Lecture Notes in Computer Science 1470

Edited by G. Goos, J. Hartmanis and J. van Leeuwen

W0106563

Springer-Verlag Berlin Heidelberg GmbH

David Pritchard   Jeff Reeve (Eds.)

Euro-Par'98
Parallel Processing

4th International Euro-Par Conference
Southampton, UK, September 1-4, 1998
Proceedings

Springer-Verlag
Berlin Heidelberg GmbH

Series Editors

Gerhard Goos, Karlsruhe University, Germany
Juris Hartmanis, Cornell University, NY, USA
Jan van Leeuwen, Utrecht University, The Netherlands

Volume Editors

David Pritchard
Jeff Reeve
University of Southampton
Department of Electronics and Computer Science
Southampton SO17 1BJ, UK
E-mail: {djp,jsr}@ecs.soton.ac.uk

Cataloging-in-Publication data applied for

Die Deutsche Bibliothek - CIP-Einheitsaufnahme

Parallel processing : proceedings / Euro-Par '98, 4th International
Euro-Par Conference, Southampton, UK, September 1 - 4, 1998.
David Pritchard ; Jeff Reeve (ed.). - Berlin ; Heidelberg ; New York ;
Barcelona ; Budapest ; Hong Kong ; London ; Milan ; Paris ;
Singapore ; Tokyo : Springer, 1998
  (Lecture notes in computer science ; Vol. 1470)
  ISBN 978-3-540-64952-6

CR Subject Classification (1991): C.1-4, D.1-4, F.1-2, G.1-2, E.1, H.2

ISSN 0302-9743
ISBN 978-3-540-64952-6   ISBN 978-3-540-49920-6 (eBook)
DOI 10.1007/978-3-540-49920-6

© Springer-Verlag Berlin Heidelberg 1998
Originally published by Springer-Verlag Berlin Heidelberg New York in 1998

Typesetting: Camera-ready by editors from LaTeX by authors
SPIN 10638643   06/3142 – 5 4 3 2 1 0   Printed on acid-free paper

Preface

Euro-Par

Euro-Par is an international conference dedicated to the promotion and advancement of all aspects of parallel computing. The major themes can be divided into the broad categories of hardware, software, algorithms and applications for parallel computing. The objective of Euro-Par is to provide a forum within which to promote the development of parallel computing both as an industrial technique and an academic discipline, extending the frontier of both the state of the art and the state of the practice. This is particularly important at a time when parallel computing is undergoing strong and sustained development and experiencing real industrial take up. The main audience for and participants in Euro-Par are seen as researchers in academic departments, government laboratories and industrial organisations. Euro-Par's objective is to become the primary choice of such professionals for the presentation of new results in their specific areas. Euro-Par is also interested in applications which demonstrate the effectiveness of the main Euro-Par themes.

There is now a permanent Web site for the series which can be reached at http://brahms.fmi.uni-passau.de/cl/europar where the history of the conference is described. Euro-Par is now sponsored by the Association of Computer Machinery and the International Federation of Information Processing.

Euro-Par'98

The format of Euro-Par'98 follows that of the past two conferences and consists of a number of workshops each individually monitored by a committee of four. There were 23 original workshops for this year's conference. The call for papers attracted 238 submissions of which 129 were accepted. Of the papers accepted 5 were judged as distinguished, 70 as regular and 54 as short papers. Distinguished papers are allowed 12 pages in the proceedings and 30 minutes for presentation, regular papers are allowed 8 pages and 15 minutes for presentation, short papers are allowed 4 pages and 15 minutes for presentation. Two extra pages could be purchased. There were on average 3.6 reviews per paper. Submissions were received from 33 countries 26 of which are represented at the conference. The principal contributors by country are the UK with 23 papers, Germany 22, France 19 and the USA 16.

The Web site for the conference is at http://www.europar98.ecs.soton.ac.uk.

Acknowledgements

Knowing the quality of past Euro-Par conferences makes the task of organising one daunting indeed and we have many people to thank. Ron Perrott, Christian

Lengauer and Luc Bougé have given us the benefit of their experience and helped us generously throughout the past 18 months. The workshop structure of the conference means that we must depend on the goodwill and enthusiasm of all the 77 programme committee members listed below. Their professionalism makes this the most academically rigorous conference in the field worldwide. The programme committee meeting at Southampton in April was well attended and thanks to sound preparation by everyone and Ron Perrott's guidance resulted in a coherent, well structured conference. The smooth running of the organisation of the conference can be attributed to a few individuals. Firstly the software for the submission and refereeing of the papers that we inherited from Lyons via Passau was significantly enhanced by Flavio Bergamaschi. This attracted many compliments from those who benefited. Panagiotis Melas ably assisted by Duncan Simpson spent copious hours checking, printing and correcting papers. Finally Lesley Courtney, secretary to the conference and the research group, has been invaluable in monitoring the conference organisation and seeing to the myriad of tasks that invariably arise, including organising the social programme.

Southampton, June 1998   Jeff Reeve and David Pritchard.

Euro-Par Steering Committee

Chair
Ron Perrott   Queen's University Belfast, UK
Vice Chair
Emilio Zapata   University of Malaga, Spain
Committee
Luc Bougé   ENS Lyon, France
Agnes Bradier   EC, Belgium
Helmar Burkhart   University of Basel, Switzerland
Paul Feautrier   University of Versailles, France
Ian Foster   Argonne National Lab, USA
Seif Haridi   SICS, Sweden
Peter Kacsuk   KFKI, Hungary
Christian Lengauer   University of Passau, Germany
Jeff Reeve   University of Southampton, UK
Paul Spirakis   CTI, Greece
Marian Vajtersic   Slovak Academy, Slovakia
Jens Volkert   Johannes Kepler University, Austria
Makoto Amamiya   Kyushu University, Japan

Euro-Par'98 Local Organisation

The conference has been organised by the Concurrent Computation Group of the Department of Electronics & Computer Science, University of Southampton

Chairs
Tony Hey
Jeff Reeve
Committee
Alistair Dunlop
Hugh Glaser
Luc Moreau
Mark Papiani
David Pritchard
Secretary   Lesley Courtney
Technical Support
Flavio Bergamaschi
Panagiotis Melas
Duncan Simpson

Euro-Par'98 Programme Committee

Workshop 1: Support Tools and Environments

Global Chair

Helmar Burkhart   University of Basel, Switzerland

Local Chair

Chris Wadsworth   Rutherford Appleton Lab, UK

Vice Chairs

Peter Kacsuk   KFKI, Hungary

Vaidy Sunderam   Emory University, USA

Workshop 2+8: Performance Evaluation and Prediction

Global Chairs

Allen Malony   University of Oregon

Rajeev Alur   University of Pennsylvania and Bell Labs

Local Chairs

Wolfgang Gentzsch   GENIAS Software, Germany

Eric Rogers   University of Southampton

Vice Chairs

Daniel Reed   University of Illinois, Urbana-Champaign

Aad van der Steen   University of Utrecht, The Netherlands

Hans-J.Siegert   TU Munich, Germany

Guenter Hommel   University of Berlin, Germany

Workshop 3: Scheduling and Load Balancing

Global Chair

Susan Flynn Hummel   IBM T.J.Watson Research Center

Local Chair

Graham Riley   University of Manchester

Vice Chairs

Rizos Sakellariou   University of Manchester

Wolfgang Gentzsch   GENIAS Software, Germany

Workshop 4: Automatic Parallelization and High-Performance Compilers

Global Chair

Jean-François Collard   CNRS and U. of Versailles, France

Local Chair

Thomas Brandes   IASC, Germany

Vice Chairs

Martin Rinard   MIT, USA

Martin Griebl   University of Passau, Germany

Workshop 5+9+15: Distributed Systems and Database Systems

Global Chairs

Andreas Reuter	International University in Germany
Ernst Mayr	TU Munich, Germany

Local Chairs

Lionel Brunie	ENS Lyon, France
Pavlos Spirakis	CTI, Greece
Kam-Fai Wong	Hong Kong University

Vice Chairs

Harald Kosch	University of Klagenfurt, Austria
Usama Fayyad	Microsoft
Arbee Chen	National Tsing Hua University, Taiwan
Friedemann Mattern	TU Darmstadt, Germany
Marios Mavronicolas	University of Cyprus, Cyprus

Workshop 6+16+18: Languages

Global Chairs

Denis Caromel	University de Nice - INRIA Sophia Antipolis
Christian Lengauer	University of Passau, Germany
Mike Quinn	Oregon State University

Local Chairs

Antonio Corradi	University of Bologna, Italy
Henk Sips	Technical University, Delft, Netherlands
Murray Cole	University of Edinburgh, UK

Vice Chairs

Gul Agha	University of Illinois
Geoffrey Fox	University of Syracuse
Phil. Hatcher	University of New Hampshire, USA
Luc Bougé	ENS Lyon, France
Beverly Sanders	University of Florida, Gainesville
Gaétan Hains	University of Orléans, France

Workshop 7+20: Numerical and Symbolic Algorithms

Global Chairs

Wolfgang Küchlin	University of Tubingen, Germany
Maurice Clint	Queen's University Belfast, UK

Local Chairs

Ali Abdallah	University of Reading, UK
Marian Vajtersic	Slovak Academy, Slovakia

Vice Chairs

Krzysztof Apt	CWI, Netherlands
Kevin Hammond	University of St. Andrews, Scotland
Michael Thune	University of Upsala, Sweden
Peter Arbenz	Institute for Sci. Computing, Switzerland

Workshop 10+17+21+22:
Theory and Applications of Parallel Computation

Global Chairs

Bill McColl	Oxford University, UK
Mike Brady	University of Oxford, UK
Paul Messina	California Institute of Technology
Freidel Hossfeld	Forschungszentrum Julich GmbH, Germany

Local Chairs

Michel Cosnard	INRIA, Nancy, France
Paul Lewis	University of Southampton, UK
Ed Zaluska	University of Southampton, UK

Vice Chairs

Andrea Pietracaprina	University of Padova, Italy
Frank Dehne	Carlton University, Canada
Patrice Quinton	IRISA-CNRS, Rennes, France
Hartmut Schmeck	University of Karlsruhe, Germany
Rajeev Thakur	Argonne National Laboratory, USA
David Walker	University of Wales, UK

Workshop 13+14: Architectures and Networks

Global Chairs

Abhiram Ranade	Indian Institute of Technology, Bombay
Mateo Valero	UPC Barcelona, Spain

Local Chairs

Kieran Herley	University College Cork, Eire
David Snelling	FECIT, UK

Vice Chairs

Sanguthevar Rajasekaran	University of Florida
Geppino Pucci	University of Padova, Italy
Olivier Temam	University of Versailles, France
Nigel Topham	University of Edinburgh, UK
Rupert Ford	University of Manchester, UK

Esprit Workshop

Global Chair

Ron Perrott	Queen's University Belfast

Local Chair

Colin Upstill	PAC,University of Southampton

Vice Chairs

Jesus Labarta	Universitat Politecnica de Catalunya, Spain
Karl Solchenbach	PALLAS, Germany

Other Referees

Arbab, Farhad
Barthou, Denis
Bischof, Stefan
Buffo, Mathieu
Detert, Ulrich
Ellmenreich, Nils
Erlebach, Thomas
Esser, Ruediger
Flammini, Michele
Foisy, Christian
Formenti, Enrico
Friedetzky, Tom
Gerndt, Michael
Gorlatch, Sergei
Harmer, Terence
Hascoet, Laurent
Hege, Hans-Christian
Herrmann, Christoph
Hogstedt, Karin
Irigoin, Francois
Jerraya, Ahmed
Lazure, Dominique
Lim, Amy
Ludwig, Thomas

Massingill, Berna
Mattern, Friedemann
McKendrick, Rob
Mehaut, Jean-Francois
Mery, Dominique
Mountjoy, Jon
Namyst, Raymond
Pazat, Jean-Louis
Perez, Christian
Rauber, Thomas
Redon, Xavier
Robert, Yves
Scales, Dan
Schwabe, Eric
Stadtherr, Hans
Stewart, Alan
Surridge, Mike
Thompson, Simon
Trinder, Phil
Utard, Gil
Vivien, Frederic
Wedler, Christoph
Wonnacott, David
Zait, Mohamed

Contents

Invited Talks
Random Number Generation and Simulation on Vector and Parallel Computers 1
Richard P. Brent

Heterogeneous HPC Environments 21
Marco Vanneschi

Quantum Cryptography on Optical Fiber Networks 35
Paul D. Townsend

Very Distributed Media Stories: Presence, Time, Imagination 47
Glorianna Davenport

HPcc as High Performance Commodity Computing on Top of Integrated Java,
CORBA, COM and Web Standards 55
Geoffrey C. Fox, W. Furmanski, T. Haupt, E. Akarsu and H. Ozdemir

Workshop 1
Support Tools and Environments 75
Chris Wadsworth and Helmar Burkhart

Process Migration and Fault Tolerance of *BSPlib* Programs Running on
Networks of Workstations 80
Jonathan M.D. Hill, Stephen R. Donaldson and Tim Lanfear

A Parallel-System Design Toolset for Vision and Image Processing 92
M. Fleury, N. Sarvan, A.C. Downton and A.F. Clark

Achieving Portability and Efficiency Through Automatic Optimisation:
An Investigation in Parallel Image Processing 102
*D. Crookes, P.J. Morrow, T.J. Brown, G. McAleese, D. Roantree and
I.T.A. Spence*

EDPEPPS: A Toolset for the Design and Performance Evaluation of Parallel
Applications 113
*T. Delaitre, M.J. Zemerly, P. Vekariya, G.R. Justo, J. Bourgeois,
F. Schinkmann, F. Spies, S. Randoux and S.C. Winter*

Verifying a Performance Estimator for Parallel DBMSs 126
*E.W. Dempster, N.T. Tomov, J. Lü, C.S. Pua, M.H. Williams, A. Burger,
H. Taylor and P. Broughton*

Generating Parallel Applications of Spatial Interaction Models 136
John Davy and Wissal Essah

Performance Measurement of Interpreted Programs 146
Tia Newhall and Barton P. Miller

Analysing an SQL Application with a *BSPlib* Call-Graph Profiling Tool ... 157
*Jonathan M.D. Hill, Stephen A. Jarvis, Constantinos Siniolakis and
Vasil P. Vasilev*

A Graphical Tool for the Visualization and Animation of Communicating
Sequential Processes 165
Ali E. Abdallah

A Universal Infrastructure for the Run-Time Monitoring of Parallel and
Distributed Applications 173
Roland Wismüller, Jörg Trinitis and Thomas Ludwig

Net-dbx: A Java Powered Tool for Interactive Debugging of MPI Programs
Across the Internet 181
Neophytos Neophytou and Paraskevas Evripidou

Workshop 2+8
Performance Evaluation and Prediction 191
Allen D. Malony and Rajeev Alur

Configurable Load Measurement in Heterogeneous Workstation Clusters .. 193
Christian Röder, Thomas Ludwig and Arndt Bode

Exploiting Spatial and Temporal Locality of Accesses: A New Hardware-Based
Monitoring Approach for DSM Systems 206
*Robert Hockauf, Wolfgang Karl, Markus Leberecht, Michael Oberhuber and
Michael Wagner*

On the Self-Similar Nature of Workstations and WWW Servers Workload . 216
Olivier Richard and Franck Cappello

White-Box Benchmarking 220
Emilio Hernández and Tony Hey

Cache Misses Prediction for High Performance Sparse Algorithms 224
Basilio B. Fraguela, Ramón Doallo and Emilio L. Zapata

h-Relation Models for Current Standard Parallel Platforms 234
C. Rodríguez, J.L. Roda, D.G. Morales and F. Almeida

Practical Simulation of Large-Scale Parallel Programs and Its Performance
Analysis of the NAS Parallel Benchmarks 244
Kazuto Kubota, Ken'ichi Itakura, Mitsuhisa Sato and Taisuke Boku

Assessing LogP Model Parameters for the IBM-SP 255
Iskander Kort and Denis Trystram

Communication Pre-evaluation in HPF 263
Pierre Boulet and Xavier Redon

Modeling the Communication Behavior of Distributed Memory Machines by
Genetic Programming 273
L. Heinrich-Litan, U. Fissgus, St. Sutter, P. Molitor and Th. Rauber

Representing and Executing Real-Time Systems 279
Rafael Ramirez

Fixed Priority Scheduling of Age Constraint Processes 288
Lars Lundberg

Workshop 3

Scheduling and Load Balancing 297
Susan Flynn Hummel, Graham Riley and Rizos Sakellariou

Optimizing Load Balance and Communication on Parallel Computers with
Distributed Shared Memory 299
Rudolf Berrendorf

Performance Analysis and Portability of the PLUM Load Balancing System 307
Leonid Oliker, Rupak Biswas and Harold N. Gabow

Experimental Studies in Load Balancing 318
Azzedine Boukerche and Sajal K. Das

On-Line Scheduling of Parallelizable Jobs 322
Christophe Rapine, Isaac D. Scherson and Denis Trystram

On Optimal k-linear Scheduling of Tree-Like Task Graphs for
LogP-Machines 328
Wolf Zimmermann, Martin Middendorf and Welf Löwe

Static Scheduling Using Task Replication for LogP and BSP Models 337
Cristina Boeres, Vinod E.F. Rebello and David B. Skillicorn

Aspect Ratio for Mesh Partitioning 347
Ralf Diekmann, Robert Preis, Frank Schlimbach and Chris Walshaw

A Competitive Symmetrical Transfer Policy for Load Sharing 352
Konstantinos Antonis, John Garofalakis and Paul Spirakis

Scheduling Data–Parallel Computations on Heterogeneous and Time–Shared
Environments 356
Salvatore Orlando and Raffaele Perego

A Lower Bound for Dynamic Scheduling of Data Parallel Programs 367
Fabricio Alves Barbosa da Silva, Luis Miguel Campos and Isaac D. Scherson

A General Modular Specification for Distributed Schedulers 373
Gerson G. H. Cavalheiro, Yves Denneulin and Jean-Louis Roch

Feedback Guided Dynamic Loop Scheduling: Algorithms and Experiments 377
J. Mark Bull

Load Balancing for Problems with Good Bisectors, and Applications in Finite
Element Simulations 383
Stefan Bischof, Ralf Ebner and Thomas Erlebach

An Efficient Strategy for Task Duplication in Multiport Message-Passing
Systems 390
Dingchao Li, Yuji Iwahori, Tatsuya Hayashi and Naohiro Ishii

Evaluation of Process Migration for Parallel Heterogeneous Workstation
Clusters 397
M.A.R. Dantas

Using Alternative Schedules for Fault Tolerance in Parallel Programs on a
Network of Workstations 401
Dibyendu Das

Dynamic and Randomized Load Distribution in Arbitrary Networks 405
J. Gaber and B. Toursel

Workshop 4
Automatic Parallelisation and High Performance Compilers 411
Jean-François Collard

Data Distribution at Run-Time: Re-using Execution Plans 413
Olav Beckmann and Paul H.J. Kelly

Enhancing Spatial Locality via Data Layout Optimizations 422
M. Kandemir, A. Choudhary, J. Ramanujam, N. Shenoy and P. Banerjee

Parallelization of Unstructured Mesh Computations Using Data Structure
Formalization 435
Rainer Koppler

Parallel Constant Propagation 445
Jens Knoop

Optimization of SIMD Programs with Redundant Computations 456
Jörn Eisenbiegler

Exploiting Course Grain Parallelism from FORTRAN by Mapping it to IF1 463
Adrianos Lachanas and Paraskevas Evripidou

A Parallelization Framework for Recursive Tree Programs 470
Paul Feautrier

Optimal Orthogonal Tiling 480
Rumen Andonov, Sanjay Rajopadhye and Nicola Yanev

Enhancing the Performance of Autoscheduling in Distributed Shared Memory
Multiprocessors 491
*Dimitrios S. Nikolopoulos, Eleftherios D. Polychronopoulos and
Theodore S. Papatheodorou*

Workshop 5+15
Distributed Systems and Databases 503
Lionel Brunie and Ernst Mayer

Collection-Intersect Join Algorithms for Parallel Object-Oriented Database
Systems 505
David Taniar and J. Wenny Rahayu

Exploiting Atomic Broadcast in Replicated Databases 513
Fernando Pedone, Rachid Guerraoui and André Schiper

The Hardware/Software Balancing Act for Information Retrieval on Symmetric
Multiprocessors 521
Zhihong Lu, Kathryn S. McKinley and Brendon Cahoon

The Enhancement of Semijoin Strategies in Distributed Query Optimization 528
Faza Najjar and Yahya Slimani

Virtual Time Synchronization in Distributed Database Systems Using a Cluster
of Workstations 534
Azzedine Boukerche, Timothy E. LeMaster, Sajal K. Das and Ajoy Datta

Load Balancing and Processor Assignment Statements 539
C. Rodríguez, F. Sande, C. León, I. Coloma and A. Delgado

Mutual Exclusion Between Neighboring Nodes in a Tree that Stabilizes Using
Read/Write Atomicity 545
Gheorghe Antonoiu and Pradip K. Srimani

Irreversible Dynamos in Tori 554
P. Flocchini, E. Lodi, F. Luccio, L. Pagli and N. Santoro

MPI-GLUE: Interoperable High-Performance MPI Combining Different
Vendor's MPI Worlds 563
Rolf Rabenseifner

High Performance Protocols for Clusters of Commodity Workstations 570
P. Melas and E. J. Zaluska

Significance and Uses of Fine-Grained Synchronization Relations 578
Ajay D. Kshemkalyani

A Simple Protocol to Communicate Channels over Channels 591
Henk L. Muller and David May

SciOS: Flexible Operating System Support for SCI Clusters 601
Povl T. Koch and Xavier Rousset de Pina

Indirect Reference Listing: A Robust Distributed GC 610
José M. Piquer and Ivana Visconti

Active Ports: A Performance-Oriented Operating System Support to Fast LAN
Communications 620
G. Chiola and G. Ciaccio

Workshop 6+16+18

Languages **625**
Henk Sips, Antonio Corradi and Murray Cole

A Tracing Protocol for Optimizing Data Parallel Irregular Computations ..629
Thomas Brandes and Cécile Germain

Contribution to Better Handling of Irregular Problems in HPF2 639
Thomas Brandes, Frédéric Brégier, Marie Christine Counilh and Jean Roman

OpenMP and HPF: Integrating Two Paradigms 650
Barbara Chapman and Piyush Mehrotra

Towards a Java Environment for SPMD Programming 659
Bryan Carpenter, Guansong Zhang, Geoffrey Fox, Xiaoming Li, Xinying Li and Yuhong Wen

Language Constructs and Run-Time Systems for Parallel Cellular
Programming 669
Giandomenico Spezzano and Domenico Talia

Task Parallel Skeletons for Irregularly Structured Problems 676
Petra Hofstedt

Synchronizing Communication Primitives for a Shared Memory Programming
Model 682
Vladimir Vlassov and Lars-Erik Thorelli

Symbolic Cost Analysis and Automatic Data Distribution for a Skeleton-Based
Language 688
Julien Mallet

Optimising Data-Parallel Programs Using the BSP Cost Model 698
D.B. Skillicorn, M. Danelutto, S. Pelagatti and A. Zavanella

A Parallel Multigrid Skeleton Using BSP 704
Femi O. Osoba and Fethi A. Rabhi

Flattening Trees 709
Gabriele Keller and Manuel M.T. Chakravarty

Dynamic Type Information in Process Types 720
Franz Puntigam

Generation of Distributed Parallel Java Programs 729
Pascale Launay and Jean-Louis Pazat

An Algebraic Semantics for an Abstract Language with Intra-Object-
Concurrency 733
Thomas Gehrke

An Object-Oriented Framework for Managing the Quality of Service of
Distributed Applications 738
Stéphane Lorcy and Noël Plouzeau

A Data Parallel Java Client-Server Architecture for Data Field Computations
over Z^n 742
Jean-Louis Giavitto, Dominique De Vito and Jean-Paul Sansonnet

Workshop 7+20
Numerical and Symbolic Algorithms **747**
Maurice Clint and Wolfgang Kreuchlin

On the Influence of the Orthogonalization Scheme on the Parallel Performance
of GMRES 751
Valérie Frayssé, Luc Giraud and Hatim Kharraz-Aroussi

A Parallel Solver for Extreme Eigenpairs 763
Leonardo Borges and Suely Oliveira

Parallel Solvers for Large Eigenvalue Problems Originating from Maxwell's
Equations 771
Peter Arbenz and Roman Geus

Waveform Relaxation for Second Order Differential Equation $y'' = f(x, y)$. 780
Kazufumi Ozawa and Susumu Yamada

The Parallelization of the Incomplete LU Factorization on AP1000 788
Takashi Nodera and Naoto Tsuno

An Efficient Parallel Triangular Inversion by Gauss Elimination with
Sweeping 793
Ayşe Kiper

Fault Tolerant QR-Decomposition Algorithm and its Parallel
Implementation 798
Oleg Maslennikow, Juri Kaniewski and Roman Wyrzykowski

Parallel Sparse Matrix Computations Using the PINEAPL Library:
A Performance Study 804
Arnold R. Krommer

Using a General-Purpose Numerical Library to Parallelize an Industrial
Application: Design of High-Performance Lasers 812
Ida de Bono, Daniela di Serafino and Eric Ducloux

Fast Parallel Hermite Normal Form Computation of Matrices over $F[x]$ ... 821
Clemens Wagner

Optimising Parallel Logic Programming Systems for Scalable Machines ... 831
Vítor Santos Costa and Ricardo Bianchini

Experiments with Binding Schemes in LOGFLOW 842
Zsolt Németh and Péter Kacsuk

Experimental Implementation of Parallel TRAM on Massively Parallel
Computer 846
Kazuhiro Ogata, Hiromichi Hirata, Shigenori Ioroi and Kokichi Futatsugi

Parallel Temporal Tableaux 852
R.I. Scott, M.D. Fisher and J.A. Keane

Workshop 10+17+21+22
Theory and Algorithms for Parallel Computation 863

Bill McColl and David Walker

BSP, LogP, and Oblivious Programs 865
Jörn Eisenbiegler, Welf Löwe and Wolf Zimmermann

Parallel Computation on Interval Graphs Using PC Clusters: Algorithms and
Experiments 875
A. Ferreira, I. Guérin Lassous, K. Marcus and A. Rau-Chaplin

Adaptable Distributed Shared Memory: A Formal Definition 887
Jordi Bataller and José M. Bernabéu-Aubán

Parameterized Parallel Complexity 892
Marco Cesati and Miriam Di Ianni

Asynchronous (Time-Warp) versus Synchronous (Event-Horizon) Simulation
Time Advance in BSP 897
Mauricio Marín

Scalable Sharing Methods Can Support a Simple Performance Model 906
Jonathan Nash

Long Operand Arithmetic on Instruction Systolic Computer Architectures and
Its Application in RSA Cryptography 916
Bertil Schmidt, Manfred Schimmler and Heiko Schröder

Hardware Cache Optimization for Parallel Multimedia Applications 923
C. Kulkarni, F. Catthoor and H. De Man

Parallel Solutions of Simple Indexed Recurrence Equations 933
Yosi Ben-Asher and Gady Haber

Scheduling Fork Graphs under LogP with an Unbounded Number of
Processors 940
Iskander Kort and Denis Trystram

A Data Layout Strategy for Parallel Web Servers 944
Jörg Jensch, Reinhard Lüling and Norbert Sensen

ViPIOS: The Vienna Parallel Input/Output System 953
Erich Schikuta, Thomas Fuerle and Helmut Wanek

A Performance Study of Two-Phase I/O 959
Phillip M. Dickens and Rajeev Thakur

Workshop 13+14
Architectures and Networks 967
Kieran Herley and David Snelling

Predictable Communication on Unpredictable Networks: Implementing BSP
over TCP/IP 970
Stephen R. Donaldson, Jonathan M.D. Hill and David B. Skillicorn

Adaptive Routing Based on Deadlock Recovery 981
Nidhi Agrawal and C.P. Ravikumar

On the Optimal Network for Multicomputers: Torus or Hypercube? 989
Mohamed Ould-Khaoua

Constant Thinning Protocol for Routing h-Relations in Complete Networks 993
Anssi Kautonen, Ville Leppänen and Martti Penttonen

NAS Integer Sort on Multi–threaded Shared Memory Machines 999
Thomas Grün and Mark A. Hillebrand

Analysing a Multistreamed Superscalar Speculative Instruction Fetch
Mechanism 1010
Rafael R. dos Santos and Philippe O.A. Navaux

Design of Processor Arrays for Real-time Applications 1018
Dirk Fimmel and Renate Merker

Interval Routing & Layered Cross Product: Compact Routing Schemes for
Butterflies, Mesh of Trees and Fat Trees 1029
Tiziana Calamoneri and Miriam Di Ianni

Gossiping Large Packets on Full-Port Tori 1040
Ulrich Meyer and Jop F. Sibeyn

Time-optimal Gossip in Noncombining 2-D Tori with Constant Buffers ...1047
Michal Šoch and Pavel Tvrdík

Divide-and-Conquer Algorithms on Two-Dimensional Meshes 1051
Miguel Valero-García, Antonio González, Luis Díaz de Cerio and Dolors Royo

All-to-all Scatter in Kautz Networks 1057
Petr Salinger and Pavel Tvrdík

Reactive Proxies: A Flexible Protocol Extension to Reduce ccNUMA Node
Controller Contention 1062
Sarah A.M. Talbot and Paul H.J. Kelly

Handling Multiple Faults in Wormhole Mesh Networks 1076
Tor Skeie

Shared Control — Supporting Control Parallelism Using a SIMD-like
Architecture 1089
Nael B. Abu-Ghazaleh and Philip A. Wilsey

Workshop 23

ESPRIT Projects **1101**
Ron Perrott and Colin Upstill

Parallel Crew Scheduling in PAROS 1104
*Panayiotis Alefragis, Christos Goumopoulos, Efthymios Housos, Peter Sanders,
Tuomo Takkula and Dag Wedelin*

Cobra: a CORBA-compliant Programming Environment for High-Performance
Computing 1114
Thierry Priol and Christophe René

OCEANS: Optimising Compilers for Embedded ApplicatioNS 1123
*Michel Barreteau, François Bodin, Peter Brinkhaus, Zbigniew Chamski, Henri-
Pierre Charles, Christine Eisenbeis, John Gurd, Jan Hoogerbrugge, Ping Hu,
William Jalby, Peter M.W. Knijnenburg, Michael O'Boyle, Erven Rohou,
Rizos Sakellariou, André Seznec, Elena A. Stöhr, Menno Treffers and
Harry A.G. Wijshoff*

Industrial Stochastic Simulations on a European Meta-Computer 1131
Ken Meacham, Nick Floros and Mike Surridge

Porting the SEMC3D Electromagnetics Code to HPF 1140
Henri Luzet and L.M. Delves

HiPEC: High Performance Computing Visualization System Supporting
Networked Electronic Commerce Applications 1149
Reinhard Lüling and Olaf Schmidt

Index of Authors 1153

Workshop 6+16+18
Languages

Henk Sips, Antonio Corradi and Murray Cole

Co-chairmen

Three Workshops(6, 16, and 18) have programming languages and models as central theme. Workshop 6 focuses on the use of object oriented paradigms in parallel programming; Workshop 16 has the design of parallel languages as primary focus, and Workshop 18 deals with programming models and methods. Together they present a nice overview of current research in these areas.

Object Oriented Programming.

The Object-Oriented (OO) technology has received a renovated stimulus by the ever-increasing usage of the Web and Internet technology. The globalisation has enlarged the number of the potentially involved users to such an extent to suggest a reconsideration of the available environments and tools. This has also motivated the attempt to clarify all debatable and ambiguous points, not all of which of practical and immediate application. On the one hand, several issues connected to program correctness and semantics are still unclear. In particular, the introduction of concurrency and parallelism within an object framework is still subject to discussion, in its verification and modelling. The same is for the aggregation of different objects and behaviour in predetermined patterns. How to accommodate several execution capacities and resources into the same object or pattern requires still work and new proposals. In any case, there is much to be done, both in the abstract area and in the applied field. On the other hand, the distributed framework has not only introduced examples in need of practical solutions and environments, but also forced to reflect on all applicable models, starting from traditional ones, such as the client-server one and RPCs, to less traditional ones, such as the agent models. The growth of Java, CORBA, and their capacity of attracting implementors and resources produce unifying perspectives, with the possibility of offering a new integrated framework in which to solve most common problems. This starts to produce the possibility of creating generally available components to be really employed, by reducing the convenience of the redesign from scratch.

This conference session is an occasion to expose to an enlarged audience working into parallelism some of the hot topics and researches going on in the OO area. And, even if the OO community has many occasions of meeting and many forums to exchange opinions, EUROPAR seems a particular opportunity of both presenting experiences and receiving contributions with possibilities of cross-fertilisation. The five papers presented in the conference explore several of the most strategic directions of evolution of the OO area.

The first paper, "Dynamic Type Information in Process Types" by Puntigam, uses the process as an example of objects with dynamic type. The goal is to make possible all the checks typical of static types even in the dynamic case: the process is modelled as an active object that, depending on its state, is capable of accepting different messages from different clients. The presented model is a refinement of a previous work of the same author and is based on a calculus of objects that communicate with asynchronous message passing. The third paper of the session, by Gehrke, "An Algebraic Semantics for an Abstract Language with Intra-Object-Concurrency," addresses the problems of intra-object concurrency, working with a process algebra method. The goal is to introduce the formal semantics for intra-object concurrency in OO frameworks where active processes can be distinguished from passive object. Let us recall that this is the Java assumption. The topics of the other papers are all connected to Java. The paper by Launay and Pazat, "Generation of distributed parallel Java programs" and the fifth paper, by Giavitto, De Vito and Sansonnett, "A Data Parallel Java Client-Server Architecture for Data Field Computations" addresses the point of enlarging the usage of the Java Framework. Launay and Pazat propose a framework capable of transparently distributing Java components of an application onto the available target architecture. Giavitto, De Vito and Sansonnett apply their effort to make Java usable in the data- parallel paradigm, for client-server applications. The fourth paper, by Lorcy and Plouzeau, "An object-oriented framework for managing the quality of service of distributed applications", addresses the quality of service problem for interactive applications. The authors elaborate on the known concept of 'contract', with new considerations and insight.

Programming Languages.

As up to now data parallel languages have been the most succesful attempt to bring parallel programming closer to the application programmer. The most seriuos attempt has been the definition of High Performance Fortran (HPF) as an extension of Fortran 90. Several commercial compilers are available today. However, user experience indicates that for irregularly structured problems, the current definition is often inadequate. The first two papers in the HPF session deal with this problem. The paper by Brandes and Germain, called "A tracing protocol for optimizing data parallel irregular computations" describes a dynamic approach by allowing the user to specify which data has to be traced for modifications. The second paper by Brandes, Bregier, Counilh, and Roman proposes a programming style for irregular problems close to regular problems. In this way compile-time and run-time techniques can be readily combined.

A recent language extension for shared-memory programming that has caught a lot of attention is OpenMP. OpenMP is a kind of reincarnation of the old PCF programming model. The paper by Chapman and Mehrotra describes various ways how HPF and OpenMP can be combined to form a combined powerful programming system. Also the Java language can be fruitfully used as a basis for parallel programming. The paper by Carpenter, Zhang, Fox, Li, Li, and Wen outlines a conservative set of language extensions to Java to support SPMD style of programming.

General parallel programming languages have a hard time in obtaining optimal performance for specific cases. If the application domain is restricted, better performance can be obtained by using a domain specific parallel programming language. The paper by Spezzano and Talia describes the langauge CARPET intended for programming cellular automata systems.

Finally, the paper by Hofstadt presents the integration of task parallel extensions into a functional programming language. This approach is illustrated by a branch and bound problem example.

Programming Models and Languages.

Producing correct software is already a difficult task in the sequential context. The challenge is compounded by the conceptual complexity of parallelism and the requirement for high performance. Building on the foundations laid in Workshop 7 in the preceding instantiation of Euro-Par, Workshop 18 focuses on programming and design models that abstract from low-level programming techniques, present software developers with interfaces that reduce the complexity of the parallel software construction task, and support correctness issues. It is also concerned with methodological aspects of developing parallel programs, particularly transformational and calculational approaches, and associated ways of integrating cost information into them.

The majority of papers this year work from a "skeletal" programming perspective, in which syntactic restrictions are used both to raise the conceptual level at which parallelism is invoked and to constrain the resulting implementation challenge. Mallet's work links the themes of program transformation (here viewed as a compilation strategy) and cost analysis, using symbolic methods to choose between distribution strategies. His source language is the by now conventional brew of nested vectors and the map, fold, scan skeleton family, while the cost analysis borrows from the polytope volume techniques of the Fortran parallelization world, an interesting and encouraging hybrid. Skillicorn and colleagues work with the P3L language as source and demonstrate that the use of BSP as an implementation mechanism enables a significant simplification of the underlying optimisation problem. The link is continued in the work of Osoba and Rabhi, in which a skeleton abstracting the essence of the multigrid approach benefits from the portability and costability of BSP. In contrast, Keller and Chakravarty work with the well know data-parallel language NESL, introducing techniques which allow the existing concept of "flattening transformations" (which allow efficient implementation of the nested parallel structures expressible in the source language) to be extended to handle user-defined recursive types, and in particular parallel tree structures. Finally, Vlassov and Thorelli apply the ubiquitous principle of simplification through abstraction to the design of a shared memory programming model.

In summary, we expect from these session a fruitful discussion of the hot topics in concurrency and parallelism in several areas. This enlarged exchange of ideas can impact on advances of the discipline in the whole distributed and concurrency field.

A Tracing Protocol for Optimizing Data Parallel Irregular Computations

Thomas Brandes[1] and Cécile Germain[2]

[1] Institute for Algorithms and Scientific Computing (SCAI)
GMD, Schloß Birlinghoven, D-53754 St. Augustin, Germany
e-mail: brandes@gmd.de
[2] Laboratoire de Recherche en Informatique (LRI-CRNS)
Université Paris-Sud, F-91405 Orsay Cedex, France
e-mail: cecile@lri.fr

Abstract. High Performance Fortran (HPF) is the de facto standard language for writing data parallel programs. In case of applications that use indirect addressing on distributed arrays, HPF compilers have limited capabilities for optimizing such codes on distributed memory architectures, especially for optimizing communication and reusing communication schedules between subroutine boundaries.

This paper describes a dynamic approach for optimizing unstructured communication in codes with indirect addressing. The basic idea is that runtime data reflecting the communication patterns will be reused if possible. The user has only to specify which data in the program has to be traced for modifications. The experiments and results show the effectiveness of the chosen approach.

1 Introduction

Data parallel languages as High Performance Fortran [9, 10] have been designed to make possible efficient and high-level parallel programming for distributed memory machines. As calculations in computationally intensive codes mostly occur in loops or intrinsic calls involving loops, a challenging problem is to handle efficiently loops with indirectly referenced distributed arrays. For irregular parallel loops, complex data structures, called *schedules*, are required for accessing remote items of distributed arrays and for communication optimization. The schedules cannot be worked out at compile time, but only at run-time, when the values of the indirection arrays are known. The code design which first builds the schedule, then uses it to carry out the actual communication and computation, has been coined as the *inspector/executor* scheme that was pioneered very early [16, 12], and developed in the PARTI [6, 2] and CHAOS [17] libraries. It is the state-of-the-art design for academic compilers [13, 11, 18, 4] as well as commercial ones [15].

The preprocessing associated with the inspector creates a large run-time overhead, because it involves a large amount of computation and all-to-all communications. Hence the inspector/executor scheme targets irregular loops which exhibit especially temporal locality where the schedule can be reused. The schedule reusing problem can be addressed at compile-time, through data-flow analysis [6, 8]. The main drawbacks of this approach are the limits of inter-procedural analysis and the wall of separate compilation. More precisely, real codes with indirect addressing often require inter-procedural analysis [1, 7, 5].

At the other extreme, language support makes the concept of schedule visible to the end-user. The HPF+ project [3] proposes *gather* and *scatter* clauses referring to explicit *schedule variables*, or *reuse* clause for independent loop. The user is responsible for tracking the modifications, but through high level conditional constructs. Reusing schedules across procedure calls relies on the Fortran 90 **save** attribute for the schedule variable.

The seminal paper [14] proposes a schedule reusing method based on time-stamps. Each modification of an array is mirrored by an update of the time-stamp, and the time-stamps are recorded at inspector invocation. A schedule may be reused if all relevant arrays have the same time-stamps at the point of reuse as the recorded ones. As no directive is provided, all array modifications have to be tracked. Moreover, array descriptors are shared through all arrays having the same structure parameters, so that an indirection array may be considered modified without necessity.

In this paper, we propose a new protocol to handle temporal locality. It has been implemented and evaluated in the framework of the ADAPTOR HPF compilation system [4] that already supported the inspector/executor scheme. The protocol is called URB, because the run-time system can either Use, Refresh or Build a schedule for each parallel loop. With some help from the user, URB is able to track changes in the irregular loops exactly: a new schedule is built only when either an irregular parallel loop is structurally different from the previously executed ones, or when the indirection arrays have been modified. The structural information includes the distribution, shape and size of the arrays, and is completely handled by the compiler and the run-time system. The intervention from the user is limited to provide an attribute for each indirection array, through a directive. This attribute indicates that all modifications of this array have to be recorded at run-time, and allows to handle procedure calls without inter-procedural analysis. Finally, the tracking scheme is conservative in the sense that, when tracing directives are missing, the generated code simply rebuilds a new schedule for each execution of an irregular loop. Hence, codes using the trace directive and other codes (e.g. library routines) can securely be mixed.

The rest of the paper is organized as follows. Section 2 describes the realization of indirect addressing of distributed arrays via the inspector/executor scheme. Section 3 introduces the protocol that is needed for tracking indirection arrays and how it can be realized. Section 4 presents performance results for kernels and a real application, and we conclude in section 5.

2 Unstructured Communication

In the following, we introduce a formal definition of the *communication schedule* that is built up during the inspector phase. We discuss it for a typical gather operation. The array T specifies the home of the iteration where the data is needed.

$$\text{forall } (I \in I_T, M(I)) \; T(I) = A(L(I))$$

Such an indirect addressing will be noted as $\mathcal{I}_{T,A,L,M} = \{I_T, d_T, I_A, d_A, L, M\}$, the inspector info, where I_T and I_A specify the index space of the arrays T and A, d_T and d_A their distributions on the available processors P. M is a predicate defining the subset S_T that is the iteration space of the parallel loop. L stands for the indirect addressing of the array A. All indices and arrays can be multi-dimensional ($I = I_1, \ldots, I_n$), and L stands for a set of integer arrays $L_1, \ldots, L_m$.

$$d_T : I_T \to P \qquad\qquad d_A : I_A \to P$$
$$M : I_T \to \{true, false\}, \qquad S_T := \{I \;\mid\; M(I) = true\} \subseteq I_T$$
$$L : S_T(\subseteq I_T) \to I_A, \qquad\qquad I \mapsto L(I)$$

We assume that the function L is specified by an integer array or by a set of integer arrays L_j and that they are aligned with the array T. In other words, $L(I)$ is available on the processor that owns $T(I)$. Also the predicate M, if available as a logical array, will not involve any communication.

For every processor pair (p, q), $p \neq q$, the set $C_{p \leftarrow q}$ specifies which elements of I_A processor p needs from processor q:

$$C_{p \leftarrow q} := \{(I, L(I)) | I \in S_T, d_T(I) = p, d_A(L(I)) = q\}$$

$C_{p \leftarrow q} \subseteq d_T^{-1}(p) \times d_A^{-1}(q)$, so for every $(I, K) \in C_{p \leftarrow q}$, I is a local index of T on processor p and K is a local index of A on processor q. Data movement from processor q to processor p will be necessary if $C_{p \leftarrow q}$ is not empty. Processor q has to send the corresponding data and processor p must receive it. In the following, the collection of sets $C_{p \leftarrow q}$, $S_{T,A,L,M} = \{C_{p \leftarrow q}\}$, is called the *communication schedule*. Due to the fact that L is not a function, the computation cannot be done by closed formulas as it might be possible in case of structured communication.

for $q \in P$: $C_{p \leftarrow q} := \{\}$ end for for I in $d_T^{-1}(p) \cap S_T$ $\quad q := d_A(L(I))$ $\quad C_{p \leftarrow q} := C_{p \leftarrow q} + \{(I, L(I))\}$ $\quad$ end for end for	for $q \in P, p \neq q$ $\quad$ send $C_{p \leftarrow q}$ to processor q $\quad$ recv $C_{q \leftarrow p}$ from processor q end for

Fig. 1. Schedule building algorithms

for $q \in P$, $p \neq q$, $C_{q \leftarrow p} \neq \{\}$
  compute $V_{q \leftarrow p}$ (local on p)
  send $V_{q \leftarrow p}$ to processor q
end for
for $q \in P$, $p \neq q$, $C_{p \leftarrow q} \neq \{\}$
  recv $V_{p \leftarrow q}$ from processor q
  for $(I, v) \in V_{p \leftarrow q}$ set $T(I) = v$
end for
	for $q \in P$, $p \neq q$, $C_{p \leftarrow q} \neq \{\}$
	  compute $W_{p \leftarrow q}$ (local on p)
	  send $W_{p \leftarrow q}$ to processor q
	end for
	for $q \in P$, $p \neq q$, $C_{q \leftarrow p} \neq \{\}$
	  recv $W_{q \leftarrow p}$ from processor q
	  for $(v, K) \in W_{q \leftarrow p}$ set $A(K) = v$
	end for

Fig. 2. Executor algorithms

Each processor p has to compute the sets $C_{p \leftarrow q}$ for all processors q, as shown in Fig. 1, left part. By an all-to-all communication, the processors exchange the information which data they need from each other (*index exchange*, Fig. 1, right part). This is necessary in order to make sure that the other processors know which values they have to send. The inspector info together with its schedule builds the inspector data.

In the *executor* phase, the processors use the communication schedule of the inspector data to carry out the communication for gathering or scattering data. Therefore the values of T and A belonging to a domain X are transferred.

$$val_A : I_A \to X \qquad V_{p \leftarrow q} := \{(I, val_A(K)) | (I, K) \in C_{p \leftarrow q}\}$$

Figure 2 (left part) shows the algorithm which is executed on each processor p to gather the values of the distributed array A. For a scatter operation, $forall(I \in I_T, M(I))A(L(I)) = T(I)$, the communication schedule can be used in the same way to realize the necessary communication (Fig.2, right part).

$$val_T : I_T \to X \qquad W_{p \leftarrow q} := \{(val_T(I), K) | (I, K) \in C_{p \leftarrow q}\}$$

3 The URB Protocol

3.1 Dynamic Reuse of Communication Schedules

Let be $\mathcal{I}_{T,A,L,M} = \{I_T, d_T, I_A, d_A, L, M\}$ an indirect addressing for which a communication schedule $\mathcal{S}_{T,A,L,M}$ has been computed. This communication schedule can be reused if the sets $C_{p \leftarrow q}$ have not changed. This is the case if I_T, I_A, d_T, d_A, L and M have not changed. This model points out that the inspector does not depend on the actual values of the arrays T and A, but only on their shape and distribution. Thus, an inspector has the potential to be reused for instance for different sections of the same array, or for aligned arrays.

Our URB protocol is based on run-time analysis. For this dynamic approach, it will be decided at runtime whether a communication schedule can be reused or not. Therefore, inspectors are kept in a database. Two problems must be addressed: at first, tracing the modifications of the indirection and mask arrays, and secondly, avoiding the explosion of the data base while keeping useful information.

The penalty of this approach is that modifications of the indirection arrays must be traced. In the presence of a mask, this is also necessary for the corresponding arrays within the mask.

3.2 Principle of Dynamic Tracing

The previous formulation shows the inspector database management as an instance of a cache management problem. The URB protocol is based on this analogy.

To allow tracing, the descriptor of an indirection array L includes a *dirty flag*. Each write to L sets its dirty flag. The schedule can thus be reused if and only if the dirty flag of L is clear. The approach has to guarantee that every update or every possible update of L will set its dirty flag.

Furthermore, if the dirty flag of an array L is set, all inspector for which this array is involved, become invalid. The entries for these inspectors can be freed avoiding the explosion of the database.

In fact, the previous scheme fails if an indirection array L_j is involved in more than one indirect assignment. The reason is simply that the flag of the L_j may be shared by many inspectors, but that their status with respect to these inspectors should not be shared. Hence, the descriptor of a traced array actually includes not only one dirty flag, but a *dirty mask*, with a flag for each possible inspector. Each write to the array sets *all* flags of the dirty mask, while each inspector update clears only the corresponding flag in the dirty mask.

The overall run-time algorithm for building or retrieving schedules is sketched in Fig. 3. Even when the schedule is reusable, the actual schedule can need refreshing. This means that all the hard work of computing ownership and local addresses is done, but that a translation term from the base address has to be taken into account. The inspector data then is updated with the refreshed schedule.

```
Schedule GetSchedule (InspectorInfo I)
begin
var InspectorData : P       /* InspectorData = InspectorInfo + Schedule */
        P := DBSearch (I)
        L := IndirectionArrayDescriptor(I)
        if (P == NULL) or DirtyMask(L, P) then
          DBFree (P)
          P := InspectorBuild (I)
          DirtyMaskClear(L, P)
        elseif not EqualBaseAddress (P, I) then
          ScheduleRefresh (P, I)
        endif
        return (Schedule(P))
    end
```

Fig. 3. The URB protocol

3.3 Tracing of Arrays

Tracing clearly belongs to the compiler, which has to insert the code setting the dirty flags as a side-effect of a write of some arrays. The requirement for tracing is described by the user directive !ADP$ TRACE. This directive has a Fortran attribute semantics, and follows Fortran 90 rules of scoping. By default, arrays are not traced. To insure correctness, this case must lead to the conservative scheme where the schedule is computed at each access. If at least one indirection array is not traced, no entry in the data base is created. The schedule is built, used and destroyed immediately after its use.

An array must be flagged dirty when it is changed. The syntactic constructs that can modify an array are limited, and can be analyzed at compile-time: assignments with this array as a left-hand-side, redistributions, allocations and deallocations. Thus, the trace attribute is compatible with the DYNAMIC and ALLOCATABLE attributes, but not with the POINTER or TARGET attribute, because the compiler will not be able to record modification of such arrays.

3.4 Procedure Handling

We want the URB protocol to be compatible with the general HPF framework that does not include the tracing directive at the present time. Thus, all combination of traced and not traced for actual and dummy arguments must be considered.

The run-time system copies information from the descriptor of the actual argument into the descriptor of the dummy argument at the entry of the subprogram, and vice versa when leaving the subprogram. Four cases must be considered, following the fact that the actual and the dummy arguments have or have not the trace attribute. If none has, we are done. If the actual and the dummy arguments both have, the modifications of the array will be traced through the subroutine boundaries, by copying forth and back the dirty mask. If the actual does not have the attribute and the dummy has, the modifications of the dirty mask of the dummy inside the procedure have no impact on the actual: reusing of schedules where this array is involved is only possible during the lifetime of this subroutine call.

Finally, procedures without explicit or implicit interface, or with an explicit or implicit interface that does not give the trace attribute to an output dummy argument, are assumed to modify the actual argument. Thus, on the calling side, the compiler will set the dirty flag of the corresponding array when the procedure returns. With this scheme, the case of library subroutines that modify an indirection array, and that are not aware of the trace attribute, is correctly handled.

4 Performance Results

All experiments were performed on an IBM SP2 equipped with P2SC nodes, and a 80MB/s maximal bandwidth per node. We used the user space protocol, and the hardware timer of the Power2 architectures. Measurements are averages.

4.1 Basic Performance

We studied the basic performance of the URB protocol for a typical gather operation, $A(I) = B(L(I))$, in order to make evident the various components of the inspector-executor scheme. Three schemes of indirect addressing L have been considered: identity (i), a cyclic shift (c), and uniform random values (r). The random and shift distributions are standing here for two typical degrees of irregularity, heavy and light. Identity is useful to measure pure overhead. It also describes *data-dependent locality*, where no communication is needed because the indirection array operates inside disjoint index subsets. Domain decomposition methods are a typical example of this situation.

First, a breakdown of the various components of the inspector and of the executor, for the random scheme, on an 8 processors configuration and a 4K array of real (Table 1) shows that the only potential overhead of the URB protocol, namely searching the database, is negligible. It also highlights the cost of the inspector step.

Table 1. Timing of unstructured communication (seconds, 8 processors).

	inspector step	executor step
no reuse	0.16 sec	0.07 sec
reuse	0.25e-4 sec	0.07 sec

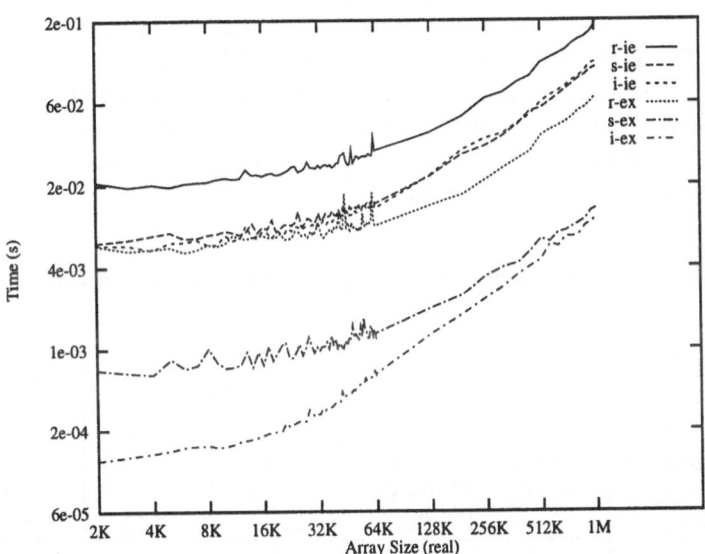

Fig. 4. Performance effects of schedule reusing: 16 processors. Curves labeled *ie* are the non-reuse case, and curves labeled *ex* are the reuse case.

Figure 4 presents a detailed comparison of timings with and without schedule reuse, on a 16 processors configuration. The random scheme displays a nearly linear difference (constant speedup) between the reuse and non reuse case, because of the index exchange, which in this case has exactly the same volume as the data movement. The identity scheme, in the reuse case, describes the minimal cost of an indirect operation, which includes reading the schedule and the local copy between the user arrays. The shift scheme, in the reuse case, has the only additional overhead of each processor sending one 4-bytes message. This overhead is very significant for small arrays, but becomes negligible for large arrays, where the performance converges with the one of the identity case. The shift and identity schemes have very similar performance in the non-reuse case, and display the overhead of building the schedule (ownership and offset computations).

4.2 Performance Results for a Real Application

The AEROLOG computational fluid dynamic software developed by MATRA is devoted to the study of compressible fluid flows around complex geometries. The kernel of the numerical solver (called CLASS for Conservative LAws Systematic Solver) is common to all physical models and is based on a second order accurate time and space finite volume scheme. Within the European Esprit project PHAROS, this code has been ported to HPF [5].

The HPF parallelization strategy is based on the coarse grain parallelism given by the decomposition of the mesh data into sub-domains. In one time step, the time increments for the physical variables are computed. The routines called within one time step can be sorted into two groups, the *local* routines and the *boundary* routines. The local routines are called independently over sub-domains. This group represents typically up to 90 % of the total CPU time of the serial code. The local computations scale well as no communication is necessary. The boundary routines perform the boundary conditions. These routines make an intensive use of indirect addressing and involve dependences between data belonging to different sub-domains.

Table 2. Execution times in seconds (304200 mesh points, 20 iterations).

	P=1	P=2	P=4	P=8	P=16
local	190.46	95.84	46.68	23.43	12.86
boundary replicated	11.24	35.99	38.53	42.97	52.19
compress	12.03	11.49	11.56	13.24	14.05
notrace	165.65	96.88	52.43	28.66	21.12
trace	34.42	18.72	12.16	7.73	5.88

In the initial HPF version, the whole mesh data involved in the boundary computations as well as the boundary computations itself are replicated on all processors. This avoids unstructured communication but the replication is very expensive. In the tuned HPF version, the mesh data is packed before the replication. This reduces the communication volume, but the scalability is still limited.

The third version distributes the mesh data thus requires unstructured communication. Performance results are very poor if the communication schedule is not reused (notrace). Only the reuse of communication schedules (trace) gives acceptable and scalable performance results. With ADAPTOR, the user gets this performance gain only by inserting some TRACE directives in his HPF code. Table 2 shows the performance results for a medium industrial test case with 16 sub-domains of 304200 total mesh points executing for 20 iterations.

5 Conclusions

In this paper, we have presented a strategy to implement effectively an inspector-executor paradigm where communication schedules can be reused. It requires no more data-flow and inter-procedural analysis at compile time. It will also reuse communication schedules in situations where even best compile time analysis might fail, e.g. if the indirection arrays are modified under a mask. Although the method described here requires a new language construct, the TRACE directive, the user does not have to track the multiple execution paths. This should be task of the compiler and of the run-time system.

The realization of our concept in the ADAPTOR HPF compilation system showed that it can be easily integrated in an existing HPF compiler that supports the inspector/executor scheme. The compiler itself has only to deal with the new TRACE attribute. The runtime system requires the support of reusing communication schedules and the implementation of a data base for these schedules. For the tracing of arrays with the TRACE attribute, the array descriptor has to be extended by a dirty mask. Tracing can be tracked along subroutine boundaries if an array or a section of an array is passed via such a descriptor. This is generally the case for ADAPTOR, but tracking fails if the HPF compiler passes an array only by a pointer.

The work described here demonstrates two new ideas. The first is that a schedule can be associated with more than one parallel indirect assignment, allowing aggressive optimization even in dynamic applications. The second is the analogy to schedule reusing with cache management.

We believe that the ideas of the URB protocol might also be useful for other optimizations in an HPF compiler, especially in the context of indirect and dynamic distributions, and for the efficient support of DOACROSS loops with irregular dependencies. In particular, the fact that the URB protocol is very efficient on the identity case has an important application: the HPF 2.0 Approved Extensions deal with data-dependent locality through the coupling of the ON HOME and RESIDENT directives. Schedule reusing is simpler, for the user and the compiler, and more versatile, because the same code can be kept unmodified whether data-dependent locality exists or not. The improvement of the current strategy as well as the investigation of benefits for other optimizations will be part of our future work.

Acknowledgments

We acknowledge the CRI and the CNUSC for access to their IBM SP2.

References

1. G. Agrawal and J. Saltz. Interprocedural communication optimizations for distributed memory compilation. In *Language and Compilers for Parallel Computing*, pages 1–16, Aug. 1994.

2. G. Agrawal, A. Sussman, and J. Saltz. Compiler and run-time support for structured and block-structured applications. In *Supercomputing*, pages 578–587. IEEE, 1993.

3. S. Benkner and H. Zima. Definition of HPF+ Rel.2. Technical report, HPF+ Consortium, 1997.

4. T. Brandes and F. Zimmermann. ADAPTOR - A Transformation Tool for HPF Programs. In K. Decker and R. Rehmann, editors, *Programming Environments for Massively Parallel Distributed Systems*, pages 91–96. Birkhäuser Verlag, Apr. 1994.

5. T. Brandes, F. Zimmermann, C. Borel, and M. Brédif. Evaluation of High Performance Fortran for an Industrial Computational Fluid Dynamics Code. In *Proceedings of VECPAR 98, Porto, Portugal*, June 1998. Accepted for publication.

6. R. Das et al. Communication optimization for irregular scientific computations on distributed memory architecturess. *Journal of Parallel and Distributed Computing*, (22):462–478, 1994.

7. C. Germain, J.Laminie, M. Pallud, and D. Etiemble. An HPF Case Study of a Domain-Decomposition Based Irregular Application. In *PACT'97*. LNCS, Sep 1997.

8. M. Gupta, E. Schonberg, and H. Srinivasan. A unified data-flow framework for optimizing communication in data-parallel programs. *IEEE Trans. on Parallel and Distributed Systems*, 7(7):689–704, 1996.

9. High Performance Fortran Forum. *High Performance Fortran Language Specification*. Rice Univ., Nov. 1994. *Version 1.1*.

10. High Performance Fortran Forum. *High Performance Fortran Language Specification*, Oct. 1997. *Version 2.0*.

11. S. Hiranandani, K. Kennedy, and C. Tseng. Compiling Fortran D for MIMD Distributed-Memory machines. *CACM*, 35(8):66–80, Aug. 1992.

12. S. Hiranandani, J. Saltz, P. Mehrotra, and H. Berryman. Performance of hashed cache migration schemes on multicomputers. *Journal of Parallel and Distributed Computing*, 12:415–422, 1991.

13. C. Koelbel, P. Mehrotra, and J. V. Rosendale. Supporting shared data structures on distributed memory architectures. In *2nd. Symp. on Principles and Practice of Parallel Programming*, pages 177–186. ACM, 1990.

14. R. Ponnusamy, J. Saltz, and A. Choudhary. Runtime compilation techniques for data partitioning and communication schedule reuse. In *ACM Int. Conf. on Supercomputing*, pages 361–370, 1993.

15. Portland Group, Inc. PGHPF User's Guide. Manual Release 2.2, PGI, 1997.

16. R. Mirchandaney et al. Principles of run-time support for parallel processing. In *ACM Int. Conf. on Supercomputing*, pages 140–152, 1988.

17. S. D. Sharma and al. Run-time and Compile-time Support for Adaptive Irregular Problems. In *Supercomputing'94*, pages 99–106. IEEE, 1994.

18. H. Zima and B. Chapman. Compiling for distributed-memory systems. *Proceedings of the IEEE*, Feb. 1993.

Contribution to Better Handling of Irregular Problems in HPF2

Thomas Brandes[1], Frédéric Brégier[2], Marie Christine Counilh[2], Jean Roman[2]

[1] GMD/SCAI, Institute for Algorithms and Scientific Computing,
German National Research Center for Computer Science, Schloss Birlinghoven,
PO Box 1319, 53754 St. Augustin, Germany
[2] LaBRI, ENSERB and Université Bordeaux I, 33405 Talence Cedex, France

Abstract. In this paper, we present our contribution for handling irregular applications with HPF2. We propose a programming style of irregular applications close to the regular case, so that both compile-time and run-time techniques can be more easily performed. We use the well-known *tree data structure* to represent irregular data structures with hierarchical access, such as sparse matrices. This algorithmic representation avoids the indirections coming from the standard irregular programming style. We use derived data types of Fortran 90 to define trees and some approved extensions of HPF2 for their mapping. We also propose a run-time support for irregular applications with loop-carried dependencies that cannot be determined at compile-time. Then, we present the TriDenT library, which supports distributed trees and provides run-time optimizations based on the inspector/executor paradigm. Finally, we validate our contribution with experimental results on IBM SP2 for a sparse Cholesky factorization algorithm.

1 Introduction

High Performance Fortran (**HPF**) [10] is the current standard language for writing data parallel programs for shared and distributed memory parallel architectures. HPF is well suited for regular applications; however, many scientific and engineering applications (fluid dynamics, structural mechanics, ...) use irregular data structures such as sparse matrices. Irregular data need a special representation in order to save storage requirements and computation time, and the data accesses use indirect addressing through pointers stored in index arrays (see for example SPARSKIT [14]). In these applications, patterns of computations are generally irregular and special data distributions are required to both provide high data locality for each processor and good load balancing. Then, compile-time analysis is not sufficient to determine data dependencies, location of data and communication patterns; indeed, they are only known at run-time because they depend on the input data. Therefore, it is currently difficult to achieve good performance with these irregular applications.

The second version of the language, **HPF2** [10], introduces new features for efficient irregular data distributions and for data locality specification in irregular computations. To deal with the lack of compile-time information on irregular codes, run-time compilation techniques based on the inspector/executor method have been proposed and widely used.

Major works include PARTI [16] and CHAOS [12] libraries used in Vienna Fortran and Fortran90D [13] compilers, and PILAR library [11] used in PARADIGM compiler [2]. They are efficient for solving iterative irregular problems in which communication and computation phases alternate. RAPID [9] is another run-time system based on a computation specification library for specifying irregular data objects and tasks manipulating them. It is not limited to iterative computations and can address algorithms with *loop-carried* dependencies.

In order to reduce the overhead due to these run-time techniques, the technique called sparse-array rolling (SAR) [15] exploits compile-time information about the representation of distributed sparse matrices for optimizing data accesses. In the Vienna Fortran Compilation System [6], a new directive (SPARSE directive) [15] specifies the sparse matrix structure used in the program to improve the performance of existing sparse codes.

Bik and Wijshoff [3] propose another approach based on a "sparse compiler" that converts a code operating on dense matrices and annoted with sparsity related information into an equivalent sparse code.

Our approach consists of combining compile-time analysis and run-time techniques, but in a way that differs from the previously cited works. We propose a programming style of irregular applications close to the regular case so that both compile-time and run-time techniques can be more easily performed. In order to do so, we use the well-known *tree data structure* to represent sparse matrices and, more generally, irregular data structures with a hierarchical access. This algorithmic representation masks the implementation structure and avoids the indirections coming from the standard irregular programming style. Like in the SAR approach [15], the compiler knows more information about the distributed data, but unlike this approach, information naturally result from the representation of the irregular data structure itself. Our approach is not restricted to sparse matrices but it necessitates a coding of irregular applications using our tree structure. In order to increase portability, we use Derived Data Types (DDT) of Fortran 90 to define trees and hence irregular data structures, and we use some approved extensions of HPF2 for their mapping. In all the following of this paper, we will refer to our language proposition as **HPF2/Tree**. We also propose a run-time support for irregular applications with loop-carried dependencies that cannot be determined at compile-time. Each iteration is performed on a subset of processors only known at run-time. This support is based on an inspector/executor scheme which builds the processor sets, the loop indices for each processor and generates the necessary communications. So, the first step of our work has been to develop a library, called **TriDenT**, which supports distributed trees and implements the above-mentioned optimizations. The performance of this library have been demonstrated on irregular applications such as sparse Cholesky factorization. The following step of this work is to provide an HPF2/Tree compiler; we are currently working on the integration of the TriDenT library in the **ADAPTOR** compilation platform [5].

This paper (see [4] for a more detailed version) is organized as follows. In Section 2, we give a short background on a sparse column Cholesky factorization algorithm which will be our illustrative and experimental example for the validation of our approach. Section 3 presents HPF2/Tree and section 4 describes the TriDenT library. Section 5 gives an experimental study on IBM SP2 for sparse Cholesky factorization applied on real size problems, and provides an analysis of the performance. Finally, section 6 gives some perspectives of this work.

2   An Illustrative Example

The Cholesky factorization of sparse symmetric positive definite matrices is an extremely important computation arising in many scientific and engineering applications. However, this factorization step is quite time-consuming and is frequently the computational bottleneck in these applications. Consequently, it is a significant interesting example for our study. The goal is to factor a sparse symmetric positive definite $n \times n$ matrix A into the form $A = LL^T$, with L lower triangular. Two steps are typically performed for this computation.

First, we perform a *symbolic factorization* to compute the non-zero structure of L from the *ordering* of the unknowns in A; this ordering, for example using a *nested dissection strategy* must reduce the *fill in* and increase the parallelism in the computations. This (irregular) data structure is allocated and its initial non-zero coefficients are those of A. In the pseudo-code given below, this step is implicitly contained in the instruction $L = A$. From this symbolic factorization, one can deduce for each k the two following sets defined as the sparsity structure of row k and the sparsity structure of column k of L :

$$Struct(L_{k*}) = \{j < k \text{ such that } l_{kj} \neq 0\}$$
$$Struct(L_{*k}) = \{i > k \text{ such that } l_{ik} \neq 0\}.$$

Second, the *numerical factorization* computes the non-zero coefficients of L in the data structure. This step, which is the most time-consuming, can be performed by the following sparse column-Cholesky factorization algorithm (see for example [7, 1] and included references).

```
1. L = A
2. for k = 1 to n do
3.    for j ∈ Struct(L_k*) do
4.       for i ∈ Struct(L*j), i ≥ k do % cmod(k,j,i) %
5.          l_ik = l_ik - l_kj * l_ij
6.    l_kk = √(l_kk) % cdiv(k) %
7.    for i ∈ Struct(L*k) do
8.       l_ik = l_ik / l_kk
```

where $cmod(k, j, i)$ represents the modification of the rows of the column k by the corresponding terms in column j, and $cdiv(k)$ represents the division of the column k by the scalar $\sqrt{l_{kk}}$.

This algorithm is said to be a *left-looking* algorithm, since at each stage it accesses needed columns to the left of the current column in the matrix. It is also referred to as a *fan-in* algorithm, since the basic operation is to combine the effects of multiple previous columns on a single subsequent column.

If we consider now the parallelism induced by sparsity and achieved by distributing the columns on processors (so, we exploit the parallelism of the outer loop of instruction 2.), we can see that a given column k depends only of columns belonging to $Struct(L_{k*})$; so, we have *loop-carried dependencies* in this algorithm. In order to take advantage of this structural parallelism, we use an irregular distribution called *subtree-to-subcube* mapping and which leads to an efficient reduction of communication while keeping a good load balance between processors; this mapping is computed algorithmically from the sparse data structure for L (see for example [7] and included references).

3 HPF2/Tree

3.1 Representation of Irregular Data Structures with Trees

In this paper, we consider irregular data structures with a hierarchical access and we propose the use of a tree data structure for their representation. For example, the sparse matrix used in the previous oriented column Cholesky algorithm must be a data structure with a hierarchical column access and the user can represent it by a tree with three levels numbered from 0 to 2. The k-th node on level 1 represents the k-th column. Its p-th son represents the p-th nonzero element defined by its value and row number. Fig. 1 gives a sparse matrix and its corresponding tree. In order to distinguish the levels, we number the columns with roman numbers and the rows with arabic numbers.

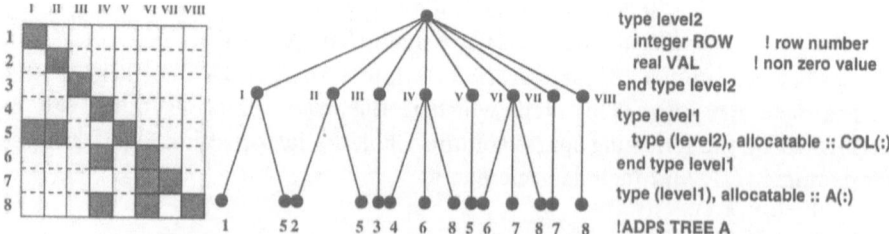

Fig. 1. A Sparse Matrix and its HPF2/Tree Declaration.

3.2 Tree Representation and TREE Directive in HPF2

Trees are embedded in the HPF2 programming language by using the Derived Data Type (DDT) of Fortran 90. For every level of a tree (except level 0), a DDT must be defined. It contains all the data type definitions for this level and (except for the level with the greatest number) the declaration of an array variable whose type is the one associated with the following level. A tree is then declared as an array of the type the DDT associated with the level 1 (cf. Fig. 1).

The TREE directive distinguishes tree variables from other variables using DDT because these variables are used in a restricted way; in particular, pointer attributes of Fortran 90 and recursions are not allowed. So, our tree data structures must not be recursively defined and cannot be used for example to represent quadtree decompositions of matrices [8]. However, this limitation leads to an efficient implementation of trees by the compiler (cf. 4.1).

The accesses to a specific tree element or to a subtree are performed according to the usual Fortran 90 notation. The advantage of this programming style is the analogy between the level notion for trees and the classical dimension notion for arrays used in regular applications. Then, the compiler can perform the classical optimizations based on dependence analysis much more easily than with indirection arrays currently used in irregular applications.

3.3 Distribution of Trees

In order to support irregular applications, HPF2/Tree includes the **GEN_BLOCK** and **INDIRECT** distribution formats and allows the mapping of the components of DDT according to HPF2.0 approved extensions [10]. The constraints for the mapping of derived type components allow the mapping of structure variables at only one level, the *reference* level. For the two distributions of tree A given at Fig. 2, the reference level is level 1.

In HPF2/Tree, the levels preceding the reference level are replicated while the levels following it are distributed according to the distribution of this reference level. This distribution implies an implicit alignment of data with the data of the reference level. This data locality is an important advantage of tree distribution and a compiler can exploit it to generate an efficient code. This is clearly more difficult to obtain when a sparse storage format, such as CSC format [14], is used because the indirection and data arrays are of different sizes and their distribution formats also differ.

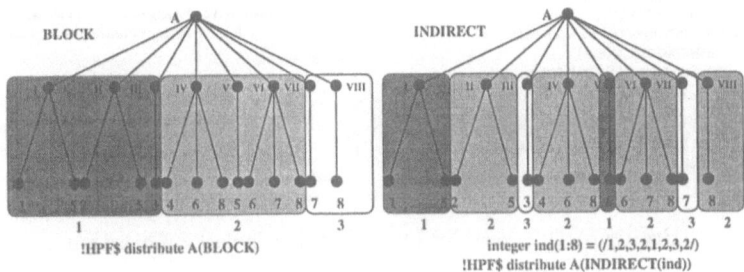

Fig. 2. Tree Distributions with 3 Processors.

3.4 Application to the Sparse Fan-In Cholesky Algorithm

This section describes an HPF2/Tree code (cf. Program 3 (a)) for the sparse fan-in Cholesky algorithm presented in section 2. This code uses the tree A declared in Fig. 1 and another tree, named B, used to store the dependencies between columns. Then, the first value of B(K)%COL(:)%DEP is K, and the other ones are the column numbers given by $Struct(L_{k*})$. In section 4.3, we will show that these data can be computed during an appropriate inspection phase so that this tree B can be avoided. The distribution of tree A uses a specific INDIRECT distribution of nodes on level 1 according to a subtree-to-subcube mapping (cf. section 2). Then, the tree B is aligned with the tree A using the HPF ALIGN directive.

In the outer K loop, the ON directive asserts that the statements at iteration K are performed by the set of processors that own at least one column with a non-zero value in row K. The internal L loop performs a reduction inside this set of processors using the NEW variable TMP_VAL specified in the ON directive. Each iteration L is performed by the processor which owns the column with number J = B(K)%COL(L)%DEP; so, the computation of the contributions for the column K is distributed. Moreover, a compiler might identify that this reduction uses an all-to-one scheme due to the instruction following the loop. The tree notation which avoids indirection in the code, and the data locality and alignment provided by the distribution of the tree make the compiler able to extract the needed communications outside this internal loop (for A(K)%COL(:)%ROW and B(K)%COL(:)%DEP variables). This would not be possible by using a sparse storage format (such as CSC) due to the indirections.

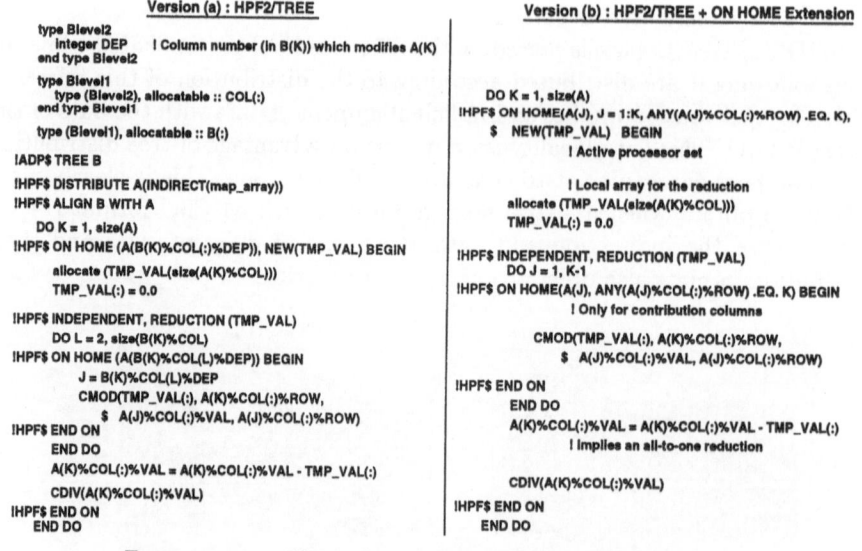

Program 3. 2 Versions of the Fan-In Cholesky Algorithm.

However, this example also shows that run-time techniques are required because the set of active processors specified by the ON directive cannot be known until run-time. In the following section, we describe the TriDenT run-time system which provides supports for tree and for such irregular active processor sets.

4   The TriDenT Library

TriDenT is a library which consists of two parts: a set of routines for the manipulation of trees, and another one for the optimization of computations and communications in irregular processor sets based on the inspector/executor paradigm.

We currently use this library to validate the tree approach and our optimization proposals by writing HPF2 codes with explicit calls to TriDenT primitives. Furthermore, the TriDenT library has been designed to be integrated as easily as possible in the compilation platform ADAPTOR [5] in order to automatically translate HPF2/Tree codes in SPMD codes with calls to TriDenT primitives. ADAPTOR also supports HPF2 features and is based on the **DALIB** (Distributed Array LIBrary) run-time system for distributed arrays. The following sections present the two components of the TriDenT library.

4.1 Support for Distributed Trees

TriDenT implements trees by using an array representation. The **TREE** directive allows the compiler to generate one-dimensional arrays for the implementation of the Derived Data Types associated with the tree; this implementation consists of one array for each scalar variable of a DDT and of some pointer arrays (cf. Fig. 4). Hence, TriDenT provides efficient access functions to tree elements and to subtrees; moreover, the array implementation allows to take advantage of the potential cache effects.

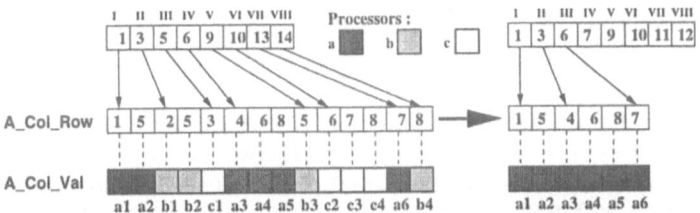

Fig. 4. Indirect Distribution and Local Data for Processor a.

The distribution of trees (cf. 3.3) is performed by using the technique described in [13]. It consists in distributing all the data arrays, for the levels from the reference level, with **GEN_BLOCK** distribution format according to the tree distribution format. Then, the processor local data are stored in a contiguous memory space (cf. Fig. 4 for the indirect distribution of Fig. 2).

4.2 Support for Irregular Processor Subsets

TriDenT is a run-time system which integrates the active processor set notion introduced in HPF2 by the ON directive. This directive specifies how computation is partitioned among processors and therefore may affect the efficiency of computation. Then TriDenT includes context push and pop as well as reduction and broadcast for active processor sets. Moreover, if the ON directive includes an indirection (e.g. ON HOME (A(INDIR(1:4)))), the set can be only determined at run-time and so an inspection phase is necessary. TriDenT provides further optimizations for handling ON directive used inside a global DO loop with loop-carried dependencies which can be determined before this loop.

Our main goal is to improve the efficiency of the global computation (inspection and execution), even if the executor is used only once. The inspector consists of three parts.

The **Set Inspector** analyzes the ON directive of the global loop in order to create in advance the processor subsets associated with each iteration. This will avoid during the execution of the global loop the synchronizations due to the set building messages. In order to allow an automatic generation of these irregular sets, we propose to extend the ON directive by specifying the algorithmic properties to be verified by the variables in the HOME clause (cf. Program 3 (b)). The **Loop Index Inspector** determines the useful loop indices for each processor. The **Communication Inspector** inspects two kinds of communications: those to extract out of the global loop and across-iterations communications. Moreover, the executor optimizes the communications by splitting the send and receive operations and integrating them in the code so as to minimize the synchronisations.

4.3 Application to the Sparse Fan-In Cholesky Algorithm

We briefly described the optimizations achieved by inpection mechanisms on the HPF2/Tree code for the sparse fan-in Cholesky algorithm given in Program 3 (b). This code uses the HOME clause extension introduced in section 4.2 and avoids the effective storage of the dependence tree B used in Program 3 (a) (whose size is of the same order of magnitude as A(:)%COL(:)%ROW).

The Set Inspector involved by the extended ON directive for the global loop directly generates from the A(:)%COL(:)%ROW data (and so without the tree B) the processor set for each iteration K (ACTIVE_PROC(K)).

In this example, the Loop Index Inspection for the inner loop allows to avoid the scan of the A(J)%COL(:)%ROW, and to directly use the valid iteration index sets for each processor.

The compiler can identify A(K)%COL(:)%ROW as loop invariant, and can detect that these data must be sent to the ACTIVE_PROC(K) set; then the Communication Inspector can extract the emission outside the global loop and keep the reception inside iterations.

5 Experimental Results

We validate our contribution with experimental results for a sparse fan-in Cholesky factorization on IBM SP2 with 16 processors. The test matrix is a $n \times n$ sparse matrix with $n = 65024$ achieved from a 2D-grid finite element problem. This matrix is distributed with an INDIRECT mapping according to the *subtree-to-subcube* distribution (cf. section 2). We compare the three versions described at Fig. 5. The **A** version uses the tree B (cf. Program 3 (a)), whereas the **B** and **C** versions use the proposed extended HOME clause (cf. Program 3 (b)). In all the following, the *global time* consists in the summation of *inspection* and *execution times*. All these versions use the Processor Set Inspection. We have verified the effectiveness of this basic inspection by comparing the execution of **A** with another version which doesn't use the processor subsets (35 seconds against 460 seconds for the global time with 16 processors).

Name	Irregular Set Insp.	Loop Index Insp.	Communication Insp.	Dep. (T)ree B or (O)N HOME Ext.
A	Yes	No	No	T
B	Yes	Yes	No	O
C	Yes	Yes	Yes	O

Fig. 5. Characteristics of the Three Versions of the Fan-In Cholesky Factorization

The Fig. 6 presents the relative efficiency with regard to version **A**, first for the execution time only, and second for the global time. The first graph demonstrates the improvment obtained : from 2 to 8% for the Loop Index Inspection (**B**), from 3 to 13% with Communication Inspection added (**C**). The second graph illustrates that the global time, even for only one executor phase, can be better than for the reference version **A**. It appears that the Loop Index Inspection leads to an interesting improvment (4% for **B**). However, the cost of the Communication Inspection is just recover by the executor on 16 processors; nevertheless, one can expect a greater improvement with more processors.

In all our experimentations, we can notice that inspection times are proportional to global time. This property is important for the scalability of our execution support. For example, this time represents between 9% and 15% for version **C** according to the number of processors.

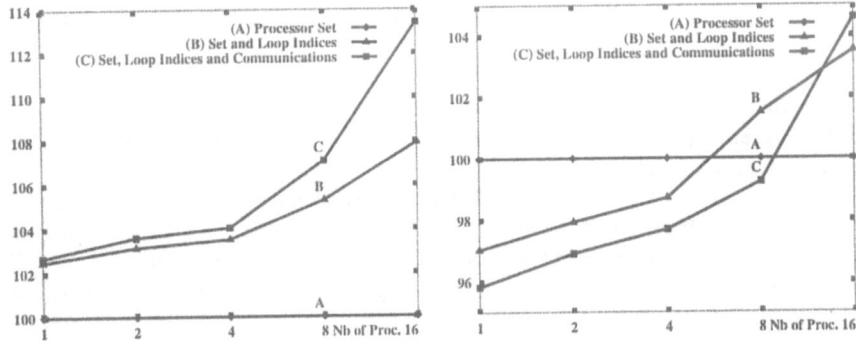

Fig. 6. Relative Efficiencies (Execution Time and Global Time).

So, these experimental results validate our run-time system. Of course, if the factorization step is executed many times, then the inspector cost will be negligible compared with the gain achieved by multiple executions. See [4] for other measures which validate this approach.

6   Conclusion and Perspectives

In this paper, we present a contribution to better handling irregular problems in a HPF2 context. We use a tree representation to take data or distribution irregularity into account. This can help the compiler to perform standard analysis and optimizations. We also introduce an extension for the **ON HOME** directive in order to enable automatic irregular processor set creations, and we propose a run-time support based on inspection/execution paradigm for active processor sets, loop indices and communications for algorithms with loop-carried dependencies. Our approach is validated by experimental results for sparse Cholesky factorization.

Our future work is to extend the properties of our trees (geometrical distributions as BRS or MRD [15], flexible trees in order to enable reallocations and redistributions), and to improve the performance of our Inspector/Executor system (use of the task parallelism induced by the loops). Finally, we are currently working on the integration of TriDenT in the compilation platform ADAPTOR.

References

1. C. Ashcraft, S. C. Eisenstat, J. W.-H. Liu, and A. H. Sherman. A Comparison of Three Column Based Distributed Sparse Factorization Schemes. In *Fifth SIAM Conference on Parallel Processing for Scientific Computing*, 1991.
2. P. Banerjee, J. A. Chandy, M. Gupta, J. G. Holm, A. Lain, D. J. Palermo, S. Ramaswamy, and E. Su. The PARADIGM Compiler for Distributed-Memory Message Passing Multicomputers. In *the First International Workshop on Parallel Processing*, Bangalore, India, December 1994.
3. A. J.C. Bik and H. A.G. Wijshoff. Simple Quantitative Experiments with a Sparse Compiler. In A. Ferreira J. Rolim Y. Saad and T. Yang editors, editors, *Proc. of Third International Workshop, IRREGULAR'96*, volume 1117 of *Lecture Notes in Computer Science*, pages 249–262, Springer, August 1996.
4. T. Brandes, F. Brégier, M.C. Counilh, and J. Roman. Contribution to Better Handling of Irregular Problems in HPF2. Technical Report RR 120598, LaBRI, May 1998.
5. T. Brandes and F. Zimmermann. *Programming Environments for Massively Parallel Distributed Systems*, chapter ADAPTOR - A Transformation Tool for HPF Programs, pages 91–96. In K.M. Decker and R.M. Rehmann, editors, Birkhauser Verlag, April 1994.
6. B. Chapman, S. Benkner, R. Blasko, P. Brezany, M. Egg, T. Fahringer, H.M. Gerndt, J. Hulman, B. Knaus, P. Kutschera, H. Moritsch, A. Schwald, V. Sipkova, and H. Zima. *Vienna Fortran Compilation System, User's Guide Edition*, 1993.
7. K. A. Gallivan et al. *Parallel Algorithms for Matrix Computations*. SIAM, Philadelphia, 1990.
8. J. D. Frens and D. S. Wise. Auto-blocking Matrix-Multiplication or Tracking BLAS3 Performance with Source Code. Technical Report 449, Computer Scince Department, Indiana University, December 1996.
9. C. Fu and T. Yang. Run-Time Techniques for Exploiting Irregular Task Parallelism on Distributed Memory Architectures. *Journal of Parallel and Distributed Computing*, 42:143–156, 1997.
10. HPF Forum. *High Performance Fortran Language Specification*, January 1997. Version 2.0.
11. A. Lain. *Compiler and Run-time Support for Irregular Computations*. PhD thesis, Illinois, 1996.
12. S.S. Mukherjee, S. D. Sharma, M. D. Hill, J. R. Larus, A. Rogers, and J. Saltz. Efficient Support for Irregular Applications on Distributed-Memory Machines. In *ACM SIGPLAN Symposium on Principles & Practice of Parallel Programming (PPoPP)*, July 1995.
13. R. Ponnusamy, Y.S. Hwang, R. Das, J. Saltz, A. Choudhary, and G. Fox. Supporting Irregular Distributions in Fortran 90D/HPF Compilers. *IEEE Parallel and Distributed Technology*, 1995. Technical Report CS-TR-3268 and UMIACS-TR-94-57.

14. Y. Saad. SPARSKIT : a Basic Tool Kit for Sparse Matrix Computations - Version 2. Technical report, CSRD, University of Illinois, June 1994.
15. M. Ujaldon, E. L. Zapata, B. M. Chapman, and H. P. Zima. New Data-Parallel Language Features for Sparse Matrix Computations. In *Proc. of 9th IEEE International Parallel Processing Symposium*, Santa Barbara, California, April 1995.
16. J. Wu, R. Das, J. Saltz, H. Berryman, and S. Hiranandani. Distributed Memory Compiler Design for Sparse Problems. *IEEE Transactions on Computers*, 44(6), 1995.

OpenMP and HPF: Integrating Two Paradigms*

Barbara Chapman[1] and Piyush Mehrotra[2]

[1] VCPC, University of Vienna, Vienna, Austria.
[2] ICASE, MS 403, NASA Langley Research Center, Hampton VA 23681.
barbara@vcpc.univie.ac.at, pm@icase.edu

Abstract. High Performance Fortran is a portable, high-level extension of Fortran for creating data parallel applications on non-uniform memory access machines. Recently, a set of language extensions to Fortran and C based upon a fork-join model of parallel execution was proposed; called OpenMP, it aims to provide a portable shared memory programming interface for shared memory and low latency systems. Both paradigms offer useful features for programming high performance computing systems configured with a mixture of shared and distributed memory. In this paper, we consider how these programming models may be combined to write programs which exploit the full capabilities of such systems.

1 Introduction

Recently, high performance systems with physically distributed memory, and multiple processors with shared memory at each node, have come onto the market. Some have relatively low latency and may support at least moderate levels of shared memory parallelism. This has motivated a group of hardware and compiler vendors to define OpenMP, an interface for shared memory programming which they hope to see adopted by the community. OpenMP [2] supports a model of shared memory parallel programming which is based to some extent on PCF [3], an earlier effort to define a standard interface, but increases the power of its parallel constructs by permitting parallel regions to include subprograms.

HPF was designed to exploit data locality and does not provide features for utilizing shared memory in a system; OpenMP provides the latter, however it does not support data locality. In this paper we consider how these paradigms might be combined to program distributed memory multiprocessing systems (DMMPs) which may require both of these. We assume that such a system consists of a number of interconnected **processing nodes**, or simply **nodes**, each of which is associated with a specific physical memory module and one or more general-purpose **processors** which share this memory and execute a program's instructions at run time: we emphasize this two-level structure by calling such systems SM-DMMPs.

---

* This research was supported by the National Aeronautics and Space Administration under NASA Contract No. NAS1-97046 while both authors were in residence at the Institute for Computer Applications in Science and Engineering (ICASE), NASA Langley Research Center, Hampton, VA 23681-0001.

We first consider interfacing the paradigms, enabling each of them to be used where appropriate. However, there is also potential for a tighter integration of at least some features of HPF with those offered by OpenMP. We therefore examine an approach which combines HPF data mappings with OpenMP constructs, permitting exploitation of shared memory whilst coordinating the mapping of data and work to the target machine in a manner which preserves data locality.

This paper is organized as follows. Section 2 briefly describes the features of each paradigm, and in Section 3 we consider an interface between them based upon the HPF extrinsic mechanism. We speculate on the potential for a closer integration of the two, and illustrate each of them with a multiblock application example.

2 A Brief Comparison of HPF and OpenMP

High Performance Fortran (HPF) is a set of extensions to Fortran, designed to facilitate data parallel programming on a wide range of architectures [1]. It presents the user with a conceptual single thread of control. HPF directives allow the programmer to specify the distribution of data across processors, thus providing a high level description of the program's data locality, while the compiler generates the actual low-level parallel code for communication and scheduling of instructions. HPF also defines directives for expressing loop parallelism and simple task parallelism, where there is no interaction between tasks. Mechanisms for interfacing with other languages and programming models are provided.

OpenMP is a set of directives, with bindings for Fortran 77 and C/C++, for explicit shared memory parallel programming ([2]). It is based upon the *fork-join* execution model, in which a single master thread begins execution and spawns worker threads to perform computations in parallel as required. The user must specify the parallel regions, and may explicitly coordinate threads. The language thus provides directives for declaring such regions, for sharing work among threads, and for synchronizing them. An order may be imposed on variable updates and critical regions defined. Threads may have private copies of some of the program's variables. Parallel regions may differ in the number of threads which execute them, and the assignation of work to threads may also be dynamically determined. Parallel sections easily express other forms of task parallelism; interaction is possible via shared data. The user must take care of any potential race conditions in the code, and must consider in detail the data dependences which need to be respected.

Both paradigms provide a parallel loop with private data and reduction operations, yet these differ significantly in their semantics. Whereas the HPF INDEPENDENT loop requires data independence between iterations (except for reductions), OpenMP requires only that the construct be specified within a parallel region in order to be executed by multiple threads. This implies that the user has to handle any inter-iteration data dependencies explicitly. HPF permits private variables only in this context; otherwise, variables may not have different values in different processes. OpenMP permits private variables – with potentially different values on each thread – in all parallel regions and work-sharing constructs. This extends to common blocks, private copies of which may be local to threads. Even subobjects of shared common blocks may be private. Sequence association is permitted for shared common block data under OpenMP; for HPF, this is the case only if the common blocks are declared to be sequential.

The HPF programming model allows the code to remain essentially sequential. It is suitable for data-driven computations, and on machines where data locality dominates performance. The number of executing processes is static. In contrast, OpenMP creates tasks dynamically and is more easily adapted to a fluctuating workload; tasks may interact in non-trivial ways. Yet it does not enable the binding of a loop iteration to the node storing a specific datum. Thus each programming model has specific merits with respect to programming SM-DMMPs, and provides functionality which cannot be expressed within the other.

3 Combining HPF with OpenMP

In this section we examine two ways of combining HPF and OpenMP such that we can exploit the features of both approaches. Note that the logical processors of HPF can be associated with either the *nodes* or with the individual *processors* of an SM-DMMP. In the former case, the user must rely on automatic shared memory parallelization to exploit all processors on a node. In the latter case, a typical implementation would create separate address spaces for each individual processor, and even communication between two processors on the same node will require an explicit data transfer.

3.1 OpenMP as an HPF Extrinsic Kind

It is a stated goal of HPF to interoperate with other programming paradigms. The language specification defines the *extrinsic* mechanism for interfacing to program units which are written in other languages or do not assume the HPF model of computation. The programming language and computation model must be defined. An interface specification must be provided for each *extrinsic* procedure. This approach keeps the two models completely separate, which implies that the compilers can also be kept distinct; in particular, the OpenMP compiler does not need to know anything about HPF data distributions. For the HPF calling program, execution should proceed just as if an HPF subprogram was invoked. HPF provides language interfaces for Fortran (90/95), Fortran 77 and C. We consider only Fortran programs and OpenMP constructs in this paper.[1]

HPF provides the pre-defined programming models *global*, *serial* and *local* for use with extrinsics. It also permits additional models so long as conceptually one copy of the called procedure executes initially. That is the case for an OpenMP program. We define an *extrinsic* model below which allows us to invoke OpenMP subroutines and functions from within an HPF program. We also indicate how an alternative model might be defined.

---

[1] The two Fortran language versions differ in the assumptions on sequence and storage association, particularly when arguments are passed to subroutines.

The OpenMP extrinsic model We base the *OpenMP* extrinsic kind upon the predefined local model. It assumes that processors defined in an HPF program are associated with nodes of the target system (the number of nodes, not the total number of processors, is returned by HPF's NUMBER_OF_PROCESSORS function).

An OpenMP extrinsic procedure called under this model is realized by creating a single local procedure on each node (or HPF processor). These execute concurrently and independently, exclusively within their respective node, until they return to the HPF calling program upon completion. The calling program blocks until all local procedures have terminated. With the extrinsic kind *OpenMP*, therefore, each local procedure is an independent OpenMP routine, initially executed by a master thread on a processor of the node it is invoked on. It will create worker threads on the same node when a parallel region is encountered; no communication is possible with a local OpenMP procedure on another node. Only the processors and memory on that node are available to it.

An explicit interface will describe the HPF mapping of data required by the routine; any remapping required is performed before the *OpenMP* extrinsic procedure is called. Each invocation will access the local part of distributed data only. Scalar data arguments are available on each node. Only data visible to the master thread will be returned: if they are scalar, the value must be the same on each node. The OpenMP routine may not return to the caller from a parallel region or from within a synchronization construct, and may not have alternate entries. An OpenMP library routine will return information related to the procedure executing on the node where it is invoked only.

This programming model might be very useful for systems with high latency between nodes, such as clusters of workstations, where it is important to map data to the node which needs them. It permits use of a finer grain of parallelism to exploit the processors within each such node once the data is in place. However, it is also restrictive: an *OpenMP* extrinsic routine can only exploit parallelism within a single node, since other nodes, and their data, are not visible to it.

An alternative extrinsic model Other extrinsic kinds may be defined which conform to the requirements of HPF. In particular, an alternative can be defined which permits an OpenMP routine to utilize an entire platform, i.e. it begins with a single master thread which is cognizant of all processors executing the program. The team of threads created to execute a parallel region in the OpenMP code may thus span the entire machine. Since the OpenMP routine will not understand HPF data mappings, all distributed arguments must be mapped to the processor which will initiate the routine during its setup.

This model enables the user to switch between programming models, depending upon which is most suitable for some region of the program. However, it does not enable OpenMP to take advantage of HPF's data locality, and the initial overhead might be large. Its practical value is therefore not clear.

3.2   Merging HPF and OpenMP

An interface between two paradigms obviously does not permit full exploitation of their features. The *OpenMP* extrinsic may be satisfactory for SM-DMMP architectures with high latency. However, the alternative model is less appealing as an approach to programming large low-latency SM-DMMPs, which may benefit from both data locality and the flexibility of a shared memory model. We thus consider augmenting the OpenMP directives with HPF directives directly, so that both shared-memory parallelism and data locality can be expressed within the same routines. OpenMP directives can then describe the parallel computation, while HPF directives may map data to the nodes where it is used and bind the execution of loop iterations to nodes storing distributed data.

Data Mappings The HPF data mapping directives were carefully defined to affect a program's performance, but not its semantics. Thus they can be easily integrated into OpenMP without changing the semantics of OpenMP programs.

It is natural to associate logical HPF processors with processing nodes under this model, since data will be shared within a node unless it is thread-private. Techniques from HPF compilers may be applied to prefetch data in bulk transfers where permitted by the semantics of the code. The resulting language might allow privatization of distributed variables just as HPF does inside INDEPENDENT loops. In general, however, we expect that HPF directives will be used to explicitly map large data objects, which are likely to be shared rather than being private to each thread.

HPF does not support sequence or storage association for explicitly mapped variables. This restriction must carry over to the integrated language so that the compiler may exploit the mapping information. Common blocks with distributed constituents must follow rules of both HPF and OpenMP. That is, if they are named in a THREADPRIVATE directive, or in a PRIVATE clause, each private copy inherits the original distribution. Storage for the explicitly mapped constituents is dissociated from the storage for the common block itself.

In HPF programs, mapped data objects may be implicitly remapped across procedure boundaries to match the prescriptive mapping directives for the formal arguments. On the other hand, if they have been declared DYNAMIC, objects can be explicitly remapped using a REDISTRIBUTE or REALIGN directive. In order to facilitate remapping within an OpenMP program, we must impose the following constraint: *Any code which implicitly or explicitly remaps a data objects must be encountered by all the threads that share that object.* This ensures that if the object is remapped, the new mapping is visible to all the threads which have access to the data.

Additional HPF Constructs HPF provides the ON directive to specify the locus of computation of a block of statements, a loop iteration or subroutine call. This might be used as an alternative to the OpenMP scheduling options to bind the execution of code in a work-sharing construct to one or more nodes, or logical **HPF** processors. An **ON** clause enables the user to specify, for example, the node where a **SINGLE** region or a **SECTION** of a parallel sections construct is to be executed. Similarly, it may bind each iteration of a **DO** construct to the node where a specific data item is stored.

Finally, in contrast to the **INDEPENDENT** directive of **HPF**, parallel loops in **OpenMP** may have inter-iteration data dependencies. Thus an additional work-sharing construct could be based upon the **INDEPENDENT** loop, which permits the prefetching of data and may eliminate expensive synchronizations.

4 An Example

Scientific and engineering applications sometimes use a set of grids, instead of a single grid, to model a complex problem domain. Such *multiblock* codes use tens, hundreds, or even thousands of such grids, with widely varying shapes and sizes.

Structure of the Application The main data structure in such an application stores the values associated with each of the grids in the multiblock domain. It is realized by an array, *MBLOCK*, whose size is the number of grids *NGRID* which is defined at run time. Each of its elements is a Fortran 90 derived type containing the data associated with a single grid.

```
TYPE GRID
   INTEGER NX, NY
   REAL, POINTER :: U(:, :), V(:, :), F(:, :), R(:, :)
END TYPE GRID
TYPE (GRID), ALLOCATABLE :: MBLOCK(:)
```

The decoupling of a domain into separate grids creates internal grid boundaries. Values must be correctly transferred between these at each step of the iteration. The connectivity structure is also input at run time, since it is specific to a given computational domain. It may be represented by an array *CONNECT* (not shown here) whose elements are pairs of sections of two distinct grids.

The structure of the computation is as follows:

```
DO ITERS = 1, MAXITERS
   CALL UPDATE_BDRY (MBLOCK, CONNECT)
   CALL IT_SOLVER (MBLOCK, SUM)
   IF ( SUM .LT. EPS) THEN finished
END DO
```

The first of these routines transfers intermediate boundary results between grid sections which abut each other. We do not discuss it further. The second routine comprises the solution step for each grid. It has the following main loop:

```
SUBROUTINE IT_SOLVER( MBLOCK, SUM )
```

```
   SUM = 0.0
   DO I = 1, NGRID
    CALL SOLVE_GRID( MBLOCK(I)%U, MBLOCK(I)%V, ...)
    CALL RESID_GRID( ISUM, MBLOCK(I)%R, MBLOCK(I)%V, ...)
    SUM = SUM + ISUM
   END DO
   END
```

Such a code generally exhibits two levels of parallelism. The solution step is independent for each individual grid in the multiblock application. Hence each iteration of this loop may be invoked in parallel. If the grid solver admits parallel execution, then major loops in the subroutine it calls may also be parallelized.

Finally the routine also computes a residual; for each grid, it calls a procedure *RESID_GRID* to carry out a reduction operation. The sum of the results across all grids is produced and returned in *SUM*. This would require a parallel reduction operation if the iterations are performed in parallel.

Distributing the Grids We again presume that an HPF processor maps onto a node of the underlying machine rather than to an individual physical processor. We map each grid to a single node (which may "own" several grids) by distributing the array *MBLOCK*. Such a mapping permits levels of parallelism to be exploited. It implies that the boundary update routine may need to transfer data between nodes, whereas the solver routine can run in parallel on the processors of a node using the shared memory in the node to access data. A simple BLOCK distribution, as shown below, may not ensure load balance.

```
   !HPF$ DISTRIBUTE MBLOCK( BLOCK )
```

The INDIRECT distribution of HPF may be used to provide a finer grain of control over the mapping.

4.1 Using an OpenMP Extrinsic Routine

This version calls an *OpenMP* extrinsic to perform the solution step on each grid as well as compute local contributions to the residual. The boundary updates and the final residual summation are performed in the calling HPF program.

An interface declaration is required for the *OpenMP* extrinsic routine:

```
   EXTRINSIC('OPENMP', 'LOCAL')
    SUBROUTINE OPEN(MBLOCK, M, LSUM)
      INTEGER M(:)
      REAL LSUM(:)
      TYPE (GRID), ALLOCATABLE :: MBLOCK(:)
```

Only the *local* segment of *MBLOCK* will be passed to the local procedure, *M* is its size and *LSUM* is the residual value for the local grids on the node.

The extrinsic subroutine is invoked on each node independently. The master thread creates the required number of threads locally. It will return when each of the threads has terminated. The calling HPF routine waits until all extrinsic procedures have terminated.

```
      SUBROUTINE OPEN( LOC_MBLOCK, NLOC, LOC_SUM )
          SUM = 0.0
          CALL GET_THREAD(NLOC, LOC_MBLOCK, N)
!$OMP CALL OMP_SET_NUM_THREADS(N)
!$OMP PARALLEL DO SCHEDULE (DYNAMIC), DEFAULT(SHARED)
!$OMP REDUCTION (+: LOC_SUM)
          DO I = 1, NLOC
            CALL SOLVE_GRID( LOC_MBLOCK(I)%U, ...)
            CALL RESID_GRID( ISUM, LOC_MBLOCK(I)%R, ... )
            LOC_SUM = LOC_SUM + ISUM
          END DO
          RETURN
      END
```

This approach is suitable for systems where the latency between nodes is comparatively high. It permits direct experimentation with, and exploitation of, threads in order to balance work despite large differences in grid sizes.

4.2 Combining HPF and OpenMP

The combined model is very similar to the above; however, it is simpler to write in the sense that there is no need to construct an extrinsic interface. The data declarations and distribution are identical to those of the HPF program. The *IT_SOLVER* routine can be directly written with OpenMP directives in a manner very similar to the code shown above:

```
      SUBROUTINE IT_SOLVER( MBLOCK, SUM )
!HPF$ DISTRIBUTE MBLOCK( BLOCK )
          SUM = 0.0
          CALL GET_THREAD(NGRID, MBLOCK, N)
!$OMP CALL OMP_SET_NUM_THREADS(N)
!$OMP PARALLEL DO SCHEDULE (DYNAMIC), DEFAULT(SHARED)
!$OMP REDUCTION (+: SUM)
          DO I = 1, NGRID
!$HPF     ON (HOME (MBLOCK(I)), RESIDENT
            CALL SOLVE_GRID( MBLOCK(I)%U, ...)
            CALL RESID_GRID( ISUM, MBLOCK(I)%R, ... )
!$HPF     END ON
          SUM = SUM + ISUM
          END DO
          RETURN
      END
```

This version again deals with the full array $MBLOCK$, rather than with local segments. Thus, there may be a need to specify the locus of computation so that each iteration is performed on the processor which stores its data. This can be done using an HPF-style ON-block along with the RESIDENT directive, as shown.

The execution differs. This time, the initialization of SUM will be performed once only, on a single master thread which controls the entire machine. The number of threads, N, is computed for the entire system, and the work is distributed across all processors. This may enable an improved load balance. We have avoided a two-step computation of SUM.

This approach is appropriate for a system with comparatively low latency, which allows a single master thread on the machine and a global scheduling policy, permitting a compiler/runtime system to do global load balancing if appropriate.

5   Summary

OpenMP was recently proposed for adoption in the community by a group of major vendors. It is not yet clear what level of acceptance it will find, yet there is certainly a need for portable shared-memory programming. The functionality it provides differs strongly from that of HPF, which considers the needs of distributed memory systems. Since many current architectures have requirements for both data locality and efficient shared memory parallel programming, there are potential benefits in combining elements of these two paradigms. We have shown in this paper that the HPF extrinsic mechanism enables usage of each within a single program. This may facilitate the programming of workstation clusters with standard interconnection technology, for example. It is less clear if a large low latency machine will benefit from a simple interface. For these, there appears to be added value in providing some level of integration. Given the orthogonality of many concepts in the two programming models, they are fundamentally highly compatible and a useful integrated model can be specified. In particular, a combination of HPF data locality directives with OpenMP parallelization constructs extends the scope of applicability of each of them. Since at least one major vendor has already provided data mapping options together with a shared memory parallel programming model [4], we expect that this avenue will be closely explored by the OpenMP consortium in the near future.

References

1. High Performance Fortran Forum: High Performance Fortran Language Specification. Version 2.0, January 1997
2. OpenMP Consortium: OpenMP Fortran Application Program Interface, Version 1.0, October, 1997
3. B. Leasure (Ed.): Parallel Processing Model for High Level Programming Languages, Draft Proposed National Standard for Information Processing Systems, April 1994
4. Silicon Graphics, Inc.: MIPSpro Fortran 77 Programmer's Guide, 1996

Towards a Java Environment for SPMD Programming

Bryan Carpenter, Guansong Zhang, Geoffrey Fox
Xiaoming Li*, Xinying Li and Yuhong Wen

NPAC at Syracuse University
Syracuse, New York,
NY 13244, USA
{dbc,zgs,gcf,lxm,xli,wen}@npac.syr.edu

Abstract. As a relatively straightforward object-oriented language, Java is a plausible basis for a scientific parallel programming language. We outline a conservative set of language extensions to support this kind of programming. The programming style advocated is Single Program Multiple Data (SPMD), with parallel arrays added as language primitives. Communications involving distributed arrays are handled through a standard library of collective operations. Because the underlying programming model is SPMD programming, direct calls to other communication packages are also possible from this language.

1 Introduction

Java boasts a direct simplicity reminiscent of Fortran, but also incorporates many of the important ideas of modern object-oriented programming. Of course it comes with an established track-record in the domains of Web and Internet programming. The idea that Java may enable new programming environments, combining attractive user interfaces with high performance computation, is gaining increasing attention amongst computational scientists [7, 8].

This article will focus specifically on the potential of Java as a language for scientific parallel programming. We envisage a framework called *HPJava*. This would be a general environment for parallel computation. Ultimately it should combine tools, class libraries, and language extensions to support various established paradigms for parallel computation, including shared memory programming, explicit message-passing, and array-parallel programming. This is a rather ambitious vision, and the current article only discusses some first steps towards a general framework. In particular we will make specific proposals for the sector of HPJava most directly related to its namesake: High Performance Fortran.

For now we do not propose to import the full HPF programming model to Java. After several years of effort by various compiler groups, HPF compilers are still quite immature. It seems difficult justify a comparable effort for Java

---

* Current address: Peking University

before success has been convincingly demonstrated in Fortran. In any case there are features of the HPF model that make it less attractive in the context of the integrated parallel programming environment we envisage. Although an HPF program *can* interoperate with modules written in other parallel programming styles through the HPF extrinsic procedure interface, that mechanism is quite awkward. Rather than follow the HPF model directly, we propose introducing some of the characteristic ideas of HPF—specifically its distributed array model and array intrinsic functions and libraries—into a basically SPMD programming model. Because the programming model is SPMD, direct calls to MPI [1] or other communication packages are allowed from the HPJava program.

The language outlined here provides HPF-like distributed arrays as language primitives, and new *distributed control* constructs to facilitate access to the local elements of these arrays. In the SPMD mold, the model allows processors the freedom to independently execute complex procedures on local elements. All access to *non-local* array elements must go through library functions—typically collective communication operations. This puts an extra onus on the programmer; but making communication explicit encourages the programmer to write algorithms that exploit locality, and simplifies the task of the compiler writer. On the other hand, by providing distributed arrays as language primitives we are able to simplify error-prone tasks such as converting between local and global array subscripts and determining which processor holds a particular element. As in HPF, it is possible to write programs at a natural level of abstraction where the meaning is insensitive to the detailed mapping of elements. Lower-level styles of programming are also possible.

2 Distributed Arrays

HPJava adds class libraries and some additional syntax for dealing with *distributed arrays*. These arrays are viewed as coherent global entities, but their elements may be divided across a set of cooperating processes. The new objects represent true multidimensional arrays. They allow regular section subscripting, similar to Fortran 90 arrays. These properties all appear very attractive within the scientific parallel programming community, and have become established in HPF and related languages. But they imply a kind of array structurally very different to Java's builtin arrays. A design decision was made not to attempt integration of the new arrays with standard Java array types: instead they are a new kind of entity that coexists in the language with ordinary Java arrays. The type-signatures and constructors of the multidimensional array use double brackets to distinguish them from ordinary arrays.

In this example:

```
Procs2 p = new Procs2(3, 2) ;

Range x = new BlockRange(100, p.dim(0)) ;
Range y = new BlockRange(200, p.dim(1)) ;

float [[,]] a = new float [[x, y]] on p ;
```

a is created as a 100×200 array, block-distributed over the 6 processes in **p**. The ideas will be discussed in more detail in the following paragraphs, but the fragment is essentially equivalent to the HPF declarations

```
!HPF$ PROCESSORS P(3, 2)

      REAL A(100, 200)
!HPF$ DISTRIBUTE A(BLOCK, BLOCK) ONTO P
```

Before examining the HPJava syntax for declaration of distributed arrays we will discuss the *process groups* over which their elements are scattered. A base class **Group** describes a general group of processes. It has subclasses **Procs1**, **Procs2**, . . . , representing one-dimensional process grids, two-dimensional process grids, and so on. In the example p represents a 3 by 2 grid of processes. At the time the constructor of p is executed the program should be running on six or more processes. Declaration of p in the Java program corresponds directly to declaration of the processor arrangement P in the HPF fragment. In HPJava additional objects describing individual dimensions of a process grid are accessed through the inquiry member **dim**.

Some or all of the dimensions of a multi-dimensional array can be declared as *distributed ranges*. In general a distributed range is represented by an object of class **Range**. A **Range** object defines a range of integer subscripts, and specifies how they are mapped into a process grid dimension. For example, the class **BlockRange** is a subclass of **Range** describing a simple block-distributed range of subscripts. Like **BLOCK** distribution format in HPF, it maps blocks of contiguous subscripts to each element of its target process dimension[1]. The constructor of **BlockRange** usually takes two arguments: the extent of the range and a **Dimension** object defining the process dimension over which the new range is distributed.

In general the constructor of the distributed array must be followed by an **on** clause, specifying the process group over which the array is distributed. Distributed ranges of the array must be distributed over distinct dimensions of this group. The **on** clause can be omitted in some circumstances—see Sect. 3.

A distributed array can also have ordinary sequential dimensions, in which case the range slot in the constructor is replaced by an integer extent, and the corresponding slot in the type signature should contain an asterisk:

```
float [[,*]] b = new float [[x, 10]] on p ;
```

---

[1] Other range subclasses include CyclicRange, which produces the equivalent of CYCLIC distribution format in HPF.

The asterisk is not mandatory (array types decorated in this way are regarded as subtypes of the undecorated ones) but it allows some extra flexibility in subscripting sequential dimensions[2].

With one or two extra constructors for groups and ranges, the framework described here captures all the distribution and alignment features of HPF.

Because arrays like a are declared as a collective objects we can apply collective operations to them. For example a standard library called Adlib (one of a number of existing communication libraries that may eventually be brought into the HPJava framework) provides many of the array functions of Fortran 90.

```
float [[,]] c = new float [[x, y]] on p ;

Adlib.shift(a, c, -1, 0, CYCL) ;
```

At the edges of the local segment of a the **shift** operation causes the local values of a to be overwritten with values of c from a processor adjacent in the x dimension.

Subscripting operations on distributed arrays are subject to certain restrictions. The HPJava model is a formalization of the distributed memory SPMD programming style, and all memory accesses must refer to the memory of the local processor. In general therefore an array reference like

```
a [17, 23] = 13 ;
```

is illegal because it may imply access to an element held on a different processor. The language provides several *distributed control* constructs to alleviate the inconvenience of this restriction. These will be discussed in the next few sections.

3   The *on* Construct and the Active Process Group

The concept of *process groups* has an important role in the HPJava model. The class **Group** (process grid classes are special cases) has a member function called **local**. This returns a boolean value **true** if the local process is a member of the group, **false** otherwise. In

```
if(p.local()) {
  ...
}
```

the code inside the conditional is executed only if the local process is a member p. HPJava provides a short way of writing this construct

```
on(p) {
  ...
}
```

---

[2] Incidentally, because b is has no range distributed over p.dim(1) it is understood to be *replicated* over this dimension of the grid.

The *on* construct provides some extra value. The language incorporates a formal idea of the *active process group* (APG). At any point of execution some process group is singled out as the APG. An on(p) construct specifically changes the value of the APG to p. On exit from the construct, the APG is restored to its value on entry.

Elevating the APG to a part of the language allows some simplifications. For example, it provides a natural default for the on clause in array constructors. More importantly, formally defining the APG simplifies the statement of various rules about what operations are legal *inside* distributed control constructs like *on*. A set of formal rules defines how these constructs can be nested and what data accesses are legal inside them. These rules have been presented elsewhere.

4 Locations and the *at* Construct

As noted at the end of Sect. 2, a mechanism is needed to ensure that the array reference

```
a [17, 23] = 13 ;
```

is legal, because the local process holds the element in question. In general determining whether an element is local may be a non-trivial task.

Rather than use integer values directly as local subscripts in distributed array dimensions, the idea of a *location* is introduced. A location can be viewed as an abstract element, or "slot", of a distributed range. An individual location is described by an object of the class Location. Each Location element is mapped to a particular slice of a process grid. In general two locations are identical only if they come from the same position in the same range. A subscripting syntax is used to represent location n in range x:

```
Location i = x [n]
```

Locations are used to parametrize a second distributed control construct called the *at* construct. This is similar to *on*, except that its body is executed only on processes that hold a specified location. For distributed array dimensions, locations rather than integers are used as subscripts. Access to element a [17, 23] can be safely written as:

```
Location i = x [17], j = y [23] ;

at(i)
  at(j)
    a [i, j] = 13 ;
```

Locations used as array subscripts must be elements of the corresponding ranges of the array.

5 Distributed Loops

The *at* mechanism of the previous section is often useful, but a more urgent requirement is a mechanism for *parallel* access to distributed array elements.

The last and most important distributed control construct in the language is called *overall*. It implements a distributed parallel loop. Conceptually it is quite similar to the **FORALL** construct of Fortran, except that the *overall* construct specifies exactly where its parallel iterations are to be performed. The argument of *overall* is a member of the special class **Index**. This class is a subclass of **Location**, so it is syntactically correct to use an index as an array subscript. Here is an example of a pair of nested *overall* loops:

```
float [[,]] a = new float [[x, y]], b = new float [[x, y]] ;
...
Index i, j ;
overall(i = x | :)
  overall(j = y | :)
    a [i, j] = 2 * b [i, j] ;
```

The body of an *overall* construct executes, conceptually in parallel, for every location in the range of its index. An individual "iteration" executes on just those processors holding the location associated with the iteration. The net effect of the example above should be reasonably clear. It assigns twice the value of each element of b to the corresponding element of a. Because of the rules about *where* an individual iteration iterates, the body of an *overall* can usually only combine elements of arrays that have some simple alignment relation relative to one another. The **idx** member of range can be used in parallel updates to yield expressions that depend on global index values.

The ":" in the *overall* examples above can be replaced by a general Fortran-like *subscript triplet* of the form l : u : s, where l, u and s are respectively lower bound upper bound and a stride. The ": s" part can be omitted, in which case s defaults to 1; l or u can be omitted, in which case the default to 0 or $N - 1$ respectively, where N is the size of the range specified before the "|". Hence the degenerate isolated ":" selects the whole of the range.

With the *overall* construct we can give some useful examples of parallel programs. Figure 1 gives a parallel implementation of red-black relaxation in the extended language. To support the important stencil-update paradigm, *ghost regions* are allowed on distributed arrays. Ghost regions are extensions of the locally held block of a distributed array, used to cache values of elements held on adjacent processors. In our case the width of these regions is specified in a special form of the **BlockRange** constructor. The ghost regions are explicitly brought up to date using the library function **writeHalo**. Its arguments are an array with suitable extensions and a vector defining in each dimension the width of the halo that must actually be updated.

Note that the new range constructor and **writeHalo** function are *library* features, not new language extensions. One new piece of syntax is needed: the addition and subtraction operators are overloaded so that integer offsets can be added or subtracted to locations, yielding new, shifted, locations. This kind of shifted does not imply access to off-processor data. It only works if the subscripted array has suitable ghost extensions.

Figure 2 gives a parallel implementation of Cholesky decomposition in the extended language. The first dimension of a is sequential ("collapsed" in HPF parlance). The second dimension is distributed (cyclically, to improve load-balancing). This a column-oriented decomposition.

The example introduces Fortran-like *array sections*. These are distinguished from local subscripting operations by the use of double brackets. The subscripts are typically scalar (integers) or triplets [3]. The example also involves one new operation from the Adlib library. The function **remap** copies the elements of one distributed array or section to another of the same shape. The two arrays can have any, unrelated decompositions. In the current example **remap** is used to implement a broadcast. Because b has no range distributed over p, it implicitly has *replicated* mapping; **remap** accordingly copies identical values to all processors.

We have covered most of the important *language* features we are implementing. Two additional features that are quite important in practice but have not been discussed are *subranges* and *subgroups*. A subrange is simply a range which is a regular section of some other range, created by syntax like x [0 : 49]. Subranges can be used to create distributed arrays with general HPF-like alignments. A *subgroup* is some slice of a process array, formed by restricting process coordinates in one or more dimensions to single values. Subgroups formally describe the state of the active process group inside *at* and *overall* constructs. For a more complete description of a slightly earlier version of the proposed language, see [3].

Note that if distributed array accesses are genuinely irregular, the necessary subscripting cannot usually be *directly* expressed in our language, because subscripts cannot be computed randomly in parallel loops without violating the fundamental SPMD restriction that all accesses be local. This is not regarded as a shortcoming: on the contrary it forces explicit use of an appropriate library package for handling irregular accesses (such as CHAOS [6]). Of course a suitable binding of such a package is needed in our language.

6 Discussion

We have described a conservative set of extensions to Java. In the context of an explicitly SPMD programming environment with a good communication library, we claim these extensions provide much of the concise expressiveness of HPF, without relying on very sophisticated compiler analysis. The object-oriented features of Java are exploited to give an elegant parameterization of the distributed arrays in the extended language. Because of the relatively low-level programming model, interfacing to other parallel-programming paradigms is more natural than in HPF. With suitable care, it is possible to make direct calls to, say, MPI from within the data parallel program (in [2] we suggest a concrete Java binding for MPI).

It is necessary to address various questions:

Why introduce language extensions and additional syntax—why not work entirely with class libraries for distributed arrays? In fact we consciously put

---

[3] In the example upper bounds of various triplets default to $N - 1$, determined by the extent of the corresponding array range.

```
Procs2 p = new Procs2(P, P) ;

on(p) {
  Range x = new BlockRange(N, p.dim(0), 1) ;  // ghost width 1
  Range y = new BlockRange(N, p.dim(1), 1) ;  // ghost width 1

  float [[,]] u = new float [[x, y]] ;

  int [] widths = {1, 1} ;          // Widths updated by 'writeHalo'

  // ... some code to initialise 'u'

  for(int iter = 0 ; iter < NITER ; iter++) {
    for(int parity = 0 ; parity < 2 ; parity++) {

      Adlib.writeHalo(u, widths) ;

      Index i, j ;
      overall(i = x | 1 : N - 2)
        overall(j = y | 1 + (x.idx(i) + parity) % 2 : N - 2 : 2)
          u [i, j] = 0.25 * (u [i - 1, j] + u [i + 1, j] +
                             u [i, j - 1] + u [i, j + 1]) ;
    }
  }
}
```

Fig. 1. Red-black iteration using **writeHalo**.

```
Procs1 p = new Procs1(P) ;
on(p) {
  Range x = new CyclicRange(N, p.dim(0));

  float [[*,]] a = new float [[N, x]] ;

  float [[*]]  b = new float [[N]] ;  // buffer

  // ... some code to initialise 'a'

  Location l ;
  Index m ;

  for(int k = 0 ; k < N - 1 ; k++) {

    at(l = x [k]) {
      float d = Math.sqrt(a [k, l]) ;

      a [k, l] = d ;
      for(int s = k + 1 ; s < N ; s++)
        a [s, l] /= d ;
    }

    Adlib.remap(b [[k + 1 : ]], a [[k + 1 : , k]]);

    overall(m = x | k + 1 : )
      for(int i = x.idx(m) ; i < N ; i++)
        a [i, m] -= b [i] * b [x.idx(m)] ;
  }

  at(l = x [N - 1])
    a [N - 1, l] = Math.sqrt(a [N - 1, l]) ;
}
```

Fig. 2. Cholesky decomposition.

as much of the framework into libraries as seemed practical, based on extensive experimentation (in Java and C++). All communication operations and collective operations are library functions. Although we introduce a basic syntactic framework for creating distributed arrays[4], all variants of distribution format are captured in runtime libraries, in the Range hierarchy. Similary all variants of processor arrangement are captured in the Group hierarchy. Unfortunately it seems to be difficult to obtain a pure library implementation of the basic data-parallel loops and local array subscripting operations that is flexible, convenient, *and* efficient. Our suggestion is that, while it would certainly be unrealistic to expect the syntax extensions outlined here to be incorporated in the standard Java language, relatively simple preprocessors can convert the extended language to ugly but efficient standard Java plus class-library calls, as required.

Why does the language seem to have awkward restrictions on subscripting, compared with HPF? The answer is: because HPJava is supposed to be a thin layer on top of a straightforward distributed SPMD program with calls to class libraries. In general the restrictions make it much easier to implement efficiently than HPF.

Why adopt the distributed memory model—why not do parallel programming with Java threads? Mainly because our target hardware is clusters of commodity processors, not expensive SMPs.

The language extensions described were devised partly to provide a convenient interface to a distributed-array library developed in the PCRC project [5, 4]. Hence most of the run-time technology needed to implement the language is available "off-the-shelf". The existing library includes the run-time descriptor for distributed arrays and a comprehensive array communication library. The HPJava compiler itself is being implemented initially as a translator to ordinary Java, through a compiler construction framework also developed in the PCRC project [12].

A complementary approach to communication in a distributed array environment is the one-sided-communication model of Global Arrays (GA) [9]. For task-parallel problems this approach is often more convenient than the schedule-oriented communication of CHAOS (say). Again, the language model we advocate here appears quite compatible with GA approach—there is no obvious reason why a binding to a version of GA could not be straightforwardly integrated with the the distributed array extensions of the language described here.

Finally we mention two language projects that have some similarities. Spar [11] is a Java-based language for array-parallel programming. There are some similarities in syntax, but semantically Spar is very different to our language. Spar expresses parallelism but not explicit data placement or communication—it is a higher level language. ZPL [10] is a new programming language for scientific computations. Like Spar, it is an array language. It has an idea of performing computations over a *region*, or set of indices. Within a compound statement

---

[4] Even this framework could be eliminated if Java provided a feature like C++ templates.

prefixed by a *region specifier*, aligned elements of arrays distributed over the same region can be accessed. This has similarities to our *overall* construct.

References

1. Bryan Carpenter, Yuh-Jye Chang, Geoffrey Fox, Donald Leskiw, and Xiaoming Li. Experiments with HPJava. *Concurrency: Practice and Experience*, 9(6):633, 1997.
2. Bryan Carpenter, Geoffrey Fox, Xinying Li, and Guansong Zhang. A draft Java binding for MPI. http://www.npac.syr.edu/projects/pcrc/doc.
3. Bryan Carpenter, Guansong Zhang, Geoffrey Fox, Xinying Li, and Yuhong Wen. Introduction to Java-Ad. http://www.npac.syr.edu/projects/pcrc/doc.
4. Bryan Carpenter, Guansong Zhang, and Yuhong Wen. NPAC PCRC run-time kernel definition. Technical Report CRPC-TR97726, Center for Research on Parallel Computation, 1997. Up-to-date version maintained at http://www.npac.syr.edu/projects/pcrc/doc.
5. Parallel Compiler Runtime Consortium. Common runtime support for high-performance parallel languages. In *Supercomputing '93*. IEEE Computer Society Press, 1993.
6. R. Das, M. Uysal, J.H. Salz, and Y.-S. Hwang. Communication optimizations for irregular scientific computations on distributed memory architectures. *Journal of Parallel and Distributed Computing*, 22(3):462–479, September 1994.
7. Geoffrey C. Fox, editor. *Java for Computational Science and Engineering—Simulation and Modelling*, volume 9(6) of *Concurrency: Practice and Experience*, June 1997.
8. Geoffrey C. Fox, editor. *Java for Computational Science and Engineering—Simulation and Modelling II*, volume 9(11) of *Concurrency: Practice and Experience*, November 1997.
9. J. Nieplocha, R.J. Harrison, and R.J. Littlefield. The Global Array: Non-uniform-memory-access programming model for high-performance computers. *The Journal of Supercomputing*, 10:197–220, 1996.
10. Lawrence Snyder. A ZPL programming guide. Technical report, University of Washington, May 1997.
    http://www.cs.washington.edu/research/projects/zpl/.
11. Kees van Reeuwijk, Arjan J. C. van Gemund, and Henk J. Sips. Spar: A programming language for semi-automatic compilation of parallel programs. *Concurrency: Practice and Experience*, 9(11):1193–1205, 1997.
12. Guansong Zhang, Bryan Carpenter, Geoffrey Fox, Xiaoming Li, Xinying Li, and Yuhong Wen. PCRC-based HPF compilation. In *10th International Workshop on Languages and Compilers for Parallel Computing*, 1997. To appear in Lecture Notes in Computer Science.

Language Constructs and Run-Time System for Parallel Cellular Programming

Giandomenico Spezzano and Domenico Talia

ISI-CNR c/o DEIS,
Università della Calabria,
87036 Rende (CS), Italy
{spezzano, talia}@si.deis.unical.it

Abstract. This paper describes the main features of CARPET, a high-level programming language based on cellular automata theory. The language has been designed for supporting the development of parallel high-performance software abstracting from the parallel architecture on which programs run. A CARPET user can write cellular programs to describe the actions of a very large number of simple active agents interacting locally. The CARPET run-time system allows a user to observe, also in a graphical format, the global results that arises from their parallel execution.

1 Introduction

The lack of high-level languages, tools, and application-oriented environments often limits the design and implementation of parallel algorithms that are portable, efficient, and expressive. The *restricted-computation structures* represent one of the most important models of parallel processing [8]. The interest for these models is due to the possibility to restrict the form of computations so as to restrict communication volume achieving high performance. *Restricted-computation* models offer a user a structured paradigm of parallel programming and improve the performance of the parallel algorithms reducing the overheads due to the communication *latency*. Further, tools can be designed to estimate the performance of various constructs of a high-level language on a specific parallel architecture.

Cellular processing languages based on the cellular automata (CA) model [10] represent a significant example of restricted-computation that it is used to model parallel computation for a large number of applications in biology, physics, geophysics, chemistry, economics, artificial life, and engineering. A cellular automaton consists of one-dimensional or multi-dimensional lattice of *cells*, each of which is connected to a finite neighborhood of cells that are nearby in the lattice. Each cell in the regular spatial lattice can take any of a finite number of discrete state values. Time is discrete, as well, and at each time step all the cells in the lattice are updated by means of a local rule, called *transition function*, that determines the cell's next state based upon the states of its neighbors. That is, the state of a cell at a given time depends only on its own state and the states of its nearby neighbors at the previous time step. Different neighborhoods can be defined for the cells. The most common neighborhoods in the two-dimensional case are the von Neumann neighborhood consisting of the North, South, East, West neighbors and the Moore neighborhood composed of eight neighbor cells. The global behavior of the an automaton is defined by the evolution of the states of all cells as a result of multiple interactions.

CA are intrinsically parallel so they can be simulated onto parallel computers running the cell transition functions in parallel with high efficiency, as the communication flow between processors can be kept low. In fact, in our approach, a cellular algorithm is composed of all the transition functions of cells that compose the lattice. Each transition function generally contains the same local rule, but it is also possible to define some cells with different transition functions (inhomogeneous cellular automata).

According to this approach, we designed and implemented a high-level programming language, called CARPET (CellulAR Programming EnvironmenT) [9], that allows a user to design cellular algorithms. In particular, CARPET has been used for programming cellular algorithms in the CAMEL (Cellular Automata environMent for systEms ModeLing environment) [3]. A user can design cellular programs by CARPET describing the actions of many simple active agents (implemented by the cells) interacting locally, then the CAMEL system runs cell transition functions in parallel allowing a user to observe the global complex evolution that arises from all the local interactions. A number of cellular programming languages such as Cellang [4], CDL [6], and CARP [7] have been defined in the last decade. However, none of those contains all the features of CARPET neither a parallel run-time support for them has been implemented till today.

2 CARPET

The rational of CARPET is to make parallel computers available to application-oriented users hiding the implementation issues coming from their architectural complexity. A CARPET user can program complex problems that may be represented as discrete across a lattice. Parallelism inherent to its programming model is not apparent to the programmer.

CARPET implements a cellular automaton as an SPMD program. CA are implemented as a number of processes each one mapped on a distinct PE that executes the same code on different data. According to this approach, a user must specify by CARPET only the transition function of a single cell of the system he wants to simulate. The language uses the control structures, the types, the operators of the C language. A CARPET program is composed by a declaration part that appears only once in the program and must precede any statement (except those of C pre-processor) and by a program body. The program body has the usual C statements, without I/O instructions, and a set of statements to deal with the state of a cell and its neighborhood. Further, CARPET users may use C functions or procedures to improve the structure of programs.

The declaration section includes constructs to define the dimensions of the automaton (dimension), the radius of the neighborhood (radius), the type of the neighborhood (neighbor), and to specify the state of a cell (state) as a set of typed substates that can be *char, shorts, integers, floats, doubles* and *arrays* of these basic types.

The dimension declaration defines the number of dimensions of the automaton in a discrete Cartesian space. The maximum number of dimensions allowed in the current implementation is 3. Radius defines the set of cells that can compose the neighborhood of a cell.

In CARPET the state of a cell is composed of a record of typed substates, unlike classical cellular automata where the cell state is represented by a few bits. The typification of the substates extends the range of the applications that can be coded in CARPET simplifying the writing the programs and improving their readability. In the following example the state is composed of three substates:

state (short direction, float mass, speed);

A substate of the current cell can be referred by the variable cell_substate (eg., cell_speed). To guarantee the semantics of cell updating in cellular automata the value of one substate of a cell can be modified only by the update operation. After an execution of the update statement, the value of the substate, in the current iteration, is unchanged. The new value does take effect in the next iteration.

The neighbor declaration assigns a name to a set of specified neighbor cells (of the current cell). This mechanism allows a cell to access by name the values of the substates of its neighbor cells. Neighborhoods can be asymmetrical or have any other special topological properties (e.g., hexagonal). Furthermore, in the neighbor declaration it must be defined the name of a vector that has as length the number of elements composing the logic neighborhood. The name of the vector can be used as an alias in referring to the neighbor cell. For instance, the von Neumann neighborhood shown in figure 1, can be defined as follows:

neighbor Neum[4]([0,-1]North,[-1,0]West,[0,1]South,[1,0]East);

A substate of a neighbor cell is referred, for instance, as North_speed. Using the vector name the same substate can be referred also as Neum[0]_speed. This referring way makes simpler to write loops in CARPET programs.

CARPET allows the program to know the number of iterations that have been executed by the predefined variable step. Step is silently updated by the run-time system. Initially its value is 0 and it is incremented by 1 each time all the cells of the automata are updated. This feature allows a user to define also time-dependent neighborhoods. The step variable permits also to change dynamically the values of the substates dependent upon the iterations.

In modeling a complex system, it is often necessary to describe some global features of the system. CARPET allows a user to define global parameters and initialize them to specific values. The value of a parameter is the same in every cell of the automaton. For this reason, the value of each parameter cannot be

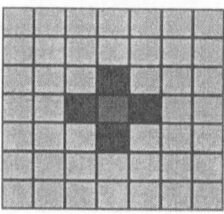

Fig. 1. The von Neumann neighborhood in a two-dimensional CA lattice.

changed in the program but it can only be modified, during the simulation, by the UI. An example of parameter declaration in CARPET is

parameter (permeability 0.9) ;

Input and output of a CARPET program can be performed by files or by the edit function of the UI. A file can contain the values of one substate of all cells, so these values can be loaded at the step 0 to initialize the automaton. The output of a CARPET program can automatically be saved in files at regular intervals to be post-processed by mathematical or visualization tools. CARPET offers a user the possibility to define non-deterministic rules using of a random function. Finally, CARPET allows a user to define cells with different transition functions by means of the Getx, Gety, Getz operations that return the value of coordinates X, Y, and Z of a cell in the automaton.

Differently form other cellular languages, CARPET does not provide statements to configure the automata, to visualize the cell values or to define the process-to processor mapping. These features can be defined by means of the GUI of its runtime system. The CAMEL GUI allows a user to define the size of cellular automata, the number of the processors onto which an automaton must be executed, and choose the colors to be assigned to the cell substates to support the graphical visualization of their values. By excluding constructs for configuration and visualization from the language, it is possible to execute the same CARPET compiled code using different configurations.

3   A Parallel Cellular Run Time System

Parallel computers are the best practical support for the effective implementation of high-performance CA [1]. According to this basic idea has been developed CAMEL. It is a parallel software system based on the cellular automata model that constitutes the parallel run-time system of CARPET. CAMEL has been implemented on a parallel computer composed of a mesh of Transputers connected to a host node [2]. CAMEL provides a GUI to configure a program, to monitor the parameters of a simulation and to dynamically change them at run time. The parallel execution of cellular algorithms is implemented by the parallel execution of the transition function of each cell in the Single Program Multiple

Data (SPMD) way. A portable implementation of CAMEL for MIMD parallel computers based on the MPI communication library is developing.

The CAMEL system is composed by a set of *Macrocell* processes, each one running on a single processing element of the parallel machine, and by a *Controller* process running on a master processor. Each *Macrocell* process implements an automaton partition that includes several elementary cells, and it makes use of a communication system that handles the data exchange among cells. All the *Macrocells* of the lattice execute in parallel the local rule that results in a global transformation of the whole lattice state. CAMEL uses a load balancing strategy for assigning lattice partitions to the processors of a parallel computer [2]. This load balancing is a domain decomposition strategy similar to the scattered decomposition technique.

4   A Simple Program

This section shows a simple program written by CARPET. This example should familiarize the reader with the CARPET approach. The program in figure 2 shows how the CARPET constructs can be used to implement the *parity rule* algorithm. The parity rule is a very simple example of "self-reproduction" in cellular automata; an initial pattern is replicated at some specific iteration that is a power of two. The cells can have 0 or 1 values only. A cell takes the sum of its neighbors and goes to 1 if the sum is odd, or to 0 if the sum is even. Let us call N the number of '1' cells among the four nearest neighbors of a given cell. The transition rule is the following: given a cell, if N is odd, the new value of the cell will be 0; if N is even the cell's value does not change.

```
caef
{ dimension 2; /*bidimensional lattice */
  radius 1;
  state (short value);
  neighbor cross[4]([0,-1]North,[-1,0]West,
                    [0,1]South,[1,0]East);
}
{int i; short N = 0;
 for (i = 0; i<4; i++)
  N = cross_value[i] + N;
 if (N%2 == 1)
  update (cell_value,0); /*updating the state of a cell*/
}
```

Fig. 2. The parity rule algorithm written in CARPET.

5   Performance

Together with the minimization of elapsed time, scalability is a major goal in the design of parallel computing applications. Scalability shows the potential ability of parallel computers to speedup their performance by adding more processing elements. The scalability of CARPET programs has been measured increasing the number of PEs used to solve the same problem, and using the same number of processing elements to solve bigger problems [2].

The speedup obtained on 32 Transputers with an automaton composed of 224x70 cells which simulated the Ontake mountain (Japan) landslide was 25.0 (78.1% efficient). Table 1 shows the times (in seconds) and the speed-up obtained running the landslide simulation on CAMEL using 1, 2, 4, 8, 16 and 32 PEs. The second column shows the number of cells, which are mapped, on each PE. Similar performance results have been obtained in other applications developed by CARPET [3].

Table 1. Execution times and speed-up.

Number of PEs	Number Cells/PE	Time for 2000 steps	Speed-up
1	15680	10231.59	1
2	7840	6392.48	1.6
4	3920	3181.40	3.2
8	1960	1447.96	7.0
16	980	728.11	14.0
32	490	407.84	25.0

6   Final Remarks and Future Work

As stated also by M. J. Flynn [5], the cellular automata model is a new mathematical way to represent problems that allows to effectively use parallel computers achieving scalable performance.

In this paper, we described the CARPET programming language designed for programming cellular algorithms on parallel computers. CARPET has been used successfully to implement several real life applications such as landslide simulation, lavaflow models, freeway traffic simulation, image processing, and genetic algorithms. The experience during the design, implementation, and use of the CARPET language showed us that high-level languages are very useful in the development of cellular algorithms for solving complex problems in science and engineering.

Currently, the CARPET language is used in the COLOMBO (Parallel COmputers improve cLean up of sOils by Modelling BiOremediation) project within the ESPRIT framework. The COLOMBO main objective is the use of CA models for the bioremediation of contaminated soils. This project is developing a portable MPI-based implementation of CARPET and CAMEL on MIMD parallel computers such as the Meiko CS-2, the CRAY T3E, and workstation clusters. The new implementation will make the CARPET programs portable on a large number of MIMD machines.

Acknowledgements This research has been partially funded by the CEC ESPRIT project n° 24,907.

References

1. B. P. Brinch Hansen, Parallel Cellular Automata: a Model for Computational Science. Concurrency: Practice and Experience, 5:425-448, 1993.
2. M. Cannataro, S. Di Gregorio, R. Rongo, W. Spataro, G. Spezzano, and D. Talia, A Parallel Cellular Automata Environment on Multicomputers for Computational Science. Parallel Computing, 21:803-824, 1995.
3. S. Di Gregorio, R. Rongo, W. Spataro, G. Spezzano, and D. Talia., A Parallel Cellular Tool for Interactive Modeling and Simulation. IEEE Computational Science & Engineering 3:33-43, 1996.
4. J. D. Eckart, Cellang 2.0: Reference Manual. ACM Sigplan Notices, 27:107-112, 1992.
5. M.J. Flynn, Parallel Processors Were the Future and May Yet Be. IEEE Computer 29:152, 1996.
6. C. Hochberger and R. Hoffmann, CDL - a Language for Cellular Processing. In: Proc. 2nd Intern. Conference on Massively Parallel Computing Systems, IEEE Computer Society Press, 1996.
7. G. Junger, Cellular Automaton Tool User Manual. GMD, Sankt Augustin, Germany, 1994.
8. D.B. Skillicorn and D. Talia, Models and Languages for Parallel Computation. ACM Computing Survey, 30, 1998.
9. G.Spezzano and D. Talia, A High-Level Cellular Programming Model for Massively Parallel Processing. In: Proc. 2nd Int. Workshop on High-Level Programming Models and Supportive Environments (HIPS97), IEEE Computer Society, pages 55-63, 1997.
10. J. von Neumann, Theory of Self Reproducing Automata. University of Illinois Press, 1966.

Task Parallel Skeletons for Irregularly Structured Problems

Petra Hofstedt*

Department of Computer Science, Dresden University of Technology,
hofstedt@inf.tu-dresden.de

Abstract. The integration of a task parallel skeleton into a functional programming language is presented. Task parallel skeletons, as other algorithmic skeletons, represent general parallelization patterns. They are introduced into otherwise sequential languages to enable the development of parallel applications. Into functional programming languages, they naturally are integrated as higher-order functional forms.

We show by means of the example branch-and-bound that the introduction of task parallel skeletons into a functional programming language is advantageous with regard to the comfort of programming, achieving good computation performance at the same time.

1 Introduction

Most parallel programs were and are written in imperative languages. In many of these languages, the programmer has to use low-level constructs to express parallelism, synchronization and communication. To support platform-independent development of parallel programs standards and systems have been invented, e.g. MPI and PVM. In functional languages, such supporting libraries have been added in rudimentary form only recently. Hence, the advantages of functional programs, such as their ability to state powerful algorithms in a short, abstract and precise way, cannot be combined with the ability to control the parallel execution of processes on parallel architectures.

Our aim is to remedy that situation. A functional language has been extended by constructs for data and task parallel programming. We want to provide comfortable tools to exploit parallelism for the user, so that she is burdened as few as possible with communication, synchronization, load balancing, data and task distribution, reaching at the same time good performance by exploitation of parallelism. The extension of functional languages by algorithmic skeletons is a promising approach to introduce data parallelism as well as task parallelism into these languages.

As demonstrated for imperative languages, e.g. by Cole [3], there are several approaches how to introduce skeletons into functional languages as higher-order

---

* The work of this author was supported by the 'Graduiertenkolleg Werkzeuge zum effektiven Einsatz paralleler und verteilter Rechnersysteme' of the German Research Foundation (DFG) at the Dresden University of Technology.

parallel forms. However, most authors concentrated on data parallel skeletons, e.g. [1], [4], [5]. Hence, our aim has been to explore the promising concept of task parallel skeletons for functional languages by integrating them into a functional language. A major focus of our work is on reusability of methods implemented as skeletons.

Algorithmic skeletons are integrated into otherwise sequential languages to express parallelization patterns. In our approach, currently skeletons are implemented in a lower-level imperative programming language, but presented as higher-order functions in the functional language. Implementation-details are hidden within the skeletons. In this way, it is possible to combine expressiveness and flexibility of the sequential functional language with the efficiency of parallel special purpose algorithms. Depending on the type of parallelism exploited, skeletons are distinguished in data and task parallel ones. Data parallel skeletons apply functions on multiple data at the same time. Task parallel skeletons express which elements of a computation may be executed in parallel. The implementation in the underlying system determines the number and the location of parallel processes that are generated to execute the task parallel skeleton.

2 The Branch-and-Bound Skeleton in a Functional Language

Branch-and-bound methods are systematic search techniques for solving discrete optimization problems. Starting with a set of variables with a finite set of discrete values (a domain) assigned to each of the variables, the aim is to assign a value of the corresponding domain to each variable in such a way that a given objective function reaches a minimum or a maximum value and several constraints are satisfied. First, mutually disjunct subproblems are generated from a given initial problem by using an appropriate branching rule (**branch**). For each of the generated subproblems an estimation (**bound**) is computed. By means of this estimation, the subproblem to be branched next is chosen (**select**) and decomposed (branched). If the chosen problem cannot be branched into further subproblems, its solution (if existing) is an optimal solution. Subproblems with non-optimal or inadmissible variable assignments can be eliminated during the computation (**elimination**). The four rules **branch, bound, select** and **elimination** are called basic rules.

The principal difference between parallel and sequential branch-and-bound algorithms lies in the way of handling the generated knowledge. Subproblems generated from problems by decomposition and knowledge about local and global optima belong to this knowledge. While with sequential branch-and-bound one processor generates and uses the complete knowledge, the distribution of work causes a distribution of knowledge, and the interaction of the processors working together to solve the problem becomes necessary.

Starting point for our implementation was the functional language DFS ('Datenparallele funktionale Sprache' [6]), which already contained data parallel skeletons for distributed arrays. DFS is an experimental programming language to be used on parallel computers. The language is strict and evaluates DFS-programs in a call-by-value strategy accordingly.

To give the user the possibility to exploit parallelism in a very comfortable way, we have extended the functional language DFS by task parallel skeletons. One of them was a branch-and-bound skeleton. The user provides the basic rules **branch, bound, select** and **elimination** using the functional language. Then she can make a function call to the skeleton as follows:

`branch&bound branch bound select elimination problem.`

A parallel abstract machine (PAM) represents the runtime environment for DFS. The PAM consists of a number of interconnected nodes communicating by messages. Each node consists of three units: the message administration unit, which handles the incoming and outgoing messages, the skeleton unit, which is responsible for skeleton processing, and the reduction unit, which performs the actual computation. Skeletons are the only source of parallelism in the programs.

To implement the parallel branch-and-bound skeleton, several design decisions had to be made with the objective of good computation performance and high comfort. In the following, the implementation is characterized according to Trienekens' classification ([7]) of parallel branch-and-bound algorithms.

Table 1. Classification by Trienekens

knowledge sharing	global/local knowledge base complete/partial knowledge base update strategy
knowledge use	access strategy reaction strategy
dividing the work	units of work load balancing strategy
synchronicity	synchronicity of each process
basic rules	**branch, bound, select, elimination**

Each process uses a *local partial knowledge base* containing only a part of the complete generated knowledge. In this way, the bottleneck arising from the access of all processes to a shared knowledge base is avoided, but at the expense of the actuality of the knowledge base. A process stores newly generated knowledge at its local knowledge base only; if a local optimum has been computed, the value is broadcasted to all other processes (*update strategy*).

When a process has finished a subtask, it accesses its local knowledge base, to store the results at the knowledge base and to get a new subtask to solve. If a process receives a message containing a local optimum, the process compares this optimum with its actual local optimum and the bounds of the subtasks still to be solved (*access strategy*). A process receiving a local optimum from another process first finishes its actual task and then reacts according to the received message (*reaction strategy*). This may result in the execution of unnecessary work. But the extent of this work is small because of the high granularity of the distributed work.

A *unit of work* consists of branching a problem into subproblems and computing the bounds of the newly generated subproblems. The *load balancing strategy* is simple and suited to the structure of the computation, because new subproblems are generated during computation. If a processor has no more work, it asks its neighbours one after the other for work. If a processor receives a request for work, it returns a unit of work – if one is available – to the asking processor. The processor sends that unit of work which is nearest to the root of the problem tree and has not been solved yet.

The implemented distributed algorithm works *asynchronously*.

The basic rules are provided by the user using the functional language.

3 Performance Evaluation

To evaluate the performance of task parallel skeletons, we implemented branch-and-bound for the language DFS as a task parallel skeleton in C for a GigaCluster GCel1024 with 1024 transputers T805(30 MHz) (each with 4 MByte local memory) running the operating system Parix.

Performance measurements for several machine scheduling problems – typical applications of the branch-and-bound method – were made to demonstrate the advantageous application of skeletons. In the following, three cases of a machine scheduling problem for 2 machines and 5 products have been considered. The number of possible orders of machine allocation is 5! = 120. This very small problem size is sufficient to demonstrate the consequences for the distribution of work and the computation performance, if the part of the problem tree, which must be computed, has a different extent. In case (a) the complete problem tree had to be generated. In case (b) only one branch of the tree had to be computed. Case (c) is a case where a larger part of the problem tree than in case (b) had to be computed. Each of the machine scheduling problems has been defined first in the standard functional way and second by use of the branch-and-bound skeleton. The measurements were made using different numbers of processors.

First we counted branching steps, i.e. we measured the average overall number of decompositions of subproblems of all processors working together in the computation of the problem. These measurements showed the extend of the computed problem tree working sequentially and parallelly. It became obvious that the overall number of branching steps is increasing in the case of a small number of to be branched subproblems. The local partial knowledge bases and the asynchronous behaviour of the algorithm cause the execution of unnecessary work. If the whole problem tree had to be generated, we observed a decrease of the average overall number of branching steps with increasing number of processors. This behaviour is called *acceleration anomaly* ([2]). Acceleration anomalies occur if the search tree generated in the parallel case is smaller than the one generated in the sequential case. This can happen in the parallel case because of branching several subproblems at the same time. Therefore it is possible to find an optimum earlier than in the sequential case. Acceleration anomalies cause a disproportional decrease of the average maximum number of branching steps per processor with increasing number of processors, a super speedup.

Table 2. Average overall number of reduction steps

	1 proc. functional	1 proc. skeleton	4 proc. skeleton	6 proc. skeleton	8 proc. skeleton	16 proc. skeleton
(a)	17197	20727	1848,2	1792,9	1074,9	709,3
(b)	54	47	138,0	218,6	299,0	451,6
(c)	138	118	179,8	261,8	258,6	484,8

Table 3. Average maximum number of reduction steps per processor

	1 proc. functional	1 proc. skeleton	4 proc. skeleton	6 proc. skeleton	8 proc. skeleton	16 proc. skeleton
(a)	17197	20727	896,0	710,2	390,5	113,1
(b)	54	47	43,2	44,7	47,0	46,4
(c)	138	118	65,5	66,1	60,9	51,7

To compare sequential functional programs with programs defined by means of skeletons, we counted reduction steps. The reduction steps include besides branching a problem, the computation of a bound of the optimal solution of a subproblem, the comparison of these bounds for selection and elimination of a subproblem from a set of to be solved subproblems, and the comparison of bounds to determine an optimum. Table 2 shows the *average overall number of reduction steps* of all processors participating in the computation. In Table 3 the *average maximum numbers of reduction steps per processor* are given. Table 2 and Table 3 clearly show the described effect of an acceleration anomaly. Because the set of reduction steps contains comparison steps of the operation 'selection of subproblems from a set of to be solved subproblems', the distribution of work causes a decrease of the number of comparison steps for this operation at each processor.

Looking at both Table 2 and Table 3 it becomes apparent that the numbers of reduction steps in the sequential cases of (a), (b), and (c) of the computation of the problem first defined in the standard functional way and second using the skeleton differ. That is caused by different styles of programming in functional and imperative languages. In case (a) an obvious decrease of the average maximum number of reduction steps per processor (Table 3) caused by the distribution of the subproblems onto several processors is observable. At the same time the average overall number of reduction steps (Table 2) is also decreasing as explained before. The distribution of work onto several processors yields a large increase of efficiency in cases when a large part of the problem tree must be computed. In case (b) the average maximum number of reduction steps per processor nearly does not change while the overall number of reduction steps is increasing, because, firstly, subproblems, which are to be branched, are distributed, and secondly, a larger part of the problem tree is computed. Because in case (b) the solution can be found in a short time, working parallelly as well as sequentially, the use of several processors produces overhead only. In case (c) the same phenomena as in case (b) are observable. Moreover, the average maximum number of reduction steps decreases to nearly 50% in case of parallel computation in comparison to the sequential computation.

4 Conclusion

The concept, implementation, and application of task parallel skeletons in a functional language were presented. Task parallel skeletons appear to be a natural and elegant extension to functional programming languages. This has been shown using the language DFS and a parallel branch-and-bound skeleton as an example. Performance evaluations showed that using the implemented skeleton for finding solutions for a machine scheduling problem is performance better, especially if a large part of the problem tree has to be generated. Also in the case of the necessity to compute a smaller part of the problem tree only, a distribution of work is advantageous.

Acknowledgements The author would like to thank Herbert Kuchen and Hermann Härtig for discussions, helpful suggestions and comments.

References

1. Botorog, G.H., Kuchen, H.: Efficient Parallel Programming with Algorithmic Skeletons. In: Boug, L. (Ed.): Proceedings of Euro-Par'96, Vol.1. LNCS 1123. 1996.
2. de Bruin, A., Kindvater, G.A.P., Trienekens, H.W.J.M.: Asynchronous Parallel Branch and Bound and Anomalies. In: Ferreira, A.: Parallel algorithms for irregularly structured problems. Irregular '95. LNCS 980. 1995.
3. Cole, M.: Algorithmic Skeletons: Structured Management of Parallel Computation. MIT Press. 1989.
4. Darlington, J., Field, A.J., Harrison, P.G., Kelly, P.H.J., Sharp, D.W.N., Wu, Q., While, R.L.: Parallel Programming Using Skeleton Functions. In: Bode, A. (Ed.): Parallel Architectures and Languages Europe : 5th International PARLE Conference. LNCS 694. 1993.
5. Darlington, J., Guo, Y., To, H.W., Yang, J.: Functional Skeletons for Parallel Coordination. In: Haridi, S. (Ed.): Proceedings of Euro-Par'95. LNCS 966. 1995.
6. Park, S.-B.: Implementierung einer datenparallelen funktionalen Programmiersprache auf einem Transputersystem. Diplomarbeit. RWTH Aachen 1995.
7. Trienekens, H.W.J.M.: Parallel Branch and Bound Algorithms. Dissertation. Universität Rotterdam 1990.

Synchronizing Communication Primitives for a Shared Memory Programming Model

Vladimir Vlassov and Lars-Erik Thorelli

Royal Institute of Technology, Electrum 204, S-164 40 Kista, Sweden
{vlad, le}@it.kth.se

Abstract. In this article we concentrate on semantics of operations on synchronizing shared memory that provide useful primitives for shared memory access and update. The operations are a part of a new shared-memory programming model called mEDA. The main goal of this research is to provide a flexible, effective and relatively simple model of inter-process communication and synchronization via synchronizing shared memory.

1 Introduction

There is a belief [7], and we share it, that dataflow computing in combination with multithreading is a general answer on two fundamental issues in multi-processing: memory latency and synchronization [3]. Synchronization of parallel processes [11] is a subject of considerable interest for various research group, both in industry and academia [8, 9, 12–14]. This article presents some aspects of a new shared-memory programming model that is based on dataflow principles and provides a unified approach for communication and synchronization of parallel processes via synchronizing shared memory. The model is called mEDA, Extended Dataflow Actor, where the prefix "m" is used to distinguish a new revision of the model from the EDA model presented by Wu [17, 15]. The model was inspired from different computation and memory models, such as the data flow model [6], the Actor model [1], I-structures [3] and M-structures [5].

In this article we concentrate on the semantics of synchronizing shared memory operations introduced in the model. The model states that a shared-memory operation (store, fetch) holds a synchronization type which specifies special synchronization requirements for both the accessing process and the accessed cell. In this way, common patterns of parallel programs such as data dependency, shared protected data, barriers, stream communication, AND- and OR-parallelism, can be conveniently expressed. The notion of processes accessing shared "boxes" (cells) using such operations provides a natural way of communication and synchronization.

Section2 informally presents the semantics of mEDA synchronizing memory operations. Section3 outlines synchronization and communication properties of the operations. Section4 shortly illustrates how mEDA allows parallel processes to communicate and be synchronized via shared memory. Our conclusions are given in Section5.

2 Synchronizing Memory Operations

Assume that a parallel application consists of a dynamic set of processes (or threads) and a set of cells which are shared, uniquely identified and known to the processes of the application. The processes interact with one another in a one-sided fashion storing data to, or fetching data from, shared cells. The processes can also transfer values from one cell to another. A process may access cells only by special store/fetch/transfer operations defined in the model. Since each process has a notion of local state, on this level of abstraction, a shared cell can be regarded as a special kind of process. An mEDA operation sends to the shared memory an access request with the following attributes: address of requested cell, operation code, value to be stored (in the case of store operation), and identifier of the requesting process. The latter is used to direct a fetch response or a store acknowledgment when needed. A shared cell may contain stored data (a value) and a FIFO queue that holds the postponed access-requests which can not be satisfied because of inappropriate state of the requested cell. We extend the set of values, which can be stored to, or fetched from, the shared cell, by an "empty" value. We call a shared cell *full* if it contains a normal value. The cell is *empty* if it contains the empty value.

mEDA recognizes four synchronization types of store operations (denoted by X, S, I, and U-store), and four synchronization types of fetch operations (X, S, I, and U-fetch). In addition the model defines 16 transfer operations briefly described in Section 2.3. The semantics of store and fetch operations is based on considerations given below. See also Fig. 1.

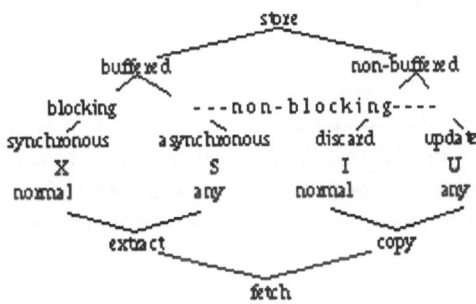

Fig. 1. Categorization of EDA operations

2.1 Store Operations

First, we state that a store operation to an empty cell always succeeds. If the cell is full, the store request either can be postponed until the variable is emptied (*buffered store*), or be served somehow as soon as it arrives (*non-buffered store*). Next we assume that a buffered store can be either blocking (synchronous) or non-blocking (asynchronous). A blocking store, denoted by X in Fig. 1, requires an acknowledgment that a value has been stored. The requesting process is suspended until the acknowledgment arrives. On the other hand, the process executing a non-blocking buffered store (S), continues its computation as soon as the request is sent to shared cell. A non-buffered store-request to a full cell can be either ignored (I), or served by updating the requested cell independently of its state (U). In both cases, a synchronizing acknowledgment is not needed since non-buffered requests are served without delay. Based on the above observations we arrive at four types of store operations: X-store ("eXclusive"), S-store ("Stream"), I-store ("Ignore"), and U-store ("Update").

2.2 Fetch Operations

We distinguish two kinds of fetch operations categorized in Fig. 1: (i) *extracting* operations which empty a shared cell after its value has been copied to the accessing process, and (ii) *copying* operations which copy a value from the cell to the process and leave the cell intact. We also distinguish the cases that a fetch operation may fetch (extract or copy) either only a normal value or any value (normal or empty). The request to fetch a normal value is postponed on an empty cell until the latter is filled. In all cases, a requesting process is blocked until the requested value (any or only normal) is fetched from the cell. The operations requesting any value can be classified as non-blocking. Thus, mEDA includes four types of fetch operations: X-fetch, S-fetch, I-fetch, and U-fetch, where the first two are extracting and the first and third one are blocking.

2.3 Transfer Operations

A transfer operation combines a pair fetch/store in one operation. Consider the expression `store(s_s, B, fetch(s_f, A))` that requires to fetch a value from cell A and store the value to cell B. Here `s_f` and `s_s` specify synchronization types of fetch and store to cell A and cell B, respectively. To implement the expression efficiently, we introduce special operations (XX-trans, SX-trans, etc.) which transfer data between shared cells. The operations allow increasing the efficiency of the model. All transfer operations are non-blocking and do not affect the requesting process: the value fetched from A, is directly stored to B.

3 Synchronization and Communication Properties of mEDA Operations

The mEDA operations on a shared cell issued by cooperating processes trigger each other. When a process empties a full cell or fills an empty cell, one or more postponed requests (if any) queued on the cell, are served. This may cause resuming of blocked processes waiting for the completion of X-store, X-fetch or I-fetch. Storing a normal value to an empty cell allows all first pending copy requests (I-fetch) and one following pending extract request (X-fetch), if any, to be resumed. Extracting a normal value from a full cell (X-fetch, S-fetch) or storing the empty value to the full cell (U-store) allows resuming and servicing one postponed store request, if any, queued on the cell. It is important to note that the shared memory may not service incoming requests until all postponed requests which can be resumed, are served.

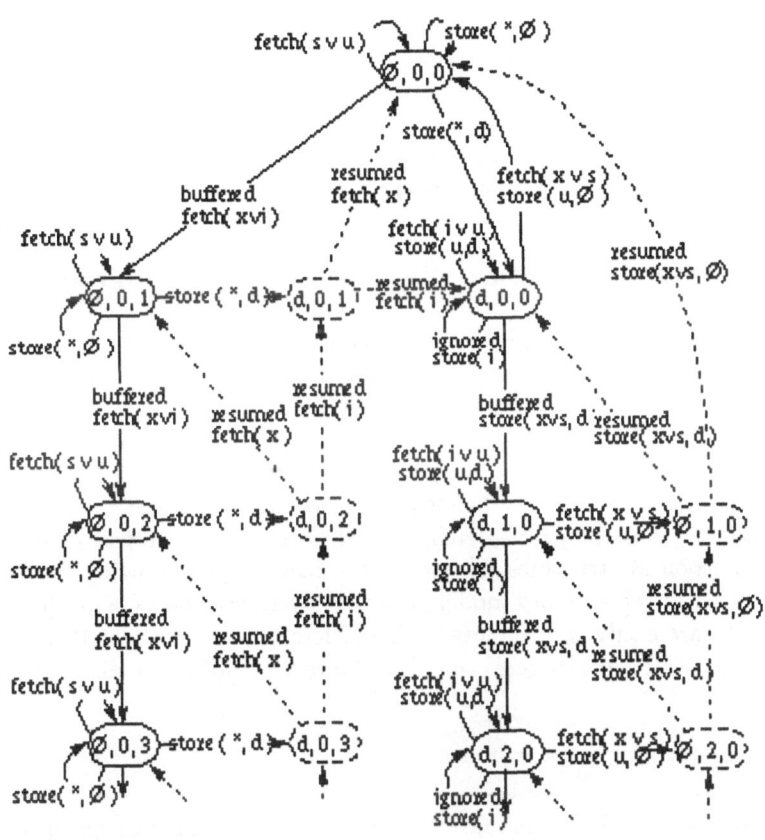

Fig. 2. State Diagram of a Shared Cell

Figure 2 illustrates a state diagram of a shared cell. Each state is marked by a triple index (C, q_s, q_f), where:

– C - indicates which value, normal or empty, is stored in the cell:

$$C = \begin{cases} \emptyset \text{ if the cell is empty} \\ d \text{ if the cell is full} \end{cases}$$

– $q_s \in \{0, 1, 2, \ldots\}$ - the number of store-requests (X or S) queued on the cell.
– $q_f \in \{0, 1, 2, \ldots\}$ - the number of fetch-requests (X or I) queued on the cell.

Each transaction is marked by incoming or resumed access-requests causing the transaction. For instance,

- fetch$(i \vee u)$ means a fetch request with either I or U synchronization.
- store$(*, \emptyset)$ means store the empty value $\emptyset$ with any synchronization.
- store(u, d) means store a normal value d with U synchronization.

Intuitively it is easy to understand that a queue of requests postponed on a shared cell, cannot contain a mixture of storing and fetching requests. If the queue is not empty, it contains either only store requests of X and/or S types, queued on the empty cell, or only fetch requests (X and/or I) queued on the full cell. This consideration is confirmed by the diagram in Fig. 2. While resuming postponed requests, the cell passes through the states depicted in a dash shape until a request (if any) that heads the queue, can be resumed.

4 Process Interaction via Synchronizing Shared Memory

The store and fetch operations introduced in mEDA can be used in different combinations in order to provide a wide range of various communication and synchronization patterns for parallel processes such as data dependency, locks, barriers, and OR-parallelism. For example, X-fetch in combination with any buffered store (X- or S-store) supports data-driven computation where a process accessing shared location by X-fetch is blocked until requested data are available. One can see that the pair X-fetch/X-store is similar to the pair of *take* and *put* operations on M-structures [5], and to the pair of synchronizing *read* and *write* introduced in the memory model presented by Dennis and Gao in [8].

Locks are easily implemented using X-fetch/X-store, and OR-parallelism is achieved with I-fetch/I-store, since an I-store to a full cell is discarded.

5 Conclusions

We have presented the semantics of shared memory operations introduced in the mEDA model, and demonstrated how these operations allow implementing a wide variety of inter-process communication and synchronization patterns in a convenient way. Orthogonal mEDA operations provide a unified approach for communication and synchronization of parallel processes via synchronizing shared memory supporting different access and synchronization needs.

The model was implemented on top of the PVM system [10]. Two software implementation of mEDA were earlier reported in [2] and [16]. In the last version [16], the synchronizing shared memory for PVM is implemented as distributed virtual shared variables addressed by task and variable identifiers. The values stored to and fetched from the shared memory are PVM messages. The mEDA library can be used in combination with PVM message passing. This allows a wide variety of communication and synchronization patterns to be programmed in a convenient way.

Finally, we believe that the memory system of a multiprocessor supporting mEDA synchronizing memory operations, can be relatively efficiently and cheaply implemented. This issue may be of interest to computer architects.

References

1. Agha, G.: Concurrent Object-Oriented Programming. Communications of the ACM **33** (1990) 125–141
2. Ahmed, H., Thorelli, L.-E., Vlassov, V.: mEDA: A Parallel Programming Environment. Proc. of the 21st EUROMICRO Conference: Design of Hardware/Software Systems, Como, Italy. IEEE Computer Society Press. (1995) 253–260
3. Arvind and Thomas, R. E.: I-structures: An Efficient Data Structure for Functional Languages. TM-178. LCS/MIT (1980)
4. Arvind and Iannucci, A.: Two fundamental issues in multiprocessing. Parallel computing in Science and Engineering. Lecture Notes in Computer Science, Vol. 295. Springer-Verlag (1978) 61–88
5. Barth, P. S., Nikhil, R. S., Arvind.: M-Structures: Extending a Parallel, Non-Strict, Functional Language with States. TM-237. LCS/MIT (1991)
6. Dennis, J. B.: Evolution of "Static" dataflow architecture. In: Gaudiot, J.-L., Bic L. (eds.): Advanced Topics in Dataflow Computing. Prentice-Hall (1991) 35–91
7. Dennis, J. B.: Machines and Models for Parallel Computing. International Journal of Parallel Programming **22** (1994) 47–77
8. Dennis, J. B., Gao, G. R.: On memory models and cache management for shared-memory multiprocessors. In: Proc. of the IEEE Symposium on Parallel and Distributed Processing, San Antonio (1995)
9. Fillo, M., Keckler, S.W., Dally, W. J., Carter, N. P., Chang, A., Gurevich, Ye., Lee, W. S.: The M-Machine Multicomputer. In: Proc. of the 28th IEEE/ACM Annual Int. Symp. on Microarchitecture, Ann Arbor, MI (1995)
10. Geist, A., Beguelin, A., Dongarra, J., Jiang, W., Manchek, R., Sunderam, V.: PVM: Parallel Virtual Machine. A Users' Guide and Tutorial for Networked Parallel Computing. MIT Press (1994)
11. Mellor-Crummey, J. M., Scott, M. L.: Algorithms for Scalable Synchronization on Shared-Memory Multiprocessors. ACM Trans. on Computer Systems **9** (1991) 21–65
12. Ramachandran, U., Lee, J.: Cache-Based Synchronization in Shared Memory Multiprocessors. Journal of Parallel and Distributed Computing **32** (1996) 11–27
13. Scott, S. L.: Synchronization and Communication in the T3E Multiprocessor. In: Proc. 7th International Conference on Architectural Support for Programming Languages and Operating Systems (ASPLOS VII) (1996) 26–36
14. Skillicorn, D. B., Hill, J. M. D., McColl, W.F.: Questions and Answers about BSP. Tech. Rep. PRG-TR-15-96. Oxford University. Computing Laboratory (1996)
15. Thorelli, L.-E.: The EDA Multiprocessing Model. Tech. Rep. TRITA-IT-R 94:28. Dept. of Teleinformatics, Royal Institute of Technology, Stockholm, Sweden (1994).
16. Vlassov, V., Thorelli, A Synchronizing Shared Memory: Model and Programming Implementation. In: Recent Advances in PVM and MPI. Proc.of the 4th European PVM/MPI Users' Group Meeting. Lecture Notes on Computer Science, Vol. 964. Springer-Verlag (1995) 288–293
17. Wu, H.: Extension of Data-Flow Principles for Multiprocessing. Tech. Rep. TRITA-TCS-9004 (Ph D thesis). Royal Institute of Technology, Stockholm, Sweden (1990)

Symbolic Cost Analysis and Automatic Data Distribution for a Skeleton-Based Language

Julien Mallet

IRISA, Campus de Beaulieu, 35042 Rennes, France
mallet@irisa.fr

Abstract. We present a skeleton-based language which leads to portable and cost-predictable implementations on MIMD computers. The compilation process is described as a series of program transformations. We focus in this paper on the step concerning the distribution choice. The problem of automatic mapping of input vectors onto processors is addressed using symbolic cost evaluation. Source language restrictions are crucial since they permit to use powerful techniques on polytope volume computations to evaluate costs precisely. The approach can be seen as a cross-fertilization between techniques developed within the FORTRAN parallelization and skeleton communities.

1 Introduction

A good parallel programming model must be portable and cost predictable. General sequential languages such as FORTRAN achieve portability but cost estimations are often very approximate. The approach described in this paper is based on a restricted language which is portable and allows an accurate cost analysis. The language enforces a programming discipline which ensures a predictable performance on the target parallel computer (there will be no "performance bugs").

Language restrictions are introduced through skeletons which encapsulate control and data flow in the sense of [6], [7]. The skeletons act on vectors which can be nested. There are three classes defined as sets of restricted skeletons: *computation skeletons* (classical data parallel skeletons), *communication skeletons* (data motion over vectors), and *mask skeletons* (conditional data parallel computation). The target parallel computers are MIMD computers with shared or distributed memory. Our compilation process can be described as a series of program transformations starting from a source skeleton-based language to an SPMD-like skeleton-based language. There are three main transformations: in-place updating, making all communications explicit and distribution. Due to space concerns, we focus in this paper only on the step concerning the choice of the distribution.

We tackle the problem of automatic mapping of input vectors onto processors using a fixed set of classical distributions (e.g. row block, block cyclic,...). The goal is to determine mechanically the best distribution for each input vector through cost evaluation. The restrictions of the source language and the fixed set of standard data distributions guarantee that the parallel cost (computation + communication) can be computed accurately for all source programs. Moreover, these restrictions allow us to evaluate and compare accurate execution times in a symbolic form. So, the results do not depend on specific vector sizes nor on a specific number of processors.

Even if our approach is rooted in the field of data-parallel and skeleton-based languages, one of its specificities is to reuse techniques developed for FORTRAN parallelization (polytope volume computation) and functional languages (single-threading analysis) in a unified framework.

The article is structured as follows. Section 2 is an overview of the whole compilation process. Section 3 presents the source language and the target parallel language. In Section 4, we describe the symbolic cost analysis through an example: LU decomposition. We report some experiments done on a MIMD distributed memory computer in Section 5. We conclude by a review of related work.

The interested reader will find additional details in an extended version of this paper ([12]).

2 Compilation Overview

The compilation process consists of a series of program transformations:

$$\mathcal{L}_1 \longrightarrow \mathcal{L}_2 \longrightarrow \mathcal{L}_3 \rightrightarrows \mathcal{L}_4 \longrightarrow \mathcal{L}_5$$

Each transformation compiles a particular task by mapping skeleton programs from one intermediate language into another. The source language $(\mathcal{L}_1)$ is composed of a collection of higher-order functions (skeletons) acting on vectors (see Section 3). It is primarily designed for a particular domain where high performance is a crucial issue: numerical algorithms. $\mathcal{L}_1$ is best viewed as a parallel kernel language which is supposed to be embedded in a general sequential language (e.g. C).

The first transformation $(\mathcal{L}_1 \rightarrow \mathcal{L}_2)$ deals with in-place updating, a standard problem in functional programs with arrays. We rely on a type system in the spirit of [17] to ensure that vectors are manipulated in a single-threaded fashion. The user may have to insert explicit copies in order for his program to be well-typed. As a result, any vector in a $\mathcal{L}_2$ program can be implemented by a global variable.

The second transformation $(\mathcal{L}_2 \rightarrow \mathcal{L}_3)$ makes all communications explicit. Intuitively, in order to execute an expression such as map $(\lambda \mathtt{x}.\mathtt{x+y})$ in parallel, y must be broadcast to every processor before applying the function. The transformation makes this kind of communication explicit. In $\mathcal{L}_3$, all communications are expressed through skeletons.

The transformation $\mathcal{L}_3 \rightarrow \mathcal{L}_4$ concerns automatic data distribution. We restrict ourselves to a small set of standard distributions. A vector can be distributed cyclicly, by contiguous blocks or allocated to a single processor. For a matrix (vector of vectors), this gives 9 possible distributions (cyclic cyclic, block cyclic, line cyclic, etc.). The transformation constructs a single vector from the input vectors according to a given data distribution. This means, in particular, that all vector accesses have to be changed according to the distributions. Once the distribution is fixed, some optimizations (such as copy elimination) become possible and are performed on the program. The transformation yields an SPMD-like skeleton program acting on a vector of processors.

In order to choose the best distribution, we transform the $\mathcal{L}_3$ program according to all the possible distributions of its input parameters. The symbolic cost of each transformed program can then be evaluated and the smallest one chosen (see Section 4). For most numerical algorithms, the number of input vectors is small and this approach is practical. In other cases, we would have to rely on the programmer to prune the search space.

The $\mathcal{L}_4 \rightarrow \mathcal{L}_5$ step is a straightforward translation of the SPMD skeleton program to an imperative program with calls to a standard communication library. We currently use C with the MPI library.

Note that all the transformations are automatic. The user may have to interact only to insert copies ($\mathcal{L}_1 \rightarrow \mathcal{L}_2$ step) or, in $\mathcal{L}_4$, when the symbolic formulation does not designate a single best distribution (see 4.4).

3   The Source and Target Languages

The source language $\mathcal{L}_1$ is basically a first-order functional language without general recursion, extended with a collection of higher-order functions (the skeletons). The main data structure is the vector which can be nested to model multi-dimensional arrays. An $\mathcal{L}_1$ program is a main expression followed by definitions. A definition is either a function definition or the declaration of the (symbolic) size of an input vector. Combination of computations is done through function composition (.) and the predefined iterator iterfor. iterfor n f a acts like a loop applying n times its function argument f on a. Further, it makes the current loop index accessible to its function argument.

The $\mathcal{L}_1$ program implementing LU decomposition is:

```
LU(M) where
    M :: Vect n (Vect n Float)
    LU(a)                   = iterfor (n-1) (calc.fac.apivot.colrow) a
    calc(i,a,row,piv)       = (i,(map (map first)
                                .rect i (n-1) (rect i (n-1) fcalc)
                                .map zip3.zip3)(a,row,piv))
    fcalc(a,row,piv)        = (a-row*piv,row,piv)
    fac(i,a,row,col,piv)    = (i,a,row,(map(map /).map zip.zip)(col,piv))
    apivot(i,a,row,col)     = (i,a,row,col,map (brdcast (i-1)) row)
    colrow(i,a)             = (i,a,brdcast (i-1) a,map (brdcast (i-1)) a)
    zip3(x,y,z)             = map p2t (zip(x,zip(y,z)))
    p2t(x,(y,z))            = (x,y,z)
```

For handling vectors, there are three classes of skeletons: computation, communication, and mask skeletons. The *computation skeletons* include the classical higher order functions map, fold and scan. The *communication skeletons* describe restricted data motion in vectors. In the definition of colrow in LU, brdcast (i-1) copies the (i-1)th vector element to all the other elements. Another communication skeleton is gather i M which copies the ith column of the matrix M to the ith row of M. There are seven communication skeletons which have been chosen because of their availability on parallel computers as hard-wired or optimized communication routines. The *mask skeletons* are data-parallel skeletons which apply their function argument only to a selected set of vector elements. In the definition of calc in LU, rect i (n-1) fcalc applies fcalc on vector elements whose index ranges from i to (n-1). Note that our approach is not restricted to these sole skeletons: even if it requires some work and caution, more iterators, computation or mask skeletons could be added.

In order to enable a precise symbolic cost analysis, additional syntactic restrictions are necessary. The scalar arguments of communication skeletons, the iterfor operator and mask skeletons must be linear expressions of iterfor indexes and variables of vector size. We rely on a type system with simple subtyping (not described here) to ensure that each variable appearing in such expression, are only iterfor and vector size variables.

It is easy to check that these restrictions are verified in LU. The integer argument of iterfor is a linear expression of an input vector size. The arguments of rect and brdcast skeletons are made of linear expressions of iterfor variable i or vector size variable n.

The target language $\mathcal{L}_4$ expresses SPMD (Single Program Multiple Data) computations. A program acts on a single vector whose elements represent the data spaces of the processors. Communications between processors are made explicit by data motion within the vector of processors in the spirit of the source parallel language of [8]. $\mathcal{L}_4$ introduces new versions of skeletons acting on the vector of processors.

An $\mathcal{L}_4$ program is a composition of computations, with possibly a communication inserted between each computation. More precisely, a program *FunP* has the following form:

$$FunP ::= (\text{pimap } Fun \mid \text{piterfor } E\ FunP)\ (\ .\ Com)^{0/1}\ (\ .\ FunP)^{0/1}$$

A computation is either a parallel operation pimap *Fun* which applies *Fun* on each processor, or an iteration piterfor *E FunP* which applies *E* times the parallel computation *FunP* to the vector of processors. The functions *Fun* are similar to functions of $\mathcal{L}_1$. The only differences are the skeleton arguments which may include modulo, integer division, minimum and maximum functions. *Com* denotes the communication between processors. It describes data motion in the same way as the communication skeletons of $\mathcal{L}_1$ but on the vector of processors. For example, the corresponding parallel communication skeleton of brdcast is pbrdcast which broadcasts values from one processor of the vector of processors to each other.

4 Accurate Symbolic Cost Analysis

After transformation of the program according to different distributions, this step aims at automatically evaluating the complexity of each $\mathcal{L}_4$ program obtained in order to choose the most efficient. Our approach to get an accurate symbolic cost is to reuse results on polytope volume computations ([16], [5]). It is possible because the restrictions of the source language and the fixed set of data distributions guarantee that the abstracted cost of all transformed source programs can be translated into a polytope volume description.

First, an abstraction function $\mathcal{CA}$ takes the program and yields its symbolic parallel cost. The cost expression is transformed so that it can be seen as the definition of a polytope volume. Then, classic methods to compute the volume of a polytope are applied to get a symbolic cost in polynomial form. Finally, a symbolic math package (such as Maple) can be used to compare symbolic costs and find the smallest cost among the $\mathcal{L}_4$ programs corresponding to the different distribution choices.

4.1 Cost Abstraction and Cost Language $\mathcal{LC}$

The symbolic cost abstraction $\mathcal{CA}$ is a function which extracts cost information from parallel programs. The main rules are shown below. We use a non-standard notation for indexed sums: instead of $\sum_{i=1}^{n}$, we note $\sum_i \{ {}^{1 \leq i}_{i \leq n} \}$. Communication costs are expressed as polynomials whose constants depend on the target computer. For example, the cost of pbrdcast involves the parameters α_{br} and β_{br} which denote respectively the time of one-word transfer between two processors and the message startup time on the parallel computer considered. One just has to set those constants to adapt the analysis for a specific parallel machine.

$$
\begin{aligned}
\mathcal{CA} \, [\![\text{f1 . f2}]\!] &= \mathcal{CA} \, [\![\text{f1}]\!] + \mathcal{CA} \, [\![\text{f2}]\!] \\
\mathcal{CA} \, [\![\text{piterfor e } (\lambda i.f)]\!] &= \sum_i \{ {}^{1 \leq i}_{i \leq e} \} \, \mathcal{CA} \, [\![\text{f}]\!] \\
\mathcal{CA} \, [\![\text{pimap } (\lambda i.f)]\!] &= \max_{i=0}^{p-1} \mathcal{CA} \, [\![\text{f}]\!] \qquad \text{where } p \text{ is the number of processors} \\
\mathcal{CA} \, [\![\text{pbrdcast}]\!] &= (\alpha_{\text{br}} * b + \beta_{\text{br}}) * p \quad \text{where } \begin{cases} p \text{ is the number of processors} \\ b \text{ is the size of broadcast data} \end{cases} \\
\mathcal{CA} \, [\![\text{rect e1 e2 f}]\!] &= \sum_i \{ {}^{e1 \leq i}_{i \leq e2} \} \, \mathcal{CA} \, [\![\text{f}]\!]
\end{aligned}
$$

The cost abstraction yields accurate costs expressed in a simple language $\mathcal{LC}$. $\mathcal{LC}$ is made of sums of generalized sums (G-sums): $\sum_i \{...\} \, (...)$ and generalized maxs (G-maxs): $\max_{i=e}^{e'} (...)$ possibly nested. A G-max denotes the maximum expression one can get when the max variable ranges over the integer interval. Inequalities in G-sums involve only linear expressions and maximum and minimum function of them. There may be polynomial included in the body of G-sums and G-maxs. The variables of polynomials are only denoting vector sizes or the number of processors. In order to illustrate the analysis, we consider here only the two best distribution choices for LU: row block and row cyclic. The abstracted costs are given below; note that only the part of the costs which differs is detailed:

$$\mathcal{C}^1_{bloc} \equiv \sum_i \{ {\scriptstyle 1 \le i \atop 1 \le n-1} \} \left(\max_{i_p=0}^{p-1} \sum_j \left\{ {\scriptstyle \max(0, i-i_p*b) \le j \atop j \le \min(b-1, n-1-i_p*b+b)} \right\} \left(\sum_k \{ {\scriptstyle 1 \le k \atop k \le n-1} \} 1 \right) \right) + C$$

$$\mathcal{C}^1_{cyc} \equiv \sum_i \{ {\scriptstyle 1 \le i \atop 1 \le n-1} \} \left(\max_{i_p=0}^{p-1} \sum_j \left\{ {\scriptstyle \max(0, \lfloor \frac{i-i_p}{p} \rfloor) \le j \atop j \le \min(b, \lceil \frac{n-1-i_p}{p} \rceil)} \right\} \left(\sum_k \{ {\scriptstyle 1 \le k \atop k \le n-1} \} 1 \right) \right) + C$$

We emphasize that writing the source program in $\mathcal{L}_1$ is crucial to get an accurate symbolic cost. First of all, without a severe limitation of the use of recursion no precise cost could be evaluated in general. Further, the restrictions imposed by $\mathcal{L}_1$ ensure that every abstract parallel cost belongs to $\mathcal{LC}$. The mask skeletons (which limit conditional application to index intervals), the communication and the computation skeletons all have a complexity which depends polynomially on the vector size. Their costs can be described by nested sums. Another important restriction is that expressions involving iteration indexes (mask skeletons and iterfor bounds) are linear. This restriction, in addition to the form of the standard distributions which keep the linearity of the vector accesses, ensure that the inequalities occurring in $\mathcal{LC}$ are linear.

4.2 Transformation to Descriptions of Polytope Volume

Polytope volumes are only defined in terms of nested G-sums with linear inequalities containing neither max nor min ([5]). In order to apply methods to compute polytope volumes, we must remove the G-max, min, max occurring in the cost expressions.

The transformation takes the form of a structural recursive scan of the cost expression. The min and max appearing in inequalities are transformed into conjunctions of inequalities. For example, the expression $\max(0, i - i_p * b) \le j$, in $\mathcal{C}^1_{bloc}$, becomes $0 \le j \wedge i - i_p * b \le j$. G-max expressions are propagated through their subexpressions. For example, in $\mathcal{C}^1_{bloc}$, $\max_{i_p=0}^{p-1} \sum_j \left\{ {\scriptstyle 0 \le j \wedge i - i_p * b \le j \atop j \le b-1 \wedge j \le n-1-i_p*b+b} \right\} (\ldots)$

is temporarily transformed into $\sum_j \left\{ {\scriptstyle 0 \le j \wedge \min_{i_p=0}^{p-1}(i - i_p*b) \le j \atop j \le b-1 \wedge j \le \max_{i_p=0}^{p-1}(n-1-i_p*b+b)} \right\} \max_{i_p=0}^{p-1}(\ldots)$. The

max value of a G-sum is a sum where the upper bounds are maximized and the lower ones minimized. For a sum inequality $i \le e$, we propagate the G-max inside e until it reaches a variable or a constant. We can finally use the facts that $\max_{i_p=0}^{p-1} i_p = p - 1$ and $\max_{i_p=0}^{p-1} k = k$ if $k \not\equiv i_p$. For an inequality $e \le i$, the transformation propagates a G-min analogously. Since no polynomial in $\mathcal{LC}$ may contain a G-max index, the G-max of a polynomial is just this polynomial. In the case of distributed LU, the cost expressions are transformed into

$$\mathcal{C}^1_{bloc} = \sum_{(i,j,k)} \left\{ {\scriptstyle 1 \le i \le n-1 \atop {0 \le j \wedge i-(p-1)*b \le j \atop j \le b-1 \wedge j \le n-1+b \wedge k \le k \le n-1}} \right\} 1 + C \equiv \mathcal{C}^2_{bloc}$$

$$\mathcal{C}^1_{cyc} = \sum_{(i,j,k)} \left\{ {\scriptstyle 1 \le i \le n-1 \wedge 0 \le j \wedge \lfloor \frac{i-p+1}{p} \rfloor \le j \atop j \le b-1 \wedge j \le \lceil \frac{n}{p} \rceil - 1 \wedge i \le k \le n-1} \right\} 1 + C \equiv \mathcal{C}^2_{cyc}$$

The technique described to remove G-max expressions may maximize the real cost. This is due to the duplication of G-max in G-sums. A more complicated but accurate symbolic solution also exists. The idea is to evaluate (as described in the next section) the nested G-sums first. When the G-sum is reduced to a polynomial the problem amounts to computing symbolic maximum of polynomials over symbolic intervals. This can be done using symbolic differentiation, solution and comparison of polynomials. At this point, we have only used the simpler approximated solution because the approximation is minor for our examples (e.g. none for LU) and the accurate symbolic solution is not implemented as such in existing symbolic math packages.

4.3 Parametrized Polytope Volume Computation

A parametrized polytope is a set of points whose coordinates satisfy a system of linear inequalities with possibly unknown parameters. After the previous transformation, the cost expression denotes a polytope volume, that is the number of points in the associated polytope. The works of [16] and [5] describe algorithms for computing symbolically the volume of parametrized polytopes. The result is a polynomial whose variables are the system parameters. Further, in [5], Clauss presents an extension for systems of non-linear inequalities with floor ($\lfloor \rfloor$) and ceiling ($\lceil \rceil$) functions as appearing in $\mathcal{C}^2_{cyc}$.

Thus, the algorithm of [5] allows one to transform each parallel cost expression into a polynomial. This method applied to LU yields:

$$\mathcal{C}^2_{bloc} = n^3 \left(\frac{3p^2-1}{6p^3} \right) - \frac{n^2}{2p} + \frac{n}{6p} + C \equiv \mathcal{C}^3_{bloc}$$
$$\mathcal{C}^2_{cyc} = \frac{n^3}{3p} + n^2 \left(\frac{p-2}{2p} \right) + n \left(\frac{-3p^2+6p-2}{6p} \right) + \frac{p^2-3p+2}{6} + C \equiv \mathcal{C}^3_{cyc}$$

4.4 Symbolic Cost Comparison

The last step is to compare the symbolic costs of different distribution choices. It amounts to computing the symbolic intervals where the difference of cost (polynomial) is positive or negative. Symbolic math packages such as Maple can be used for solving this problem. In the case of LU, Maple produces the following condition:

$$\mathcal{C}^3_{bloc} - \mathcal{C}^3_{cyc} \geq 0 \Leftrightarrow n \geq p$$

The programmer may have to indicate if the relations given by Maple are satisfied or not. In our example, he must indicate if $n \geq p$. Another (automatic) solution is to use the relations as run-time tests which choose between several versions of the program.

In our example, the difference in the two costs can be explained by the fact that the cyclic distribution provides a much better load balancing than the block distribution whereas communications are identical.

5 Experiments

We have performed experiments on an Intel Paragon XP/S with a handful of standard linear algebra programs (LU, Cholesky factorization, Householder, Jacobi elimination, ...). Our implementation is not completed and some compilation steps, such as the destructive update step and part of the symbolic cost computation, were done manually. Below, a table gathers the execution times obtained for LU decomposition. They are representative of the results we got for the other few programs. For all programs, the distribution chosen by the cost analysis proved to be the best one in practice.

Processors	Skel. cyclic	Skel. bloc	C Seq.	HPF cyclic	ScaLAPACK cyclic
1	14.77	15.07	13.61	15.36	3.78
4	5.25	6.75	×	5.41	1.84
16	2.97	5.33	×	3.06	1.50
32	2.57	5.58	×	2.67	1.41

We compared the sequential execution of skeleton programs with standard (and portable) C versions and our parallel implementation with High Performance Fortran (a manual distribution approach). No significant sequential or parallel runtime penalty seems to result from programming using skeletons, at least for such regular algorithms.

We compared our code with the parallel implementation of NESL, a skeleton-based language [1]. The work on the implementation of NESL has mostly been directed towards SIMD machines. On the Paragon, the NESL compiler distributes vectors uniformly on processors and communications are not optimized. Not surprisingly, the parallel code is very inefficient (at least fifty times slower than our code).

We also compared our implementation with ScaLAPACK, an optimized library of linear algebra programs designed for distributed memory MIMD parallel computers [4]. In ScaLAPACK, the user may explicitly indicate the data distribution. So, we indicated the best distribution found by the cost analysis in each ScaLAPACK program considered. If our code on 1 processor is much slower than its ScaLAPACK equivalent (between 3 to 5 times slower), the difference decreases as the number of processors increases (typically, 1.8 times slower on 32 processors). Much of this difference comes from the machine specific routines used by ScaLAPACK for performing matrix operations (the BLAS library). This suggests a possible interesting extension of our source language. The idea would be to introduce new skeletons corresponding to the BLAS operations in order to benefit from these machine specific routines.

These preliminary results are promising but more experiments are necessary to assess both the expressiveness of the language and the efficiency of the compilation. We believe that these experiments may also indicate useful linguistic extensions (e.g. new skeletons) and new optimizations.

6   Related Work and Conclusion

The community interested in the automatic parallelization of FORTRAN has studied automatic data distribution through parallel cost estimation ([10], [3]).

If the complete FORTRAN language (unrestricted conditional, indexing with runtime value, ...) is to be taken into account, communication and computation costs cannot be accurately estimated. In practice, the approximated cost may be far from the real execution time leading to a bad distribution choice. [16], [5] focus on a subset of FORTRAN: loop bound and array indexes are linear expressions of the loop variables. This restriction allows them to compute a precise symbolic computation cost through their computations of polytope volume. Unfortunately, using this approach to estimate communication costs is not realistic. Indeed, the cost would be expressed in terms of point-to-point communications without taking into account hard-wired communication primitives. All these works estimate real costs too roughly to ensure that a good distribution is chosen.

The skeleton community has studied the transformation of restricted computation patterns into lower-level parallel primitives. [7] defines a restricted set of skeletons which are transformed using cost estimation. Only cost-reducing transformations are considered. It is, however, well known that often intermediary cost-increasing transformations are necessary to derive a globally cost optimal algorithm. [15], [11] and [14] define cost analyses for skeleton-based languages. Their skeletons are more general than ours leading to approximate parallel cost (communication or/and computation). Furthermore, the costs are not symbolic (the size of input matrices and the number of processors are supposed to be known). [9] define a precise communication cost for scan and fold skeletons on several parallel topologies (hypercube, mesh, ...) which allows them to apply optimizations of communications through cost-reducing transformations. There are also a few real parallel implementations of skeleton-based languages. [2] uses cost estimations based on profiling which does not ensure a good parallel performance for different sizes of inputs. [13] uses a finer cost estimation but implementation decisions are taken locally and no arbitration of tradeoffs is possible.

We have presented in this paper the compilation of a skeleton-based language for MIMD computers. Working by program transformations in a unified framework simplifies the correctness proof of the implementation. One can show independently for each step that the transformation preserves the semantics and that the transformed program respects the restrictions enforced by the target language. The overall approach can be seen as promoting a programming discipline whose benefit is to allow precise analyses and a predictable parallel implementation. The source language restrictions are central to the approach as well as the techniques to evaluate the volume of polytopes. We regard this work as a rare instance of cross-fertilization between techniques developed within the FORTRAN parallelization and skeleton communities. A possible research direction is to study dynamic redistributions chosen at compile-time. Some parallel algorithms are much more efficient in the context of dynamic data redistribution. A completely automatic and precise approach to this problem would be possible in our framework. However, this would lead to a search space of exponential size. A possible solution is to have the user select the redistribution places.

Acknowledgements : Thanks to Pascal Fradet, Daniel Le Métayer, Mario Südholt for commenting on an earlier version of this paper.

References

1. G. E. Blelloch, S. Chatterjee, J. C. Hardwick, J. Sipelstein, and M. Zagha. Implementation of a portable nested data-parallel language. In *4th ACM Symp. on Princ. and Practice of Parallel Prog.*, pages 102–112. 1993.
2. T. Bratvold. A Skeleton-Based Parallelising Compiler for ML. In *5th Int. Workshop on the Imp. of Fun. Lang.*, pages 23–33, 1993.
3. S. Chatterjee, J. R. Gilbert, R. Schreiber, and S. Teng. Automatic array alignment in data-parallel program. In *20th ACM Symp. on Princ. of Prog. Lang.*, pages 16–28, 1993.
4. J. Choi and J. J. Dongarra. Scalable linear algebra software libraries for distributed memory concurrent computers. In *Proc. of the 5th IEEE Workshop on Future Trends of Distributed Computing Systems*, pages 170–177, 1995.
5. P. Clauss. Counting solutions to linear and nonlinear constraints through Ehrhart polynomials: Applications to analyze and transform scientific programs. In *ACM Int. Conf. on Supercomputing*, 1996.
6. M. Cole. A skeletal approach to the exploitation of parallelism. In *CONPAR'88*, pages 667–675. Cambridge University Press, 1988.
7. J. Darlington, A. J. Field, P. G. Harrison, P. H. J. Kelly, D. W. N. Sharp, Q. Wu, and R. L. While. Parallel programming using skeleton functions. In *PARLE '93*, pages 146–160. LNCS 694, 1993.
8. J. Darlington, Y. K Guo, H. W. To, and Y. Jing. Skeletons for structured parallel composition. In *5th ACM Symp. on Princ. and Practice of Parallel Prog.*, pages 19–28, 1995.
9. S. Gorlatch and C. Lengauer. (De)Composition rules for parallel scan and reduction. In *3rd IEEE Int. Conf. on Massively Par. Prog. Models*, 1998.
10. M. Gupta and P. Banerjee. Demonstration of automatic data partitioning techniques for parallelizing compilers on multicomputers. *IEEE Transactions on Parallel and Distributed Systems*, 3(2):179–193, 1992.
11. C. B. Jay, M. I. Cole, M. Sekanina, and P. Steckler. A monadic calculus for parallel costing of a functional language of arrays. In *Euro-Par'97 Parallel Processing*, pages 650–661. LNCS 1300, 1997.
12. J. Mallet. Compilation for MIMD computers of a skeleton-based language through symbolic cost analysis and automatic data distribution. Technical Report 1190, IRISA, May 1998.
13. S. Pelagatti. *A Methodology for the Development and the Support of Massively Parallel Programs.* PhD thesis, Pise University, 1993.
14. R. Rangaswami. *A Cost Analysis for a Higher-order Parallel Programming Model.* PhD thesis, Edinburgh University, 1996.
15. D. B. Skillicorn and W. Cai. A cost calculus for parallel functional programming. Technical report, Queen's University, 1993.
16. N. Tawbi. Estimation of nested loops execution time by integer arithmetic in convex polyhedra. In *Int. Symp. on Par. Proc.*, pages 217–223, 1994.
17. P. Wadler. Linear types can change the world! In *Programming Concepts and Methods*, pages 561–581. North Holland, 1990.

Optimising Data-Parallel Programs Using the BSP Cost Model

D.B. Skillicorn[1] M. Danelutto, S. Pelagatti and A. Zavanella[2]

[1] Department of Computing and Information Science
Queen's University, Kingston, Canada
[2] Dipartimento di Informatica
Universitá di Pisa, Corso Italia 40
56125 Pisa, Italy

Abstract. We describe the use of the BSP cost model to optimise programs, based on skeletons or data-parallel operations, in which program components may have multiple implementations. BSP's view of communication transforms the problem of finding the best implementation choice for each component into a one-dimensional minimisation problem. A shortest-path algorithm that finds optimal implementations in time linear in the number of operations of the program is given.

1 Problem Setting

Many parallel programming models gain expressiveness by raising the level of abstraction. Important examples are skeletons, and data-parallel languages such as HPF. Programs in these models are compositions of moderately-large building blocks, each of which hides significant parallel computation internally.

There are typically multiple implementations for each of these building blocks, and it is straightforward to order these implementations by execution cost. What makes the problem difficult is that different implementations require different arrangements of their inputs and outputs. Communication steps must typically be interspersed to rearrange the data between steps.

Choosing the best global implementation is difficult because the cost of a program is the sum of two different terms, an execution cost (instructions), and a communication cost (words transmitted). In most parallel programming models, it is not possible to convert the communication cost into comparable units to the execution cost because the message transit time depends on which other messages are simultaneously being transmitted, and this is almost impossible to determine in practice.

The chief advantage of the BSP cost model, in this context, is that it accounts for communication (accurately) in the same units as computation. The problem of finding a globally-optimal set of implementation choices becomes a one-dimensional minimisation problem, in fact with some small adaptations, a shortest path problem.

The complexity of the shortest path problem with positive weights is quadratic in the number of nodes (Dijkstra's algorithm) or $\mathcal{O}((e+n)\log n)$, where e is the number of edges, for the heap algorithm. We show that the special structure of this particular application reduces the problem to shortest path in a layered graph, with complexity linear in the number of program steps.

BSP's cost model assumes that the bottleneck in communication performance is at the processors, rather than in the network itself. Thus the cost of communication depends on the fan-in and fan-out of data at each processor, and a single architectural parameter, g, that measures effective inverse bandwidth (in units of time per word transferred). Such a model is extremely accurate for today's parallel computers. Other programming models can use the technique described here to the extent that the BSP cost model reflects their performance. In practice, this requires being able to decide which communication actions will occur in the same time frame. This will almost certainly be possible for P3L, HPF, GoldFish, and other data-parallel models.

In Section 2 we show how the BSP cost model can be used to model the costs of skeletons or data-parallel operations. In Section 3, we present an optimisation algorithm with linear complexity. In Section 4, we illustrate its use on a nontrivial application written in the style of P3L. In Section 5, we review some related work.

2 Cost Modelling

Consider the program in Figure 1, in which program operations, BSP supersteps [7], are shown right to left in a functional style. Step A has two potential implementations, and we want to account for the total costs of the two possible versions of the program.

Step A

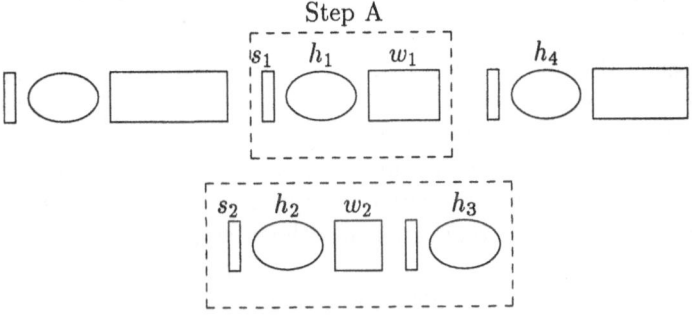

Alternate Implementation of Step A

Fig. 1. Making a Local Optimisation. Rectangles represent computation, ovals communication, and bars synchronisation. Each operation is labelled with its total cost parameter. Both implementations may contain multiple supersteps although they are drawn as containing one. The alternate implementation of step A requires extra communication (and hence an extra barrier) to place data correctly. In general, it is possible to overlap this with the communication in the previous step.

The communication stage of the operation preceding step A arranges the data in a way suitable for the first implementation of step A. Using the second implementation for step A requires inserting extra data manipulation code to change this arrangement. This extra code can be merged with the communication at the end of the preceding superstep, saving one barrier synchronisation, and potentially reducing the total communication load. For example, identities involving collective communication operations may allow some of the data movements to 'cancel' each other.

We can describe the cost of these two possible implementations of Step A straightforwardly in the BSP cost framework. The cost of the original implementation of Step A is $w_1 + h_1 g + s_1 l$ and the cost of the new implementation is $w_2 + h_2 g + s_2 l$ plus a further rearrangement cost $h_3 g + l$ To determine the extent to which the $h_3 g$ cost can be overlapped with the communication phase of the previous step we must break both costs down into fan-out costs and fan-in costs. So suppose that h_4 is the communication cost of the preceding step and

$$h_3 = \max(h_3^{out}, h_3^{in}) \qquad h_4 = \max(h_4^{out}, h_4^{in})$$

since fan-ins and fan-outs are costed additively. The cost of the combined, overlapped communication is

$$\max(h_3^{out} + h_4^{out}, h_3^{in} + h_4^{in}) g$$

The new implementation is cheaper than the old, therefore, if

$$w_2 + h_2 g + s_2 l + \max(h_3^{out} + h_4^{out}, h_3^{in} + h_4^{in}) g < w_1 + h_1 g + s_1 l + h_4 g$$

3 Optimisation

The analysis in the previous section shows how a program is transformed into a directed acyclic graph in which the nodes are labelled with the cost of computation (and synchronisation) and the edges with the cost of communication. Alternate implementation corresponds to choosing one path between two points rather than another. In what follows, we simply assume that different implementations for each program step are known, without concerning ourselves with how these are discovered.

Given a set of alternate implementations for each program step, the communication costs of connecting a given implementation of one step with a given implementation of the succeeding step can be computed as outlined above. If there are a alternate implementations for each program step, then there are a^2 possible communication patterns connecting them. If the program consists of n steps, then the total cost of setting up the graph is $(n-1)a^2$.

We want to find the shortest path through this layered graph. Consider the paths from any implementation block at the end of the program. There are (at most) a paths leading back to implementations of the previous step. The cost of a path is the sum of the costs of the blocks it passes through and the edges it contains. After two communication steps, there are a^2 possible paths, but only (at most) a of them can continue, because we can discard all but the cheapest of any paths that meet at a block. Thus there are only a possible paths to be extended beyond any stage. Eventually, (at most) a paths reach the start of the program, and the cheapest overall can be selected. The cost of this search is $(n-1)a^2$, and it must be repeated for each possible starting point, so the overall cost of the optimisation algorithm is $(n-1)a^3$. This is linear in the length of the program being optimised. The value of a is typically small, perhaps 3 or 4, so that the cubic term in a is unlikely to be significant in practice.

4 An Example

We illustrate the approach with a non-trivial program, Conway's game of Life, expressed in the intermediate code used by the P3L compiler[5,1]. The operations available to the P3L compiler are:

distribute :: Distr_pattern $\rightarrow$ SeqArray α $\rightarrow$ ParArray (SeqArray α)

gstencil :: ParArray α $\rightarrow$ Distr_pattern $\rightarrow$ Stencil $\rightarrow$ Distr_pattern $\rightarrow$ ParArray α

lstencil :: ParArray α $\rightarrow$ Stencil $\rightarrow$ Distr_pattern $\rightarrow$ ParArray (SeqArray α)

map :: ParArray α $\rightarrow$ $(\alpha \rightarrow \beta)$ $\rightarrow$ ParArray β

reduce :: ParArray α $\rightarrow$ $(\alpha \rightarrow \alpha \rightarrow \alpha)$ $\rightarrow$ α

reduceall :: ParArray α $\rightarrow$ $(\alpha \rightarrow \alpha \rightarrow \alpha)$ $\rightarrow$ ParArray α

gather :: ParArray α $\rightarrow$ Distr_pattern $\rightarrow$ SeqArray α

where distribute distributes an array to a set of worker processes according to a distribution pattern (Distr_pattern), gstencil fetches data from neighbour workers according to a given stencil, lstencil arranges the data local to a worker to have correct haloes in a stencil computation, map and reduce are the usual functional operations, reduceall is a reduce in which all the workers get the result, and gather collects distributed data in an array to a single process. Possible intermediate P3L code for the game of life is:

```
1. d1 = (block_block p)
2. W = distribute d1 World
3. while NOT(C2)
4.      W1 = gstencil W d1 [(1,1),(1,1)] d1
5.      W2 = lstencil W1 [(1,1),(1,1)] d1
6.      (W3,C)= map (map update) W2
7.      C1 = map (reduce OR C)
8.      C2 = reduceall OR C
9. X = gather d1 W3
```

where d1 is the distribution adopted for the World matrix (block,block over a rectangle of p processors), and W is the ParArray of the distributed slices of the initial World matrix. Line 4 gathers the neighbours from non-local processors. Line 5 creates the required copies. Line 6 performs all the update in parallel. Array C contains the boolean values saying if the corresponding element has changed. Line 9 gathers the global results. We concentrate on the computation inside the loop. It may be divided into three phases: Phase A gathers the needed values at each processor (Line 4). Phase B computes all of the updates in parallel (Lines 5–7). Phase C checks for termination: nothing has changed in the last iteration (Line 8). There are several possible implementations for these steps:

Operation	Computation Cost	Comment
A_1	l	direct stencil implementation
A_2	$2l$	gather to a single process then distribute
A_3	$3l$	gather to a single process then broadcast
B	$(9+t_\oplus)N^2+l$	local update and reduce
C_1	$(p-1)t_{OR}$	total exchange of results, then local reductions
C_2	$(\log p - 1)(l + t_{OR})$	tree reduction

where $t_\oplus$ is the cost of a local update, t_{OR} is the time taken to compute a logical OR, and N is the length of the side of the world region held by each processor.

In A_1, all workers in parallel exchange the boundary elements $(4(N+1))$, requiring a single barrier (costing l). In A_2, all the results are gathered on a single processor and redistributed giving each worker its halo; this requires much more data exchange $(N^2p + 4(N+1))$ and two synchronisations $(2l)$. In A_3, data are collected as in A_2 and then broadcasted to all the workers. This requires three synchronisations $(3l)$, as optimal BSP broadcast is done in two steps [7]. The cost of the communication between different implementations is:

Pair	Communication Cost	Comment
(A_1, B)	$4(N+1)g$	send the halo of an $N \times N$ block
(A_2, B)	$N^2pg + 4(N+1)g$	gather and distribute the halo
(A_3, B)	$3N^2pg$	gather and broadcast, no haloes
(B, C_1)	$(p-1)g$	total exchange
(B, C_2)	$2g$	
$(C_1, *)$	0	data already placed as needed
$(C_2, *)$	$2(\log p - 1)g$	

where p is the number of target processors, and N the length of side of the block of World on each processor). The optimal solution depends on the values of g, l and s of the target parallel computer. In general, for this example, A_1 is better than A_2 and A_3, but C_1 is better than C_2 only for small numbers of processors or networks with large values of g. For instance, if we consider a CRAY T3D (with values of $g \approx 0.35\mu s$/word, $l \approx 25\mu s$ and $s \approx 12Mflops$ as in [7]), C_1 is faster than C_2 when $p \leq 128$, so the optimal implementation is A_1-B-C_1. For larger numbers of processors, A_1-B-C_2 is the best implementation. For architectures with slower barriers (Convex Exemplar, IBM SP2 or Parsytec GC) the best solution is always A_1-B-C_1.

5   Related Work

The optimisation problem described here has, of course, been solved pragmatically by a large number of systems. Most of these use local heuristics and a small number of program and architecture parameters, and give equivocal results [4,3,2]. Unpublished work on P3L compiler performance has shown that effective optimisation can be achieved, but it requires detailed analysis of the target architecture and makes compilation time-consuming. The technique presented here is expected to replace the present optimisation algorithm. A number of global optimisation approaches have also been tried. To [8] gives an algorithm with complexity n^2a to choose among block and cyclic distributions of data for a sequence of data-parallel collective operations. Rauber and Rünger [6] optimise data-parallel programs that solve numerical problems. Different implementations are homogeneous families of algorithms with tunable algorithmic parameters (the number of iterations, the number of stages, the size of systems), and the cost model is a simplification of LogP.

6 Conclusions

The contribution of this paper is twofold: a demonstration of the usefulness of the perspective offered by the BSP cost model in simplifying a complex problem to the point where its crucial features can be seen; and then the application of a shortest-path algorithm for finding optimal implementations of skeleton and data-parallel programs.

The known accuracy of the BSP cost model in practice reassures us that the simplified problem dealt with here has not lost any essential properties. Reducing the problem from a global problem with many variables to a one-dimensional problem that can be optimised sequentially makes a straightforward optimisation algorithm, with linear complexity, possible.

Acknowledgement. We gratefully acknowledge the input of Barry Jay and Fabrizio Petrini, in improving the presentation of this paper.

References

1. B. Bacci, B. Cantalupo, M. Danelutto, S. Orlando, D. Pasetto, S. Pelagatti, and M. Vanneschi. An environment for structured parallel programming. In L. Grandinetti, M. Kowalick, and M. Vaitersic, editors, *Advances in High Performance Computing*, pages 219–234. Kluwer, Dordrecht, The Netherlands, 1997.
2. Z. Bozkus, A. Choudhary, G. Fox, T. Haupt, S. Ranka, and M.-Y. Wu. Compiling Fortran 90D/HPF for distributed memory MIMD computers. *Journal of Parallel and Distributed Computing*, 21(1):15–26, April 1994.
3. S. Ciarpaglini, M. Danelutto, L. Folchi, C. Manconi, and S. Pelagatti. ANACLETO: a template-based p3l compiler. In *Proceedings of the Seventh Parallel Computing Workshop (PCW '97)*, Australian National University, Canberra, 1997.
4. M. Danelutto, F. Pasqualetti, and S. Pelagatti. Skeletons for data parallelism in p3l. In C. Lengauer, M. Griebl, and S. Gorlatch, editors, *Proc. of EURO-PAR '97, Passau, Germany*, volume 1300 of *LNCS*, pages 619–628. Springer-Verlag, August 1997.
5. S. Pelagatti. *Structured development of parallel programs*. Taylor&Francis, London, 1997.
6. T. Rauber and G. Rünger. Deriving structured parallel implementations for numerical methods. *The Euromicro Journal*, (41):589–608, 1996.
7. D.B. Skillicorn, J.M.D. Hill, and W.F. McColl. Questions and answers about BSP. *Scientific Programming*, 6(3):249–274, 1997.
8. H.W. To. *Optimising the Parallel Behaviour of Combinations of Program Components*. PhD thesis, Imperial College, 1995.

A Parallel Multigrid Skeleton Using BSP

Femi O. Osoba and Fethi A. Rabhi

Department of Computer Science, University of Hull,
Hull HU6 7RX
{B.O.Osoba,F.A.Rabhi}@dcs.hull.ac.uk

Abstract. Skeletons offer the opportunity to improve parallel software development by providing a template-based approach to program design. However, due to the large number of architectural models available and the lack of adequate performance prediction models, such templates have to be optimised for each architecture separately. This paper proposes a programming environment based on the Bulk Synchronous Parallel(BSP) model for multigrid methods, where key implementation decisions are made according to a cost model.

1   Introduction

Programming environments and CASE tools are increasingly becoming popular for developing parallel software. Skeleton-based programming environments offer known advantages of intellectual abstraction and development of structured programs [1,9]. Most skeleton-based systems are implemented at a low level using functional languages [3,12] or imperative languages with parallel constructs e.g. PVM/C, dataparallel languages etc. [1,9]. However such models suffer from having no simple cost calculus, thereby hindering the functionality of the skeleton. To generate the required target code for different architectures, such skeleton-based systems have to be optimsed for each architecture. This paper describes a skeleton-based programming environment that is implemented at the low-level with a programming model that posseses a cost calculus, namely the Bulk Synchronous Parallelism (BSP) model.

The next section presents our case study skeleton and Section 3 presents an overview of the proposed programming environment. The next sections present an overview of the user interface, the specification analysis phase and the code generation process. Finally the last section presents some conclusions and future directions of our work.

2   A Case Study : Multigrid Methods

Most scientific computing problems represent a physical system by a mathematical model equation. Presently we are considering a model differential equation of the form $\nabla . K \nabla u = f$ in the 2D space. To make it suitable for a computer solution, the continuous physical domain of the system being modelled is discretised and this leads to a set of linear equations of the form $Ax = b$ with

n unknowns, which needs to be solved for. A particular class of methods for solving such equations is known as *multigrid(MG) methods*. Such methods have been selected as a case study on the basis of their usefulness in an important application area (CFD) and their similarity with SIT algorithms which were the focus of earlier work [9, 12].

The principle behind a multigrid method is to use a relaxation method (e.g. Gauss Seidel or Jacobi) on a sequence of m discretisation grids $G^1 \ldots G^m$ (G^1 represents the finest grid and G^m the coarsest). By employing several levels of discretisation, the multigrid method is able to accelerate the convergence rate of the relaxation method on the finest grid. One iteration of the MG method from finest to coarsest grid and back is called a *cycle*. During a cycle, the different stages in an MG algorithm are *relaxation* which involves the use of an iterative method on the current grid level, *restriction* which involves the transfer of the residual from one grid to the next coarser grid, *coarsest grid solution* which provides an exact solution on the coarsest grid and *interpolation* which is concerned with the transfer of the solution obtained from a coarser grid to the next finer grid. Additional details on multigrid methods can be found in [2].

3 Overview of a Skeleton-Based Programming Environment

The overall structure of the system is very similar to other skeleton-based programming environments [9] as figure 1 illustrates. The primary aim is separating the role of the user and the system developer. The user decides on how the multigrid method should be used and what parameters are to determine the optimal performance irrespective of the underlying architecture, while the system developer addresses issues concerned with generating efficient code and determining suitable or recommended parameters for different architectures based on performance prediction and cost modelling. The main components of the system as shown in figure 1 are described in turn in the next sections.

3.1 User Interface

The role of the User Interface is to allow the user to enter and edit the parameters of the multigrid algorithm. The User Interface also displays the results of the computation which consist of a visual representation of the solution. Figure 2 shows three sets of parameters: *multigrid parameters*, *physical parameters* and *implementation parameters*.

Multigrid parameters The multigrid parameters include the number of grids, the relaxation method, the stencil (e.g. 4 or 8 points) and the cycling strategy (V or W cycles). Another parameter is the number of relaxation iterations performed at various grid levels. Some parameters (e.g. the relaxation method) are expected to be changed after the Specification Analyser has been invoked. When selecting values, the user primarily attempts to reduce the overall execution time by minimizing the number of cycles required to converge to the solution.

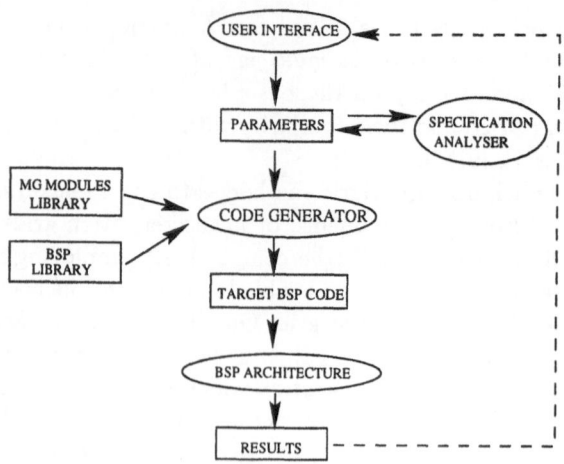

Fig. 1. Overall description of the system

PARAMETERS

MULTIGRID	PHYSICAL	IMPLEMENTATION
-Number of grids -Number of points -Relaxation method -Stencil -Number of iterations -Cycling strategy	-Initial data -Boundary data -Number of cycles -K	-BSP architecture (p,g,l) -Partitioning strategy -Tiles shape

Fig. 2. User Interface parameters

Physical parameters These parameters are dependent on the physical attributes of the problem e.g. the constant K describes the anisotropy in the problem.

Implementation parameters These parameters represent the characteristics of the parallel implementation. First, the BSP architecture is specified as a triplet (p, g, l) [6]. Another parameter represents the partitioning strategy, which is generally based on grid partitioning techniques. Finally, the tiles shape parameter refers to the shape of sub-partitions (e.g. square or rectangular tiles).

3.2 Specification Analyser

The role of the Specification Analyser is to predict the optimal value of some of the parameters given a minimal subset of values provided by the user. These predictions are made based on estimates on the computation and communication requirements for parallel execution. We base our domain partitioning theoretical cost model on a simple analytic model presented in [11]. Because BSP removes the notion of network topology [6], the goal is to build solutions that are optimal with respect to total computation, total communication, and total number of *supersteps*[6]. For our system this is represented by a parameter set. Each set represents one variation of the algorithm. A parameter set consists of Number of Cycles(Cy), Relaxation Method(RM), Stencil type(St), and Tileshape(Ts). Designing a particular program then becomes a matter of choosing a parameter set that is optimal for the range of machine sizes envisaged for the application. Our BSP theoretical cost model for a standard MG algorithm for a regular grid is of the form:

$$Cy[C.(n^2/p) + g.(n(1/t_r + 1/t_c).M)] + l.S \tag{1}$$

where C is the total number of computation steps, M is the total amount of communication, p is the number of processors (and also the number of tiles), t_r is the number of tiles in a row, t_c the number of tiles in a column, and S is the total number of supersteps. C is a function of (RM, St, n^2, p), M is a function of (Ts, RM, St) and S is a function of (RM, Ts). So for a given number of cycles (Cy), the analyser produces a parameter set (Cy, RM, St, Ts) which results in the lowest execution time for the particular architecture.

3.3 The Code Generator

The role of the Code Generator is to produce BSP code according to the current values of the parameters. The Code Generator uses a set of reusable library modules which form the *MG Module Library* and the BSP library (BSPlib) [4] for handling communication and synchronisation. The MG modules are coded in C and implement each of the phases identified in Section 2.

For further details on the multigrid programming environment, the reader is referred to the Ph.D thesis in [8].

4 Conclusion and Future Work

This paper described a skeleton-based programming environment for multigrid methods which generates BSP code. The user is offered an interface that abstracts from the underlying hardware thus ensures portability and intellectual abstraction. By combining the skeletons idea with BSP, accurate analysis and predictions of the performance of the code on architectures can be made. Since the system is aware of the global properties of the underlying BSP architecture, the system can decide which aspects of the MG algorithm may need to be changed to provide optimal performance and then performs the code generation.

Future work will refine the performance prediction model according to practical results, develop a systematic framework for automatic code generation and increase the parameter set to cater for more realistic problems. Finally, we will also investigate a suitable notation for specifying and modifying parameters and evaluate the proposed system with real users who work with such methods.

5 Acknowledgements

We would like to thank members of the BSP team at Oxford University Computing Laboratory U.K, for their continuous support on the use of the BSP library, and also for making the SP-2 available to us.

References

1. G.H. Botorog and H. Kuchen, Algorithmic skeletons for adaptive multigrid methods, In *Proceedings of Irregular '95*, LNCS 980, Springer Verlag , 1995, pp. 27-41.
2. J.H. Bramble, *Multigrid methods*, Pitman Research Notes in Mathematics, Series 294, Longman Scientific & Technical, 1993.
3. J. Darlington, A. Field, P. Harrison,P. Kelly, D. Sharp, Q. Wu, R. While, *Parallel Programming using Skeleton Functions*, In *Proceedings of PARLE '93*, LNCS, Munich, Springer Verlag, June 1993.
4. M. W. Goudreau, J.M.D. Hill, K. Lang, W.F. McColl, S. B. Rao, D.C. Stefanescu, T. Suel, and T. Tsantilas, A proposal for the BSP worldwide standard library, July 1996, available on WWW http://www.bsp-worldwide.org/.
5. O.A. McBryan, P.O. Frederickson, J. Linden, A. Schuller, K. Solchenbach, K. Stuben, C.A. Thole, and U. Trottenberg, Multigrid methods on parallel computers - a survey of recent developments, *IMPACT Comput. Sci. Eng.*, vol, 3, pp. 1-75, 1991.
6. W.F. McColl, Bulk synchronous parallel computing, In *Abstract Machine Models for Highly Parallel Computers*, J.R. Davy and P.M. Dew (eds), Oxford University Press, 1995, pp. 41-63.
7. M. Nibhanupudi, C. Norton, and B. Szymanski, Plasma Simulation On Networks of Workstations using the Bulk Synchronous Parallel mode, in *Proceedings of the International Conference on Parallel and Distributed Processing Techniques and Applications*, Athens, GA, November 1995.
8. B.O. Osoba, *Design of a Parallel Multigrid Skeleton-Based System using BSP*, Ph.D thesis in preparation in the Department of Computer Science, University of Hull, 1998.
9. P.J. Parsons and F.A. Rabhi, Generating parallel programs from paradigm-based specifications, to appear in the *Journal of Systems Architectures*, 1998.
10. F.A. Rabhi, A Parallel Programming Methodology Based on Paradigms, In *Transputer and Occam Developments*, P. Nixon (Ed.), IOS Press, 1995, pp. 239-252.
11. P. Ramanathan and S. Chalasani, Parallel multigrid algorithms on CM-5. In *IEE Proc. Computers and Digital Techniques*, vol. 142, no 3, May 1995.
12. J. Schwarz and F.A. Rabhi, A skeleton-based implementation of iterative transformation algorithms using functional languages, In *Abstract Machine Models for Parallel and Distributed Computing*, M. Kara *et al.* (eds), IOS Press, 1996,

Flattening Trees

Gabriele Keller[1] and Manuel M. T. Chakravarty[2]

[1] Fachbereich Informatik, Technische Universität Berlin, Germany
keller@cs.tu-berlin.de
[2] Inst. of Inform. Sciences and Electronics, University of Tsukuba, Japan
chak@is.tsukuba.ac.jp

Abstract. Nested data-parallelism can be efficiently implemented by mapping it to flat parallelism using Blelloch & Sabot's *flattening* transformation. So far, the only dynamic data structure supported by flattening are vectors. We extend it with support for user-defined recursive types, which allow parallel tree structures to be defined. Thus, important parallel algorithms can be implemented more clearly and efficiently.

1 Introduction

The *flattening* transformation of Blelloch & Sabot [6, 4] implements *nested* data-parallelism by mapping it to *flat* data-parallelism. Compared to flat parallelism, nested parallelism allows algorithms to be expressed on a higher level of abstraction while providing a language-based performance model [5]; in particular, algorithms operating on irregular data structures and divide-and-conquer algorithms benefit from nested parallelism. Efficient code can be generated for a wide range of parallel machines [2, 11, 12, 10]. However, flattening supports only vectors (i.e., homogeneous, ordered sequences) as dynamic data structures. Thus, important parallel algorithms, like hierarchical n-body codes and other adaptive algorithms based on tree structures, are awkward to program—the tree structure has to be mapped to a vector structure, implying explicit index calculations, to keep track of the parent-child relation, and leading to a suboptimal data distribution on distributed-memory machines.

In this paper, we propose user-defined *recursive types* to tackle the mentioned problems. We extend flattening such that it maps the new structures to efficient, flat data-parallel code. Our extension fits easily into existing formalizations and implementations of flattening; in particular, the optimization techniques of previous work [11, 12, 10, 7, 9] remain applicable. This paper makes the following three main contributions: (1) It demonstrates the usefulness of recursive types for nested data-parallel languages (Section 2), (2) it formally specifies our extension of flattening including user-defined recursive types (Section 3), and (3) it provides experimental results gathered with the resulting code on a Cray T3E (Section 4). Regarding point (2), as a side-effect of our extension, we contribute to a rigorous specification of flattening by formalizing the instantiation of polymorphic primitives. Thereby, we also introduce a new kind of primitives, so-called *chunkwise operations*, for more efficient data redistribution on distributed-memory machines. We use the functional language NESL [5] throughout this paper, but the discussed techniques also work for imperative languages [1]. Many details and discussions have been omitted in this paper due to shortage of space—this material can be found in [8].

Section 2 discusses the benefit of recursive types for tree-based algorithms in a purely vector-oriented language. Section 3 formalizes our extended flattening transformation. Section 4 presents benchmarks. Finally, Section 5 concludes.

2 The Problem: Encoding Trees by Vectors

NESL [5] is strict functional language featuring *nested vectors* as its central data structure. In addition to built-in parallel operations on vectors, the *apply-to-each* construct is used to express parallelism. In its general form $\{e : x_1$ in $e_1, \ldots, x_n$ in $e_n \mid f\}$ we call e the *body*, the x_i in e_i the *generators*, and f (which is optional) the *filter*. The body e is evaluated for each element of the vectors e_i and the result is included in the result vector if f evaluates to T (true); the vectors e_i are required to be of equal length and are processed in lock-step. For example, $\{x + y : x$ in $[1, -2, 3]$, y in $[4, 5, 6] \mid x > 0\}$ evaluates to $[5, 9]$. Nested parallelism occurs where parallel operations appear in the body of an apply-to-each, e.g., $\{plus\_scan (x) : x$ in $xs\}$, which for $xs = [[1, 2, 3], [4], [5, 6]]$ yields $[[0, 1, 3], [0], [0, 5]]$. The built-in plus_scan is a prescan using addition [4].

The implementation of a tree-based algorithm in such a language implies representing trees by nested vectors, obscuring the code with explicit index calculations, to keep track of the parent-child relation. Moreover, in an implementation based on flattening, these nested vectors are represented by a set of flat vectors in the target code. All *data elements* of the tree are mapped to a single vector and are uniformly distributed to achieve load balancing. This, however, leads to superfluous redistributions as those algorithms usually traverse the trees breadth-first, i.e., level by level, all nodes on one level are processed in parallel.

We illustrate these problems at the example of Barnes & Hut's hierarchical *n*-body code [3]. It minimizes the number of force calculations by grouping particles hierarchically into *cells* according to their spatial position. The hierarchy is represented by a tree. This allows approximating the accelerations induced by a group of particles on distant particles by using the centroid of that group's cell. The algorithm has two phases: (1) The tree is constructed from a particle set, and (2) the acceleration for each particle is computed in a down-sweep over the tree. The following NESL function outlines the tree construction:

```
function bhTree (ps) : [MassPnt] -> [Cell] =
   if #ps == 1 then [Cell (ps[0], [])]            — cell has only one particle
   else let ⟨split particles ps into spatial groups pgs⟩
           subtrees  = {bhTree (pg): pg in pgs};
           children  = {subtree[0] : subtree in subtrees};          (⋆)
           cd        = centroid ({mp : Cell (mp, is) in children});
```

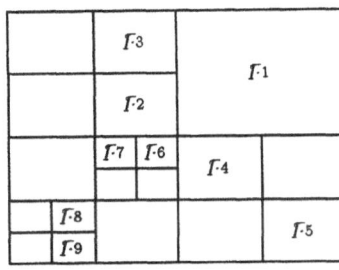 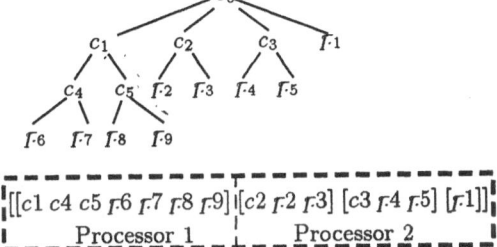

Fig. 1. Example of a Barnes-Hut tree and its representation as a vector.

```
        sizes    = {#subtree : subtree in subtrees}                (*)
in [Cell (cd, sizes)] ++ flatten (subtrees)   — whole tree as flat vector (*)
```

We represent the tree as a *vector of cells*. Each cell of the form `Cell` (mp, szs) corresponds to a node of the tree and is a pair of a mass point mp and the sizes of the subtrees szs (tuples can be named in NESL). Given such a vectorized **tree**, the acceleration of a set of mass points `mps` is computed by

```
function accels (tree, mps) : ([Cell], [MassPnt]) -> [Vec] =
  if #mps == 0 then []
  else let Cell (cd, chNos) = tree[0];              — get root         (*)
            (split mps into closeMps and farMps (direct force calculation))
            farAcs    = {accel (cd, mp) : mp in farMps};
            subTrees  = partition (drop (tree, 1), chNos);            (*)
            closeAcss = {accels (tree, closeMps) : tree in subTrees};
  in (combine farAcs and closeAcss)
```

It computes the acceleration for the mass points in **farMps** directly (using the function **accel**) and recurses into the tree for those in **closeMps**. The function **drop** omits the first element of a vector and **partition** forms a nested vector according to the lengths passed in the second argument (it is the same as $\mathcal{P}$ in the next section). The type **Vec** represents 'vectors' in the sense of physics.

The lines marked by (*) in the functions are (partially) artifacts of maintaining the tree as a vector. Figure 1 depicts the grouping of an exemplary particle distribution and the corresponding tree. The tree is both built and traversed level by level, i.e., all nodes in one level of the tree are processed in a parallel step. Let us consider the data layout (over two processors) for the example tree in Figure 1. To ensure proper load balancing, all cells of the already constructed subtrees have to be redistributed in each recursive step of **bhTree**. Similarly, **accels**, while descending the tree, has to redistribute those cells that correspond to one level of the tree. We will quantify these costs experimentally in Section 4.

It should be clear that it is more suitable to store the nodes of the tree in a distinct vector for each level of the tree, and then, to chain these vectors to represent the whole tree. At the source level, such a structure corresponds to regarding each node as being composed from some node data plus a vector of tree nodes that have exactly the *same* structure; in other words, we need a *recursive type*. For our example, we can use **datatype Node (MassPnt, [Node])**. Then, we simplify the function **bhtree** as follows: We omit computing **children** and sizes, we compute cd by centroid ({mp : Node (mp, chs) in trees}), and change the body of the let into Node (cd, subtrees). In accels, computing subtrees becomes superfluous, and closeAcss is computed by accels (tree, closeMps) : tree in subtrees} where Node (cd, subtrees) = tree.

Generally, we extend NESL with named tuples that refer to themselves, but only in the element type of a vector (to avoid infinite types), as we can terminate recursion only by empty vectors—there are no arbitrary sum types. Handling sum types efficiently in flattening seems much harder than recursive types.

3 Flattening Trees by Program Transformation

After the source language extension, we proceed to the implementation of user-defined recursive types by flattening. The flattening transformation serves three aims: First, it replaces all nested parallel expressions by flat parallel expressions that contain the same degree of parallelism, second, it changes the representation of data structures such that vectors contain only elements of base type, and third, it replaces all polymorphic vector primitives with monomorphic instances.

In this section, we begin by introducing the flat data-parallel kernel language FKL, which is the target language of the first part of the flattening transformation. We continue with a discussion of the target representation of data structures, and then, describe an instantiation procedure, which implements the change of the representation of data structures as well as the generation of all necessary monomorphic instances of the primitives. Due to changing the representation of data structures, the instantiation of the polymorphic primitives becomes technically challenging—especially so, in the presence of recursive types. The representation of recursive types together with the instantiation procedure are the central technical contributions of this paper.

Although the instantiation of the polymorphic primitives—excluding recursive types—is already implicit in previous work, it was never described in detail (for example, Blelloch [4] merely gives some examples and leaves the non-trivial induction of a general algorithm to the inclined reader); in particular, we provide the first formal specification. Our treatment also leads to more efficient code on distributed-memory machines than previous approaches (which concentrated on vector and shared-memory machines). A central idea of our method is the fact that only the *primitive* vector operations actually access or manipulate elements of nested vectors. Therefore, we can regard nested vectors as an abstract data type with some freedom in the concrete implementation.

We will not discuss the first step of the flattening, as it is already detailed in previous work [4, 11, 10] and not affected by the addition of recursive types—however, a complete specification of the flattening transformation can be found in an unabridged version of this paper [8].

3.1 The Flat Kernel Language

A kernel language (FKL) program consists of a list of declarations produced by the rule $D \rightarrow V (V_1, \ldots, V_n) = E$ with variables V and expressions produced by

$E \to C \mid V \mid V(E_1, \ldots, E_n) \mid$ let $V = E_1$ in $E_2 \mid$ if E_1 then E_2 else $E_3 \mid [E_1, \ldots, E_n]$

where C are constants. We assume programs are typed, with types from

$T \to Int \mid Bool \mid V \mid (T_1, \ldots, T_n) \mid [T] \mid \mu V.T$

For brevity, we only have Int and $Bool$ as primitive types. In a recursive type $\mu x.T$, all occurrences of x in T must be within element types of vectors (to get a finite type). For example, the type Node of Section 2 is represented as $\mu x.(MassPnt, [x])$, where $MassPnt$ abbreviates the tuple of a mass point.

The second component of FKL are its primitive operations. Among the most important are the usual arithmetic and logic operations as well as construction of tuples $\tau^n : \alpha_1 \times \cdots \times \alpha_n \to (\alpha_1, \ldots, \alpha_n)$ and the corresponding projections $\pi_n^i : (\alpha_1, \ldots, \alpha_n) \to \alpha_i$ (the function type is given to the right of the colon). The vector operations include operations length $\# : [\alpha] \to Int$, concatenation $+\!\!+ : [\alpha] \times [\alpha] \to [\alpha]$, and indexing $ind : [\alpha] \times Int \to \alpha$. Moreover, we have distribution $dist : \alpha \times Int \to [\alpha]$, where $dist\,(a, n)$ yields $[a, \ldots, a]$ with length n, permutation $perm : [\alpha] \times [Int] \to [\alpha]$, where $perm\,(xs, is)$ permutes xs according to the index vector is, packing $pack : [\alpha] \times [Bool] \to [\alpha]$, where $pack\,(xs, fs)$ removes all elements from xs that correspond to a $false$ value in fs. Furthermore, there are families of reduction $\oplus_\tau\text{-}reduce : [\tau] \to \tau$ and prescan $\oplus_\tau\text{-}scan : [\tau] \to [\tau]$ functions for associative binary primitives $\oplus_\tau$ operating on data of basic type τ. Finally, FKL contains three functions that form the basis for handling nested vectors: $\mathcal{F} : [[\alpha]] \to [\alpha]$ correspond to $+\!\!+$ -$reduce$ and removes one level of nesting, e.g., $\mathcal{F}\,([[1, 2, 3], [], [4, 5]]) = [1, 2, 3, 4, 5]$; $\mathcal{S} : [[\alpha]] \to [Int]$ corresponds to $\{\#xs : xs \leftarrow xss\}$ and returns the toplevel nesting structure, e.g., $\mathcal{S}\,([[1, 2, 3], [], [4, 5]]) = [3, 0, 2]$; and $\mathcal{P} : [\alpha] \times [Int] \to [[\alpha]]$ reconstructs a nested vector from the results of the previous two function, e.g., $\mathcal{P}\,([1, 2, 3, 4, 5], [3, 0, 2]) = [[1, 2, 3], [], [4, 5]]$. For each nested vector xs, we have $\mathcal{P}\,(\mathcal{F}\,(xs), \mathcal{S}\,(xs)) = xs$. We write the application of primitives like $+\!\!+$ infix. Some of the primitives are polymorphic (those with α in the type) and, as said, we discuss their instantiation later, but we assume that in an FKL program all polymorphism in user-defined functions is already removed (by code duplication and type specialization).

To compensate the lack of general nested parallelism (i.e., no apply-to-each construct), FKL supports all primitive functions $p : T_1 \times \cdots \times T_n \to T$ in a vectorized form $p^\uparrow : [T_1] \times \cdots \times [T_n] \to [T]$, which applies p in parallel to all the elements of its argument vectors (which must be of the same length). For example, we have $[1, 2, 3] +_{Int}^\uparrow [4, 5, 6] = [5, 7, 9]$. In general, a primitive and its vectorized form relate through $p^\uparrow(xs_1, \ldots, xs_n) = \{p(x_1, \ldots, x_n) : x_1 \leftarrow xs_1, \ldots, x_n \leftarrow xs_n\}$. Note, the part of the transformation generating FKL guarantees that nothing like $(p^\uparrow)^\uparrow$ is needed. The next subsection introduces two additional sets of primitives: One handles recursive types and the other, although not strictly necessary, handles some operations on nested vectors more efficiently. We delay the discussion of these primitives as it benefits from knowing the target data representation.

We choose FKL as the target language as there are optimizing code-generation techniques mapping it on different parallel architectures [2, 7, 9].

3.2 Concrete Data Representation

Before presenting the instantiation procedure for polymorphic primitives, we discuss an efficient target representation of nested vectors, vectors of tuples, and vectors of recursive types.

Nested vectors. We can represent nested vectors of basic type using only flat vectors by separating the data elements from the nesting structure. For example, $[[1, 2, 3], [], [4, 5]]$ is represented by a pair consisting of the data vector $[1, 2, 3, 4, 5]$ and the *segment descriptor* $[3, 0, 2]$ containing the lengths of the subvectors. The primitives $\mathcal{F}$ and $\mathcal{S}$ extract these two components, whereas $\mathcal{P}$ combines them into an *abstract* compound structure representing a nested vector. In general, this representation requires a single data vector and one segment descriptor per nesting level of the represented vector. Instances of polymorphic primitives operating on these nested vectors can be realised by combining primitives on vectors of basic type with the functions $\mathcal{F}$, $\mathcal{S}$, and $\mathcal{P}$, as we will see below.

This representation allows an optimized treatment of costly reordering primitives, such as permutation. Consider the expression $perm'\ (as, is)$, where as is of type $[[Int]]$. Both the data vector and the segment descriptor of as have to be permuted. We have $perm'\ (as, is) = \mathcal{P}\ (perm\ (\mathcal{F}\ (as), is'), perm\ (\mathcal{S}\ (as), is))$, where $perm$ operates on vectors of type Int and is' is a new permutation vector computed from is and $\mathcal{S}\ (as)$. So, for example, $perm'\ ([[3, 4, 5], [1, 3]], [1, 0]) = \mathcal{P}\ (perm\ ([3, 4, 5, 1, 3], [2, 3, 4, 0, 1]), perm\ ([3, 2], [1, 0]))$.

This scheme is expensive, because (a) a new index vector is' is computed for each level of nesting, and moreover, (b) the data vector is permuted elementwise, whereas the original expression allows to deduce that several continuous blocks of elements (the subvectors) are permuted, i.e., we lose information about the structure of communication. We can prevent this behaviour by employing an additional set of primitive functions, the so-called *chunkwise operations* $distC$, $permC$, and $indC$. The operation $distC$ gets a vector together with a natural number and yields a vector that repeats the input vector as often as specified by the second argument. The operations $permC$ and $indC$ both get three arguments: (1) a flat data vector, (2) a segment descriptor, and (3) an index vector or index position (depending on the function). They permute and index blocks of the data vector chunkwise, where the chunk size is specified by the segment descriptor. Their semantics is defined as $permC\ (xs, s, is) = perm'\ (\mathcal{P}\ (xs, s), is)$ and $indC\ (xs, s, i) = ind'\ (\mathcal{P}\ (xs, s), i)$, respectively, where $perm'$ and ind' operate on vectors of type $[[s]]$ when xs is of type $[s]$ and s is a basic type. Their implementation using blockwise communication is straightforward.

Vectors of tuples. Vectors of tuples are represented by tuples of vectors. Accordingly, applications of vector primitives operating on such vectors are pulled inside the tuple, as proposed by [4].

Recursive Types. The most complicated case is the representation of vectors of recursive types by vectors of basic type in such a way that the nodes of each level of the represented tree structure are stored in a separate vector (as discussed in Section 2). In Subsection 3.1 we required that, in a recursive type $\mu x . T$, each occurrence of x in T has the form $[x]$. Let us consider the possible contexts of these occurrences. If the outermost type constructor of T is a primitive type (e.g., Int), there is no recursion and we represent T as usual. But, if the outermost type constructor of T is a vector, we have $[[x]]$, i.e., a nested vector of recursive type. Similar to nested vectors of basic type, we regard them as an abstract structure manipulated by the three functions $\mathcal{F}$, $\mathcal{S}$, and $\mathcal{P}$. Next, if the outermost type constructor of T is a tuple, we have to represent it as a tuple of vectors like in the non-recursive case, while treating the components of the resulting tuple in the same way as T itself. Finally, if the outermost type constructor of T is a recursive type of the form $\mu x' . T'$, no special treatment is necessary, apart from handling T' in the same way as T.

Overall, any occurrence of a recursive type variable x (except in the root of a tree) is represented by $[[x]]$, due to the propagation of vectors inside tuples and the requirement that x occurs only as $[x]$. Thus, these occurrences are manipulated by $\mathcal{F}$, $\mathcal{S}$, and $\mathcal{P}$. These functions allow us to represent a tree structure of potentially unbounded depth using only flat vectors by separating data from nesting structure. Nevertheless, a problem remains: For nested vectors, the nesting depth of the data is statically known (due to strong typing), but not so for the depth of a recursive type. Hence, we get a sequence of segment descriptors of statically unknown length. Each of these is accompanied by the data vector for one level of the tree. A value of recursive type $\mu x . T$ is always terminated by empty vectors of a recursive occurrence x in T, but due to the propagation of vectors inside tuples, we need a special value to identify the recursion termination. Hence, we wrap the representation of vectors of recursive type into a *maybe* type $\langle \alpha \rangle$, which is either $\langle x \rangle$ (where x is of type α) or $\langle \rangle$. To process values of type $\langle \alpha \rangle$, we add two primitives $(\cdot ?) : \langle \alpha \rangle \to Bool$ and $(\cdot \uparrow) : \langle \alpha \rangle \to \alpha$. Operation $(\cdot ?)$ yields true if the argument has the form $\langle x \rangle$, in this case, $(\cdot \uparrow)$ returns x (it is undefined, otherwise). The tree of Figure 1 is represented by (c_0, v) with

$$v = \langle ([c_1, c_2, c_3, p_1]) \rangle, \qquad \mathcal{P}(w, [2, 2, 2, 0]))$$
$$w = \langle ([c_4, c_5, p_2, p_3, p_4, p_5], \; \mathcal{P}(x, [2, 2, 0, 0, 0]))) $$
$$x = \langle ([p_6, p_7, p_8, p_9], \; \mathcal{P}(y, [0, 0, 0, 0]))) $$
$$y = \langle ([], \qquad \mathcal{P}(\langle \rangle, []))) $$

3.3 Instantiation of Polymorphic Functions

Now, we are in a position to define the instantiation of the polymorphic uses of primitives. We denote the type of an instance of a primitive by annotating the type *substituted for the type variable* (α in the signatures from Subsection 3.1) as a subscript. For example, $+\!\!+_{[Int]}$ has type $[[Int]] \times [[Int]] \to [[Int]]$, and we have to generate instances for all occurrences apart from $+\!\!+_{Int}$ and $+\!\!+_{Bool}$. The generic primitives τ^n, π_i^n, $\#$, $\mathcal{F}$, $\mathcal{S}$, and $\mathcal{P}$ are not instantiated as their code is independent of the type instance.

The transformation algorithm. In the presence of recursive types, we cannot transform the program by replacing uses of polymorphic primitives with expressions that merely use primitives on basic type; instead, we add new declarations to a program, until all uses of primitives are either directly supported or covered by a previously added declaration. The addition of a declaration may introduce primitives occurring at new instances, which in turn triggers the addition of declarations for these instances. We discuss the termination of this process later.

Each equation says that if a primitive occurs at a type matching the left hand-side, add a declaration that is an instance of the right hand-side for that type. We start with the distribution *dist*, together with its chunkwise variant:

$$dist_{(T_1,\ldots,T_n)}\,(x,n) = (dist_{T_1}\,(\pi_1^n(x),n),\ldots,dist_{T_n}\,(\pi_n^n(x),n))$$
$$dist_{\mu x.T}\,(x,n) \quad = \langle dist_{T\{\mu x.T/x\}}\,(x,n)\rangle$$
$$dist_{[T]}\,(x,n) \quad = \mathcal{P}(distC_{[T]}\,(x,n), dist_{Int}\,(\#x,n))$$
$$distC_{\mu x.T}\,(xs,n) \quad = \textbf{if } xs? \textbf{ then } distC_{T\{\mu x.T/x\}}\,(xs\!\uparrow,n) \textbf{ else } \langle\rangle$$
$$distC_{[T]}\,(xs,n) \quad = \mathcal{P}(distC_{[T]}\,(\mathcal{F}\,(xs),n), distC_{Int}\,(\mathcal{S}\,(xs),n))$$

The rules for tuples propagate the function inside. Vectors are distributed using the chunkwise version, of which the second rule demonstrates the handling of the maybe type $\langle\alpha\rangle$. We omit the rules for tuples in the following, as they have the same structure as above. The next set of rules covers chunkwise concatenation:

$$xs \mathbin{+\!\!+}_{\mu x.T} ys = \textbf{if } xs? \textbf{ then } (\textbf{if } ys? \textbf{ then } xs\!\uparrow \mathbin{+\!\!+}_{T\{\mu x.T/x\}} ys\!\uparrow \textbf{ else } xs) \textbf{ else } ys$$
$$xs \mathbin{+\!\!+}_{[T]} ys \quad = \mathcal{P}(\mathcal{F}(xs) \mathbin{+\!\!+}_T \mathcal{F}(ys)), \mathcal{S}(xs) \mathbin{+\!\!+}_{Int} \mathcal{S}(ys))$$

We continue with plain and chunkwise indexing. Note that indexing an empty vector of recursive type leads to an error as the index is out of bounds.

$$ind_{\mu x.T}\,(xs,i) \quad = \textbf{if } xs? \textbf{ then } ind_{T\{\mu x.T/x\}}\,(xs,i) \textbf{ else error}$$
$$ind_{[T]}\,(xs,i) \quad = indC_T\,(\mathcal{F}(xs),\mathcal{S}(xs),i)$$
$$indC_{\mu x.T}\,(xs,s,i) = \textbf{if } xs? \textbf{ then } indC_{T\{\mu x.T/x\}}\,(xs,s,i) \textbf{ else } \langle\rangle$$
$$indC_{[T]}\,(xs,s,i) \quad = \mathcal{P}(indC_T\,(\mathcal{F}(xs), +_{Int}\text{-}reduce^\uparrow\,(\mathcal{P}\,(\mathcal{S}(xs),s)),i),$$
$$indC_{Int}\,(\mathcal{S}(xs),s,i))$$

For $indC_{[T]}$, the segment descriptor passed to the recursive call is computed by summing over the segments of the current level. For space reasons, we omit the rules for *perm*, *pack*, and *combine*. They are similar to *ind*, but, e.g., permuting an empty vector does not raise an error, but returns the empty vector. Finally, considering vectorized primitives, the rules for types $(T_1,\ldots,T_n)$ and $\mu x.T$ are as above; for $[T]$, the vectorized version is generated by vectorizing the above rules—this vectorization is essentially the same as the first step of the flattening transformation that was mentioned in the beginning of the present section. The details of this and of handling *perm*, *pack*, and *combine* can be found in [8].

Termination. Considering the termination of the above process, we see that for each primitive p, the rules for instances $p_{(T_1,\ldots,T_n)}$ and $p_{[T]}$ decrease the structural depth of the type subscript. Merely the rules for $p_{\mu x.T}$ may be problematic as they recursively unfold the type in their right hand-side. However, in

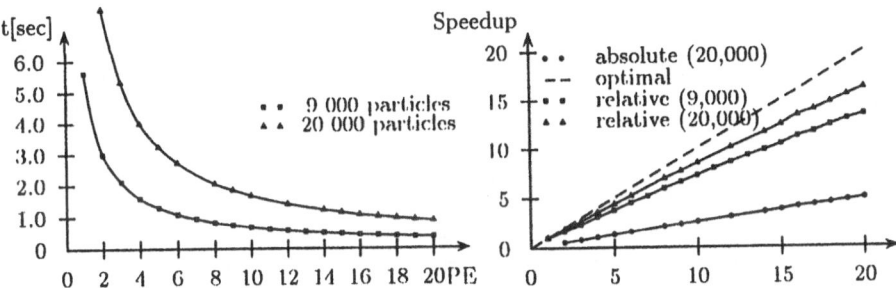

Fig. 2. Absolute time and relative speed-up for Barnes-Hut on a Cray T3E.

the definition of the kernel languages, we required that type variables bound by μ occur only as the element type of a vector. So, for all occurrences where the type unfolding substitutes the recursive type, we get an expression that requires $p_{[\mu x. T]}$. If p is a chunk operation or $+\!\!+$, the right hand-side of an instance of the rule for $p_{[T']}$ with $T' = \mu x. T$ requires $p_{\mu x. T}$, which is exactly the instance we started with and which, therefore, is already defined. If p is neither a chunk operation nor $+\!\!+$, it requires the corresponding chunkwise operation and we already discussed the termination for chunkwise operations.

4 Experimental Results

We measured the accumulated runtime of the expressions in the `let`-binding of `bhtree` (Section 2) for the *original* NESL-*program* using 15,000 particles and CMU-NESL [2] (on a single processor, 200MHz Pentium/Linux machine), to quantify the overhead of mapping the tree structure to a vector. From the overall runtime of 1.75 seconds, 0.5 seconds (29%) are spend on the operations introduced by the mapping. We gathered this data on a sequential machine, due to inaccurate profiling information from CMU-NESL on the Cray T3E. The inefficiencies would even be worse in a parallel run, since these operations include algorithmically unnecessary reordering operations, which cause communication.

To measure our proposed techniques for implementing recursive types, we timed an implementation of Barnes-Hut generated using the presented rules.[1] Figure 2 shows the timing of a single simulation step on a Cray T3E for 9000 and 20000 particles as well as the relative speedup and the absolute speed up (we only had access to 20 processors). The relative speedup is already quite close to the theoretical optimal speedup, but the absolute speedup shows that there is still room for improvement and we already know some possible optimizations. We did not compare CMU-NESL and our implementation on the Cray T3E directly, since our code generation techniques for the flat language already outperform CMU-NESL by more than an order of magnitude [9].

---

[1] The code was 'hand-generated'; we are currently implementing a compiler.

5 Conclusion

After its introduction by Blelloch & Sabot, flattening has received considerable attention, but to the best of our knowledge, we are the first to extend flattening to tree structures. There are other approaches to implementing nested data-parallelism, and of course, there is a wealth of literature on trees in parallel computing, but we do not have the space to discuss this work here—some of it is discussed in [8]. Our extension is easily integrated into existing systems and the gathered benchmarks support the efficiency of the generated code.

Acknowledgements. We are indebted to Martin Simons and Wolf Pfannenstiel for fruitful discussions and helpful comments on earlier versions of this paper. Furthermore, we are grateful to the anonymous referees of EuroPar'98 for their valuable comments and the members of the CACA seminar (University of Tokyo) for lively discussions. The first author receives a PhD scholarship from the DFG (German Research Council). She thanks the SCORE group of the University of Tsukuba, especially, Tetsuo Ida, for the hospitality while being a guest at SCORE.

References

1. P. K. T. Au, M. M. T. Chakravarty, J. Darlington, Yike Guo, Stefan Jähnichen, G. Keller, M. Köhler, W. Pfannenstiel, and M. Simons. Enlarging the scope of vector-based computations: Extending Fortran 90 with nested data parallelism. In W. K. Giloi, editor, *Proc. of the Intl. Conf. on Advances in Parallel and Distributed Computing*. IEEE Computer Society Press, 1997.
2. G. E. Blelloch, S. Chatterjee, J. C. Hardwick, J. Sipelstein, and M. Zagha. Implementation of a portable nested data-parallel language. In *4th ACM SIGPLAN Symp. on Principles and Practice of Parallel Programming*, 1993.
3. J. Barnes and P. Hut. A hierarchical $O(n \log n)$ force calculation algorithm. *Nature*, 324, December 1986.
4. G. E. Blelloch. *Vector Models for Data-Parallel Computing*. The MIT Press, 1990.
5. G. E. Blelloch. Programming parallel algorithms. *Communications of the ACM*, 39(3):85–97, 1996.
6. G.E. Blelloch and G. W. Sabot. Compiling collection-oriented languages onto massively parallel computers. *Journal of Parallel and Distributed Computing*, 8:119–134, 1990.
7. S. Chatterjee. Compiling nested data-parallel programs for shared-memory multiprocessors. *ACM Trans. on Prog. Lang. and Systems*, 15(3), 1993.
8. Gabriele Keller and Manuel M. T. Chakravarty. Flattening trees—unabridged. Forschungsbreicht 98-6, Technical University of Berlin, 1998. http://cs.tu-berlin.de/cs/ifb/TechnBerichteListe.html.
9. G. Keller. *Transformation-based Implementation of Nested Parallelism for Parallel Computers with Distributed Memory*. PhD thesis, Technische Universität Berlin, Fachbereich Informatik, 1998. Forthcoming.
10. G. Keller and M. Simons. A calculational approach to flattening nested data parallelism in functional languages. In J. Jaffar, editor, *The 1996 Asian Computing Science Conference*, LNCS. Springer Verlag, 1996.

11. J. Prins and D. Palmer. Transforming high-level data-parallel programs into vector operations. In *Proceedings of the Fourth ACM SIGPLAN Symposium on Principles and Practice of Parallel Programming*, pages 119–128, San Diego, CA., May 19-22, 1993. ACM.

12. D. Palmer, J. Prins, and S. Westfold. Work-efficient nested data-parallelism. In *Proceedings of the Fifth Symposium on the Frontiers of Massively Parallel Processing (Frontiers 95)*. IEEE, 1995.

Dynamic Type Information in Process Types

Franz Puntigam

Technische Universität Wien, Institut für Computersprachen, Argentinierstr. 8,
A-1040 Vienna, Austria.
franz@complang.tuwien.ac.at

Abstract. Static checking of process types ensures that each object accepts all messages received from concurrent clients, although the set of acceptable messages can depend on the object's state. However, conventional approaches of using dynamic type information (e.g., checked type casts) are not applicable in the current process type model, and the typing of self-references is too restrictive. In this paper a refinement of the model is proposed. It solves these problems so that it is easy to handle, for example, heterogeneous collections.

1 Introduction

The process type model was proposed as a statically checkable type model for concurrent and distributed systems based on active objects [10, 11]. A process type specifies not only a set of acceptable messages, but also constraints on the sequence of these messages. Type safety can be checked statically by ensuring that each object reference is associated with an appropriate *type mark* which specifies all message sequences accepted by the object via this reference. The object accepts messages sent through different references in arbitrary interleaving. A type mark is a limited resource that represents a "claim" to send messages. It can be used up, split into several, more restricted type marks or forwarded from one user to another, but it must not be exceeded by any user.

Support of using dynamic type information shall be added. But, checked type casts break type safety: An object's dynamic type is not a "claim" to send messages. This problem is solved by checking against dynamic type marks.

Each object has better knowledge about its own state than about the other objects' states. The additional knowledge can be used in changing the self-references' type marks in a very flexible and still type-safe way.

The rest of the paper is structured as follows: An improved version of the typed process calculus presented in [11] is introduced in Sect. 2. Support of using dynamic type information is added in Sect. 3, a more flexible typing of self-references in Sect. 4. Static type checking is dealt with in Sect. 5.

2 A Typed Calculus of Active Objects

The proposed calculus describes systems composed of active objects that communicate through asynchronous message passing. An object has its own thread of execution, a behavior, an identifier, and an unlimited buffer of received messages. According to its behavior, an object accepts messages from its buffer, sends messages to other objects, and creates new objects. All messages are received and accepted in the same logical order as they were sent.

We use following notation. *Constants* (denoted by $x, y, z, \ldots$) are used as message selectors and in types. *Identifiers* $(u, v, w, \ldots)$ in the infinite set $\mathbb{V}$ are used as object and procedure identifiers as well as formal parameters. For each symbol e, $\tilde{e}$ is an abbreviation of $e_1, \ldots, e_n$; e.g., $\tilde{x}$ is a sequence of constants. $\{\tilde{e}\}$ denotes the smallest set containing $\tilde{e}$, and $|\tilde{e}|$ the length of the sequence. For each symbol $\cdot$, $\tilde{e} \cdot g$ stands for $e_1 \cdot g, \ldots, e_n \cdot g$, and $\tilde{e} \cdot \tilde{g}$ for $e_1 \cdot g_1, \ldots, e_n \cdot g_n$ $(|\tilde{e}| = |\tilde{g}|)$.

Processes $(p, q, \ldots)$ specify object behavior. Their syntax is:

$$
\begin{aligned}
p &::= \{\tilde{s}\} \mid u.m; p \mid \underline{u} := (\langle \tilde{u} \rangle p{:}\varphi); q \mid \underline{v}\$u\langle \tilde{\alpha}, \tilde{u} \rangle; p \mid u\langle \tilde{\alpha}, \tilde{u} \rangle \mid u{=}v\,?\,p \mid q \\
s &::= x\langle \tilde{u} \rangle; p \\
m &::= x\langle \tilde{\alpha}, \tilde{u} \rangle \\
\alpha &::= \tau \mid \varphi \\
\tau &::= \{\tilde{a}\}[\tilde{x}] \mid *\underline{u}\{\tilde{a}\}[\tilde{x}] \mid \sigma + \tau \mid u \\
a &::= x\langle \tilde{u}{:}\tilde{\mu}, \tilde{\alpha} \rangle [\tilde{y}] \triangleright [\tilde{z}] \\
\varphi &::= \langle \tilde{u}{:}\tilde{\mu}, \tilde{\alpha} \rangle \tau \mid *\underline{u}\langle \tilde{u}{:}\tilde{\mu}, \tilde{\alpha} \rangle \tau \mid u \\
\mu &::= \mathsf{t} \mid \mathsf{ot} \mid \mathsf{pt}
\end{aligned}
$$

A *selector* $\{\tilde{s}\}$ specifies a possibly empty, finite set of *guarded processes* $(r, s, \ldots)$. Each s_i is of the form $x\langle \tilde{u} \rangle; p$, where x is a message selector, $\tilde{u}$ a list of formal parameters, and p a process to be executed if s_i is selected. An s_i is selectable if the first received message in the object's buffer is of the form $x\langle \tilde{\alpha}, \tilde{v} \rangle$, where $\tilde{\alpha}, \tilde{v}$ are arguments to be substituted for the parameters $\tilde{u}$ $(|\tilde{\alpha}, \tilde{v}| = |\tilde{u}|)$; $\tilde{\alpha}$ is a list of types and $\tilde{v}$ a list of object and procedure identifiers. A process $u.m; p$ sends a message m to the object with identifier u, and then behaves as p. A *procedure definition* $\underline{u} := (\langle \tilde{u} \rangle p{:}\varphi); q$ introduces an identifier u of a procedure of type φ and then behaves as q; the procedure takes $\tilde{u}$ as parameters and specifies the behavior p. A process $\underline{v}\$u\langle \tilde{\alpha}, \tilde{u} \rangle; p$ creates a new object with identifier v and then behaves as p; the new object behaves as $u\langle \tilde{\alpha}, \tilde{u} \rangle$. A *call* $u\langle \tilde{\alpha}, \tilde{u} \rangle$ behaves as specified by the procedure with identifier u. A *conditional expression* $u{=}v\,?\,p \mid q$ behaves as p if the identifiers u and v are equal, otherwise as q.

There are two kinds of types, object types $(\pi, \varrho, \sigma, \tau, \ldots)$ and procedure types $(\varphi, \psi, \ldots)$. A type is denoted by $\alpha, \beta, \gamma, \ldots$ if its kind does not matter. Furthermore, there are meta-types $(\mu, \nu, \ldots)$: "ot" is the type of all object types, "pt" the type of all procedure types, and "t" the type of all types of any kind.

The *activating set* $[\tilde{x}]$ of an object type $\{\tilde{a}\}[\tilde{x}]$ is a multi-set of constants, the *behavior descriptor* $\{\tilde{a}\}$ a finite collection of *message descriptors* $(a, b, c, \ldots)$. A message descriptor $x\langle \tilde{u}{:}\tilde{\mu}, \tilde{\alpha} \rangle [\tilde{y}] \triangleright [\tilde{z}]$ describes a message with selector x, type parameters $\tilde{u}$ of meta-types $\tilde{\mu}$, and parameters of types $\tilde{\alpha}$. The $\tilde{u}$ can occur in $\tilde{\alpha}$. The message descriptor is *active* if the activating set $[\tilde{x}]$ contains all constants in the multi-set $[\tilde{y}]$ (*in-set*). When a corresponding message is sent (or accepted), the type is updated by removing the constants in the in-set from the activating set and adding those in the multi-set $[\tilde{z}]$ (*out-set*). Using type updating repeatedly for all active message descriptors, the type specifies a set of acceptable message sequences. An expression $*\underline{u}\{\tilde{a}\}[\tilde{x}]$ is a recursive version of an object type. A type combination $\sigma + \tau$ specifies a set of acceptable message sequences that includes all arbitrary interleavings of acceptable message sequences specified by σ and τ. An identifier u can be used as type parameter.

A procedure type $\langle \tilde{\underline{u}}:\tilde{\mu}, \tilde{\alpha}\rangle \tau$ specifies that a procedure of this type takes type parameters $\tilde{u}$ of meta-types $\tilde{\mu}$ and parameters of types $\tilde{\alpha}$. Objects behaving according to the procedure accept messages as specified by τ.

For example, a procedure $B := (\langle \underline{u}\rangle \{\mathrm{put}\langle \underline{v}\rangle; \{\mathrm{get}\langle \underline{w}\rangle; w.\mathrm{back}\langle v\rangle; B\langle u\rangle\}\}{:}\varphi_B)$ specifies the behavior of a buffer accepting "put" and "get" in alternation and sending "back" to the argument of "get" after receiving "get". The type of B is given by $\varphi_B =_{\mathrm{def}} \langle \underline{u}{:}t\rangle \{\mathrm{put}\langle u\rangle[e]\triangleright[f], \mathrm{get}\langle \{\mathrm{back}\langle u\rangle[\mathrm{once}]\triangleright[]\}[\mathrm{once}]\rangle[f]\triangleright[e]\}[e]$. The type of a procedure specifying the behavior of an infinite buffer that accepts as many "get" messages as there are elements in the buffer can be given by $\varphi_{BI} =_{\mathrm{def}} \langle \underline{u}{:}t\rangle \{\mathrm{put}\langle u\rangle[]\triangleright[f], \mathrm{get}\langle \{\mathrm{back}\langle u\rangle[\mathrm{once}]\triangleright[]\}[\mathrm{once}]\rangle[f]\triangleright[]\}[]$.

Each underlined occurrence of an identifier binds this and all following occurrences of the identifier. An occurrence is free if it is not bound. *Free(e)* denotes the set of all identifiers occurring free in e. Two processes are regarded as equal if they can be made identical by renaming bound identifiers (α-conversion) and repeatedly applying the equations $\{\tilde{r}, s\} = \{\tilde{r}, s, s\}$ and $\{\tilde{r}, s, \tilde{s}\} = \{\tilde{r}, \tilde{s}, s\}$ to selectors. Two types are regarded as equal if they can be made identical by renaming bound identifiers (α-conversion) and applying these equations:

$$\{\tilde{a}, b\} = \{\tilde{a}, b, b\} \qquad \{\tilde{a}, c, \tilde{c}\} = \{\tilde{a}, \tilde{c}, c\} \qquad\qquad [\tilde{x}, y, \tilde{y}] = [\tilde{x}, \tilde{y}, y]$$

$$\sigma + \tau = \tau + \sigma \qquad (\varrho + \sigma) + \tau = \varrho + (\sigma + \tau) \qquad\qquad \tau + \{\}[] = \tau$$

$$*\underline{u}\alpha = [*\underline{u}\alpha/u]\alpha \qquad \{\tilde{a}\}[\tilde{x}, \tilde{y}] = \{\tilde{a}\}[\tilde{x}] \quad (\tilde{y} \notin \mathit{Eff}\{\tilde{a}\})$$

$$\{x\langle \tilde{\underline{u}}:\tilde{\mu}, \tilde{\alpha}\rangle [\tilde{x}]\triangleright[\tilde{y}, \tilde{y}'], \tilde{a}\}[\tilde{z}] = \{x\langle \tilde{\underline{u}}:\tilde{\mu}, \tilde{\alpha}\rangle [\tilde{x}]\triangleright[\tilde{y}], \tilde{a}\}[\tilde{z}] \quad (\tilde{y}' \notin \{\tilde{x}\}\cup\mathit{Eff}\{\tilde{a}\})$$

$$\{\tilde{a}\}[\tilde{x}] + \{\tilde{c}\}[\tilde{y}] = \{\tilde{a}, \tilde{c}\}[\tilde{x}, \tilde{y}] \quad (\tilde{x} \in \mathit{Eff}\{\tilde{a}\};\ \tilde{y} \in \mathit{Eff}\{\tilde{c}\})$$

where $\mathit{Eff}\{x_1\langle \tilde{\underline{u}}_1:\tilde{\mu}_1, \tilde{\alpha}_1\rangle[\tilde{x}_1]\triangleright[\tilde{y}_1], \ldots, x_i\langle \tilde{\underline{u}}_i:\tilde{\mu}_i, \tilde{\alpha}_i\rangle[\tilde{x}_i]\triangleright[\tilde{y}_i]\} = \{\tilde{x}_1\} \cup \cdots \cup \{\tilde{x}_i\}$.

Subtyping on meta-types is defined by $\mathrm{ot} \le \mathrm{t}$ and $\mathrm{pt} \le \mathrm{t}$. Subtyping for types depends on an environment Π containing typing assumptions of the forms $u \le \alpha$ and $\alpha \le u$. The subtyping relation $\le$ on types and the redundancy elimination relation $\doteqdot$ on object types are the reflexive, transitive closures of the rules:

$$\Pi \cup \{u \le \alpha\} \vdash u \le \alpha \qquad\qquad \Pi \cup \{\alpha \le u\} \vdash \alpha \le u$$

$$\frac{\Pi \vdash \tau + \varrho \doteqdot \sigma}{\Pi \vdash \sigma \le \tau} \qquad \frac{\Pi \vdash \pi \le \varrho \quad \Pi \vdash \sigma \le \tau}{\Pi \vdash \pi + \sigma \le \varrho + \tau} \qquad \frac{\Pi \vdash \tilde{\gamma} \le \tilde{\alpha} \quad \Pi \vdash \sigma \le \tau \quad \tilde{\nu} \le \tilde{\mu}}{\Pi \vdash \langle \tilde{\underline{u}}:\tilde{\mu}, \tilde{\alpha}\rangle \sigma \le \langle \tilde{\underline{u}}:\tilde{\nu}, \tilde{\gamma}\rangle \tau}$$

$$\Pi \vdash \{\tilde{a}, \tilde{c}\}[\tilde{x}] \doteqdot \{\tilde{a}\}[\tilde{x}] \quad (\forall c \in \{\tilde{c}\} \bullet \exists a \in \{\tilde{a}\} \bullet$$
$$c = x\langle \tilde{\underline{u}}:\tilde{\mu}, \tilde{\alpha}\rangle[\tilde{y}, \tilde{y}']\triangleright[\tilde{z}, \tilde{z}'] \ \wedge\ \tilde{z}' \notin \mathit{Eff}\{\tilde{a}\} \ \wedge$$
$$a = x\langle \tilde{\underline{u}}:\tilde{\nu}, \tilde{\gamma}\rangle[\tilde{y}]\triangleright[\tilde{z}, \tilde{z}''] \ \wedge\ \tilde{\mu} \le \tilde{\nu} \ \wedge\ \Pi \vdash \tilde{\alpha} \le \tilde{\gamma})$$

($\alpha \le \beta$ and $\sigma \doteqdot \tau$ are simplified notations for $\emptyset \vdash \alpha \le \beta$ and $\emptyset \vdash \sigma \doteqdot \tau$.) The relation $\doteqdot$ is used for removing redundant message descriptors from behavior descriptors. Using these definitions we can show, for example, $\varphi_{BI} \le \varphi_B$.

A type $\{\tilde{a}\}[\tilde{x}]$ is *deterministic* if all message descriptors in $\{\tilde{a}\}$ have pairwise different message selectors. For each deterministic type σ and message m there is only one way of updating σ according to m.

A *system configuration* C contains expressions of the forms $u \mapsto \langle p, \tilde{m} \rangle$ and $u \mapsto \langle \tilde{u} \rangle p{:}\varphi$. The first expression specifies an object with identifier u, behavior p and sequence of received messages $\tilde{m}$. The second expression specifies a procedure of identifier u. C contains at most one expression $u \mapsto e$ for each $u \in \mathbb{V}$. $C[u_1 \mapsto e_1, \ldots, u_n \mapsto e_n]$ is a system configuration with expressions added to C, where $u_1, \ldots, u_n$ are pairwise different and do not occur to the left of $\mapsto$ in C.

For each $u \in \mathbb{V}$ the relation $\xrightarrow{u}$ on system configurations (defined by the rules given below) specifies all possible execution steps of object u. ($[\tilde{e}/\tilde{u}]f$ is the simultaneous substitution of the expressions $\tilde{e}$ for all free occurrences of $\tilde{u}$ in f.)

$$C[u \mapsto \langle \{x\langle \tilde{\underline{u}} \rangle; p, \tilde{s}\}, x\langle \tilde{\alpha}, \tilde{v} \rangle, \tilde{m} \rangle] \xrightarrow{u} C[u \mapsto \langle ([\tilde{\alpha}, \tilde{v}/\tilde{u}]p), \tilde{m} \rangle]$$

$$C[u \mapsto \langle v.m; p, \tilde{m} \rangle, v \mapsto \langle q, \tilde{m}' \rangle] \xrightarrow{u} C[u \mapsto \langle p, \tilde{m} \rangle, v \mapsto \langle q, \tilde{m}', m \rangle]$$

$$C[u \mapsto \langle v := (\langle \tilde{\underline{u}} \rangle p{:}\varphi); q, \tilde{m} \rangle] \xrightarrow{u} C[u \mapsto \langle q, \tilde{m} \rangle, v \mapsto \langle \tilde{\underline{u}} \rangle p{:}\varphi]$$

$$C[u \mapsto \langle v\$w\langle \tilde{\alpha}, \tilde{u} \rangle; p, \tilde{m} \rangle] \xrightarrow{u} C[u \mapsto \langle p, \tilde{m} \rangle, v \mapsto \langle w\langle \tilde{\alpha}, \tilde{u} \rangle \rangle]$$

$$C[u \mapsto \langle v\langle \tilde{\alpha}, \tilde{u} \rangle, \tilde{m} \rangle, v \mapsto \langle \tilde{\underline{v}} \rangle p{:}\varphi] \xrightarrow{u} C[u \mapsto \langle ([\tilde{\alpha}, \tilde{u}/\tilde{v}]p), \tilde{m} \rangle, v \mapsto \langle \tilde{\underline{v}} \rangle p{:}\varphi]$$

$$C[u \mapsto \langle v{=}v\,?\,p\,|\,q, \tilde{m} \rangle] \xrightarrow{u} C[u \mapsto \langle p, \tilde{m} \rangle]$$

$$C[u \mapsto \langle v{=}w\,?\,p\,|\,q, \tilde{m} \rangle] \xrightarrow{u} C[u \mapsto \langle q, \tilde{m} \rangle] \quad (v \neq w)$$

$\longrightarrow$ denotes the closure of these relations over all $u \in \mathbb{V}$, and $\xrightarrow{*}$ the reflexive, transitive closure of $\longrightarrow$. $\xrightarrow{*}$ defines the operational semantics of object systems.

3 Dynamic Type Information

It seems to be easy to extend the calculus with a dynamic type checking concept: Processes of the form $\alpha{\leq}\beta\,?\,p\,|\,q$ are introduced, where α or β initially is a type parameter to be replaced with a (dynamic) type during computation. The typing assumption $\alpha \leq \beta$ (for α or β in $\mathbb{V}$) is statically known to hold in p. The relation $\xrightarrow{u}$ (for each $u \in \mathbb{V}$) is extended by the rules:

$$C[u \mapsto \langle \alpha{\leq}\beta\,?\,p\,|\,q, \tilde{m} \rangle] \xrightarrow{u} C[u \mapsto \langle p, \tilde{m} \rangle] \quad (\alpha \leq \beta)$$

$$C[u \mapsto \langle \alpha{\leq}\beta\,?\,p\,|\,q, \tilde{m} \rangle] \xrightarrow{u} C[u \mapsto \langle q, \tilde{m} \rangle] \quad (\alpha \not\leq \beta)$$

For example, assume that a reference w is of type mark $\mathrm{St}[\mathrm{f}]$, where $\mathrm{St} =_{\mathrm{def}}$ $\{\mathrm{put}\langle \underline{u}{:}t, u \rangle[\mathrm{e}]{\triangleright}[\mathrm{f}], \mathrm{get}\langle \mathrm{R}[\mathrm{one}] \rangle[\mathrm{f}]{\triangleright}[\mathrm{e}]\}$ and $\mathrm{R} =_{\mathrm{def}} \{\mathrm{back}\langle \underline{v}{:}t, v \rangle[\mathrm{one}]{\triangleright}[\,]\}$. The process $w.\mathrm{get}\langle w' \rangle; \{\mathrm{back}\langle \underline{u}, \underline{v} \rangle; u{\leq}\mathrm{St}[\mathrm{e}]\,?\,v.\mathrm{put}\langle \mathrm{St}[\mathrm{f}], v \rangle; \{\}\,|\,\{\}\}$ (w' denotes the object's self-reference) is type-safe. First, "get" is sent to w; then "back" is accepted. If the type mark of v is a subtype of $\mathrm{St}[\mathrm{e}]$, v is put into v. The type used as argument of "put" is $\mathrm{St}[\mathrm{f}]$ because the type is updated when sending "put".

Type information is lost when types are updated. $\mathrm{St}[\mathrm{e}]$ is a supertype of v's type mark u; but, $\mathrm{put}\langle u, v \rangle$ cannot be sent to v: So far it is not possible to specify that v's type mark is equal to u, except that u is updated. To solve the problem, processes of the form $\sigma \leq \tau ? \underline{v}; p \mid q$ are introduced, where σ initially is a type parameter to be replaced with an object type. The relation $\xrightarrow{u}$ is extended:

$$C[u \mapsto \langle \sigma \leq \tau ? \underline{v}; p \mid q, \tilde{m} \rangle] \xrightarrow{u} C[u \mapsto \langle ([\varrho/v]p), \tilde{m} \rangle] \quad (\tau + \varrho \doteq \sigma)$$

$$C[u \mapsto \langle \sigma \leq \tau ? \underline{v}; p \mid q, \tilde{m} \rangle] \xrightarrow{u} C[u \mapsto \langle q, \tilde{m} \rangle] \quad (\sigma \not\leq \tau)$$

v in p is replaced with a dynamically determined object type ϱ which specifies the difference between σ and τ. This version of the above process keeps type information: $w.get\langle w' \rangle; \{back\langle \underline{u}, \underline{v} \rangle; u \leq St[e] ? \underline{u}'; v.put\langle St[f] + u', v \rangle; \{\} \mid \{\}\}$. The type $St[f] + u'$ equals u after removing e and adding f to the activating set.

4 Typing Self-References

Without special treatment, static typing of self-references is very restrictive: If an object behaves according to a type τ, the type mark σ of a self-reference is a subtype of τ. Both, σ and τ specify all messages that may return results of possibly needed services. In general, it is impossible to determine statically which services are needed. Hence, σ and τ usually specify much larger sets of message sequences than needed. This problem can be solved because each object has much more knowledge about its own state than other objects: Before a message with a self-reference as argument is sent to a server, this reference's type mark is selected so that it supports the messages returned by the server. At the same time the object's type is extended correspondingly. The object's type and self-reference's type mark are determined dynamically as needed in the computation.

We extend the calculus: The distinguished name "self" can be used as arguments in procedures; "self" is replaced with a self-reference when the procedure is called. The fifth rule in the definition of $\xrightarrow{u}$ has to be replaced with:

$$C[u \mapsto \langle v \langle \tilde{\alpha}, \tilde{u} \rangle, \tilde{m} \rangle, v \mapsto \langle \tilde{v} \rangle p{:}\varphi] \xrightarrow{u} C[u \mapsto \langle ([u/self][\tilde{\alpha}, \tilde{u}/\tilde{v}]p), \tilde{m} \rangle, v \mapsto \langle \tilde{v} \rangle p{:}\varphi]$$

The type mark of an occurrence of "self" as argument is equal to the type of the corresponding formal parameter. (Elements of $\mathbb{V} \cup \{self\}$ are denoted by $o, \ldots$)

An example demonstrates the use of self and dynamic type comparisons:

$$\underline{w} := ((\langle \underline{u}, \underline{u}', \underline{v}, \underline{w}' \rangle u \leq St[f] ? v.get\langle self \rangle; \{back\langle \underline{u}, \underline{v} \rangle; w\langle u, u', v, w' \rangle\}$$
$$\mid w'\langle u, v \rangle) \quad : \quad \langle \underline{u}{:}t, \underline{u}'{:}ot, u, \langle \underline{u}{:}t, u \rangle u' \rangle u'); \ldots$$

The procedure w takes as parameters a type u, an object type u', an identifier v of type u and a procedure identifier w' of type $\langle \underline{u}{:}t, u \rangle u'$. If v is a non-empty store, an element is got from the store and w is called recursively with the element and its type as arguments. Otherwise w' is called with u and v as arguments. The type mark of self is R[one], as specified by St. The corresponding message "back" is accepted if the expression containing self is executed.

5 Static Type Checking

An important result is that the process type model with the proposed extensions supports static type checking. Type checking rules are given in Appendix A.

Theorem 1. *Let* $C = \{u_1 \mapsto \langle p_1 \rangle, \ldots, u_n \mapsto \langle p_n \rangle\}$ *be a system configuration and* $\sigma_{i,j}$ *and* τ_i $(1 \leq i, j \leq n)$ *object types such that* $\tau_i \leq \sigma_{i,1} + \cdots + \sigma_{i,n}$ *and* $\emptyset, \{u_1:\sigma_{1,i}, \ldots, u_n:\sigma_{n,i}\} \vdash \langle p_i:\tau_i \rangle$ *as defined by the type checking rules. Then, for each system configuration* D *with* $C \xrightarrow{*} D$, *if* D *contains an expression* $u \mapsto \langle p, m, \tilde{m} \rangle$, *there exists a system configuration* E *with* $D \xrightarrow{u} E$.

(The proof is omitted because of lack of space.) Provided that type checking succeeds for an initial system configuration, if an object has a nonempty buffer of received messages, the execution of this object cannot be blocked. Especially, the next message in the buffer is acceptable. Theorem 1 also implies that separate compilation is supported: If a process p_i in C is replaced with another p_i' satisfying the conditions, the consequences of the theorem still hold.

6 Related Work

Much work on types for concurrent languages and models was done. The majority of this work is based on Milner's π-calculus [3,4] and similar calculi. Especially, the problem of inferring most general types was considered by Gay [2] and Vasconcelos and Honda [14]. Nierstrasz [5], Pierce and Sangiorgi [8], Vasconcelos [13], Colaco, Pantel and Sall'e [1] and Ravara and Vasconcelos [12] deal with subtyping. But their type models differ in an important aspect from the process type model: They cannot represent constraints on message sequences and ensure statically that all sent messages are acceptable; the underlying calculus does not keep the message order.

The proposals of Nielson and Nielson [6] can deal with constraints on message sequences. As in the process type model, a type checker updates type information while walking through an expression. However, their type model cannot ensure that all sent messages are understood, and dealing with dynamic type information is not considered.

The process type model can ensure statically that all sent messages are understood, although the set of acceptable messages can change. Therefore, this model is a promising approach to strong, static types for concurrent and distributed applications based on active objects. However, it turned out that the restrictions caused by static typing are an important difficulty in the practical applicability of the process type model. So far it was not clear how to circumvent this difficulty without breaking type safety. New techniques for dealing with dynamic type information (as presented in this paper) had to be found.

The approach of Najm and Nimour [7] has a similar goal as process types. However, this approach is more restrictive and cannot deal with dynamic type information, too.

The work presented in this paper refines previous work on process types [9–11]. The major improvement is the addition of dynamic type comparisons and a special treatment of self-references. Some changes to earlier versions of the process type model were necessary. For example, the two sets of type checking rules presented in [11] had to be modified and combined into a single set so that it was possible to introduce "self".

Dynamic type comparisons exist in nearly all recent object-oriented programming languages. However, the usual approaches cannot be combined with process types because these approaches do not distinguish between object types and type marks and cannot deal with type updates. Conventional type models do not have problems with the flexibility of typing self-references.

7 Conclusions

The process type model is a promising basis for strongly typed concurrent programming languages. The use of dynamic type information can be supported in several ways so that the process type model becomes more flexible and useful. An important property is kept: A type checker can ensure statically that clients are coordinated so that all received messages can be accepted in the order they were sent, even if the acceptability of messages depends on the server's state.

References

1. J.-L. Colaco, M. Pantel, and P. Sall'e. A set-constraint-based analysis of actors. In *Proceedings FMOODS '97*, Canterbury, United Kingdom, July 1997. Chapman & Hall.
2. Simon J. Gay. A sort inference algorithm for the polyadic π-calculus. In *Conference Record of the 20th Symposium on Principles of Programming Languages*, January 1993.
3. Robin Milner. The polyadic π-calculus: A tutorial. Technical Report ECS-LFCS-91-180, Dept. of Comp. Sci., Edinburgh University, 1991.
4. R. Milner, J. Parrow, and D. Walker. A calculus of mobile processes (parts I and II). *Information and Computation*, 100:1–77, 1992.
5. Oscar Nierstrasz. Regular types for active objects. *ACM SIGPLAN Notices*, 28(10):1–15, October 1993. Proceedings OOPSLA'93.
6. Flemming Nielson and Hanne Riis Nielson. From CML to process algebras. In *Proceedings CONCUR'93*, number 715 in Lecture Notes in Computer Science, pages 493–508. Springer-Verlag, 1993.
7. E. Najm and A. Nimour. A calculus of object bindings. In *Proceedings FMOODS '97*, Canterbury, United Kingdom, July 1997.
8. Benjamin Pierce and Davide Sangiorgi. Typing and subtyping for mobile processes. In *Proceedings LICS'93*, 1993.
9. Franz Puntigam. Flexible types for a concurrent model. In *Proceedings of the Workshop on Object-Oriented Programming and Models of Concurrency*, Torino, June 1995.

10. Franz Puntigam. Types for active objects based on trace semantics. In Elie Najm et al., editor, *Proceedings FMOODS '96*, Paris, France, March 1996. IFIP WG 6.1, Chapman & Hall.

11. Franz Puntigam. Coordination requirements expressed in types for active objects. In Mehmet Aksit and Satoshi Matsuoka, editors, *Proceedings ECOOP '97*, number 1241 in Lecture Notes in Computer Science, Jyväskylä, Finland, June 1997. Springer-Verlag.

12. António Ravara and Vasco T. Vasconcelos. Behavioural types for a calculus of concurrent objects. In *Proceedings Euro-Par '97*, Lecture Notes in Computer Science. Springer-Verlag, 1997.

13. Vasco T. Vasconcelos. Typed concurrent objects. In *Proceedings ECOOP'94*, number 821 in Lecture Notes in Computer Science, pages 100–117. Springer-Verlag, 1994.

14. Vasco T. Vasconcelos and Kohei Honda. Principal typing schemes in a polyadic pi-calculus. In *Proceedings CONCUR'93*, July 1993.

A Type Checking Rules

$$\frac{\Pi \vdash \{\tilde{a}\}[\tilde{y}] \leq \tau \quad \forall 1\leq i\leq n \bullet \Pi, \Gamma[\tilde{u}_i{:}\tilde{\mu}_i, \tilde{v}_i{:}\tilde{\alpha}_i] \vdash \langle p_i{:}\{\tilde{a}\}[\tilde{z}_i]\rangle}{\Pi, \Gamma \vdash \langle \{x_1\langle \underline{\tilde{u}}_1, \underline{\tilde{v}}_1\rangle; p_1, \ldots, x_n\langle \underline{\tilde{u}}_n, \underline{\tilde{v}}_n\rangle; p_n\}{:}\tau\rangle} \quad (1) \qquad \text{SEL}$$

$$\frac{\Pi \vdash \sigma \leq \{a\}[\tilde{x}] + \varrho \quad \Pi, \Gamma[u{:}\{a\}[\tilde{y}] + \varrho] \vdash \langle \tilde{\alpha}{:}\tilde{\mu}, \tilde{o}{:}([\tilde{\alpha}/\tilde{u}]\tilde{\gamma}), p{:}\tau\rangle}{\Pi, \Gamma[u{:}\sigma] \vdash \langle (u.x\langle \tilde{\alpha}, \tilde{o}\rangle; p){:}\tau\rangle} \quad (2) \qquad \text{SEND}$$

$$\frac{\Pi, \emptyset[\tilde{w}_1{:}\tilde{\nu}, \tilde{w}_2{:}\tilde{\varphi}, u{:}\psi, \tilde{u}{:}\tilde{\mu}, \tilde{v}{:}\tilde{\alpha}] \vdash \langle \psi{:}\text{pt}, p{:}\sigma\rangle \quad \Pi, \Gamma[u{:}\psi] \vdash \langle q{:}\tau\rangle}{\Pi, \Gamma \vdash \langle (\underline{u} := (\langle \underline{\tilde{u}}, \underline{\tilde{v}}\rangle p{:}{*}\underline{u}\psi); q){:}\tau\rangle} \quad (3) \qquad \text{DEF}$$

$$\frac{\Pi \vdash \varphi \leq \langle \underline{\tilde{u}}{:}\tilde{\mu}, \tilde{\gamma}\rangle\sigma \quad \Pi, \Gamma[u{:}([\tilde{\alpha}/\tilde{u}]\sigma), v{:}\varphi] \vdash \langle \tilde{\alpha}{:}\tilde{\mu}, \tilde{o}{:}([\tilde{\alpha}/\tilde{u}]\tilde{\gamma}), p{:}\tau\rangle}{\Pi, \Gamma[v{:}\varphi] \vdash \langle (\underline{u}\$v\langle \tilde{\alpha}, \tilde{o}\rangle; p){:}\tau\rangle} \qquad \text{NEW}$$

$$\frac{\Pi \vdash \varphi \leq \langle \underline{u}{:}\tilde{\mu}, \tilde{\gamma}\rangle\sigma \quad \Pi \vdash ([\tilde{\alpha}/\tilde{u}]\sigma) \leq \tau \quad \Pi, \Gamma[u{:}\varphi] \vdash \langle \tilde{\alpha}{:}\tilde{\mu}, \tilde{o}{:}([\tilde{\alpha}/\tilde{u}]\tilde{\gamma}), \{\}{:}\{\}\square\rangle}{\Pi, \Gamma[u{:}\varphi] \vdash \langle u\langle \tilde{\alpha}, \tilde{o}\rangle{:}\tau\rangle} \qquad \text{CALL}$$

$$\frac{\Pi, \Gamma[u{:}\varrho + \sigma] \vdash \langle ([u/v]p){:}\tau\rangle \quad \Pi, \Gamma[u{:}\varrho, v{:}\sigma] \vdash \langle q{:}\tau\rangle}{\Pi, \Gamma[u{:}\varrho, v{:}\sigma] \vdash \langle (u{=}v\,?\,p\,|\,q){:}\tau\rangle} \qquad \text{EQU}_1$$

$$\frac{\Pi, \Gamma[u{:}\varphi, v{:}\psi] \vdash \langle p{:}\tau\rangle \quad \Pi, \Gamma[u{:}\varphi, v{:}\psi] \vdash \langle q{:}\tau\rangle}{\Pi, \Gamma[u{:}\varphi, v{:}\psi] \vdash \langle (u{=}v\,?\,p\,|\,q){:}\tau\rangle} \qquad \text{EQU}_2$$

$$\frac{\Pi \cup \{u \leq \alpha\}, \Gamma[u{:}\mu] \vdash \langle p{:}\tau\rangle \quad \Pi, \Gamma[u{:}\mu] \vdash \langle \alpha{:}t, q{:}\tau\rangle}{\Pi, \Gamma[u{:}\mu] \vdash \langle (u{\leq}\alpha\,?\,p\,|\,q){:}\tau\rangle} \quad (u \notin Free(\alpha)) \qquad \text{SUB}_1$$

$$\frac{\Pi \cup \{\alpha \leq u\}, \Gamma[u{:}\mu] \vdash \langle p{:}\tau\rangle \quad \Pi, \Gamma[u{:}\mu] \vdash \langle \alpha{:}t, q{:}\tau\rangle}{\Pi, \Gamma[u{:}\mu] \vdash \langle (\alpha{\leq}u\,?\,p\,|\,q){:}\tau\rangle} \quad (u \notin Free(\alpha)) \qquad \text{SUB}_2$$

$$\frac{\Pi \cup \{u \leq \sigma\}, \Gamma[u{:}\text{ot}, v{:}\text{ot}] \vdash \langle p{:}\tau\rangle \quad \Pi, \Gamma[u{:}\text{ot}] \vdash \langle \sigma{:}\text{ot}, q{:}\tau\rangle}{\Pi, \Gamma[u{:}\text{ot}] \vdash \langle (u{\leq}\sigma\,?\,\underline{v}; p\,|\,q){:}\tau\rangle} \quad (u \notin Free(\sigma)) \qquad \text{SPLIT}$$

$$\frac{\Pi \vdash \sigma \leq \varrho + \tau \quad \Pi, \Gamma[u{:}\varrho] \vdash \langle \tilde{e}{:}\tilde{g}\rangle}{\Pi, \Gamma[u{:}\sigma] \vdash \langle u{:}\tau, \tilde{e}{:}\tilde{g}\rangle} \qquad \text{OBJ}$$

$$\frac{\Pi, \Gamma \vdash \langle \tilde{e}{:}\tilde{g}, p{:}\sigma + \tau\rangle}{\Pi, \Gamma \vdash \langle \text{self}{:}\sigma, \tilde{e}{:}\tilde{g}, p{:}\tau\rangle} \qquad \text{SELF}$$

$$\frac{\Pi \vdash \psi \leq \varphi \quad \Pi, \Gamma[u{:}\psi] \vdash \langle \tilde{e}{:}\tilde{g}\rangle}{\Pi, \Gamma[u{:}\psi] \vdash \langle u{:}\varphi, \tilde{e}{:}\tilde{g}\rangle} \qquad \text{PROC}$$

$$\frac{\Pi, \Gamma[u{:}\nu] \vdash \langle \tilde{e}{:}\tilde{g}\rangle \quad \nu \leq \mu}{\Pi, \Gamma[u{:}\nu] \vdash \langle u{:}\mu, \tilde{e}{:}\tilde{g}\rangle} \qquad \text{TPAR}$$

$$\frac{\forall 1{\leq}i{\leq}n \bullet \Pi, \Gamma[u{:}\text{ot}, \tilde{u}_i{:}\tilde{\mu}_i] \vdash \langle \tilde{\alpha}_i{:}t, \{\}{:}\{\}\square\rangle \quad \Pi, \Gamma \vdash \langle \tilde{e}{:}\tilde{g}\rangle \quad \text{ot} \leq \nu}{\Pi, \Gamma \vdash \langle {*}\underline{u}\{x_1\langle \underline{\tilde{u}}_1{:}\tilde{\mu}_1, \tilde{\alpha}_1\rangle[\tilde{y}_1]{\triangleright}[\tilde{z}_1], \ldots, x_n\langle \underline{\tilde{u}}_n{:}\tilde{\mu}_n, \tilde{\alpha}_n\rangle[\tilde{y}_n]{\triangleright}[\tilde{z}_n]\}[\tilde{y}]{:}\nu, \tilde{e}{:}\tilde{g}\rangle} \qquad \text{OBJT}_1$$

$$\frac{\Pi, \Gamma \vdash \langle \sigma{:}\text{ot}, \tau{:}\text{ot}, \tilde{e}{:}\tilde{g}\rangle \quad \text{ot} \leq \mu}{\Pi, \Gamma \vdash \langle (\sigma + \tau){:}\mu, \tilde{e}{:}\tilde{g}\rangle} \qquad \text{OBJT}_2$$

$$\frac{\Pi, \Gamma[u{:}\text{pt}, \tilde{u}{:}\tilde{\mu}] \vdash \langle \tilde{\alpha}{:}t, \tau{:}\text{ot}, \{\}{:}\{\}\square\rangle \quad \Pi, \Gamma \vdash \langle \tilde{e}{:}\tilde{g}\rangle \quad \text{pt} \leq \nu}{\Pi, \Gamma \vdash \langle ({*}\underline{u}\langle \underline{\tilde{u}}{:}\tilde{\mu}, \tilde{\alpha}\rangle\tau){:}\nu, \tilde{e}{:}\tilde{g}\rangle} \qquad \text{PROCT}$$

(1) $Act\{\tilde{a}\}[\tilde{y}] = \{x_1\langle \underline{\tilde{u}}_1{:}\tilde{\mu}_1, \tilde{\alpha}_1\rangle\square{\triangleright}[\tilde{z}_1], \ldots, x_n\langle \underline{\tilde{u}}_n{:}\tilde{\mu}_n, \tilde{\alpha}_n\rangle\square{\triangleright}[\tilde{z}_n]\}$ where

$$Act\{\}[\tilde{x}] = \{\}$$
$$Act\{x\langle \underline{\tilde{u}}{:}\tilde{\mu}, \tilde{\alpha}\rangle[\tilde{x}]{\triangleright}[\tilde{y}'], \tilde{a}\}[\tilde{x}, \tilde{z}] = \{x\langle \underline{\tilde{u}}{:}\tilde{\mu}, \tilde{\alpha}\rangle\square{\triangleright}[\tilde{y}', \tilde{z}], \tilde{c}\} \ (Act\{\tilde{a}\}[\tilde{x}, \tilde{z}] = \{\tilde{c}\})$$
$$Act\{x\langle \underline{\tilde{u}}{:}\tilde{\mu}, \tilde{\alpha}\rangle[\tilde{x}]{\triangleright}[\tilde{y}'], \tilde{a}\}[\tilde{z}] = Act\{\tilde{a}\}[\tilde{z}] \qquad (\forall \tilde{x}' \bullet [\tilde{z}] \neq [\tilde{x}, \tilde{x}'])$$

and $\{\tilde{a}\}[\tilde{y}]$ is deterministic.

(2) $a = x\langle \underline{\tilde{u}}{:}\tilde{\mu}, \tilde{\gamma}\rangle[\tilde{x}]{\triangleright}[\tilde{y}]$

(3) $\psi = \langle \underline{\tilde{u}}{:}\tilde{\mu}, \tilde{\alpha}\rangle\sigma$ and $\Gamma = \{\tilde{w}_1{:}\tilde{\nu}, \tilde{w}_2{:}\tilde{\varphi}, \tilde{w}_3{:}\tilde{\varrho}\}$

Generation of Distributed Parallel Java Programs

Pascale Launay and Jean-Louis Pazat

IRISA, Campus de Beaulieu, F35042 RENNES cedex
Pascale.Launay@irisa.fr, Jean-Louis.Pazat@irisa.fr

Abstract. The aim of the Do! project is to ease the standard task of programming distributed applications using Java. This paper gives an overview of the parallel and distributed frameworks and describes the mechanisms developed to distribute programs with Do!.

1 Introduction

Many applications have to cope with parallelism and distribution. The main targets of these applications are networks and clusters of workstations (NOWs and COWs) which are cheaper than supercomputers. As a consequence of the widening of parallel programming application domains, programming tools have to cope both with task and data parallelism for distributed program generation.

The aim of the *Do!* project is to ease the task of programming distributed applications using object-oriented languages (namely Java). The *Do!* programming model is not distributed, but is explicitly parallel. It relies on structured parallelism and shared objects. This programming model is embedded in a framework described in section 2, without any extension to the Java language. The distributed programs use a distributed framework described in section 3. Program distribution is expressed through the distribution of collections of tasks and data; the code generation is described in section 4.

2 Parallel Framework

In this section, we give an overview of the parallel framework described in [8]. The aim of this framework is to separate computations from control and synchronizations between parallel tasks allowing the programmer to concentrate on the definition of tasks. This framework provides a parallel programming model without any extension to the Java language. It is based on active objects (tasks) and structured parallelism (through the class PAR) that allows the execution of tasks grouped in a collection in parallel.

The notion of active objects is introduced through task objects. A task is an object which type extends the TASK class, that represents a model of task, and is part of the framework class library. The task default behavior is inherited from the behavior described in the TASK class through its *run* method. This behavior can be re-defined to implement a specific task behavior. A task can be activated synchronously or asynchronously.

We have extended the operators design pattern [6] designed to express regular operations over COLLECTIONs through OPERATORs, in order to write data-parallel SPMD programs: a COLLECTION manages the storage and the accesses to elements; an OPERATOR represents an autonomous agents processing elements. Including the concept of active objects, we offer a parallel programming model, integrating task parallelism: active and passive objects are stored by collections; a parallel program consists in a processing over task collections, task parameters being grouped in data collections. The PAR class implements the parallel activation and synchronization of tasks, providing us with structured parallelism. Nested parallelism can be expressed, the PAR class being a task.

Figure 1 shows an example of a simple parallel program, using collections of type ARRAY; the class MY_TASK represents the program specific tasks; it extends TASK, and takes an object of type PARAM as parameter.

```
import DO.SHARED.*;

/* the task definition */                    /* the task parameter */
public class MY_TASK extends TASK {           public class MY_DATA {
    public void run (Object param) {              public void add(...) { ... }
        MY_DATA data = (MY_DATA)param;            public void remove(...) { ... }
        data.select(criterion);                   public void print(...) { ... }
        data.print(out); data.add(value); } }     public void select(...) { ... } }


    public class SIMPLE_PARALLEL {
        public static void main (String argv[ ]) {
        /* task and data array initializations */
        ARRAY tasks = new ARRAY(N); ARRAY data = new ARRAY(N);
        for (int i=0; i<N; i++)
            { tasks.add (new MY_TASK(), i); data.add (new PARAM(), i); }
        /* parallel activation of tasks */
        PAR par = new PAR (tasks,data); par.call (); } }
```

Fig. 1. A simple parallel program

3 Distributed Framework

The distributed framework [9] is used as a target of our preprocessor to distribute parallel programs expressed with our parallel framework. In the parallel framework, we use collections to manage the storage and accesses to active and passive objects. The distributed framework is based on distributed collections: a distributed collection is a collection that manages elements mapped on distinct processors, the location of the elements being masked to the user: when a client retrieves a remote element through a distributed collection, it gets a remote reference to the element, that can be invoked transparently. Distributed tasks are activated in parallel by remote asynchronous invocations. A distributed collection is composed of fragments mapped on distinct processors, each fragment managing the local subset of the collection elements. A local fragment is a non distributed collection; the distributed collection processes an access to a distributed element by remote invocation to the fragment *owning* this element (local to this element). To access an element of a distributed collection, a client identifies this element with a global identifier (relative to the whole set of elements). The distributed collection has to identify the fragment owning the element and transform the global identifier into a local identifier relevant to the local fragment. The task of converting a global identifier into the corresponding local identifier and the owner identifier devolves on a LAYOUT_MANAGER object. Different distribution policies are provided by different types of layout_managers. The program distribution is guided by the user choice of a specific layout_manager implementation.

Nevertheless the distributed framework does not manage the remote creations and accesses. They are handled by a specific runtime, and require to transform objects of the program (section 4).

4 Distributed Code Generation

We have developed a preprocessor to transform parallel programs expressed with our parallel framework into distributed programs, using our distributed framework. The parallel and distributed frameworks have the same interface, so the framework replacement is obtained by changing the imports in the program. Despite this, we have to transform some objects of the program: the objects stored in the distributed collections are distributed upon processors and need to have access to other objects. So, we have to:

- create (map) objects of the program on processors: when an object is located on a processor, its attributes are managed by the local memory and its methods run on this processor;
- have a mechanism allowing objects to be accessed remotely without any change from the caller point of view.

We have extended the object creation semantics to take into account remote creations of objects. Servers running on each host as separate threads are responsible for remote object creations. This extension is implemented using the standard Java reflection mechanism. To allow the transparent accesses to remote objects, the *Do!* preprocessor transforms a class into two classes:

- the *implementction clcss* contains the source methods implementations. An implementation object is not replicated and is located where the source object has been created (mapped). It is shared between all proxy objects.
- the *proxy clcss* has the same name and interface as the source class, but the method bodies consist in remote invocations of the corresponding methods in the implementation class. The proxy object handles a remote reference on the implementation object; it catches the invocations to the source object and redirects them to the right host. The proxy object state is never modified, so it can be replicated on each processor getting a reference on the source object.

5 Related Work

Many research projects have appeared around the Java language, aiming at filling the gaps of parallelism and distribution in Java. Some of them are presented in [1]. Some projects are based on parallel extensions to the Java language: tools [2, 4] produce Java parallel (multi-threaded) programs, relying on a standard Java runtime system using thread libraries and synchronization primitives; they do not generate distributed programs; others [7, 10] are based on distributed objects and remote method invocations. Other projects use the Java language without any extension: some environments [5] rely on a data-parallel programming model and a SPMD execution model; as in the *Do!* project, parallelism may be introduced through the notion of active objects [3].

6 Conclusion

In this paper, we have presented an overview of the *Do!* project, that aims at automatic generation of distributed programs from parallel programs using the Java language. We use the Java RMI as run-time for remote accesses; a foreseen extension of this work is to use CORBA for objects communications. Ongoing work consists in extending the parallel and distributed frameworks to handle dynamic creation of tasks, through dynamic collections (e.g. lists) and distributed scheduling.

References

1. ACM 1997 Workshop on Java for Science and Engineering Computation. *Concurrency: Practice and Experience*, 9(6):413–674, June 1997.
2. A. J. C. Bik and D. B. Gannon. Exploiting implicit parallelism in Java. *Concurrency, Practice and Experience*, 9(6):579–619, 1997.
3. D. Caromel. Towards a method of object-oriented concurrent programming. *Communications of the ACM*, 36(9):90–102, September 1993.
4. Y. Ichisugi and Y. Roudier. Integrating data-parallel and reactive constructs into Java. In *OBPDC'97*, France, October 1997.
5. V. Ivannikov, S. Gaissaryan, M. Domrachev, V. Etch, and N. Shtaltovnaya. DPJ: Java class library for development of data-parallel programs. Institute for System Programming, Russian Academy of Sciences, 1997.
6. J.-M. Jézéquel and J.-L. Pacherie. Parallel operators. In P. Cointe, editor, *ECOOP'96*, number 1098 in LNCS, Springer Verlag, pages 384–405, July 1996.
7. L. V. Kalé, M. Bhandarkar, and T. Wilmarth. Design and implementation of Parallel Java with a global object space. In *Conference on Parallel and Distributed Processing Technology and Applications*, Las Vegas, Nevada, July 1997.
8. P. Launay and J.-L. Pazat. A framework for parallel programming in Java. In *HPCN'98*, LNCS, Springer Verlag, Amsterdam, April 1998. To appear.
9. P. Launay and J.-L. Pazat. Generation of distributed parallel Java programs. Technical Report 1171, Irisa, February 1998.
10. M. Philippsen and M. Zenger. JavaParty – transparent remote objects in Java. In *PPoPP*, June 1997.

An Algebraic Semantics for an Abstract Language with Intra-Object-Concurrency*

Thomas Gehrke

Institut für Informatik, Universität Hildesheim
Postfach 101363, D-31113 Hildesheim, Germany
gehrke@informatik.uni-hildesheim.de

Object-oriented specification and implementation of reactive and distributed systems are of increasing importance in computer science. Therefore, languages like *Java* [1] and *Object REXX* [3, 7] have been introduced which integrate object-orientation and concurrency. An important area in programming language research is the definition of semantics, which can be used for verification issues and as a basis for language implementations. In recent years several semantics for concurrent object-oriented languages have been proposed which are based on the concepts of process algebras; see, for example, [2, 6, 8, 9]. In most of these semantics, the notions of processes and objects are identified: Objects are represented by sequential processes which interact via communication actions. Therefore, in each object only one method can be active at a given time. Furthermore, language implementations based on these semantics tend to be inefficient, because the simulation of message passing among local objects by communication actions is more expensive than procedure calls in sequential languages.

In contrast to this identification of objects with processes, languages like *Java* and *Object REXX* allow for intra-object-concurrency by distinguishing *passive* objects and *active* processes (in *Java*, processes are objects of a special class). Objects are data structures containing methods for execution, while system activity is done by processes. Therefore, different processes may perform methods of the same object at the same time. To avoid inconsistencies of data, the languages contain constructs to control intra-object-concurrency.

In this paper, we introduce an abstract language based on *Object REXX* for the description of the behaviour of concurrent object-based systems, i.e. we ignore data and inheritance. The operational semantics of this language is defined by a translation of systems into terms of a process calculus developed previously. This calculus makes use of process creation and sequential composition instead of the more common action prefixing and parallel composition. Due to the translation of method invocation into process calls similar to procedure calls in sequential languages, the defined semantics is appropiate as a basis for implementation.

The language $\mathcal{O}$: An $\mathcal{O}$-system consists of a set of objects and an additional sequence of statements, which describes the initial system behaviour. Each object is identified by a unique object identifier O and must contain at least one

---

* This work was supported by the DFG grant *EREAS* (Entwurf reaktiver Systeme).

```
object O1              object O2                  object O3
method m1              method m1 guarded          method m1 guarded
   a; b                   c; reply; d                a; reply; b
method m2              method m2 guarded
   a; reply; b            e; guard off; reply; f   object O4
method m3                                         method m1 guarded
   a □ reply; b                                      c; d
```

Fig. 1. $\mathcal{O}$ examples.

method. The special object identifier **self** can be used for the access of methods of the calling object (**self** is not allowed in the initial statement sequence). Methods are identified by method identifiers m and must contain at least one statement. Statements include single instructions and nondeterministic choices between sequences: $s_1 \square s_2$ performs either s_1 or s_2. We distinguish five kinds of instructions: atomic actions a representing the visible actions of systems (e.g. access to common resources), methods calls $O.m$ and the three control instructions **reply**, **guard on** and **guard off**. ; denotes sequential composition; for syntactical convenience we assume that ; has a higher priority than $\square$ (e.g., $a \square b; c$ is $a \square (b; c)$). System runs are represented by sequences $a_1...a_n$ of atomic actions, called *traces*.

During the execution of a called method, the caller is blocked until the called method has terminated. For example, the execution of the sequence $O1.m1; c$ in combination with the object $O1$ in Figure 1 generates the trace $a\,b\,c$. In some cases it is desirable to continue the execution of the caller before the called method has finished (e.g., the execution of the remaining instructions of the called method does not influence the result delivered to the caller). The instruction **reply** allows the caller to continue its execution concurrently to the remaining instructions of the called method. Therefore, the execution of $O1.m2; c$ leads to the traces $a\,b\,c$ and $a\,c\,b$. Furthermore, the execution of $O1.m3; c$ leads to the traces $a\,c$, $b\,c$ and $c\,b$.

Methods of different objects can always be executed in parallel. Concurrency within objects can be controlled by the notion of *guardedness*. If a method m of an object O is guarded, no other method of O is allowed to become active when m is taking place. If m is unguarded, other methods of O can be performed in parallel with m. The option **guarded** declares a method to be guarded. The instructions **guard on** and **guard off** allow to change the guardedness of a method during its execution. Consider the object $O2$ in Figure 1 in which both methods are declared as guarded. Although $O2.m1$ contains a **reply** instruction, the execution of the sequence $O2.m1; O2.m2$ can only generate the trace $c\,d\,e\,f$, because the guardedness of the methods prevents their concurrent execution. On the other hand the sequence $O2.m2; O2.m1$ leads to the traces $e\,f\,c\,d$ and $e\,c\,d\,f$

$$\frac{}{1 \xrightarrow{\delta} 1}\,\mathrm{T}_1 \qquad \frac{}{spawn(t) \xrightarrow{\delta} spawn(t)}\,\mathrm{T}_2 \qquad \frac{t \xrightarrow{\delta} t \quad u \xrightarrow{\delta} u}{(t;u) \xrightarrow{\delta} (t;u)}\,\mathrm{T}_3 \qquad \frac{t \xrightarrow{\delta} t}{(a:t) \xrightarrow{\delta} (a:t)}\,\mathrm{T}_4$$

$$\frac{}{\alpha \xrightarrow{\alpha} 1}\,\mathrm{R}_1 \qquad \frac{}{n \xrightarrow{\tau} \Theta(n)}\,\mathrm{R}_2 \qquad \frac{t \xrightarrow{\alpha} t'}{spawn(t) \xrightarrow{\alpha} spawn(t')}\,\mathrm{R}_3$$

$$\frac{t \xrightarrow{a\dagger} t' \quad a \neq b}{(b:t) \xrightarrow{a\dagger} (b:t')}\,\mathrm{R}_4 \qquad \frac{t \xrightarrow{\omega} t'}{t+u \xrightarrow{\omega} t'}\,\mathrm{R}_5 \qquad \frac{u \xrightarrow{\omega} u'}{t+u \xrightarrow{\omega} u'}\,\mathrm{R}_6 \qquad \frac{t \xrightarrow{\alpha} t'}{t;u \xrightarrow{\alpha} t';u}\,\mathrm{R}_7$$

$$\frac{t \xrightarrow{\delta} t' \quad u \xrightarrow{\alpha} u'}{t;u \xrightarrow{\alpha} t';u'}\,\mathrm{R}_8 \qquad \frac{t \xrightarrow{\delta} t' \quad t' \xrightarrow{\alpha} t'' \quad u \xrightarrow{\alpha'} u' \quad \{\alpha,\alpha'\} = \{a,a?\}}{t;u \xrightarrow{\tau} t'';u'}\,\mathrm{R}_9$$

Fig. 2. Transition rules.

(the trace $e\,c\,f\,d$ is prevented by the guardedness of $O2.m1$). Guardedness does not play a role in the initial statement sequence.

As a third example, consider the sequence $O3.m1; O4.m1$. This sequence leads to the possible traces $a\,b\,c\,d$, $a\,c\,b\,d$ and $a\,c\,d\,b$, because the called methods belong to different objects.

The process calculus $\mathcal{P}$: To model the behaviour of $\mathcal{O}$-systems, we introduce a process calculus $\mathcal{P}$, which is a slightly modified version of a calculus studied in [5]. We assume a countable set $\mathcal{C}$ of action names, ranged over by a, b, c. A name a can be used either for input, denoted $a?$, or for output, denoted with only the name a itself. We sometimes use $a\dagger$ as a "meta-notation" denoting either $a?$ or a. The set of actions is denoted $\mathcal{A} = \{a\dagger \mid a \in \mathcal{C}\} \cup \{\tau\}$, ranged over by α, β. τ is a special action to indicate internal behaviour of a process. $\Omega = \mathcal{A} \cup \{\delta\}$, ranged over by ω, is the set of transition labels, where δ signals successful termination. $\mathcal{N}$, ranged over by n, n', is a set of names of processes. The calculus $\mathcal{P}$, ranged over by t, u, v, is defined through the following grammar:

$$t ::= \mathbf{0} \mid \mathbf{1} \mid \alpha \mid n \mid t;t \mid spawn(t) \mid t+t \mid (a:t)$$

$\mathbf{0}$ denotes the inactive process, $\mathbf{1}$ denotes a successfully terminated term. A process name n is interpreted by a function $\Theta : \mathcal{N} \to \mathcal{P}$, called *process environment*, where n denotes a process call of $\Theta(n)$. $t;u$ denotes the sequential composition of t and u, i.e. u can perform actions when t has terminated. $spawn(t)$ creates a new process which performs t concurrently to the spawning process: $spawn(t); u$ represents the concurrent execution of t and u. The choice operator $t + u$ performs either t or u. $(a : t)$ restricts the execution of t to the actions in $\mathcal{A} \setminus \{a, a?\}$. The semantics of $\mathcal{P}$ is given by the transition rules in Figure 2.

Translation: An object system is translated into a process environment Θ and a term representing the initial statement sequence. The set $\mathcal{N}$ of process names is defined as $\mathcal{N} = \{O\_m \mid O \text{ object}, m \text{ method of } O\} \cup \{O\_mutex \mid O \text{ object}\}$. For each method m of an object O, we include a process name O_m into the

$$
\begin{array}{ll}
O1\_m1 & \mapsto a; b \\
O1\_m2 & \mapsto a; spawn(b) \\
O1\_m3 & \mapsto \tau; a + \tau; spawn(b) \\[6pt]
O2\_m1 & \mapsto o2\_l; c; spawn(d; o2\_u) \\
O2\_m2 & \mapsto o2\_l; e; o2\_u; spawn(o2\_l; f; o2\_u) \\
O2\_mutex & \mapsto o2\_l?; o2\_u?; O2\_mutex \\[6pt]
O3\_m1 & \mapsto o3\_l; a; spawn(b; o3\_u) \\
O3\_mutex & \mapsto o3\_l?; o3\_u?; O3\_mutex \\[6pt]
O4\_m1 & \mapsto o4\_l; c; d; o4\_u \\
O4\_mutex & \mapsto o4\_l?; o4\_u?; O4\_mutex
\end{array}
$$

Fig. 3. Process semantics of the objects in Figure 1.

set $\mathcal{N}$; the process term $\Theta(O\_m)$ models the behaviour of the body of m. For example, the method $m1$ of object $O1$ in Figure 1 is translated into $O1\_m1 \mapsto a; b$ (see Figure 3). The additional O_mutex processes are used for synchronizing methods. Method calls are translated into process calls of the corresponding process definitions. Therefore, the statement sequence $O1.m1; c$ is translated into $O1\_m1; c$, which is able to perform the following transitions: $O1\_m1; c \xrightarrow{\tau} a; b; c \xrightarrow{a} 1; b; c \xrightarrow{b} 1; 1; c \xrightarrow{c} 1; 1; 1$. Note that the sequence of the transition labels without τ corresponds to the trace for $O1.m1; c$.

The translation of the **reply**-instruction is realized by enclosing the remaining instructions of the method in a $spawn$-operator. For example, the method $m2$ of $O1$ is translated into the process $O1\_m2 \mapsto a; spawn(b)$. The sequence $O1.m2; c$ is translated into $O1\_m2; c$, which leads to the following transition system:

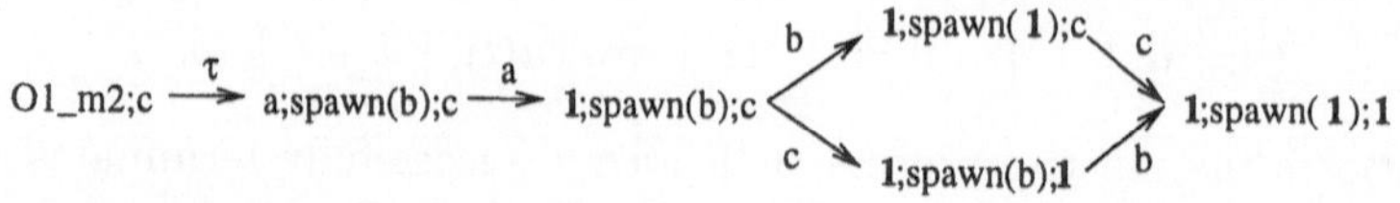

$$
O1\_m2;c \xrightarrow{\tau} a;spawn(b);c \xrightarrow{a} 1;spawn(b);c
\begin{array}{c}
\xrightarrow{b} 1;spawn(1);c \xrightarrow{c} \\
\xrightarrow{c} 1;spawn(b);1 \xrightarrow{b}
\end{array}
1;spawn(1);1
$$

Choice operators $s_1 \square s_2$ are translated into terms $\tau; t_1 + \tau; t_2$ where t_1, t_2 are the translations of s_1 and s_2. The initial τ-actions simulate the internal nondeterminism of the $\square$-operator. If an instruction sequence contains subterms of the form $(s_1 \square s_2); s_3$, we have to distribute sequential composition over choice, i.e. $(s_1 \square s_2); s_3$ has to be transformed into $s_1; s_3 \square s_2; s_3$ before the translation into the process calculus can be applied. This transformation is necessary for correct translation of the **reply**-instruction. For example, the sequence $(a; b \square \textbf{reply}; c); d$ is transformed into $a; b; d \square \textbf{reply}; c; d$, and then translated into $\tau; a; b; d + \tau; spawn(c; d)$.

Guardedness is implemented by mutual exclusion with semaphores. In the absence of data, we have to simulate semaphores by communication. For each object O using guardedness, we introduce a special semaphore process $O\_mutex \mapsto o\_l?; o\_u?; O\_mutex$. Communication on channel o_l means locking the semaphore, communication on o_u the corresponding release (unlock). To ensure that only one guarded method of an object can be active at the same time, methods of objects using guardedness have to interact with the corresponding semaphore process. For example, consider the translation of object $O2$ in Figure 3. In order to integrate unguarded actions into the synchronization mechanism, we have to enclose every unguarded action by lock and unlock actions. Therefore, in $O2.m2$ the instruction f is translated into $o2\_l; f; o2\_u$. Without the synchronization actions, unguarded actions could be performed although a guarded method has locked the semaphore. In objects without guardedness, the insertion of lock and unlock actions is not necessary and therefore omitted (see Figure 3). To enforce communication over the l and u channels, we have to restrict these actions to the translation of the initial instruction sequence. Furthermore, the semaphore process has to be spawned initially. Therefore, the initial statement sequence $O2.m1; O2.m2$ is translated into $(\{o2\_l, o2\_u\} : spawn(O2\_mutex); O2\_m1; O2\_m2)$. The semaphore process is spawned initially and runs concurrently to the method calls. $o2\_l$ and $o2\_u$ are restricted, therefore they can only be performed in communications between $O2\_m1$, $O2\_m2$ and the semaphore process. It is easy to see that the only possible trace is $c\,d\,e\,f$ (with τ-actions omitted).

In the full paper [4], the translation of O-systems is defined via translation functions. Furthermore, *weak bisimulation* is used as a equivalence relation on object systems.

References

1. K. Arnold and J. Gosling. *The Java Programming Language*. Addison-Wesley, 1996.
2. P. Di Blasio and K. Fisher. A Calculus for Concurrent Objects. In *Proceedings of CONCUR '96*, LNCS 1119. Springer, 1996.
3. T. Ender. *Object-Oriented Programming with REXX*. John Wiley and Sons, Inc., 1997.
4. T. Gehrke. An Algebraic Semantics for an Abstract Language with Intra-Object-Concurrency. Technical Report HIB 7/98, Institut für Informatik, Universität Hildesheim, May 1998.
5. T. Gehrke and A. Rensink. Process Creation and Full Sequential Composition in a Name-Passing Calculus. In *Proceedings of EXPRESS '97*, vol. 7 of *Electronic Notes in Theoretical Computer Science*. Elsevier, 1997.
6. K. Honda and M. Tokoro. An Object Calculus for Asynchronous Communication. In *Proceedings of ECOOP '91*, LNCS 512. Springer, 1991.
7. C. Michel. Getting Started with Object REXX. In *Proceedings of the SHARE Technical Conference*, March 1996.
8. E. Najm and J.-B. Stefani. Object-Based Concurrency: A Process Calculus Analysis. In *Proceedings of TAPSOFT '91 (vol. 1)*, LNCS 493. Springer, 1991.
9. D. Walker. Objects in the π-Calculus. *Information and Computation*, 116:253–271, 1995.

An Object-Oriented Framework for Managing the Quality of Service of Distributed Applications

Stéphane Lorcy, Noël Plouzeau

Irisa*
Campus de Beaulieu - 35042 Rennes Cedex - France
Email: Noel.Plouzeau@irisa.fr

Abstract. In this paper we present a framework aiming at easing the design and implementation of distributed object-oriented applications, especially interactive ones, where quality of service has to be monitored and controlled. Our framework relies on a new computation model based on a *contract* concept. Contracts are used to specify execution requirements and to monitor the remote execution of methods; they enable a good separation of concern between the application domain issues and the distribution domain ones, while promoting structured interactions between objects in these two domains, especially regarding quality of service.

1   Introduction

It is now a well-known fact that the design, implementation and testing of distributed applications pose many challenges to the computer engineer. A potential and partial cure of these problems may lie in the use of object-oriented techniques. But parallel and distributed applications have specific problems requiring the development of specific object-oriented techniques. Handling distributed objects (*ie* objects executing in execution spaces disseminated over a network) is a serious concern to the application designer. A lot of effort has been devoted to techniques for using and managing distributed objects. In this paper we focus on the quality of service management issue in object-oriented distributed applications.

Designers of distributed computations build on a huge amount of experience both on architectural side and on the algorithmic side. Concurrency control, efficiency, fault-tolerance, best use of the resources are some of the main qualities that distributed software designer strive to obtain [1, 9]. However, once obtained in some particular implementation , these qualities cannot be transposed and adapted easily to other application settings.

Cross-breeding object-oriented techniques with distributed computations ones is a fairly hot topic and many interesting works have been published [8]. Let us mention a few of major approaches.

---

* This work has been supported in part by the council of Région Bretagne

1. One of the most popular approach relies on hiding distribution. To the application program developer, all objects are local (except maybe at initialization time). The bad side is the lack of control and the fact that some important issues are hidden although they are too important to be ignored.
2. Another popular approach uses a different strategy and gives tools to the application designer to build a custom distributed object framework. The possibilities that this scheme gives are numerous and it overpowers the simple yet limited scheme of proxies. The drawback here is that using and assembling parts requires a great deal of knowledge on distributed application architectures.
3. A third scheme goes farther on the reification path. Even higher order entities such as protocols can be reified[4]. Defining a protocol between objects is done by stacking or composing existing classes of protocols. On the down side, an application needs specific protocols which are difficult to identify by the application developer.

In this paper we present another approach to the cross-breeding of object-orientedness and distributed computation domains. The main concept we use is the *contract* model. Most features derive from the semantics of this contract notion. The main aim of the contract notion is to capture the interaction properties between two or more objects. A method invocation is a common case of object interaction. By using an adequate contract, a customer object may invoke a method on a provider object *and* still get vital information on the method execution, such as delays, failures, etc.

This paper does not aim at providing an extensive definition of the contract model we designed. We rather aim at presenting the overall objectives, structure and logic of a distributed framework we built for managing quality of service issues in distributed applications. Section 2 contains a general exposition of out contract model. Section 3 gives a very brief overall description of our framework for distributed objects.

2   The Contract Concept

A contract defines the ways and means of interaction between two or more objects. Several contract models have already been defined to enhance the clarity of design and the reliability of object-oriented applications. A common form of contracts use method call precondition, postconditions, loop invariant, class invariants, etc. For instance, the Eiffel language [7] includes constructs to state explicitly this kind of contract within the code. Other forms of contracts may include the specification of valid interaction protocols between objects [5]. Also, real-time software designs include timing constraints on object interaction.

Our work on contracts and distribution aims at extending the contract notion to deal properly with many distribution related problems, such as quality of service management (eg varying network latency, network failures, site failures,...) or load balancing (eg performance evaluation of sites, selection of a site for method evaluation).

Our model relies on four important types of entities : the contract, the contractable, the customer and the command. Briefly stated, a contract is setup by a negotiation between the customer and the contractable. The contract specifies how a command will be executed, at a customer's request. The most common form of request is method invocation, ie an object (customer) wants another one (provider) to execute some method. Typical contracts may include specifications of preconditions, postconditions, execution deadlines, alternate behaviors to follow in case of contract violation. From the language point of view, a contract is an object instantiated from some class which inherits from the abstract *Contract* class. Instantiation of contract objects is performed by contractable objects only, in a way similar to the *abstract factory* design pattern [3]. As it is customary with object-oriented architectures, the exact concrete class of a contract object is hidden to the customer object, allowing sophisticated means of contract implementation selection and tuning within the contract management software.

Contract negotiation is the first phase within a contract's life. If some given customer object aims at requesting a command execution under specific conditions (eg a maximum duration of execution), the customer object must ask a known contractable object to provide a contract stating these execution conditions (which command, which delays, etc). The contractable object is fully responsible for finding out whether the conditions are reasonable or not. This implies that some estimation of the feasibility has to be performed, eg a network load evaluation. By returning a valid contract a contractable object give hints to the customer object about the current quality of service of command execution.

A contract is used by a customer object when it requests execution of a command to its contractable partner. The contractable object is responsible for finding out means to execute the command under the specifications of the contract. The contractable object is also responsible for monitoring the execution, based on the contract features.

3   Contracts Between Distributed Objects

Our contract model states that a contract negotiation starts at initiative of a customer object. In a distributed execution environment, work has to be done to find out whether a contract proposed by a customer object is feasible, which site or sites can deal with requests under the contract, what adjustments have to be made. This involves specific trading algorithms.

Our framework includes several subclasses of the *Contract* class, to provide the application developer with distributed services such as execution of command in a group of objects, etc.

The simplest contract is a *maximum delay of execution* contract. To get this kind of contract, the designer of the scene management subsystem simply needs to subclass a basic contract class of our framework, named *TimeoutContract*.

Another useful contract is the *Delta T Reachable Set Contract* (*DTRSContract* for short). A *DTRSContract* contains a list of remote sites which have to be reached in less than T units of time. *DTRSContract* objects are created and configured by the kernel of our framework, which monitors the communication between a set of sites using the *fail awareness datagram* scheme of Fetzer and Cristian [2]. This scheme uses physical clocks on each site to compute the message transmission delays and take decisions about the reachability of sites.

The *Delta Connected Group Contract* includes requirements for the connectivity of members of groups: all members of a group must follow the requirements of the same group contract. We designed a specific group management algorithm to compute a partition of the sites set [6].

4 Conclusion

Like in the Bast system [4], we use high level building blocks to help the distributed application designer. However, our contract model has more expressive power to selectively hide or expose features of the underlying distribution mechanisms. Thanks to this, distributed applications can be constructed with a limited expertise in distribution issues, without sacrifying the ways of control. This improves the separation of concern: some part of the application design does not care about distribution issues, some other part includes specific, application-specific distributed mechanisms and expose them via the contract user & contract provider metaphor.

References

1. R. Van Renesse K. P. Birman and S. Maffeis. Horus : a flexible group communication system. *Communication of the ACM*, 39(4), April 1996.
2. C. Fetzer and F. Cristian. Fail-awareness in timed asynchronous. In *15th ACM Symposium on Principles of Distributed Computing*, Philadelphia, May 1996.
3. E. Gamma, R. Helm, R. Johnson, and J. Vlissides. *Design Patterns, element of Reusable Object-Oriented Software*. Addison-Wesley, 1995.
4. B. Garbinato, P. Felber, and R. Guerraoui. Composing reliable protocols using the strategy design patterns. In *Usenix International Conference on Object-Oriented Technologies (COOTS'97)*, 1997.
5. R. Helm, I. M. Holland, and D. Gangopadhay. Contracts : Specifying behavioral compositions in object-oriented systems. In *ECOOP/OOPSLA*, pages 169–179, October 1990.
6. S. Lorcy and N. Plouzeau. A distributed algorithm for managing group membership with multiple groups. In *Proc. of the PDPTA'98 conference*, 1998.
7. B. Meyer. Applying "design by contract". *Computer (IEEE)*, 25(10):40–51, October 1992.
8. D. C. Schmidt. The adaptative communication environment : Object-oriented network, programming components for developping client/server application. *12th Sun Users Group Conference*, June 1994.
9. P. Verissimo, L. Rodriguez, F. Cosquer, H. Fonseca, and J. Frazao. An overview of the navtech system. Technical report, Department of Informatics, Faculty of Sciences of the University of Lisboa, Portugal, 1995.

A Data Parallel Java Client-Server Architecture for Data Field Computations over $\mathbb{Z}^n$

Jean-Louis Giavitto, Dominique De Vito, Jean-Paul Sansonnet

LRI u.r.a. 410 du CNRS, Bâtiment 490 – Université de Paris-Sud,
F-91405 Orsay Cedex, France.
{giavitto|devito}@lri.fr

Abstract. We describe `FieldBroker`, a software architecture, dedicated to data parallel computations on fields over $\mathbb{Z}^n$. Fields are a natural extension of the parallel array data structure. From the application point of view, field operations are processed by a field server, leading to a client/server architecture. Requests are translated successively in three languages corresponding to a tower of three virtual machines processing respectively mappings on $\mathbb{Z}^n$, sets of arrays and flat vectors in core memory. The server is itself designed as a master/multithreaded-slaves program. The aim of `FieldBroker` is to mutually incorporate approaches found in distributed computing, functional programming and the data parallel paradigm. It provides a testbed for experiments with language constructs, evaluation mechanisms, on-the-fly optimizations, load-balancing strategies and data field implementations.

1 Introduction

Collections, Data Fields and Data Parallelism. The data parallel paradigm relies on the concept of *collection*: it is an aggregate of data *handled as a whole* [6]. A *data field* is a theoretically well founded abstract view of a collection as a function from a finite index set to a value domain. Higher order functions or intensional operations on these mappings correspond to data parallel operations: point-wise applied operation (map), reduction (fold), etc. Data fields enable to represent irregular data by using a suitable index set. Another attractive advantage of the data field approach, in addition to its generality and abstraction, is that many ambiguities and semantical problems of "imperative" data parallelism can be avoided in the declarative framework of data fields.

A Distributed Paradigm for Data Parallelism. Data parallelism was motivated to satisfy the increasing needs of computing power in scientific applications. Thus, the main target of data parallel languages has been supercomputers and the privileged linguistic framework was Fortran (cf. HPF). Several factors urge to reconsider this traditional framework:

- Advances in network protocols and bandwidths have made practical the development of high performance applications whose processing is distributed over several supercomputers (metacomputing).

- The widening of parallel programming application domains (e.g. data mining, virtual reality, generalization of numerical simulations) urges to use cheaper computing resources, like NOWs (networks of workstations).
- Development in parallel compilation and run-time environments have made possible the integration of data parallelism and control parallelism, e.g. to hide the communication latency with the multithreaded execution of independent computations.
- New algorithms exhibit more and more a dynamic behavior and perform on irregular data. Consequently, new applications depend more and more on the facilities provided by a run-time (dynamic management of resources, etc.).
- Challenging applications consist of multiple heterogeneous modules interacting with each other to solve an overall design problem. New software architectures are needed to support the development of such applications.

All these points require the development of portable, robust, high-performance, dynamically adaptable, architecture neutral applications on multiple platforms in heterogeneous, distributed networks.

Many of theses attributes can be cited as descriptive characteristics of distributed applications. So, it is not surprising that distributed computing concepts and tools, which precisely face this kind of problems, become an attractive framework for supporting data parallel applications. In this perspective, we propose FieldBroker, a client server architecture dedicated to data parallel computations on data field over $\mathbb{Z}^n$. Data field operations in an application are requests processed by the FieldBroker server.

FieldBroker has been developed to provide an underlying virtual machine to the 81/2 language [5] and to compute recursive definitions of group based fields [2]. However, FieldBroker aims also to investigate the viability of client server computing for data parallel numerical and scientific applications, and the extent to which this paradigm can integrate efficiently a functional approach of the data parallel programming model. This combination naturally leads to an environment for dynamic computation and collaborative computing. This environment provides and facilitates interaction and collaboration between users, processes and resources. It also provides a testbed for experiments with language constructs, evaluation mechanisms, on-the-fly optimizations, load-balancing strategies and data field implementations.

2 A Distributed Software Architecture for Scientific Computation

The software architecture of the data field server is illustrated by Fig. 1 right. Three layers are distinguished. They correspond to three virtual machines:

- The **server** handles requests on *functions over* $\mathbb{Z}^n$. It is responsible for parallelization and synchronization between requests from one client and between different clients.

- The **master** handles operations between sets of arrays. This layer is responsible for various high-level optimizations on data field expressions. It also decides the load balancing strategy and synchronizes the computations of the slaves.
- The **slaves** implement sequential computations over contiguous data in memory (vectors). They are driven by the master requests. Master requests are of two kinds: computations to perform on the slave's data or communications (send data to other slaves; receives are implicit). Computations and communications are multithreaded in order to hide communication latency.

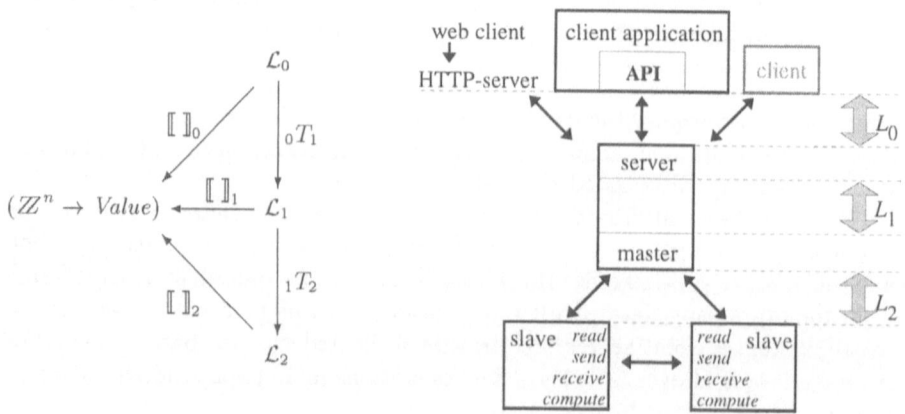

Fig. 1. *Left: Relationships between field algebras $\mathcal{L}_0, \mathcal{L}_1$ and $\mathcal{L}_2$. Right: A client/server-master/multithreaded-slaves architecture for the data parallel evaluation of data field requests.* The software architecture described on the right implements the field algebras sketched on the left. Functions $_iT_{i+1}$ are phases of the evaluation. The functions $[\![\ ]\!]_i$ are the semantic functions that map an expression to the denoted element of $\mathbb{Z}^n \to$ *Value*. They are defined such that the diagram commutes, that is $[\![e_i]\!]_i = [\![_iT_{i+1}(e_i)]\!]_{i+1}$ is true for $i \in \{0, 1\}$ and $e_i \in \mathcal{L}_i$. This property ensures the soundness of the evaluation process.

Our software architectures corresponds to a three levels language tower. Each language specifies the communications between two levels of the architecture and describes a data structure and the corresponding operations. Three languages are used, going from the more abstract $\mathcal{L}_0$ (client view on a field) to $\mathcal{L}_1$ and to the more concrete $\mathcal{L}_2$ (in core memory view on a field). The server-master and the slave programs are implemented in Java. The rationale of this design decision is to support portability and dynamic extensibility. The expected benefits of this software architecture are the following:

- **Accessibility and client independence:** requests for the data field computation are issued by a client through an API. However, because the slave is a Java program, Java applets can be easily used to communicate with the server. So, an interactive access could be provided through a web client at no further cost. In this case, the server appears as a data field desk calculator.

- **Autonomous services:** the server lifetime is not linked to the client lifetime. Thus, implementing persistence, sharing and checkpointing will be much easier with this architecture than with a monolithic SPMD program.
- **Multi-client interactions:** this architecture enables applications composition by pipelining, data sharing, etc.

The figure 1 illustrates that $\mathcal{L}_0$ terms are successively translated into $\mathcal{L}_1$ and $\mathcal{L}_2$ and that $\mathcal{L}_2$ terms are dispatched to the slaves to achieve the data parallel final processing. More details about these languages can be found in [1].

3 Conclusion

The aim of our first ongoing implementation is to evaluate the functionalities provided by such an architecture. At this stage, we have not pay attention to its performance which is certainly disappointing. A promising way to tackle this drawback consists in the use of *just-in-time* Java compiler that are able to translate Java bytecode into executable machine-dependent code.

FieldBroker integrates concepts and technics that have been developed separately. For example, relationships between the definition of functions and data fields are investigated in [4]. A proposal for an implementation is described in [3] but focuses mainly on the management of the definition domain of data fields.

One specific feature of FieldBroker is the use of heterogeneous representations, i.e. extensional and intensional data fields, to simplify field expressions. Clearly, the algebraic framework is the right one to reason about the mixing of multiple representations.

References

1. J.-L. Giavitto and D. De Vito. Data field computations on a data parallel Java client-server distributed architecture. Technical Report 1167, Laboratoire de Recherche en Informatique, Apr. 1998. 9 pages.
2. J.-L. Giavitto, O. Michel, and J.-P. Sansonnet. Group based fields. *Parallel Symbolic Languages and Systems (International Workshop PSLS'95)*, vol. 1068 of *LNCS*, pages 209–215, Beaune (France), 2-4 October 1995. Springer-Verlag.
3. J. Halén, P. Hammarlund, and B. Lisper. An experimental implementation of a higly abstract model of data parallel programming. Technical Report TRITA-IT 9702, Royal Institute of Technology, Sweden, March 1997.
4. B. Lisper. On the relation between functional and data-parallel programming languages. In *Proc. of the 6th. Int. Conf. on Functional Languages and Computer Architectures*. ACM, ACM Press, June 1993.
5. O. Michel. Introducing dynamicity in the data-parallel language 81/2. *EuroPar'96 Parallel Processing*, vol. 1123 of *LNCS*, pages 678–686. Springer-Verlag, Aug. 1996.
6. J. M. Sipelstein and G. Blelloch. Collection-oriented languages. *Proceedings of the IEEE*, 79(4):504–523, Apr. 1991.

Workshop 7+20
Numerical and Symbolic Algorithms

Maurice Clint and Wolfgang Kreuchlin

Co-chairmen

Numerical Algorithms

The workshop on numerical algorithms is organised in three sessions under the headings, Parallel Linear Algebra, Parallel Decomposition Methods and Parallel Numerical Software.

The use of numerical linear algebra in the development of scientific and engineering applications software is ubiquitous. With the rapid increase in the size of the problems which can now be solved there is a concentration on the use of iterative methods for the solution of linear systems of equations and eigenproblems. Such methods are particularly amenable to efficient implementation on many types of parallel computer. Efficient methods for the partial eigensolution of large systems include Davidson's method and the Krylov-space based method of Lanczos. An important method for the iterative solution of linear systems is another Krylov-space based method, GMRES. All of these methods incorporate an orthogonalization phase.

Frayseé, Giraud and Kharraz-Aroussi, compare the effectiveness, on a distributed memory machine, of four variants of the Gram-Schmidt orthogonalization procedure when used as components of the GMRES method. They conclude that, in a parallel environment, it is best to use ICGS (Iterative Classical Gram-Schmidt) rather than the more widely recommended modified versions of the procedure since it offers the best trade off between numerical robustness and parallel efficiency. Borges and Oliveira present a new parallel algorithm (RDME), based on Davidson's method, for computing a group of the smallest eigenvalues of a symmetric matrix. They demonstrate the effectiveness of their algorithm by comparing its performance on a Paragon with that of a parallel Lanczos-type method taken from PARPACK. Adam, Arbenz and Geus compare the performances on a HP Exemplar machine of a number of methods, including Davidson-based and Lanczos-based procedures, for computing partial eigensolutions of a large sparse symmetric matrix arising from an application of Maxwell's equations. They observe that parallel efficiencies are as might be expected from an analysis of the algorithms and that subspace iteration is not competitive with the other methods they investigate.

Many powerful algorithms for the solution of problems in numerical mathematics are based on the decomposition of linear operators into simpler forms. It is thus imperative that the potential for efficient and robust parallel implementation of such methods be studied. In addition, the use of decomposition may allow the solution of a problem to be constructed from an appropriate combination of the simultaneously computed solutions of a number of subproblems.

Osawa and Yamada present a decomposition-based waveform relaxation method, well suited to parallel implementation, for the solution of second order differential equations. The usefulness of their approach is supported by the results of experiments conducted on a KSR1 machine for both linear and non-linear equations. Incomplete LU factorization is an important operation in preconditioned iterative methods for solving sets of linear equations. This phase of the overall process is, however, the least tractable as far as parallel implementation is concerned. Nodera and Tsuma present a generalization of Bastion and Horton's parallel preconditioner and demonstrate its effectiveness on a distributed memory machine when used with the solvers, BiCGStab and GMRES. Kiper addresses the problem of computing the inverse of a dense triangular matrix and proposes an approach, based on Gaussian elimination, in which arithmetic parallelism is exploited by means of diagonal and triangular sweeps. A PRAM-based theoretical analysis is given and it is shown that, if sufficient processors are available, the method outperforms the standard algorithm of Sameh and Brent. Maslennikov, Kaniewski and Wyrzkowski present a Givens rotation-based QR decomposition method, designed for execution on a linear processor array, which permits the correction of errors arising from transient hardware faults: this is achieved at the expense of a modest increase in arithmetic complexity.

Developers of applications software intended for execution on high performance computers increasingly rely on the availability of libraries of efficient parallel implementations of algorithms from which their systems may be built. The provision of such libraries is the main goal of the ScaLAPACK project in the USA and the Parallel NAG Library project in the UK.

Krommer reports on work being undertaken within the PINAPL project, an aim of which is to extend the NAG Parallel Library to meet the needs of developers of parallel industrial applications. He reports favourably on the efficiency of execution and the scalability of a number of core sparse linear algebra routines which have been implemented on a Fujitsu AP3000 distributed memory machine. However, he sounds a note of caution by indicating that, in some cases, parallel performance may be poor. De Bono, di Serafino and Ducloux integrate a parallel 2D FFT routine from the PINEAPL Library into an industrial application concerned with the design of high performance lasers: most of the computational load in this application arises from the execution of this routine. The performance of the parallel software on an IBM SP2 is evaluated. Wagner presents an efficient parallel implementation, for execution on an IBM SP2, of an algorithm for computing the Hermite normal form of a polynomial matrix with its associated unimodular transformation matrix. The latter is used in the context of convolutional decoders to find the decoder for a given encoder. The implementation is expressed in C/C++ and uses the IBM message passing library, MPL.

Symbolic Algorithms

Symbolic computation is concerned with doing exact mathematics by computer, using the laws of Symbolic Logic or the laws of Symbolic Algebra. Consequently we are dealing with a wide spectrum, including theorem proving, logic programming, functional programming, and Computer Algebra. Symbolic computation is usually extremely resource demanding, because computation proceeds on a highly abstract level. Systems typically have to engage in some form of list processing with automated garbage collection, and they have little or no compile-time knowledge about the ensuing workloads.

After the demise of special purpose list-processing hadware, however, there is little hardware support for symbolic computation except for the use of many processors in parallel. Therefore, investigating parallelism is one of the major hopes in symbolic computation to achieve new levels of performance in order to tackle yet larger and more significant applications. At the same time the system issues in parallelisation are among the most demanding one can face: highly dynamic data structures, highly data dependend algorithms, and extremely irregular workloads. This is most obvious in logic programming, where the data form a program and parallelisation must follow the structure of the data. Hence workloads and other dynamic behaviour are even impossible to predict in theory.

Especially in the fields of Theorem Proving and Computer Algebra, there is also still the problem of finding good parallel algorithms. Frequently, naive parallelisations are possible, which however ignore sequential optimisations and perform worse than standard approaches. Parallel algorithms which outperform highly tuned sequential ones on realistic problems are still sorely needed.

The field has seen very good progress in the past years. It is a great advantage that shared memory machines and standard programming environments have taken an upswing recently. This greatly favors parallel symbolic computation applications which usually have complex code with frequent synchronization needs. We may therefore hope to see more real speedups on real problems, which still is the great overall challenge to parallel symbolic computation.

Costa and Bianchini optimise the Andorra parallel logic programming system. They analyse the caching behavior of the existing system using a simulator and develop 5 techniques for improving the system. As a result, the speedups improve significantly. Although itself based on simulation, this work is important for closing the gap between theoretical parallelisations and real architectures for which caching behaviour is critical in achieving speedup. Németh and Kacsuk's paper is concerned with the improvement of the LOGFLOW parallel logic programming system. They analyse the closed binding method and propose a new hybrid binding scheme for the logic variables. Ogata, Hirata, Ioroi and Futatsugi present an extension of earlier work on their parallel term-rewriting machine PTRAM. PTRAM was designed for shared memory multiprocessors; PTRAM/MPI in addition supports message passing computation and was implemented on a Cray T3E. Scott, Fisher and Keane investigate the parallelisation of the tableau method for theorem proving in temporal logic. Temporal logic permits reasoning about time-varying properties. Proof methods are PSPACE complete and hence provers are time and resource intensive. They present two shared memory parallel systems one of which achieves good speedups over the sequential system.

On the Influence of the Orthogonalization Scheme on the Parallel Performance of GMRES

Valérie Frayssé[1], Luc Giraud[1] and Hatim Kharraz-Aroussi[2]

[1] CERFACS, 42 av. Gaspard Coriolis, 31057 Toulouse Cedex 1, France.
fraysse,giraud@cerfacs.fr.
[2] ENSIAS, BP713, Agdal, Rabat, Morocco.

Abstract. In Krylov-based iterative methods, the computation of an orthonormal basis of the Krylov space is a key issue in the algorithms because the many scalar products are often a bottleneck in parallel distributed environments. Using GMRES, we present a comparison of four variants of the Gram-Schmidt process on distributed memory machines. Our experiments are carried on an application in astrophysics and on a convection-diffusion example. We show that the iterative classical Gram-Schmidt method overcomes its three competitors in speed and in parallel scalability while keeping robust numerical properties.

1 Introduction

Krylov-based iterative methods for solving linear systems are attractive because they can be rather easily integrated in a parallel distributed environment. This is mainly because they are free from matrix manipulations apart from matrix-vector products which can often be parallelized. The difficulty is then to find an efficient preconditioner which is good at reducing the number of iterations without degrading too much the parallel performance. We consider the solution of a large sparse linear system

$$Ax = b$$

where A is a nonsingular $n \times n$ matrix, b and x are two vectors of length n. Given a starting vector x_0, the GMRES method [10] consists in building an orthonormal basis V_m for the Krylov space

$$\mathcal{K}(A) = \left\{ x_0, Ax_0, \ldots, A^{m-1}x_0 \right\}.$$

The integer m is called the projection size. GMRES produces a solution whose restriction to this Krylov space has a minimal 2-norm residual. When one wants to limit the amount of storage for V_m, m is kept fixed and the method is restarted with a better initial guess: m is then called the restart parameter and the restarted GMRES method is denoted by GMRES(m). The construction of the basis V_m is an important step of GMRES. Several variants of the Gram-Schmidt (GS) process are available; they have different efficiencies and different numerical properties in finite precision arithmetic [1]. Classical GS (CGS) is the most efficient but is numerically unstable, modified GS (MGS) improves on CGS but can still suffer from a loss of orthogonality. The iterative MGS (IMGS) method and the iterative CGS (ICGS) have been designed to reach the numerical quality of the Householder (or Givens) factorization with a reasonable computational cost. From a point of view of computational efficiency, CGS and ICGS are the best because the scalar products can be gathered and implemented in a matrix-vector product form. However, the iterative procedures ICGS and IMGS may require more than twice as much work than their counterparts CGS and ICGS, when reorthogonalization is needed. Our implementations are the ones described in [1]. From a numerical point of view, it has been proved that MGS, IMGS and ICGS are able to produce a good enough computed Krylov basis to ensure the convergence of GMRES [4, 6].

In this paper, we wish to compare the efficiency of GMRES(m) with these four orthogonalization processes in a parallel distributed environment. It is well-known that the many scalar products arising in the orthogonalization phase are the bottleneck of Krylov based methods. Our tests are based on a restarted GM-RES code developed at CERFACS which implements the four GS variants and uses reverse communication for the preconditioner, the matrix-vector products and dot products [5]. The numerical experiments have been performed on the 128 node CRAY T3D available at CERFACS, using the MPI message passing library. We present results concerning an application in astrophysics developed in collaboration of M. Rieutord (Observatoire Midi-Pyrénées, Toulouse) and L. Valdettaro (Politecnico di Milano). In order to explain in depth these results, we study a model problem deriving from a convection-diffusion equation, on which we can test the scalability of GMRES.

2 An application in astrophysics

2.1 Description of the application

The application we are interested in belongs to the class of flow stability problems and arises in astrophysics, when modelling the internal structure of stars and planets, to study for instance the electromagnetic field in the earth kernel [8, 9]. The study of the stability of a solution of a nonlinear problem begins by solving the eigenvalue problem verified by small perturbations of the original solution. Therefore, the inertial modes (or eigenmodes) for an incompressible viscous fluid located between two concentric spheres are obtained by solving the linearized Navier Stokes continuity equation

$$\begin{cases} E\Delta\boldsymbol{\nabla} \times \boldsymbol{u} - \boldsymbol{\nabla} \times (\boldsymbol{e_z} \times \boldsymbol{u}) = \lambda\boldsymbol{\nabla} \times \times\boldsymbol{u}, \\ \text{div } \boldsymbol{u} = 0 \end{cases}$$

where $\boldsymbol{u}$ is the scaled velocity of the fluctuations, and $(\boldsymbol{e_z} \times \boldsymbol{u})$ is the non dimensional Coriolis force. The Eckman number E is equivalent to the inverse of a scaled Reynolds number and is intended to be very small. When using a spherical geometry and after projecting on the space of Chebyshev polynomials, one obtains a generalized eigenproblem $Ax = \lambda Bx$ where A and B are two non hermitian matrices, and B may be singular. The smaller E, the finer must be the discretization and thus A and B must be larger.

All the eigenvalues have an imaginary part between -1 and 1. The eigenvalues of interest are those closest to the imaginary axis. The eigenproblem is solved by the Arnoldi method with a Chebyshev acceleration [2]. The strategy for selecting the interesting eigenvalues is based on a shift and invert technique, with several imaginary shifts σ in the interval $[-i, i]$. We then solve for μ the eigenproblem

$$(A - \sigma B)^{-1} Bx = \mu x$$

with $\mu = 1/(\lambda - \sigma)$ being the largest eigenvalue of $C = (A - \sigma B)^{-1} B$. The Arnoldi method, like any Krylov-based iterative method, requires only the application of C to a vector. Clearly in this case, this operation involves the solution of linear systems such as $(A - \sigma B)x = y$. Even if A and B are sparse, the use of direct methods for solving this system imposes severe limitations on the size of A and B, and consequently limits the range of possible values of the Eckman number E. We have therefore investigated the use of GMRES(m) in a parallel distributed memory environment.

After detailing some choices about the data distribution and the preconditioning, we give some results on the performance of GMRES(m) on this example, where, for sake of simplicity, we have assumed the shift σ to be zero. In the rest of this paper, the acronym GMRES will always refer to the restarted method.

2.2 The data distribution

The matrix A is tridiagonal by blocks. The size of the diagonal blocks is usually twice the size of the off-diagonal blocks (also called coupling blocks). For example, the test matrix A_{5850} used in our experiments is 5850×5850, with 45 diagonal blocks of size 130×130 and 88 off-diagonal blocks of size 61×61. Only the blocks are stored. They are dense enough so that there is no substantial gain to hope from a sparse storage. Note that the matrix B is diagonal by blocks, with a block size compatible with the storage of A: therefore the matrices $A - \sigma B$ can be stored identically as A.

The matrix is distributed over the processors so that each processor possesses complete blocks only. The advantage of this distribution is that it allows an easy implementation of the preconditioner but the drawback is that it may induce some load imbalance when the number of blocks is not a multiple of the number of processors. In the case of two processors for instance, processor 1 will receive the first 23 diagonal blocks of A_{5850} and processor 2 the remaining 22 blocks. In other words, processor 1 (resp., processor 2) will treat a subproblem of size $130 \times 23 = 2990$ (resp., $130 \times 22 = 2860$). But out of 32 processors, 13 will have two blocks whereas 19 processors will have half of the load that is one block only. Let $n(p)$ be the size of the subproblem treated by the first processor when the matrix is distributed over p processors. If the data distribution were equilibrated, $n(p)$ would be kept proportional to $1/p$. Figure 1 shows the evolution of the ratio $n(2)/n(p)$, where the optimal distribution (dotted line) would give $n(2)/n(p) = p/2$. We see that our distribution is reasonably balanced whenever $p \leq 16$. When $p \geq 32$, the first processor is penalized by a larger amount of data than expected with the increase in processors. The opposite phenomena arises for the last processor which sees its load decrease faster than the increase in processors.

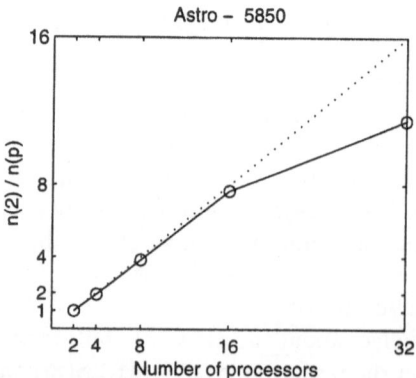

Fig. 1. Load of the first processor

2.3 Preconditioning

The restarted GMRES method applied on this problem does not converge even with large restarts. We have chosen one of the simplest preconditioner: the block Jacobi preconditioner. The preconditioner consists in the diagonal blocks of A and is well adapted to the data distribution. Each processor computes once a dense LU factorization of its blocks at the beginning of the computation. Applying the preconditioner consists in performing two triangular solves per block locally on each processor. There is no communication involved when building or applying the preconditioner.

2.4 Performance results

We describe now the results obtained for the matrix of size 5850. Since the number of processors on the CRAY T3D has to be a power of 2, we show results for 2, 4, 8, 16 and 32 processors (which is the maximum allowed for 45 blocks). The sequential code could not be run due to memory limitations: consequently, our reference for speed-ups will be the time obtained for 2 processors.

From a numerical point of view, the matrix A is very ill-conditioned (with a condition number larger than 10^{13}). GMRES with a CGS reorthogonalization does not converge on this example. In our tests, we stop the iterations when the backward error on the preconditioned system becomes lower than 10^{-7}. The original system then has a normwise backward error $\|A\tilde{x}-b\|_2/(\|A\|_2\|\tilde{x}\|_2+\|b\|_2)$ smaller than 10^{-10} (here $\tilde{x}$ denotes the computed solution). We will show the results obtained with a restart of 100 since we obtained similar behaviors for other values.

Because the matrix is ill-conditioned, the number of iterations may vary significantly with the orthogonalization strategy and the number of processors (see Figure 2). However, these variations are not monotonous and one cannot predict the preeminence of one method over the other. The variation with the number of processors is due to the different orderings of the floating-point operations in the parallel matrix-vector and dot products. Because of this sensitivity, we have chosen to evaluate the performance of the code mainly on one iteration rather than on the complete solution: the times measured over the solution are then scaled by the number of iterations required for this solution. However, to be fair, we first look at the total computational time needed to obtain the solution (see Figure 3). Regardless of the number of iterations, the fastest method is GMRES-ICGS and the slowest is GMRES-IMGS. The success of GMRES-ICGS is mainly due to the better scalability properties of the ICGS orthogonalization scheme, as we will see in the next section.

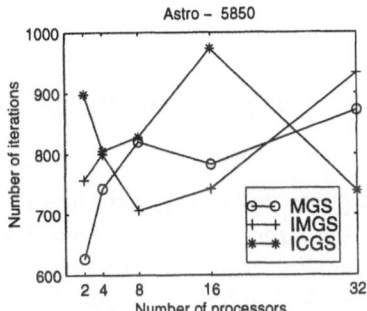

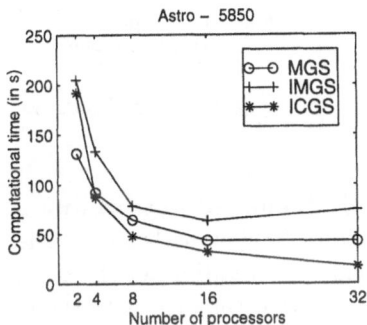

Fig. 2. Number of iterations required for convergence

Fig. 3. Time for convergence

Speed-up for the preconditioner. The speed-up for the preconditioner is taken as the ratio

$$\frac{\text{preconditioning time on } p \text{ processors}}{\text{preconditioning time on } 2 \text{ processors}}$$

and is plotted on Figure 4. We cannot compare with the sequential time because the problem is too large to fit on one processor. It departs from its optimal value (dotted line) when $p \geq 16$. This only reflects the imbalance of the processors load (see Figure 1) and, because there is no communication overhead, is the best one can expect from this data distribution.

Speed-up for the matrix-vector product. The matrix-vector product in parallel involves *a local part* corresponding to the contribution of the diagonal blocks and *a communication part* with the left and right neighbours to take into account the contribution of the coupling blocks. It is implemented so as to overlap communication and computation as much as possible. The speed-up, computed as for the preconditioning is shown on Figure 5. Here again, it reflects rather faithfully the distribution of the load on the processors (see Figure 1), which proves that the overhead due to communications does not penalize the computation.

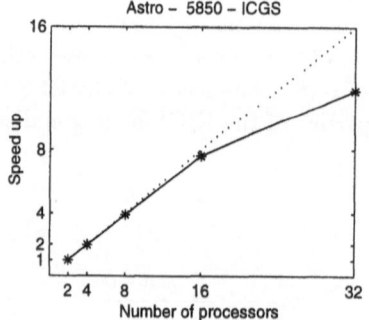 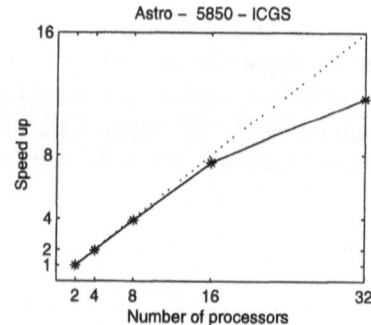

Fig. 4. Preconditioning   **Fig. 5.** Matrix-vector product

Influence of dot products. The dot products are the well-known bottleneck for Krylov methods in a distributed environment. Figure 6 shows the time spent, per iteration, in the dot products and Figure 7 gives the percentage of the solution time spent in the dot products. Both figures indicate clearly that GMRES with ICGS is the best method for avoiding the degradation of the performance generally induced by the scalar products. We even see on Figure 6 that, when $p \geq 32$, the time spent in iteration in the dot products starts increasing for IMGS and MGS whereas it continues to decrease with ICGS: this happens when the communication time overcomes the computation time in the dot products for IMGS and MGS. By gathering the scalar products, ICGS ensures more computational work to the processors.

Finally, Figure 8 gives the speed-up, for an average iteration, of the complete solution. When comparing with the best possible curve with the given processor

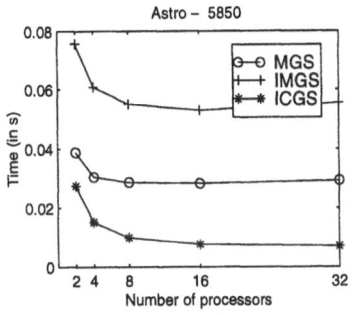
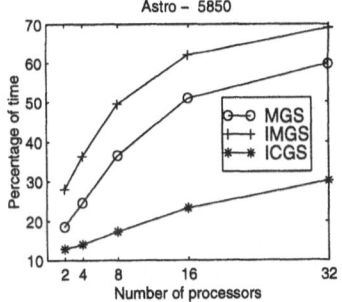

Fig. 6. Time spent in dot products per iteration

Fig. 7. Percentage of the solution time spent in dot products

load on Figure 1, we see that GMRES-MGS and GMRES-IMGS give bad performance whereas GMRES-ICGS is less affected by scalar products.

We now present a test problem for which it is easier to vary the size in order to test the parallel scalability properties of GMRES.

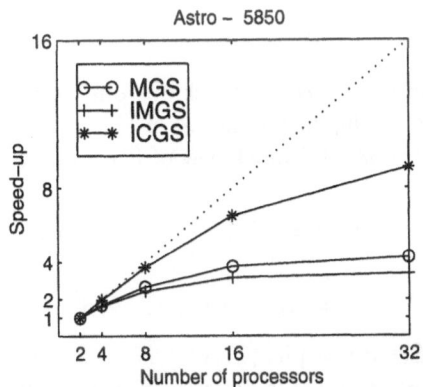

Fig. 8. Speed-up for an average iteration

3 Scalability on a model problem

We intend to investigate the parallel scalability of the GMRES code and the influence of the orthogonalization schemes. For this purpose we consider the solution, via two classic domain decomposition techniques, of an elliptic equation

$$-\left(\frac{\partial^2 u}{\partial x^2} + \frac{\partial^2 u}{\partial y^2}\right) + a\frac{\partial u}{\partial x} + b\frac{\partial u}{\partial y} = f \tag{1}$$

on the unit square $\Omega = (0,1)^2$ with Dirichlet boundary conditions.

Assume that the original domain Ω is triangulated by a set of non-overlapping coarse elements defining the N sub-domains Ω_i.

We first consider an additive Schwarz preconditioner that can be briefly described as follows. Each substructure Ω_i is extended to a larger substructure Ω_i', within a distance δ from Ω_i, where δ refers to the amount of overlap. Let A_i' denote the discretizations of the differential operator on the sub-domain Ω_i'. Let R_i^T denote the extension operator which extends by zero a function on Ω_i onto Ω, and R_i the corresponding pointwise restriction operator. With these notations the additive Schwarz preconditioner, M_{AD}, can be compactly described as

$$M_{AD} u = \sum R_i^T A_i'^{-1} R_i u.$$

We also consider the class of domain decomposition techniques that use non-overlapping sub-domains. The basic idea is to reduce the differential operator on the whole domain to an operator on the interfaces between the sub-domains.

Let I denote the union of the interior nodes in the sub-domains, and let B denote the interface nodes separating the sub-domains. Then grouping the unknowns corresponding to I in the vector u_I and the unknowns corresponding to B in the vector u_B, we obtain the following reordering of the problem:

$$Au = \begin{pmatrix} A_{II} & A_{IB} \\ A_{BI} & A_{BB} \end{pmatrix} \begin{pmatrix} u_I \\ u_B \end{pmatrix} = \begin{pmatrix} f_I \\ f_B \end{pmatrix}. \tag{2}$$

For standard discretizations, A_{II} is a block diagonal matrix where each diagonal block, A_i, corresponds to the discretization of (1) on the sub-domain Ω_i.

Eliminating u_I in the second block row of (2) leads to the following reduced equation for u_B:

$$S u_B = g_B = f_B - A_{BI} A_{II}^{-1} f_I, \tag{3}$$

where

$$S = A_{BB} - A_{BI} A_{II}^{-1} A_{IB}.$$

S is referred to as the Schur complement matrix (or also the capacitance matrix). In our experiments, Equation (1) is discretized using finite elements. The Schur complement matrix can then be written as

$$S = \sum_{i=1}^{N} S^{(i)} \tag{4}$$

where $S^{(i)}$ is the contribution from the ith sub-domain with $S^{(i)} = A_{BB}^{(i)} - A_{BI}^{(i)}(A_{II}^{(i)})^{-1} A_{IB}^{(i)}$ and $A_{XX}^{(i)}$ denotes submatrices of A_i

Without more sophisticated preconditioners, these two domain decomposition methods are not numerically scalable [12]; that is the number of iterations required grows significantly with the number of sub-domains. However, in this section we are only interested in the study of the influence of the orthogonalization schemes on the scalability of the GMRES iterations from a computer science point of view. This is the reason why we selected those two domain decomposition methods that exhibit a large amount of parallelism and we intend to see how their scalability is affected by the GMRES solver. For a detailed overview of the domain decomposition techniques, we refer to [12].

For both parallel domain decomposition implementations, we allocate one sub-domain to one processor of the target distributed computer. To reduce as much as possible the time per iteration we use an efficient sparse direct solver from the Harwell library [7] for the solution of the local Dirichlet problems arising in the Schur and Schwarz methods.

To study the scalability of the code, we keep constant the number of nodes per sub-domain when we increase the number of processors. In the experiments reported in this paper, each sub-domain contains 64×64 nodes. For the additive Schwarz preconditioner, we selected an overlap of one element (i.e. $\delta = 1$) between the sub-domains that only requires one communication after the local solution A'^{-1} while one more communication would be necessary before the solution for a larger overlap.

In Figure 9 are displayed the elapsed time for both the Schur and the Schwarz approaches observed on a 128 node Cray T3D. If the parallel code were perfectly scalable, the elapsed time would remain constant when the number of processors is increased.

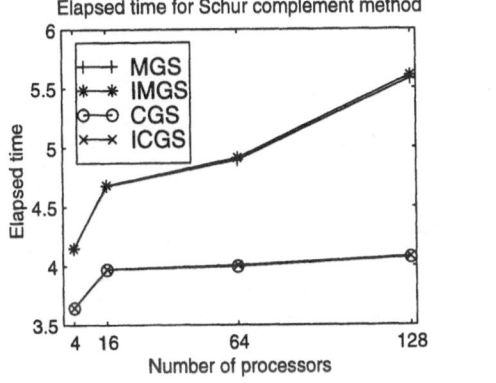
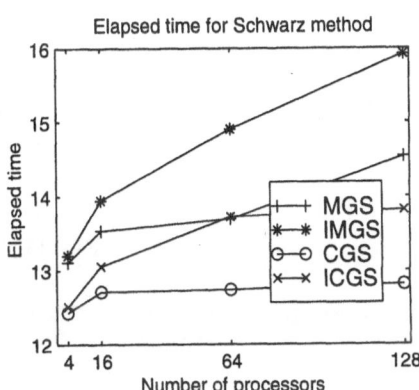

Fig. 9. Elapsed time

We define also the scaled speed-up by

$$SU_p = p \times \frac{T_4}{T_p}$$

where T_ℓ is the elapsed time to perform a complete restart step of GMRES(50) on ℓ processors. For both Schur and the Schwarz approaches we report in Figure 11 the scaled speed-ups associated with the elapsed time displayed in Figure 9.

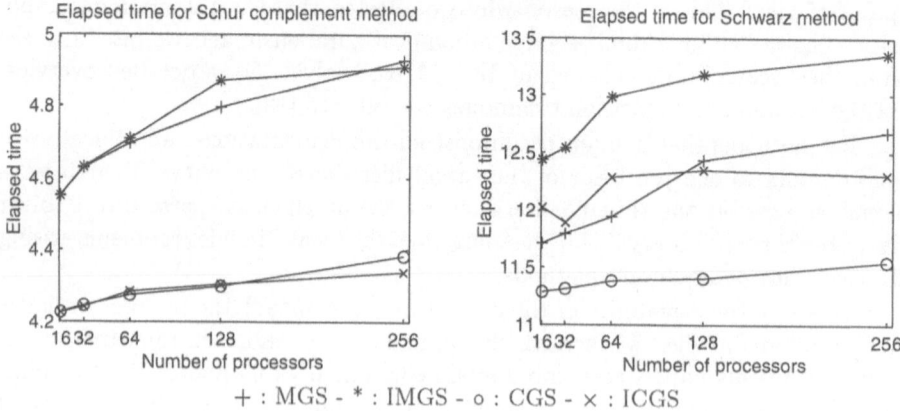

+ : MGS - * : IMGS - o : CGS - × : ICGS

Fig. 10. Elapsed time

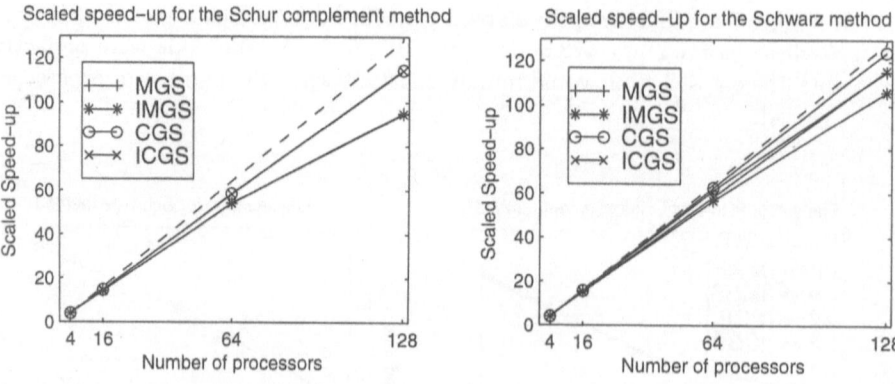

Fig. 11. Scaled speed-ups

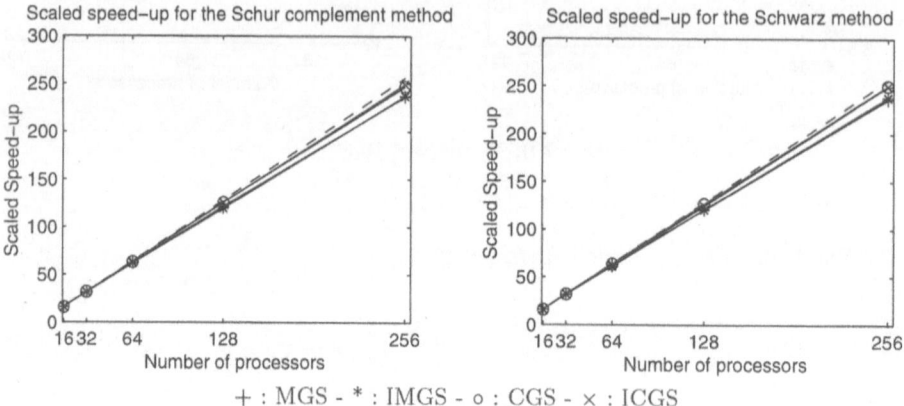

+ : MGS - * : IMGS - o : CGS - × : ICGS

Fig. 12. Scaled speed-ups

As it can be seen in Figure 9, the solution of the Schur complement system does not require iterative orthogonalization; the curves associated with CGS/ICGS and MGS/IMGS perfectly overlap each other. In such a situation, the CGS/ICGS orthogonalizations are more attractive than MGS/IMGS. They are faster and exhibit a better parallel scalability as it can be seen in the left picture of Figure 11. On 128 nodes the scaled speed-up is equal to 114 for CGS/ICGS and only 95 for MGS/IMGS. When iterative orthogonalization is required, as for the Schwarz method on our example for instance, CGS is the fastest and the most scalable but may lead to a loss of orthogonality in the Krylov basis resulting in a poor and even a loss of convergence. In that case ICGS offers the best trade-off between numerical robustness and parallel efficiency. Among the numerically reliable orthogonalization schemes, ICGS gives rise to the fastest and the most scalable iterations.

4 Conclusion

The choice of the orthogonalization scheme is crucial to obtain good performance from Krylov-based iterative methods in a parallel distributed environment. From a numerical point of view, CGS should be discarded together with MGS in some eigenproblem computations [3]. Recent works have proved that for linear systems, MGS, IMGS and ICGS ensure enough orthogonality to the computed basis so that the method converges. Finally, in a parallel distributed environment, ICGS is the orthogonalization method of choice because, by gathering the dot products, it reduces significantly the overhead due to communication.

References

1. Å. Björck. Numerics of Gram-Schmidt orthogonalization. *Linear Algebra Appl.*, 197–198:297–316, 1994.
2. T. Braconnier, V. Fraysse, and J.-C. Rioual. ARNCHEB users' guide : Solution of large non symmetric or non hermitian eigenvalue problems by the Arnoldi-Tchebycheff method. Tech. Rep. TR/PA/97/50, CERFACS, 1997.
3. F. Chaitin-Chatelin and V. Fraysse. *Lectures on Finite Precision Computations*. SIAM, Philadelphia, 1996.
4. J. Drkošová, A. Greenbaum, Z. Strakoš, and M. Rozložnik. Numerical stability of GMRES. *BIT*, 35, 1995.
5. V. Fraysse, L. Giraud, and S. Gratton. A set of GMRES routines for real and complex arithmetics. Technical Report TR/PA/97/49, CERFACS, 1997.
6. A. Greenbaum, Z. Strakoš, and M. Rozložnik. Numerical behaviour of the modified Gram-Schmidt GMRES implementation. *BIT*, 37:707–719, 1997.
7. HSL. *Harwell Subroutine Library. A Catalogue of Subroutines (Release 12)*. AEA Technology, Harwell Laboratory, Oxfordshire, England, 1995.
8. M. Rieutord. Inertial modes in the liquid core of the earth. *Phys. Earth Plan. Int.*, 90:41–46, 1995.
9. M. Rieutord and L. Valdettaro. Inertial waves in a rotating spherical shell. *J. Fluid Mech.*, 341:77–99, 1997.

762

10. Y. Saad and M. Schultz. GMRES: A generalized minimal residual algorithm for solving nonsymmetric linear systems. *SIAM J. Sci. Stat. Comput.*, 7:856–869, 1986.

11. J. N. Shadid and R. S. Tuminaro. A comparison of preconditioned nonsymmetric Krylov methods on a large-scale MIMD machine. *SIAM J. Sci. Comp.*, 14(2):440–459, 1994.

12. B.F. Smith, P. Bjørstad, and W. Gropp. *Domain Decomposition, Parallel Multilevel Methods for Elliptic Partial Differential Equations.* Cambridge University Press, New York, 1st edition, 1996.

A Parallel Solver for Extreme Eigenpairs[1]

Leonardo Borges and Suely Oliveira[2]

Computer Science Department, Texas A&M University, College Station, TX
77843-3112, USA.

Abstract. In this paper a parallel algorithm for finding a group of extreme eigenvalues is presented. The algorithm is based on the well known Davidson method for finding one eigenvalue of a matrix. Here we incorporate knowledge about the structure of the subspace through the use of an arrowhead solver which allows more parallelization in both the original Davidson and our new version. In our numerical results various preconditioners (diagonal, multigrid and ADI) are compared. The performance results presented are for the Paragon but our implementation is portable to machines which provide MPI and BLAS.

1 Introduction

A large number of scientific applications rely on the computation of a few eigenvalues for a given matrix A. Typically they require the lowest or highest eigenvalues. Our algorithm (DSE) is based on the Davidson algorithm, but calculates various eigenvalues through implicit shifting. DSE was first presented in [10] under the name RDME to express its ability to identify eigenvalues with multiplicity bigger than one. The choice of preconditioner is an important issue in eliminating convergence to the wrong eigenvalue [14] In the next section, we describe the Davidson algorithm and our version for computing several eigenvalues. In [9] Oliveira presented convergence rates for Davidson type algorithm dependent on the type of preconditioner. These results are summarized here in Section 3. Section 4 addresses parallelization strategies discussing the data distribution in a MIMD architecture, and a fast solver for the projected subspace eigenproblem. In Section 5 we present numerical and performance results for the parallel implementation on the Paragon. Further results about the parallel algorithm and other numerical results are presented in [2].

2 The Davidson Algorithm

Two of the most popular iterative methods for large symmetric eigenvalue problems are Lanczos and Davidson algorithms. Both methods solve the eigenvalue

---

[1] This research is supported by NSF grant ASC 9528912 and a Texas A&M University Interdisciplinary Research Initiative Award.

[2] Department of Computer Science, Texas A&M University, College Station, TX 77843.
email: suely@cs.tamu.edu.

problem $Au = \lambda u$ by constructing an orthonormal basis $V_k = [v_1, \ldots, v_k]$, at each kth iteration step, and then finding an approximation for the eigenvector u of A by using a vector u_k from the subspace spanned by V_k. Specifically, the original problem is projected onto the subspace which reduces the problem to a smaller eigenproblem $S_k y = \tilde{\lambda} y_k$, where $S_k = V_k^T A V_k$. Then the eigenpair $(\tilde{\lambda}_k, y_k)$ can be obtained by applying a efficient procedure for small matrices. To complete the iteration, the eigenvector y_k is mapped back as $u_k = V_k y_k$, which is an approximation to the eigenvector u of the original problem. The difference between the two algorithms consists on the way that basis V_k is built. The attractiveness of the Lanczos algorithm results from the fact that each projected matrix S_k is tridiagonal. Unfortunately, sometimes this method may require a large number of iterations. The Davidson algorithm defines a dense matrix S_k on the subspace, but since we can incorporate a preconditioner in this algorithm the number of iterations can be much lower than for Lanczos. In Davidson type algorithms, a preconditioner M_{λ_k} is applied to the current residual, $r_k = A u_k - \tilde{\lambda}_k u_k$, and the preconditioned residual $t_k = M_{\lambda_k} r_k$ is orthonormalized against the previous columns of $V_k = [v_1, v_2, \ldots, v_k]$. Although in the original formulation M_λ is the diagonal matrix $(diag(A) - \lambda I)^{-1}$ [6], the Generalized Davidson (GD) algorithm allows the incorporation of different operators for M_λ. The DSE algorithm can be summarized as follows.

Algorithm 1 – Restarted Davidson for Several Eigenvalues *Given a matrix A, a normalized vector v_1, number of eigenpairs p, restart index q, and the minimal dimension m for the projected matrix S ($m > p$), compute approximations λ and u for the p smallest eigenpairs of A.*

1. Set $V_1 \leftarrow [v_1]$.   *(initial guess)*

2. For $j = 1, \ldots, p$ (approximation for j-th eigenpair)

      While $k = 1$ or $\|r_{k-1}\| < \epsilon$ do

      (a) Project $S_k = V_k^T A V_k$.

      (b) If $(m + q) \leq dim\ S$ (restart S_k)

          Reduce $S_k \leftarrow (\Lambda_k)_{(m \times m)}$ to its m smaller eigenvectors, and update V_k for the new basis.

      (c) Compute the jth smallest eigenpair λ_k, y_k of S_k.

      (d) Compute the Ritz vector $u_k \leftarrow V_k y_k$.

      (e) Check convergence for $r_k \leftarrow A u_k - \lambda_k u_k$.

      (f) Apply preconditioner $t_k \leftarrow M r_k$.

      (g) Expand basis $V_{k+1} \leftarrow [V_k, t_k]$ using modified Gram Schmidt (MGS).

      End while

3. End For.

The core ideas of DSE (Algorithm 1) are based on the projection of A into the subspace spanned by the columns of V_k. The interation number k is not necessarily equal to $dim\ S_k$, since we have incorporated implicit restarts. The matrix S_k is obtained by adding one more column and row $V_k^T A v_k$ to matrix S_{k-1} (step **2.a**). Other important aspects of the DSE algorithm are:

(1) the eigenvalue solver for the subspace matrix S_k (step **2.c**); (2) the use of an auxiliary matrix $W_k = [w_1, \ldots, w_k]$ to provide a residual calculation $r_k = Au_k - \lambda_k u_k = w_k y_k - \lambda_k u_k$ with less computational work (step **2.e**); the choice of a preconditioner M (step **2.f**); and the use of *modified* Gram-Schmidt orthonormalization (step **2.g**) which preserves numerical stability when updating the orthonormal basis V_{k+1}. At each iteration, the algorithm expands the matrix S either until all the first p eigenvalues have been converged, or S reaches a maximum dimension $m + q$; In the latter case, restarting is applied by using the orthonormal decomposition $S_k = Y_k^T \Lambda_k Y_k$ of S. It corresponds to step **2.b** in the algorithm. Because of our choice for m, note that in step **2.c** *dim* S will be always bigger or equal to j.

3   Convergence Rate

A proof of convergence (but without a rate estimate) for the Davidson algorithm is given in Crouzeix, Philippe and Sadkane [5]. A bound on the convergence rate was first presented in [10]. The complete proof is shown in Oliveira [9]. Let A be the given matrix whose eigenvalues and eigenvectors are wanted. The preconditioner M is given for one step, and Davidson's algorithm is used with u_k being the current computed approximate eigenvector. The current eigenvalue estimate is the Rayleigh quotient $\lambda_k = \rho_A(u_k) = (u_k^T A u_k)/(u_k^T u_k)$. Let the exact eigenvector with the smallest eigenvalue of A be u, and

$$Au = \lambda u.$$

(If λ is a repeated eigenvalue of A, then we can let u be the normalized projection of u_k onto this eigenspace.)

Theorem 1. *Let P be the orthogonal projection onto $\ker(A - \lambda I)^\perp$. Suppose that A and M are symmetric positive definite. If*

$$\|P - PMP(A - \lambda I)\|_2 \leq \sigma < 1,$$

then for almost any starting value x_1, the convergence of the eigenvalue estimates $\widehat{\lambda}_k$ converge to λ ultimately geometrically with convergence factor bounded by σ^2, and the angle between the computed eigenvector and the exact eigenspace goes to zero ultimately geometrically with convergence factor bounded by σ.

A geometric convergence rate can be found for DSE (which obtains eigenvalues beyond the smallest (or largest) eigenvalue) by modifying Theorem 1. In the following theorem assume that

$$\sigma' = \|P' - P'MP'\,P'(A - \lambda_p I)P'\|_2$$

where P' is the orthogonal projection onto the orthogonal complement of the span of the first $p - 1$ eigenvectors. Then we can shown, in a similar way to Theorem 1 that the convergence factor for the new algorithm is bounded by

$(\sigma')^2$ To prove Theorem 2 we use the fact that $P's_k = s_k$, as s_k is orthogonal to the bottom p eigenvectors. and that although $(A - \lambda_p I)$ is no longer positive semi-definite, $P'(A - \lambda_p I)P'$ is.

Theorem 2. *Suppose that A and M are symmetric positive definite and that the first $p-1$ eigenvectors have been found exactly. Let P' be the orthogonal projection onto the orthogonal complement of the span of the first $p-1$ eigenvectors of A. If*

$$\|P' - P'MP'(A - \lambda_p I)P'\|_2 \le \sigma' < 1,$$

then for almost any starting value x_1, the eigenvalue estimates $\widehat{\lambda}_k$ obtained by our modified Davidson algorithm for several eigenvalues converges to λ_p ultimately geometrically with convergence factor is bounded by $(\sigma')^2$, and the angle between the exact and computed eigenvector goes to zero ultimately geometrically with convergence factor bounded by σ'.

4 Parallel Implementation

Previous implementations for the Davidson algorithm solve the eigenvalue problem in subspace S by using algorithms for dense matrices: early works [3, 4, 17] adopt EISPACK [12] routines, and later implementations [13, 15] use LAPACK [1] or reductions to tridiagonal form. Partial parallelization is obtained through the matrix-vector operations and sparse format storage for matrix A [13, 15]. Here we explore the relationship between two successive matrices S_k which allows us to represent S_k through an arrowhead matrix. The arrowhead structure is extremely sparse and the associated eigenvalue problem can be solved by a highly parallelizable method.

4.1 Data Distribution

Data partitioning significantly affects the performance of a parallel system by determining the actual degree of concurrency of the processors. Matrices are partitioned along distinct processors so that the program exploits all the best possible data parallelism: The final distribution is well balanced, and most of the computational work can be performed without communication. These two conditions make the parallel program very suited for distributed memory architectures. Both computational workload and storage requirements are the same for all processors. Communication overhead is kept as low as possible. Matrix A is split into row blocks A^i, $i = 1, \ldots, N$, each one containing $\le \lceil n/N \rceil$ rows of A. Thus processor i, $i = 1, \ldots, N$ stores A^i, the ith row block of A. Matrices V_k and W_k are stored in the same fashion. This data distribution allow us to perform many of the matrix-vector computations in place.

The orthonormalization strategy is also an important aspect in parallel environments. Recall that the modified Gram Schmidt (MGS) algorithm will be applied to the extended matrix $[V_k, t_k]$ where the current basis V_k has been previously orthonormalized. This observation reduces the computational work by eliminating the outer loop from the two nested loops in the full MGS algorithm.

4.2 The Arrowhead Relationship Between Matrices S_k

As pointed in [2, 10], the relationship between S_k and S_{k-1} can be used to show that S_k is explicitly similar to an arrowhead matrix $\tilde{S}_k$ of the form

$$\tilde{S}_k = \begin{bmatrix} \Lambda_{k-1} & \tilde{s}_k \\ \tilde{s}_k^T & s_{kk} \end{bmatrix} , \tag{1}$$

where $\tilde{s}_k = Y_{k-1}^T V_{k-1}^T w_k$, $s_{kk} = v_k^T w_k$, and the diagonal matrix Λ_{k-1} corresponds to the orthonormal decomposition $S_{k-1} = Y_{k-1} \Lambda_{k-1} Y_{k-1}^T$. In practice, the matrix S_k does not need to be stored: only a vector for Λ_k and a matrix for Y_k are required from one iteration to the next. Thus, given the eigenvalues Λ_{k-1} and eigenvectors Y_{k-1} of S_{k-1}, matrix $\tilde{S}_k$ can be used to find the eigenvalues Λ_k of S_k. Arrowhead eigensolvers [8, 11] are highly parallelizable and typically perform $\mathcal{O}(k^2)$ operations, instead of the usual $\mathcal{O}(k^3)$ effort of algorithms for dense matrices S.

5 Numerical Results

In our numerical results we employ three kind of preconditiners: diagonal preconditioner (as in the original Davidson), multigrid and ADI. A preconditioner can be expressed as the matrix which solves $Ax = b$ by applying an iterative method to $MAx = Mb$ instead. In the case of a Diagonal preconditioner this would correspond to scaling the system and then solving. Multigrid and ADI preconditioners are more complex and for that we refer the reader to [16, 18, 19]. In our implementation level 1, 2 and 3 BLAS and the Message Passing Interface (MPI) library were used for easy portability.

The computational results in this section were obtained with a finite difference approximation for

$$-\Delta u + gu = f \tag{2}$$

on a unit square domain. Here g is null inside a 0.2×0.2 square on the center of the rectangle and $g = 100$ for the remaining domain.

To compare the performance delivered by distinct preconditioners we observe the total timing and number of iterations required for the sequential DSE for finding the ten smallest eigenpairs ($p = 10$) assuming convergence for residual norms less or equal to 10^{-7}. The restart indexes were $q = 10$ and $m = 15$. This corresponds to apply restarting every time that the projected matrix S_k achieves order 25, reducing its order to 15. Table 1 presents the actual running tIMINGS In a single processor of the Intel Paragon, running three grid sizes: 31×31, 63×63, and 127×127 (matrices of orders 961, 3969 and 16129, respectively.). It reflects the tradeoff between preconditioning strategies: although the diagonal preconditioner (DIAG) is the easiest and fastest to compute, it requires an increasing number of iterations for larger matrices. Multigrid preconditioners (MG) are more expensive than DIAG, but they turn to be more effective for larger matrices. Finally, the ADI method aggregate the advantages of the previous preconditioners in the sense that it is more effective and less expensive than

MG. More details about the preconditioners used here can be found in [2] and its references.

Table 1. Sequential times and number of iterations for three preconditioners.

	matrix order 961		matrix order 3969		matrix order 16129	
	iterations	time (sec)	iterations	time (sec)	iterations	time (sec)
ADI	29	4.7	26	16.2	27	80.7
MG	34	8.8	40	43.0	40	254.1
DIAG	174	8.6	319	43.2	700	386.0

The overall behavior of the DSE algorithm (with a multigrid preconditioner) is shown in Figure 1 for matrices sizes 3969 and 16129, as a function of the number of processors. Note that the estimated optimal number of processors is not far from the actual optimal. The model for our estimates is presented in [2].

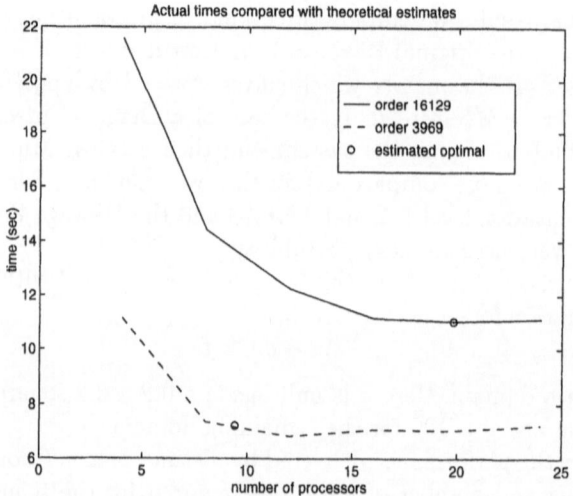

Fig. 1. Actual and estimated times for equation (2). Results for two different matrix sizes performance on the Paragon are shown.

To conclude, we compare the performance of the parallel DSE with PAR-PACK [7], a parallel implementation of ARPACK [1]. Figure 2 presents the total running times for both algorithms for the problem described above. For these runs, DSE used our parallel implementation of the ADI as its preconditioner.

---

[1] ARPACK implements an Implicitly Restarted Arnoldi Method (IRAM) which in the symmetric case corresponds to the Implicitly Restarted Lanczos algorithm. We used the regular mode when running PARPACK.

The problem was solved by using 4, 8, and 16 processors to obtain relative residuals $\|Au - \lambda u\|/\|u\|$ of order less than 10^{-5}. We show our theoretical analysis for the parallel algorithm in [2]. Other numerical results for the sequential DSE algorithm, including examples showing the behavior of the algorithm for eigenvalues with multiplicity greater than one, were presented in [10].

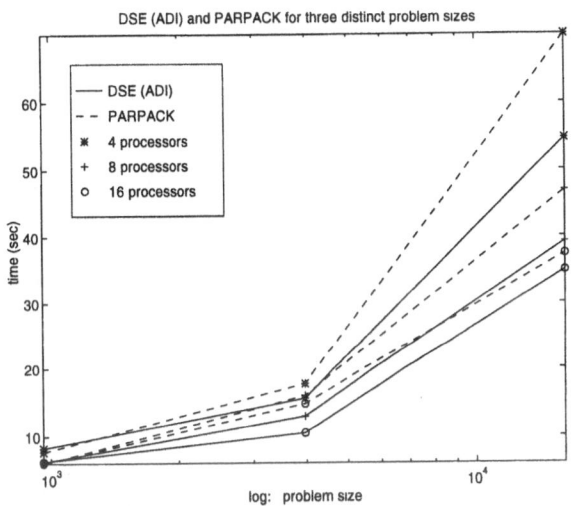

Fig. 2. Running times for DSE and PARPACK using 4, 8, and 16 processors on the Paragon.

References

1. E. Anderson, Z. Bai, C. Bischof, J. Demmel, J. Dongarra, J. Croz, A. Greenbaum, S. Hammarling, A. McKenney, S. Ostrouchov, and D. Sorensen. *LAPACK User's Guide*. SIAM, Philadelphia, 1992.
2. L. Borges and S. Oliveira. A parallel Davidson-type algorithm for several eigenvalues. *Journal of Computational Physics*. To appear.
3. G. Cisneros, M. Berrondo, and C. F. Brunge. DVDSON: A subroutine to evaluate selected sets of eigenvalues and eigenvectors of large symmetric matrices. *Compu. Chem.*, 10:281–291, 1986.
4. G. Cisneros and C. F. Brunge. An improved computer program for eigenvector and eigenvalues of large configuration iteraction matrices using the algorithm of Davidson. *Compu. Chem.*, 8:157–160, 1984.
5. M. Crouzeix, B. Philippe, and M. Sadkane. The Davidson method. *SIAM J. Sci. Comput.*, 15(1):62–76, 1994.
6. E. R. Davidson. The iterative calculation of a few of the lowest eigenvalues and corresponding eigenvectors of large real-symmetric matrices. *J. Comp. Phys.*, 17:87–94, 1975.

7. K. Maschhoff and D. Sorensen. A portable implementation of ARPACK for distributed memory parallel architectures. In *Proceedings of Copper Mountain Conference on Iterative Methods*, April 9–13 1996.

8. D. P. O'Leary and G. W. Stewart. Computing the eigenvalues and eigenvectors of symmetric arrowhead matrices. *J. Comp. Phys.*, 90:497–505, 1990.

9. S. Oliveira. On the convergence rate of a preconditioned algorithm for eigenvalue problems. Submitted.

10. S. Oliveira. A convergence proof of an iterative subspace method for eigenvalues problem. In F. Cucker and M. Shub, editors, *Foundations of Computational Mathematics Selected Papers*, pages 316–325. Springer, January 1997. (selected).

11. S. Oliveira. A new parallel chasing algorithm for transforming arrowhead matrices to tridiagonal form. *Mathematics of Computation*, 67(221):221–235, January 1998.

12. B. T. Smith, J. M. Boyle, J. J. Dongarra, B. S. Garbow, Y. Ikebe, V. C. Klema, and C. B. Moler. *Matrix eigensystem routines: EISPACK guide*. Number 6 in Lecture Notes Comput. Sci. Springer–Verlag, Berlin, Heidelberg, New York, second edition, 1976.

13. A. Stathopoulos and C. F. Fischer. A Davidson program for finding a few selected extreme eigeinpairs of a large, sparse, real, symmetric matrix. *Comp. Phys. Comm.*, 79:268–290, 1994.

14. A. Stathopoulos, Y. Saad, and C. F. Fisher. Robust preconditioning of large, sparse, symmetric eigenvalue problems. *J. Comp. and Appl. Mathematics*, 64:197–215, 1995.

15. V. M. Umar and C. F. Fischer. Multitasking the Davidson algorithm for the large, sparse eigenvalue problem. *Int. J. Supercomput. Appl.*, 3:28–53, 1989.

16. R. S. Varga. *Matrix Iterative Analysis*. Prentice-Hall, 1963.

17. J. Weber, R. Lacroix, and G. Wanner. The eigenvalue problem in configuration iteration calculations: A computer program based on a new derivation of the algorithm of Davidson. *Compu. Chem.*, 4:55–60, 1980.

18. J. R. Westlake. *A handbook of numerical matrix inversion and solution of linear equations*. Wiley, 1968.

19. D. M. Young. *Iterative Solution of Large Linear Systems*. Academic Press, 1971.

Parallel Solvers for Large Eigenvalue Problems Originating from Maxwell's Equations

Peter Arbenz[1] and Roman Geus[1]

Swiss Federal Institute of Technology (ETH), Institute of Scientific Computing,
CH-8092 Zurich,
{arbenz,geus}@inf.ethz.ch

Abstract. We present experiments with two new solvers for large sparse symmetric matrix eigenvalue problems: (1) the implicitly restarted Lanczos algorithm and (2) the Jacobi-Davidson algorithm. The eigenvalue problems originate from in the computation of a few of the lowest frequencies of standing electromagnetic waves in cavities that have been discretized by the finite element method. The experiments have been conducted on up to 12 processors of an HP Exemplar X-Class multiprocessor computer.

1 Introduction

Most particle accelerators use standing waves in cavities to produce the high voltage RF (radio frequency) fields required for the acceleration of the particles. The mathematical model for these high frequency electromagnetic fields is the eigenvalue problem solving Maxwell's equations in a bounded volume. Usually, the eigenfield corresponding to the fundamental mode of the cavity is used as the accelerating field. Due to higher harmonic components contained in the RF power fed into the cavity, and, through interactions between the accelerated particles and the electromagnetic field, an excitation of higher order modes can occur. The RF engineer designing such an accelerating cavity therefore needs a tool to compute the fundamental and about ten to twenty of the following eigenfrequencies together with the corresponding electomagnetic eigenfields.

After separation of variables depending on space and time and after elimination of the magnetic field terms the variational form of the eigenvalue problem for the electric field intensity is given by [1]

$$\begin{aligned} &\textit{Find } (\lambda, \mathbf{u}) \in \mathbb{R} \times W_0 \textit{ such that } \mathbf{u} \neq \mathbf{0} \textit{ and} \\ &(\mathbf{curl}\,\mathbf{u}, \mathbf{curl}\,\mathbf{v}) = \lambda(\mathbf{u}, \mathbf{v}), \qquad \forall \mathbf{v} \in W_0. \end{aligned} \qquad (1)$$

Let $L^2(\Omega)$ be the Hilbert space of square-integrable functions over the 3-dimensional domain Ω (the cavity) with inner product $(\mathbf{u}, \mathbf{v}) = \int_\Omega \mathbf{u}(\mathbf{x}) \cdot \mathbf{v}(\mathbf{x})\, d\mathbf{x}$ and norm $\|\mathbf{u}\|_{0,\Omega} := (\mathbf{u}, \mathbf{u})^{1/2}$. In (1), $W_0 := \{\mathbf{v} \in W \mid \operatorname{div}\mathbf{v} = 0\}$ with

$$W := \{\mathbf{v} \in L^2(\Omega)^3 \mid \mathbf{curl}\,\mathbf{v} \in L^2(\Omega)^3\,, \operatorname{div}\mathbf{v} \in L^2(\Omega)\,, \mathbf{n} \times \mathbf{v} = 0 \text{ on } \partial\Omega\}.$$

The difficulty with (1) stems from the condition $\operatorname{div} \mathbf{v} = 0$ as it is hard to find divergence-free finite elements. Therefore, ways have been looked for to get around the condition $\operatorname{div} \mathbf{v} = 0$. In this process care has to be taken in order not to introduce so-called *spurious modes*, i.e. eigenmodes that have no physical meaning [8]. We considered two approaches free of spurious modes, a penalty method and an approach based on Lagrange multipliers. In the *penalty method* approach (1) is replaced by [9, 11]

> *For fixed $s > 0$, find $(\lambda, \mathbf{u}) \in \mathbb{R} \times W$ such that $\mathbf{u} \neq 0$ and*
> $$(\mathbf{curl}\, \mathbf{u}, \mathbf{curl}\, \mathbf{v}) + s\,(\operatorname{div} \mathbf{u}, \operatorname{div} \mathbf{v}) = \lambda(\mathbf{u}, \mathbf{v}), \qquad \forall \mathbf{v} \in W. \tag{2}$$

Here, s is a positive, usually small parameter. The eigenmodes $\mathbf{u}(\mathbf{x})$ of (2) corresponding to eigenvalues $\lambda < \mu_1 s$ are eigenmodes of (1). μ_1 is the smallest eigenvalue of the negative Laplace operator $-\Delta$ on Ω.

When discretized by ordinary nodal-based finite elements, (2) leads to matrix eigenvalue problems of the form

$$A\mathbf{x} = \lambda M \mathbf{x}. \tag{3}$$

For positive s, both A and M are symmetric positive definite sparse n-by-n matrices.

In the *mixed formulation* the divergence-free condition is enforced by means of Lagrange multipliers [9].

> *Find $(\lambda, \mathbf{u}, p) \in \mathbb{R} \times H_0(\mathbf{curl}; \Omega) \times H_0^1(\Omega)$ such that $\mathbf{u} \neq 0$ and*
> (a) $(\mathbf{curl}\, \mathbf{u}, \mathbf{curl}\, \boldsymbol{\Psi}) + (\mathbf{grad}\, p, \boldsymbol{\Psi}) = \lambda(\mathbf{u}, \boldsymbol{\Psi}), \quad \forall \boldsymbol{\Psi} \in H_0(\mathbf{curl}; \Omega)$ (4)
> (b) $(\mathbf{u}, \mathbf{grad}\, q) = 0, \qquad\qquad\qquad\quad \forall q \in H_0^1(\Omega)$

Here, $H_0(\mathbf{curl}; \Omega) := \{\mathbf{v} \in L^2(\Omega)^3 \mid \mathbf{curl}\, \mathbf{v} \in L^2(\Omega)^3,\ \mathbf{n} \times \mathbf{v} = 0 \text{ on } \partial\Omega\}$. The finite element discretization of (4) yields a matrix eigenvalue problem of the form

$$\begin{bmatrix} A & C \\ C^T & O \end{bmatrix} \begin{bmatrix} \mathbf{x} \\ \mathbf{y} \end{bmatrix} = \lambda \begin{bmatrix} M & O \\ O & O \end{bmatrix} \begin{bmatrix} \mathbf{x} \\ \mathbf{y} \end{bmatrix}. \tag{5}$$

where A and M are n-by-n and C is n-by-m. M is positive definite, A is only positive semidefinite. It turns out that (5) becomes most convenient to handle if the finite elements for the vector fields are chosen to be edge elements as proposed by Nédélec [12, 8] and the Lagrange multipliers are represented by nodal-based finite elements of the same degree. Then, the columns of $M^{-1}C$ form a basis for the nullspace of A and, in principle, it suffices to compute the eigenvalues of an eigenproblem formally equal to (3) but with A and M from (5). To get rid of the high-dimensional eigenspace associated with the eigenvalue zero Bespalov [4] proposed to replace (5) by

$$\tilde{A}\mathbf{x} = \lambda M \mathbf{x}, \qquad \tilde{A} = A + CHC^T, \tag{6}$$

where H is a positive definite matrix chosen such that the zero eigenvalues are shifted to the right of the desired eigenvalues and do not disturb the computations.

In this note we consider solvers for $A\mathbf{x} = \lambda M \mathbf{x}$ with A and M from the penalty approach (3) as well as from the mixed element approach (6).

2 Algorithms

In this section we briefly survey the numerical methods that we will apply to the model problem which is a cavity of the form of a rectangular box, $\Omega = (0, a) \times (0, b) \times (0, c)$. We investigate two algorithms for computing a few of the smallest (positive) eigenvalues of the matrix eigenvalue problems originating from both the penalty method and the mixed method.

For computing a few, say p, eigenvalues of a sparse matrix eigenvalue problem

$$Ax = \lambda Mx, \qquad A = A^T, \quad M = M^T > 0, \tag{7}$$

closest to a number τ it is advisable to make a so-called *shift-and-invert* approach and apply a spectral transformation with a shift σ close to τ and solve [7]

$$(A - \sigma M)^{-1} Mx = \mu x, \qquad \mu = \frac{1}{\lambda - \sigma}. \tag{8}$$

instead of solving (7). Notice that $(A - \sigma M)^{-1} M$ is M-symmetric, i.e., it is symmetric with respect to the inner product $x^T M y$. The spectral transformation leaves the eigenvectors unchanged. The eigenvalues of (7) close to the *shift* σ become the largest absolute of (8). They are relatively well-separated which improves the speed of convergence. The cost of the improved convergence rate is the need to solve (at least approximately) systems of equations with the matrix $A - \sigma M$. In all algorithms the shift σ was chosen to be $48 < \lambda_1 \approx 48.4773$.

1. The Implicitly Restarted Lanczos algorithm (IRL). Because of the large memory consumption of the Lanczos algorithm it is often impossible to proceed until convergence. It is then necessary to *restart* the iterative process in some way with as little loss of information as possible. An elegant way to restart has been proposed by Sorensen for the Arnoldi algorithm [16], see [5] for the symmetric Lanczos case. Software is publicly available in ARPACK [10]. The algorithm is based on the spectral transformation Lanczos algorithm. The iteration process is executed until $j = p + k$, where k is some positive integer, often $k = p$. Complete reorthogonalization is done for stability reasons. This is possible since by assumption $p + k$ is not big.

In the restarting phase a clever application of the QR algorithm [14] reduces the dimension of the search space to p in such a way that the p new orthogonal basis vectors still form a Krylov sequence [16,5]. This allows the Lanczos phase to be resumed smoothly.

Here, we solved the symmetric indefinite system of equations $(A - \sigma M)x = y$ iteratively by SYMMLQ [13,3]. The accuracy of the solution of the linear system has to be at least as high as the desired accuracy in the eigenvalue calculation in order that the coefficients of the Lanczos three-term recurrence are reasonably accurate [10]. We experimented with various preconditioners. None of them was satisfactory. We obtained the best results with diagonal preconditioning. In our implementation we chose $p = k = 15$. Besides the storage for the matrices A and M, IRL requires space for two $n \times (p + k)$ arrays.

2. The Jacobi-Davidson algorithm (JDQR). In the Jacobi-Davidson algorithm the eigenpairs of $Ax = \lambda Mx$ are computed one by one. As with the implicitly restarted Lanczos procedure the search space V_j is expanded up to a certain dimension, say $j = j_{max}$. The basis of this space is constructed to be M-orthogonal but there is no three-term recurrence relation. In the expansion phase the search space is extended by a vector v orthogonal to the current eigenpair approximation $(\tilde{\lambda}, \tilde{q})$ by means of equation [15]

$$(I - \tilde{q}\tilde{q}^T M)(A - \tilde{\lambda}M)v = -(I - \tilde{q}\tilde{q}^T M)r, \qquad (I - \tilde{q}\tilde{q}^T M)v = v. \qquad (9)$$

(If eigenvectors have already been computed, this process is executed in the space M-orthogonal to them.) The solution of (9) is only needed approximately. Therefore, it can be solved iteratively. In [6] the authors propose to use a preconditioner of the form

$$(I - \tilde{q}\tilde{q}^T M)K(I - \tilde{q}\tilde{q}^T M) \qquad (10)$$

where K is a good and easily invertible approximation of $A - \tilde{\lambda}M$. They also give a generalized inverse of the matrix in (10). We solved the systems by the conjugate gradient squared (CGS) method [3] with K equal to the diagonal of $A - \tilde{\lambda}M$. The next approximate eigenvalue $\tilde{\lambda}$ and corresponding Ritz vector $\tilde{q}$ are obtained by a Ritz step for the subspace V_{j+1}. If the search space has dimension j_{max} it is shrunk to dimension j_{min} by selecting only those j_{min} Ritz vectors corresponding to Ritz values closest to the target value $\tau = 48$. In our experiments we set $j_{min} = p = 15$ and $j_{max} = 2p$ as suggested in [10].

Thus, besides the storage for the matrices A and M, memory space is needed for a $n \times j_{max}$ array and for three $n \times p$ arrays.

3   Numerical Experiments

In this section we compare IRL and JDQR for computing the 15 smallest eigenvalues of (1) with $\Omega = (0, a) \times (0, b) \times (0, c)$ where $a = 0.8421$, $b = 0.5344$, $c = 0.2187$. We apply penalty as well as mixed methods to this model problem whose eigenvalues can be computed analytically [1]. The computational results have been obtained with the HP Exemplar X-class system at the ETH Zurich.

1. Sequential results. In Figs. 1 and 2 execution times for computing the 15 lowest eigenvalues and associated eigenvectors vs. the accuracy of the computed solutions relative to the analytic solution are plotted for different mesh sizes. The largest problems sizes were $n = 65'125$ and $n = 45'226$ for the linear and quadratic edge elements, and $n = 28'278$ and $n = 118'539$ for the linear and quadratic node elements. For all experiments we used a tolerance of 10^{-8} in the stopping criterion of the eigensolvers. Loosely speaking, this means that the eigenvalues are computed to at least 8 significant digits. The execution times comprise the solution time of the eigensolver but not the time for building the matrices. For each finite element type (linear/quadratic, node element/edge element) the performance of the two algorithms is shown. Fig. 1 shows the results

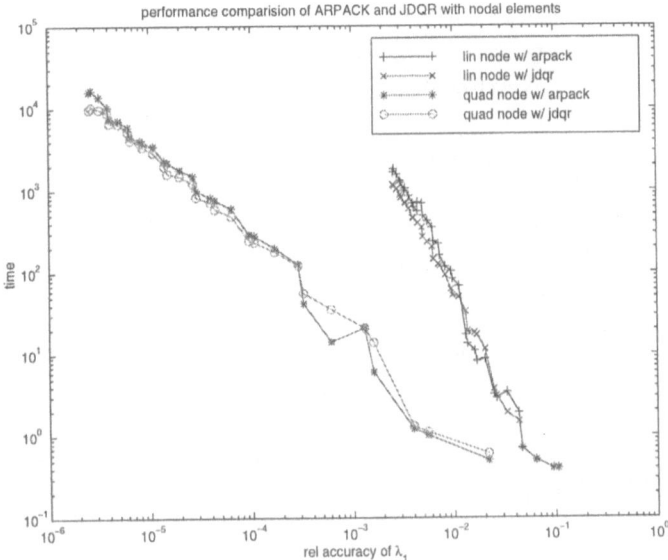

Fig. 1. Comparison of IRL and JDQR with node elements: Accuracy of the computed λ_1 relative to the analytic solution vs. computation time

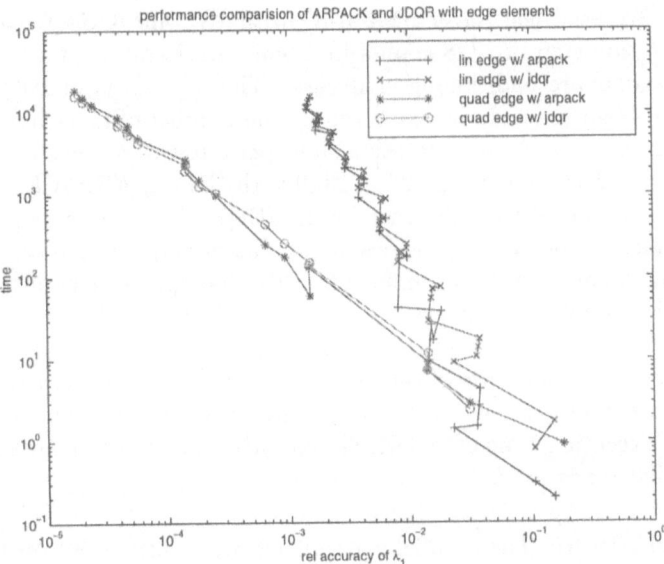

Fig. 2. Comparison of IRL and JDQR with edge elements: Accuracy of the computed λ_1 relative to the analytic solution vs. computation time

that we obtained with linear and quadratic *node* elements for λ_1. The higher eigenvalues behave similarly but are of course less accurate. Fig. 2 shows the results for the linear and quadratic *edge* elements.

For edge elements the convergence is improved by introducing the matrix C as given in (6). Experiments to that end are reported in [2]. H in (6) was heuristically chosen to be αI with $\alpha = 100/h$, where h is the mesh width.

The comparison of *linear* with *quadratic* element types reveals immediately the inferiority of the former. They give much lower accuracy for a given computation time. The node elements in turn are to be preferred to the edge elements, at least in this simplified model problem. For a given computational effort, the eigenvalues obtained with the node elements are about an order of magnitude more accurate than those obtained with the edge elements. The situation is not so evident with the linear elements.

With the quadratic elements and linear node elements JDQR is consistently faster than IRL by about 10 to 30% for the large problem sizes. For linear edge elements JDQR still is 10 to 20% ahead of IRL for most of the larger problem sizes. For the smaller problems the situation is not so clear.

We now discuss the behavior of the two algorithms in the case of the quadratic edge elements where the problem size grows from $n = 1408$ to 45226. (In the latter case A and M have each about 1'730'000 nonzero elements.) With both algorithms the number of outer iteration steps varies only little. It is ~ 65 with IRL and ~ 210 with JDQR. So, the number of restarts does not depend on the problem size. Each outer iteration step requires the solution of one system of equations. We used the solver SYMMLQ with IRL and CGS for JDQR. One (inner) iteration step of CGS counts for about two iteration steps of CGS. We applied diagonal preconditioning in all cases. The superiority of JDQR over IRL for large problem sizes can be explained by the number of inner iteration steps. The average number of inner iteration steps per outer iteration step grows from 64 to 91 with JDQR but from 423 to 1140 with IRL. In ARPACK each system of equations is solved to high accuracy. In JDQR the accuracy requirement is not so stringent and actually varies from (outer) step to step, cf. §2.2. This explains the higher iteration numbers for IRL. The projections in (9) improve the condition number of the system matrix. Further, the shift σ is updated in JDQR while it stays constant with ARPACK. These may be the reasons for the less pronounced increase of the number of inner iteration steps with JDQR. Notice that the projections made in each iteration step account for less than 10% of the execution time of JDQR. Similar observations can be made with the other element types.

2. Parallel results. The parallel experiments were carried out on the 32 processor HP Exemplar X-class system at ETH Zurich. This shared-memory CC-NUMA machine consists of two hypernodes with 16 HP PA-8000 processors and 4 GBytes memory each. The processors and memory-banks within a hypernode are connected through a crossbar-switch with a bandwidth of 960 MBytes/s per port. The hypernodes themselves are connected by a slower network with a ring

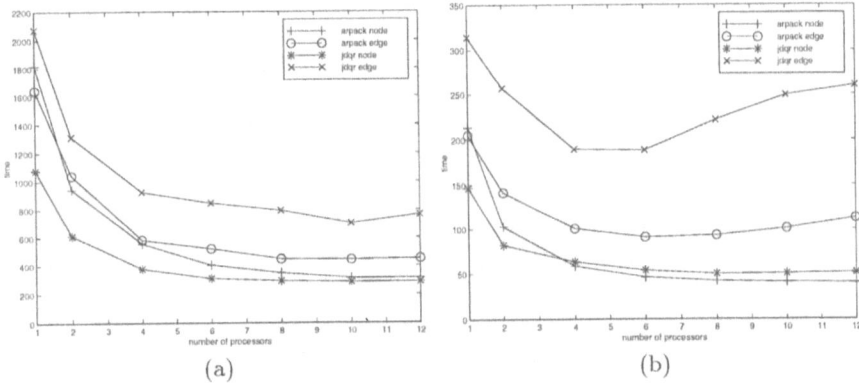

Fig. 3. Execution times (a) for the large eigenproblem ($n=22323$ for node, $n=5798$ for edge elements) and (b) for the small eigenproblem ($n=4981$ and $n=1408$, respectively)

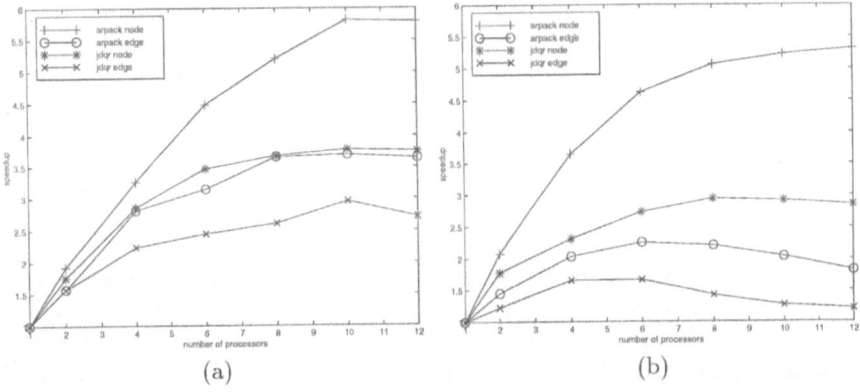

Fig. 4. Speedups (a) for the large eigenproblem and (b) for the small eigenproblem

topology. The HP PA-8000 processors have a clock rating of 180 MHz and a peak performance of 720 MFLOPS.

We carried out the parallel experiments for IRL (ARPACK) and JDQR using both quadratic node and quadratic edge elements and two different problem sizes. Linear elements are not considered because of the inferior results in the sequential case. For both edge and node elements we chose two problem sizes: the larger problems require about 1700 seconds one processor, whereas the smaller problems require only about 220 seconds. For our experiments we were allowed to use up to 12 processors.

With ARPACK about 95% of the sequential computation time is spent in the inner loops forming sparse matrix-vector products with the matrices A, M and C. JDQR spends about 90% for this task. We therefore parallelized the sparse matrix-vector product using the so called HP Exemplar advanced shared-memory programming technique, that is directive-based.

The matrix C and the strictly lower triangles of A and M are stored in the compressed sparse row format [3]. The diagonals of A and M are stored separately. We parallelized the outermost loops with the loop_parallel directives. The necessary privatization of variables was done manually by means of the loop_private directive. For the product with C and the lower triangular parts of A and M no special considerations were necesary. For the product with C^T and the upper triangular parts of A and M the processors store their "local" result into distinct vectors, that are accumulated into the global result vector afterwards. To get a well-balanced work load we distributed the triangular matrices in block-cyclic fashion over the processors.

In Fig. 3 the execution times for both problem sizes are plotted. In Fig. 4 the corresponding speedups are found. The best speedups are reached with 10 processors for the large problems. ARPACK's IRL gives speedups of 5.8 for node elements and of 3.7 for edge elements. With JDQR we get 3.8 and 3.0, respectively. These numbers are lower for the small problem size.

The reason for the better speedups with ARPACK is, that the implicit classical Gram-Schmidt orthogonalization which consumes between 5 and 10% of the sequential execution time in JDQR doesn't scale well on the HP-Exemplar. The parallelized BLAS routine DGEMV in the HP MLIB library shows no speedup even though the matrices have more than 100'000 nonzero elements! We are currently resolving this issue with HP. Using a reasonably parallelizing DGEMV routine we expect JDQR to scale as well as ARPACK.

Fig. 4 further shows that in our experiments node elements scale better than edge elements. Since we chose the problem sizes such that both edge and node elements require about the same computation time on one processor, and the linear systems arising from node elements are better conditioned, the matrices originating from edge elements are smaller. Matrices A and M stemming from edge elements together have about 7 times fewer non-zero elements than in the nodal case. So, the relative parallelization overhead is bigger for edge elements. For IRL the situation is improved because instead of A we store the shifted matrix $A - \sigma M$, which has about twice as many non-zero elements.

Our parallel experiments show the limitations of directive-based shared-memory programming. Before and after each parallel section (e.g. parallel loops) of the code, the compiler inserts global synchronization operations to ensure correct execution. However, in most cases, these global synchronization operations are unnecessary. That is why our implementation doesn't scale well. In order to remove unnecessary synchronization points, a different programming paradigm, such as message-passing or low-level shared-memory programming, has to be employed.

Our parallel results depend very much on the computer architecture, the parallel libraries and the programming paradigm we used. They can therefore not be easily generalized. Nevertheless, our experiments prove that reasonable parallel performance can be obtained on small processor numbers with only modest programming effort using directive-based shared-memory programming on the HP Exemplar.

References

1. St. Adam, P. Arbenz, and R. Geus, *Eigenvalue solvers for electromagnetic fields in cavities*, Tech. Report 275, ETH Zürich, Computer Science Department, October 1997, (Available at URL http://www.inf.ethz.ch/publications/tr.html).

2. P. Arbenz and R. Geus, *Eigenvalue solvers for electromagnetic fields in cavities*, FORTWIHR International Conference on High Performance Scientific and Engineering Computing (F. Durst et al., ed.), Springer-Verlag, 1998, (Lecture Notes in Computational Science and Engineering).

3. R. Barret, M. Berry, T. F. Chan, J. Demmel, J. Donato, J. Dongarra, V. Eijkhout, R. Pozo, Ch. Romine, and H. van der Vorst, *Templates for the solution of linear systems: Building blocks for iterative methods*, Society for Industrial and Applied Mathematics, Philadelphia, PA, 1994, (Available from Netlib at URL http://www.netlib.org/templates/index.html).

4. A. N. Bespalov, *Finite element method for the eigenmode problem of a RF cavity resonator*, Soviet Journal of Numerical Analysis and Mathematical Modelling **3** (1988), 163–178.

5. D. Calvetti, L. Reichel, and D. C. Sorensen, *An implicitely restarted Lanczos method for large symmetric eigenvalue problems*, Electronic Transmissions on Numerical Analysis **2** (1994), 1–21.

6. D. R. Fokkema, G. L. G. Sleijpen, and H. A. van der Vorst, *Jacobi-Davidson style QR and QZ algorithms for the partial reduction of matrix pencils*, Preprint 941, revised version, Utrecht University, Department of Mathematics, Utrecht, The Netherlands, January 1997.

7. R. Grimes, J. G. Lewis, and H. Simon, *A shifted block Lanczos algorithm for solving sparse symmetric generalized eigenproblems*, SIAM J. Matrix Anal. Appl. **15** (1994), 228–272.

8. J. Jin, *The finite element method in electromagnetics*, Wiley, New York, 1993.

9. F. Kikuchi, *Mixed and penalty formulations for finite element analysis of an eigenvalue problem in electromagnetism*, Computer Methods in Applied Mechanics and Engineering **64** (1987), 509–521.

10. R. B. Lehoucq, D. C. Sorensen, and C. Yang, *ARPACK users' guide: Solution of large scale eigenvalue problems by implicitely restarted Arnoldi methods*, Department of Mathematical Sciences, Rice University, Houston TX, October 1997, (Available at URL http://www.caam.rice.edu/software/ARPACK/index.html).

11. R. Leis, *Zur Theorie elektromagnetischer Schwingungen in anisotropen Medien*, Mathematische Zeitschrift **106** (1968), 213–224.

12. J. C. Nédélec, *Mixed finite elements in* $\mathbb{R}^3$, Numerische Mathematik **35** (1980), 315–341.

13. C. C. Paige and M. A. Saunders, *Solution of sparse indefinite systems of linear equations*, SIAM J. Numer. Anal. **12** (1975), 617–629.

14. B. N. Parlett, *The symmetric eigenvalue problem*, Prentice Hall, Englewood Cliffs, NJ, 1980.

15. G. L. G. Sleijpen, A. G. L. Booten, D. R. Fokkema, and H. A. van der Vorst, *Jacobi-Davidson type methods for generalized eigenproblems and polynomial eigenproblems*, BIT **36** (1996), 595–633.

16. D. Sorensen, *Implicite application of polynomial filters in a k-step Arnoldi method*, SIAM J. Matrix Anal. Appl. **13** (1992), 357–385.

Waveform Relaxation for Second Order Differential Equation $y'' = f(x, y)$

Kazufumi Ozawa and Susumu Yamada[1]

Graduate School of Information Science,
Tohoku University, Kawauchi Sendai 980-8576, JAPAN

Abstract. Waveform relaxation (WR) methods for second order equations $y'' = f(t, y)$, $y(t_0) = y_0$, $y'(t_0) = y_0'$ are studied. For linear case, the method converges superlinearly for any splittings of the coefficient matrix. For nonlinear case, the method converges quadratically only for waveform Newton method. It is shown, however, that the method with approximate Jacobian matrix converges superlinearly. The accuracy, execution times and speedup ratios of the WR methods on a parallel computer are discussed.

1 Introduction

In this paper we propose a waveform relaxation (WR) method for solving $y'' = f(x, y)$ on a parallel computer. The basic idea of this method is to solve a sequence of differential equations, which converges to the exact solution, with a starting solution $y^{[0]}(x)$. In this iteration, we can solve each component(block) of the equation in parallel.

2 Linear Differential Equation

Consider first the linear equation of the form

$$y''(x) = Qy(x) + g(x), \quad y(x_0) = y_0, \quad y'(x_0) = y_0', \quad y \in \mathbf{R}^m, \qquad (1)$$

where we assume $-Q$ is a positive definite matrix. The Picard iteration for solving (1) is given by

$$y^{[\nu+1]''}(x) = Qy^{[\nu]}(x) + g(x), \quad y^{[\nu]}(x_0) = y_0, \quad y^{[\nu]'}(x_0) = y_0'. \qquad (2)$$

Here we consider the rate of convergence of (2) and propose an acceleration of the iteration.

The solutions of (1) and (2) are given by

$$y(x) = y_0 + (x - x_0)y_0' + \int_{x_0}^{x} \int_{x_0}^{s} \{Qy(\tau) + g(\tau)\} d\tau ds, \qquad (3)$$

$$y^{[\nu+1]}(x) = y_0 + (x - x_0)y_0' + \int_{x_0}^{x} \int_{x_0}^{s} \{Qy^{[\nu]}(\tau) + g(\tau)\} d\tau ds, \qquad (4)$$

respectively. By subtracting (3) from (4) we can obtain

$$\varepsilon^{[\nu+1]}(x) = \int_{x_0}^{x} \int_{x_0}^{s} Q\varepsilon^{[\nu]}(\tau) d\tau ds, \tag{5}$$

where $\varepsilon^{[\nu]}(x) = y^{[\nu]}(x) - y(x)$. Taking the norm of the left-hand side and assuming $||\varepsilon^{[0]}(x)|| \leq K(x - x_0)$, we have by induction

$$||\varepsilon^{[\nu]}(x)|| \leq K \frac{c_1^{\nu}(x - x_0)^{2\nu+1}}{(2\nu + 1)!}, \tag{6}$$

where we set $c_1 = ||Q||$. The inequality shows that Picard iteration (2) converges on any finite interval $x \in [x_0, T]$ and the rate of convergence is superlinear. It is, however, expected that the method converges slowly if the system is stiff or the length of the interval is large.

Here we propose an acceleration of the Picard iteration using the splitting $Q = N - M$. The iterative method to be considered is

$$y^{[\nu+1]''}(x) + My^{[\nu+1]}(x) = Ny^{[\nu]}(x) + g(x), \tag{7}$$

where we assume that matrix M is symmetric and positive definite. The error of the iterate $y^{[\nu]}(x)$ satisfies the differential equation

$$\varepsilon^{[\nu+1]''}(x) + M\varepsilon^{[\nu+1]}(x) = N\varepsilon^{[\nu]}(x). \tag{8}$$

In order to obtain the explicit expression for $\varepsilon^{[\nu]}(x)$, here we define the square root and the trigonometric functions of matrices.

Let λ_k $(k = 1, \ldots, m)$ be the eigenvalues of M, and P is a unitary matrix that diagonalizes M, i.e. $M = P\mathrm{diag}(\lambda_1, \ldots, \lambda_m) P^{-1}$, then using P the square root of M can be defined by

$$\sqrt{M} = P\mathrm{diag}\left(\sqrt{\lambda_1}, \ldots, \sqrt{\lambda_m}\right) P^{-1}.$$

For any square matrix H and scalar x, the trigonometric functions can be defined by

$$\cos(Hx) = \sum_{k=0}^{\infty} \frac{(-1)^k H^{2k}}{(2k)!} x^{2k}, \quad \sin(Hx) = \sum_{k=0}^{\infty} \frac{(-1)^k H^{2k+1}}{(2k+1)!} x^{2k+1}.$$

Using the functions defined above we have

$$\varepsilon^{[\nu+1]}(x) =$$
$$(\sqrt{M})^{-1} \int_{x_0}^{x} \left(\cos(\sqrt{M}\tau)\sin(\sqrt{M}x) - \cos(\sqrt{M}x)\sin(\sqrt{M}\tau)\right) N\varepsilon^{[\nu]}(\tau) d\tau$$

$$= (\sqrt{M})^{-1} \int_{x_0}^{x} \sin\left(\sqrt{M}(x - \tau)\right) N\varepsilon^{[\nu]}(\tau) d\tau$$

$$= \int_{x_0}^{x} \int_{x_0}^{\tau} \cos\left(\sqrt{M}(x - \tau)\right) N\varepsilon^{[\nu]}(s) ds d\tau, \tag{9}$$

where the condition $\varepsilon^{[\nu+1]}(x)|_{x=x_0} = \frac{\mathrm{d}}{\mathrm{d}x}\varepsilon^{[\nu+1]}(x)|_{x=x_0} = 0$ is used. To bound $\varepsilon^{[\nu]}(x)$ if we use the Euclidean norm then we have

$$|| \cos(\sqrt{M}x)|| \le ||P|| \left|\left|\mathrm{diag}\left(\cos(\sqrt{\lambda_1}x), \ldots, \cos(\sqrt{\lambda_n}x)\right)\right|\right| \, ||P^{-1}|| \le 1, \quad (10)$$

since $||P|| = ||P^{-1}|| = 1$. Assuming $||\varepsilon^{[0]}(x)|| \le K(x - x_0)$ as before and letting $c_2 = ||N||$, we find by induction the inequality

$$||\varepsilon^{[\nu]}(x)|| \le K \frac{c_2{}^\nu (x - x_0)^{2\nu+1}}{(2\nu + 1)!}. \quad (11)$$

The result shows that although the rate of convergence of splitting method (7) remains superlinear, we can accelerate the convergence by making the value of $||N||$ small.

3 Nonlinear Differential Equation

Next we discuss the rate of convergence of the waveform relaxation for the second order nonlinear equation

$$y''(x) = f(x, y), \quad y(x_0) = y_0, \quad y'(x_0) = y_0', \quad y \in \mathbf{R}^m. \quad (12)$$

For the first order nonlinear equation $y' = f(x, y)$, the method called the waveform Newton method is proposed[5]. The method is given by

$$y^{[\nu+1]'}(x) - J_\nu y^{[\nu+1]}(x) = f(y^{[\nu]}(x)) - J_\nu y^{[\nu]}(x), \quad (13)$$

where J_ν is the Jacobian matrix of $f(y^{[\nu]})$. It is shown by Burrage[1] that the waveform Newton method converges quadratically on all finite intervals $x \in [x_0, T]$. In this section we first consider the convergence for the case that J_ν is replaced with an approximation in order to enhance the efficiency of the parallel computation.

Instead of (13) let us consider the iteration

$$y^{[\nu+1]'}(x) - \tilde{J}_\nu y^{[\nu+1]}(x) = f(y^{[\nu]}(x)) - \tilde{J}_\nu y^{[\nu]}(x), \quad (14)$$

where $\tilde{J}_\nu$ is an approximation to J_ν. Let $\varepsilon^{[\nu]} = y^{[\nu]} - y$ then we have

$$f(y^{[\nu]}) = f(y + \varepsilon^{[\nu]}) = f(y) + \frac{\partial f}{\partial y}\varepsilon^{[\nu]} + O(||\varepsilon^{[\nu]}||^2). \quad (15)$$

Substituting (15) into (14) and ignoring the term $O(||\varepsilon^{[\nu]}||^2)$, we have

$$\varepsilon^{[\nu+1]'} = \tilde{J}_\nu \varepsilon^{[\nu+1]} + \left(J_\nu - \tilde{J}_\nu\right) \varepsilon^{[\nu]}, \quad (16)$$

and taking the inner product we have

$$\left\langle \frac{\mathrm{d}}{\mathrm{d}x}\varepsilon^{[\nu+1]}, \varepsilon^{[\nu+1]} \right\rangle = \left\langle \tilde{J}_\nu \varepsilon^{[\nu+1]}, \varepsilon^{[\nu+1]} \right\rangle + \left\langle (J_\nu - \tilde{J}_\nu)\varepsilon^{[\nu]}, \varepsilon^{[\nu+1]} \right\rangle$$

$$\le \mu_1 ||\varepsilon^{[\nu+1]}||^2 + \mu_2 ||\varepsilon^{[\nu+1]}|| \, ||\varepsilon^{[\nu]}||, \quad (17)$$

where the norm $|| \cdot ||$ is defined by $||u||^2 := < u, u >$, and μ_1 and μ_2 are the subordinate matrix norms of $\tilde{J}_\nu$ and $J_\nu - \tilde{J}_\nu$, respectively. The left-hand side of (17) can be written as

$$\left\langle \frac{d}{dx} \varepsilon^{[\nu+1]}, \varepsilon^{[\nu+1]} \right\rangle = \frac{1}{2} \frac{d}{dx} ||\varepsilon^{[\nu+1]}||^2 = ||\varepsilon^{[\nu+1]}|| \frac{d}{dx} ||\varepsilon^{[\nu+1]}||,$$

so that, under the assumption that $||\varepsilon^{[\nu+1]}|| \neq 0$, we have

$$\frac{d}{dx} ||\varepsilon^{[\nu+1]}|| \leq \mu_1 ||\varepsilon^{[\nu+1]}|| + \mu_2 ||\varepsilon^{[\nu]}||. \tag{18}$$

Assuming $||\varepsilon^{[0]}(x)|| \leq K(x - x_0)$ as before and using $||\varepsilon^{[0]}(x_0)|| = 0$, we have

$$||\varepsilon^{[\nu+1]}(x)|| \leq \mu_2 e^{\mu_1 (x-x_0)} \int_{x_0}^{x} e^{-\mu_1 (s-x_0)} ||\varepsilon^{[\nu]}(s)|| \, ds$$

$$\leq \mu_2 e^{\mu_1 (x-x_0)} \int_{x_0}^{x} ||\varepsilon^{[\nu]}(s)|| \, ds, \tag{19}$$

which leads to

$$||\varepsilon^{[\nu]}(x)|| \leq K \frac{(\mu_2 e^{\mu_1 (x-x_0)})^\nu (x - x_0)^{\nu+1}}{(\nu + 1)!}, \tag{20}$$

showing that the rate of convergence of iteration (14) is not quadratic but superlinear.

By the way, the second order nonlinear equation

$$y''(x) = f(x, y), \quad y(x_0) = y_0, \quad y'(x_0) = y_0', \quad y \in \mathbf{R}^m \tag{21}$$

can be rewritten as the first order system

$$z'(x) = (y'(x), f(x, y))^T \equiv F(z(x)), \tag{22}$$

where $z(x) = (y(x), y'(x))^T$. The waveform Newton method for (22) is given by

$$z^{[\nu+1]'}(x) - J_\nu z^{[\nu+1]}(x) = F(z^{[\nu]}(x)) - J_\nu z^{[\nu]}(x). \tag{23}$$

In this case the Jacobian matrix J_ν is given by

$$J_\nu = \frac{\partial F}{\partial z} = \begin{pmatrix} \frac{\partial y'}{\partial y} & \frac{\partial y'}{\partial y'} \\ \frac{\partial f}{\partial y} & \frac{\partial f}{\partial y'} \end{pmatrix} = \begin{pmatrix} 0 & I \\ \frac{\partial f}{\partial y} & 0 \end{pmatrix}, \tag{24}$$

where I is an identity matrix. This result shows that for second order equation (12), if we define the waveform Newton method analogously by

$$\frac{d^2}{dx^2} y^{[\nu+1]}(x) - \frac{\partial f}{\partial y} y^{[\nu+1]}(x) = f(y^{[\nu]}(x)) - \frac{\partial f}{\partial y} y^{[\nu]}(x), \tag{25}$$

then the method also converges quadratically. Moreover, from the same discussion as for the first order equation, we can conclude that if an approximate Jacobian $\tilde{J}_\nu$ is applied, then the rate of convergence of the iteration

$$\frac{d^2}{dx^2} y^{[\nu+1]}(x) - \tilde{J}_\nu y^{[\nu+1]}(x) = f(y^{[\nu]}(x)) - \tilde{J}_\nu y^{[\nu]}(x) \tag{26}$$

is superlinear.

4   Numerical Experiments

4.1   Linear-Equations

In order to examine the efficiency of the waveform relaxations, we solve first the large linear system of second order differential equations. Consider the wave equation

$$\begin{cases} \dfrac{\partial^2 u}{\partial t^2} = \dfrac{\partial^2 u}{\partial x^2}, & 0 \le x \le 1, \\[2mm] u(x,0) = \sin(\pi x), & u(0,t) = u(1,t) = 0, \end{cases} \tag{27}$$

where the exact solution is $u(x,t) = \cos(\pi t)\sin(\pi x)$. The semi-discretization by 3-point spatial difference yields the system of linear equations

$$y''(t) = Qy(t), \quad y(t) = (u_1, \ldots, u_m)^T, \tag{28}$$

where

$$Q = \frac{1}{(\Delta x)^2} \begin{pmatrix} -2 & 1 & & & \text{\Large 0} \\ 1 & -2 & 1 & & \\ & \ddots & \ddots & \ddots & \\ & & 1 & -2 & 1 \\ \text{\Large 0} & & & 1 & -2 \end{pmatrix} \in \mathbf{R}^{m \times m}, \quad \Delta x = \frac{1}{m+1}.$$

In the splitting method given here we take M as the block diagonal matrix given by $M = \mathrm{diag}(M_1, M_2, \ldots, M_\mu)$, and each block M_l is the tridiagonal matrix given by

$$M_l = \frac{1}{(\Delta x)^2} \begin{pmatrix} 2 & -1 & & & \text{\Large 0} \\ -1 & 2 & -1 & & \\ & \ddots & \ddots & \ddots & \\ & & -1 & 2 & -1 \\ \text{\Large 0} & & & -1 & 2 \end{pmatrix} \in \mathbf{R}^{d \times d}, \; l = 1, 2, \ldots, \mu, \; \mu = m/d,$$

where we have assumed that m is divisible by d and, as a result, $\mu = m/d$ is an integer. In the implementation, the system is decoupled into μ subsystems and each of the subsystems is integrated concurrently on different processors, if the number of processors available are greater than that of the number of the subsystems. If this is not the case each processor integrates several subsystems sequentially. In any way our code is designed so as to have a uniform work load across the processors. The basic method to integrate the system is the 2-stage Runge-Kutta-Nyström method called the indirect collocation, which has an excellent stability property[3]. In our experiment we set $m = 256$ and integrate the system from $x = 0$ to 1 with the stepsize $h = 0.1$ under the stopping criteria $||y^{[\nu]} - y^{[\nu-1]}|| \le 10^{-7}$. The result on the parallel computer KSR1, which is a parallel computer with shared address space and distributed physical memory(see e.g. [6]), is shown in Table 1.

Table 1. Result for the linear problem on KSR1

| d | iterations | CPU time(sec) | | S_p | S'_p | E | E' | P | $\mu = m/d$ |
		serial	parallel						
1	11579	566.02	61.92	9.14	2.58	0.57	0.16	16	256
2	6012	586.73	53.79	10.91	2.97	0.68	0.19	16	128
4	3052	362.94	31.44	11.54	5.08	0.72	0.32	16	64
8	1462	240.81	19.56	12.31	8.17	0.76	0.51	16	32
16	688	181.74	13.94	13.04	11.5	0.81	0.72	16	16
32	427	201.56	27.11	7.43	5.89	0.93	0.73	8	8
64	403	369.86	97.27	3.80	1.64	0.95	0.41	4	4
128	403	791.10	407.41	1.94	0.39	0.97	0.20	2	2
256	1	159.76	–	–	1.00	–	–	1	1

$$S_p = \frac{\text{CPU time(serial)}}{\text{CPU time(parallel)}}, \quad S'_p = \frac{\text{CPU time(serial, } d = 256)}{\text{CPU time(parallel)}},$$

$$E = \frac{S_p}{P}, \quad E' = \frac{S'_p}{P}, \quad P = \text{number of processors}$$

4.2 Nonlinear Equations

Next we consider the nonlinear wave equation given by

$$v_i'' = \exp(v_{i-1}(x)) - 2\exp(v_i(x)) + \exp(v_{i+1}(x)), \quad i = 0, \pm 1, \pm 2, \ldots, . \quad (29)$$

This equation describes the behaviour of the well-known Toda lattice and has the soliton solution given by

$$v_i(x) = \log(1 + \sinh^2 \tau \, \mathrm{sech}^2(i\tau - w(x+q))), \quad i = 0, \pm 1, \pm 2, \ldots,$$

$$w = \sinh \tau,$$

where q is an arbitrary constant. The equation to be solved numerically is not (29) but its m-dimensional approximation given by

$$\begin{cases} y_i'' = \exp(y_{i-1}) - 2\exp(y_i) + \exp(y_{i+1}), & i = 2, \ldots, m-1 \\ y_1'' = 1 - 2\exp(y_1) + \exp(y_2), \\ y_m'' = \exp(y_{m-1}) - 2\exp(y_m) + 1 \end{cases} \quad (30)$$

initial conditions : $\quad y_i'(0) = v_i'(0), \; y_i(0) = v_i(0),$

where we have to choose a sufficiently large m to lessen the effect due to the finiteness of the boundary. As the parameters we take $q = 40$, $\tau = 0.5$ and $w = \sinh 0.5$. The approximate Jacobian $\tilde{J}_\nu$ used here is the block diagonal matrix given by

$$\tilde{J}_\nu(i,j) = \begin{cases} J_\nu(i,j), \; d(l-1)+1 \leq i,j \leq dl, l = 1, \ldots, m/d \\ 0, \quad \text{otherwise}, \end{cases}$$

where we have assumed m is divisible by d as before.

In our numerical experiment we use the KSR1 parallel computer, and integrate the 1024-dimensional system from $x = 0$ to 5 with stepsize $h = 0.05$ under the stopping criteria $||y^{[\nu]} - y^{[\nu-1]}|| \leq 10^{-10}$. The result is shown in Table 2.

In our algorithms, if the value of d becomes small then the degree of paral-

Table 2. Results for the nonlinear problem on KSR1

d	iterations	CPU time(sec) serial	parallel	S_p	S_p'	E	E'	P	$\mu = m/d$
1	17	53.98	6.10	8.85	3.49	0.55	0.22	16	1024
2	15	61.44	6.43	9.55	3.31	0.60	0.21	16	512
4	14	58.14	6.24	9.32	3.41	0.58	0.21	16	256
8	14	60.19	6.37	9.46	3.34	0.59	0.21	16	128
16	13	54.33	5.87	9.25	3.63	0.58	0.23	16	64
32	12	50.81	5.71	8.91	3.73	0.56	0.23	16	32
64	6	25.74	3.05	8.43	6.98	0.53	0.44	16	16
128	6	26.79	4.78	5.60	4.46	0.70	0.56	8	8
256	4	20.33	6.43	3.16	3.31	0.79	0.83	4	4
512	4	21.03	11.84	1.78	1.71	0.89	0.86	2	2
1024	4	21.30	—	—	1.00	–	–	1	1

$$S_p = \frac{\text{CPU time(serial)}}{\text{CPU time(parallel)}}, \quad S_p' = \frac{\text{CPU time(serial, } d = 1024)}{\text{CPU time(parallel)}}$$

$$P = \text{number of processors}$$

$$E = \frac{S_p}{P}, \quad E' = \frac{S_p'}{P}$$

lelism becomes large, since the number of the subsystems $\mu(= m/d)$ becomes large and, moreover, the cost of the LU decomposition that takes place in the inner iteration in each of the Runge-Kutta-Nyström schemes, which is proportional to d^3, becomes small. However, as is shown in the tables, it is in general true that the smaller the value of d, the larger the number of the WR iterations, which means the increase of the overheads such as synchronisation and communications. Therefore, small d implementation is not necessary advantageous. In our experiments we can achieve the best speedups when the number of subsystems is equal to that of processors.

References

1. Burrage, K., Parallel and Sequential Methods for Ordinary Differential Equations, Oxford University Press, Oxford, 1995.
2. Hairer, E., Nørsett S. P., and Wanner, G., Solving Ordinary Differential Equations I (Nonstiff Problems, 2nd Ed.), Springer-Verlag, Berlin, 1993.

3. van der Houwen, P. J., Sommeijer, B.P. Nguyen Huu Cong, Stability of collocation-based Runge-Kutta-Nyström methods, BIT 31(1991), 469-481.
4. Miekkala, U. and Nevanlinna, O., Convergence of dynamic iteration methods for initial value problems, SIAM J. Sci. Comput. 8(1987), pp.459-482.
5. White, J. and Sangiovanni-vincentelli, A.L., Relaxation Techniques for the Simulation of VLSI Circuits, Kluwer Academic Publishers, Boston, 1987.
6. Papadimitriou P., The KSR1 — A Numerical analyst's perspective, Numerical Analysis Report No.242, University of Mancheter/UMIST.

The Parallelization of the Incomplete LU Factorization on AP1000

Takashi NODERA and Naoto TSUNO

Department of Mathematics
Keio University
3-14-1 Hiyoshi Kohoku Yokohama 223, Japan

Abstract. Using a finite difference method to discretize a two dimensional elliptic boundary value problem, we obtain systems of linear equations $Ax = b$, where the coefficient matrix A is a large, sparse, and nonsingular. These systems are often solved by preconditioned iterative methods. This paper presents a data distribution and a communication scheme for the parallelization of the preconditioner based on the incomplete LU factorization. At last, parallel performance tests of the preconditioner, using BiCGStab(ℓ) and GMRES(m) method, are carried out on a distributed memory parallel machine AP1000. The numerical results show that the preconditioner based on the incomplete LU factorization can be used even for MIMD parallel machines.

1 Introduction

We are concerned with the solution of linear systems of equations

$$Ax = b \tag{1}$$

which arises from the discretization of partial differential equations. For example, a model problem is the two dimensional elliptic partial differential equation:

$$-u_{xx} - u_{yy} + \sigma(x, y)u_x + \gamma(x, y)u_y = f(x, y) \tag{2}$$

which defined on the unit square Ω with Dirichlet boundary conditions $u(x, y) = g(x, y)$ on $\partial\Omega$. We require $\sigma(x, y)$ and $\gamma(x, y)$ to be a bounded and sufficiently smooth function taking on strictly positive values. We shall use a finite difference approximation to discretize the equation (2) on a uniform grid points (**MESH** × **MESH**) and allow a five point finite difference molecule A. The idea of incomplete LU factorization [1, 4, 7] is to use $A = \tilde{L}\tilde{U} - R$, where

$$
A = \left\{ \begin{array}{ccc} & A_{i,j}^N & \\ A_{i,j}^W & A_{i,j}^C & A_{i,j}^E \\ & A_{i,j}^S & \end{array} \right\} \quad
\tilde{L} = \left\{ \begin{array}{ccc} & 0 & \\ L_{i,j}^W & L_{i,j}^C & 0 \\ & L_{i,j}^S & \end{array} \right\}, \quad
\tilde{U} = \left\{ \begin{array}{ccc} & U_{i,j}^N & \\ 0 & U_{i,j}^C & U_{i,j}^E \\ & 0 & \end{array} \right\} \tag{3}
$$

where $\tilde{L}$ and $\tilde{U}$ are lower and upper triangular matrices, respectively. The incomplete LU factorization preconditioner which is efficient and effective for the

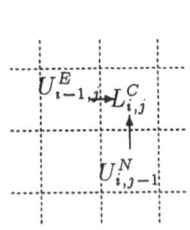 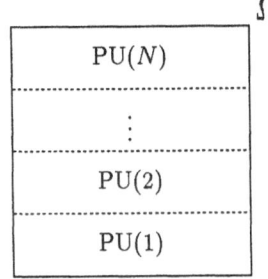 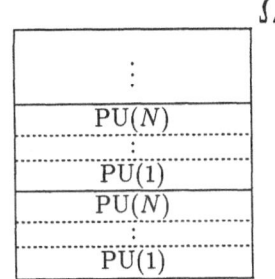

Fig. 1. The data dependency among neighborhoods.

Fig. 2. The data distribution for N processors.

Fig. 3. The cyclic data distribution for N processors.

single CPU machine, may not be appropriate for the distributed memory parallel machine just like AP1000.

In this paper, we describe some recent work on the efficient implementation of preconditioning for a MIMD parallel machine. In section 2, we discuss the parallelization of our algorithm, and then we describe the implementation of incomplete LU factorization on the MIMD parallel machine. In section 3, we present some numerical experiments in which BiCGStab(ℓ) algorithm and GMRES(m) algorithm are used to solve a two dimensional self-adjoint elliptic boundary value problem.

2 Parallelization

In section 1, as we presented the incomplete LU factorization, it involves in two parts. One is the decomposition of A into $\tilde{L}$ and $\tilde{U}$. The another is the inversion of $\tilde{L}$ and $\tilde{U}$ by forward and backward substitution. First of all, we consider the decomposition process. The data dependency of this factorization is given in Fig. 1. Namely, the entry of $L_{1,j}^C$ can be calculated as soon as $L_{i-1,j}^C$ and $L_{i,j-1}^C$ have been calculated on the grid points. The $L_{1,j}^C$ only depends on $L_{1,j-1}^C$ on the west boundary and $L_{i,1}^C$ only depends on $L_{i-1,1}^C$ on the south boundary. Next, we consider the inversion process. The calculation of $w_{i,j} = \tilde{L}^{-1}v$ and $w_{i,j} = \tilde{U}^{-1}v$ is given by forward and backward substitution as follows.

$$w_{i,j} = (v_{i,j} - L_{i,j}^S w_{i,j-1} - L_{i,j}^W w_{i-1,j})/L_{i,j}^C \tag{4}$$

$$w_{i,j} = v_{i,j} - U_{i,j}^N w_{i,j+1} - U_{i,j}^E w_{i+1,j} \tag{5}$$

We now consider brief introduction to the method of parallelization by Bastian and Horton [4]. They proposed the algorithm of dividing the domain Ω, as shown in Figure 2, when we do the parallel processing by N processors such as PU(1), ..., PU(N). Namely, PU(1) starts with its portion of the first low of grid points. After end, a signal is sent to PU(2), and then it starts to process its

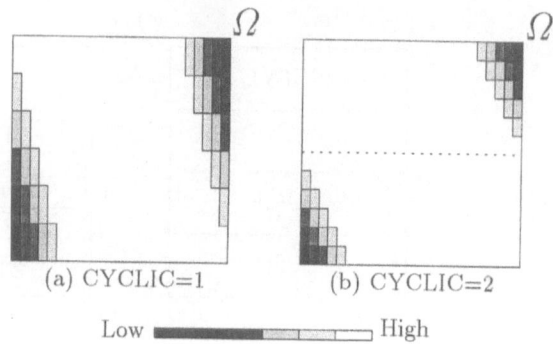

(a) CYCLIC=1   (b) CYCLIC=2

Low ▬▬▬▬▬▬▭▭ High

Fig. 4. The efficiency of parallel computation in the grid points on Ω.

section of the first row. At the time, PU(1) has started with the calculation of its component of the second row. All processors are running when PU(N) begins with the calculation of its component of the first row.

2.1 Generalization of the algorithm by Bastian and Horton

As we can show in Fig. 3, we divide the domain Ω by N processors cyclically. Here, we denote CYCLIC which denotes the number of cyclic division of the domain. The parallelization of Fig. 2 is equal to CYCLIC=1. If it divides as shown in the Fig. 3, the computation of the k-th processor PU(k) can be begun more quickly than the case of Bastian and Horton. Therefore, in this case, the efficiency of parallel processing will be higher than the ordinary one.

In Fig. 4, we show the efficiency of parallel computation in the square grid points for CYCLIC= 1, 2, respectively. In these figures, the efficiency of parallel processing is shown a deeper color of the grids is low.

The efficiency E of the parallel processing by this division is calculated as following equation 6. The detailed derivation of this equation is given in the forthcoming our paper [10, 11].

$$E = \frac{\text{CYCLIC} \times \text{MESH}}{\text{CYCLIC} \times \text{MESH} + N - 1} \tag{6}$$

On the other hand, for distributed memory parallel machine, the number of CYCLIC is increased, the communication between processors is also increased. Moreover, concealment of the communication time by calculation time also becomes difficult. Therefore, we will consider the trade-off between the efficiency of parallel processing and the overhead of communication.

3 Numerical Experiences

In this section we demonstrate the effectiveness of our proposed implementation of ILU factorization on a distribute memory machine Fujitsu AP1000 for some

Table 1. AP1000 specification

Architecture	Distributed Memory, MIMD
Number of processors	64
Inter processor networks	Broadcast network(50MB/s)
	Two-dimensional torus network
	(25MB/s/port)
	Synclronization network

Table 2. The computational time (sec.) for the multiplication of matrix $A(\tilde{L}\tilde{U})^{-1}$ and vector v on example 1.

Mesh Size	CYCLIC	$(\tilde{L}\tilde{U})^{-1}v$	Av	Total
128×128	1	2.52×10^{-2}	2.88×10^{-3}	2.81×10^{-2}
	2	5.86×10^{-2}	6.08×10^{-3}	6.47×10^{-2}
256×256	1	5.44×10^{-2}	9.18×10^{-3}	6.36×10^{-2}
	2	1.41×10^{-1}	1.41×10^{-2}	1.55×10^{-1}
	4	2.12×10^{-1}	2.68×10^{-2}	2.38×10^{-1}
512×512	1	1.56×10^{-1}	4.16×10^{-2}	1.98×10^{-1}
	2	3.31×10^{-1}	5.05×10^{-2}	3.81×10^{-1}
	4	5.05×10^{-1}	6.70×10^{-2}	5.72×10^{-1}
	8	8.26×10^{-1}	1.21×10^{-1}	9.47×10^{-1}

large sparse matrices problems. The Specification of AP1000 is given in Table 1. Each cell of AP1000 employs RISC-type SPARC or SuperSPARC processor chip. As Table 2 shows, the computational time for $A(\tilde{L}\tilde{U})^{-1}v$ for various CYCLIC, which obtained by the following example 1. From these results, it is considered suitable by the AP1000 to decrease data volume of the communication between processors, using CYCLIC= 1. So, we use 1 as the number of CYCLIC in the following numerical experiments.

[Example 1.] This example involves discretizations of boundary value problem for the partial differential equation (Joubert [5]).

$$-u_{xx} - u_{yy} + \sigma u_x = f(x, y)$$
$$u(x, y)|_{\partial\Omega} = 1 + xy$$

where the right-hand side is taken such that the true solution is $u(x, y) = 1 + xy$ on Ω. We discretize above equation 7 on a 256×256 mesh of equally spaced discretization points. By varying the constant σ, the amount of nonsymmetricity of the matrix may be changed. We utilize the initial approximation vector of $x_0 = 0$ and use simple stopping criterion $\|r_k\|/\|r_0\| \leq 10^{-12}$. We also used double precision real arithmetic to perform the runs presented here.

In Table 3, we show the computational times required to reduce the residual norm by a factor of 10^{-12} for an average over 3 trials of preconditioned or un-

Table 3. The computational time (sec.) for example 1.

Algorithm	σh									
	0	2^{-3}	2^{-2}	2^{-1}	2^0	2^1	2^2	2^3	2^4	2^5
GMRES(5)	*	*	63.25	31.74	33.51	32.15	32.98	32.82	36.68	43.91
GMRES(5)+ILU	*	48.21	25.74	27.29	22.91	18.10	12.24	9.23	7.07	6.13
GMRES(10)	*	117.04	50.52	47.75	50.41	50.89	50.75	47.82	44.42	43.54
GMRES(10)+ILU	296.18	36.25	39.58	40.95	37.02	25.45	16.08	11.94	8.31	6.63
GMRES(20)	*	109.68	89.45	88.79	94.16	92.41	92.00	90.39	84.65	79.01
GMRES(20)+ILU	197.47	60.54	76.01	70.29	54.82	36.51	29.32	15.55	9.83	5.89
BiCGStab(1)	30.11	23.83	20.34	20.66	23.18	23.04	48.42	101.04	*	*
BiCGStab(1)+ILU	30.34	21.28	17.82	17.25	15.09	10.91	8.46	6.73	4.86	3.13
BiCGStab(2)	35.16	26.78	24.14	24.67	26.38	29.65	30.07	25.64	24.69	24.90
BiCGStab(2)+ILU	31.86	22.87	19.67	21.27	16.47	10.37	7.47	6.51	5.23	3.63
BiCGStab(4)	47.28	33.40	30.93	32.86	39.23	39.51	36.47	36.75	32.33	30.96
BiCGStab(4)+ILU	34.38	24.56	22.47	23.17	18.99	12.00	8.49	7.79	5.69	4.30

* It does not converge after 3000 iterations.

preconditioned BiCGStab(ℓ) and GMRES(m) algorithm. For the preconditioner of incomplete LU factorization, in most cases both methods worked quite well.

References

1. Meijerink, J. A. and van der Vorst, H. A.: An iterative solution method for linear systems of which the coefficient matrix is symmetric M-matrix, *Math. Comp.*, Vol. 31, pp. 148-162 (1977).
2. Gustafsson, I.: A class of first order factorizations, *BIT* Vol. 18, pp. 142-156 (1978).
3. Saad, Y. and Schultz, M. H.: GMRES: A generalized minimal residual algorithm for solving nonsymmetric linear systems, *SIAM J. Sci. Stat. Comput.*, Vol. 7, No. 3, pp. 856–869 (1986).
4. Bastian, P. and Horton, G.: Parallelization of robust multigrid methods: ILU factorization and frequency decomposition method, *SIAM J. Sci. Stat. Comput.*, Vol. 12, No. 6, pp. 1457–1470 (1991).
5. Joubert, W.: Lanczos methods for the solution of nonsymmetric systems of linear equations, *SIAM J. Matrix Anal. Appl.*, Vol. 13, No. 3, pp. 926–943 (1992).
6. Sleijpen, G. L. G. and Fokkema, D. R.: BiCGSTAB(ℓ) for linear equations involving unsymmetric matrices with complex spectrum, *ETNA*, Vol. 1, pp. 11–32 (1993).
7. Bruaset, A. M.: A survey of preconditioned iterative methods, Pitman Research Notes in Math. Series 328, Longman Scientific & Technical (1995).
8. Nodera, T. and Noguchi, Y.: Effectiveness of BiCGStab(ℓ) method on AP1000, Trans. of IPSJ (in Japanese), Vol. 38, No. 11, pp. 2089-2101 (1997).
9. Nodera, T. and Noguchi, Y.: A note on BiCGStab(ℓ) Method on AP1000, *IMACS Lecture Notes on Computer Science*, to appear (1998).
10. Tsuno, N.: The automatic restarted GMRES method and the parallelization of the incomplete LU factorization, Master Thesis on Graduate School of Science and Technology, Keio University (1998).
11. Tsuno, N. and Nodera, T.: The parallelization and performance of the incomplete LU factorization on AP1000, Trans. of IPSJ (in Japanese), submitted.

An Efficient Parallel Triangular Inversion by Gauss Elimination with Sweeping

Ayşe Kiper

Department of Computer Engineering,
Middle East Technical University,
06531 Ankara Turkey
ayse@rorqual.cc.metu.edu.tr

Abstract. A parallel computation model to invert a lower triangular matrix using Gauss elimination with sweeping technique is presented. Performance characteristics that we obtain are $O(n)$ time and $O(n^2)$ processors leading to an efficiency of $O(1/n)$. A comparative performance study with the available fastest parallel matrix inversion algorithms is given. We believe that the method presented here is superior over the existing methods in efficiency measure and in processor complexity.

1   Introduction

Matrix operations have naturally been the focus of researches in numerical parallel computing. One fundamental operation is the inversion of matrices and that is mostly accompanied with the solution of linear system of equations in the literature of parallel algorithms. But it has equally well significance individually and in matrix eigenproblems. Various parallel matrix inversion methods have been known and some gave special attention to triangular types [1,2,3,5]. The time complexity of these algorithms is $O(\log^2 n)$ using $O(n^3)$ processors on the parallel random accesss machine (PRAM). No faster algorithm is known so far although the best lower bound proved for parallel triangular matrix inversion is $\Omega(\log n)$.

In this paper we describe a parallel computation model to invert a dense triangular matrix using Gauss elimination (although it is known [2] that Gauss elimination is not the fastest process) with sweeping strategy which allows maximum possible parallelisation. We believe that the considerable reduction in the processors used results in an algorithm that is more efficient than the available methods.

2   Parallel Gauss Elimination with Sweeping

A lower triangular matrix L of size n and its inverse $L^{-1} = B$ satisfy the equation

$$LB = I. \tag{1}$$

The parallel algorithm that we will introduce is based on the application of the standard Gauss elimination on the augmented matrix

$$[L \,|\, I]. \tag{2}$$

Elements of the inverse matrix B are determined by starting from the main diagonal and progressing in the subdiagonals of decreasing size. The Gauss elimination by sweeping (hereafter referred to as GES) has two main type of sweep stages:

(i) *Diagonal Sweeps* (DS) : evaluates all elements of each subdiagonal in parallel in one division step.
(ii) *Triangular Sweeps* (TS) : updates all elements of each subtriangular part in parallel in two steps (one multiplication and one subtraction).

We can describe GES as follows:

Stage 1. Initial Diagonal Sweep (IDS) : the main diagonal elements of B are all evaluated in terms of l_{ij} (are the elements of the matrix L) in parallel by

$$\alpha_{ii} = \frac{1}{l_{ii}}, \quad i = 1, 2, \ldots, n. \tag{3}$$

Stage 2. Initial Triangular Sweep (ITS) : updates the elements of the lower triangular part below the main diagonal in columnwise fashion

$$\alpha_{ij} = \frac{-l_{ij}}{l_{jj}}, \quad i = 2, 3, \ldots, n; \quad j = 1, 2, \ldots, n-1 \tag{4}$$

in one parallel step being different than the other regular triangular sweep steps.
 The IDS and ITS lead to the updated matix

$$B_{IS} = \begin{bmatrix} \alpha_{11} & & & \\ \alpha_{21} & \alpha_{22} & 0 & \\ \cdot & & \cdot & \\ \cdot & & & \cdot \\ \alpha_{n1} & & & \alpha_{nn} \end{bmatrix}. \tag{5}$$

Then the regular DS and TS are applied alternately (denoting their each application by a prime over the elements of the current matrix) until B is obtained as

$$B = \begin{bmatrix} \alpha_{11} & & & & \\ \alpha'_{21} & \alpha_{22} & & 0 & \\ \alpha''_{31} & \alpha'_{32} & \alpha_{33} & & \\ \cdot & \alpha''_{42} & \cdot & & \cdot \\ \cdot & & & \cdot & \\ \alpha^{(n-1)}_{n1} & \cdot & \cdot & \alpha''_{n,n-2} & \alpha'_{n,n-1} & \alpha_{nn} \end{bmatrix}. \tag{6}$$

If k denotes the application number of the sweeps in DSk and TSk , the stage computations can be generalised as:

Stage $(2k+1)$. DSk : $(k+1)$th subdiagonal elements of B are computed using

$$\alpha^{(k)}_{k+i,i} = \frac{\alpha^{(k-1)}_{k+i,i}}{l_{k+i,k+i}}, \quad i = 1, 2, \ldots, n-k \qquad (7)$$

leading to the intermediate matrix

$$B_{DSk} = \begin{bmatrix} \alpha_{11} & & & & & \\ \alpha'_{21} & \alpha_{22} & & & & \\ . & \alpha'_{32} \cdot & & 0 & & \\ . & & . & . & & \\ \alpha^{(k)}_{k+1,1} & & . & . & & . \\ \alpha^{(k-1)}_{k+2,1} & . & & . & . & \\ . & . & . & & . . & \\ . & . & . & . & . & \\ \alpha^{(k-1)}_{n,1} & . & . \alpha^{(k-1)}_{n,n-k-1} & \alpha^{(k)}_{n,n-k} & . . \alpha'_{n,n-1} & \alpha_{nn} \end{bmatrix} \qquad (8)$$

Stage $(2k+2)$. TSk : The constants are evaluated in the first step by

$$C^{(k)}_{k+p,i} = \alpha^{(k)}_{i+1,i} l_{k+p,i+1}, \quad p = i+1, i+2, \ldots, n-k; \quad i = 1, 2, \ldots, n-(k+1) \qquad (9)$$

After the constant evaluations are completed, new updated values of elements below the kth subdiagonal are computed columnwise using

$$\alpha^{(k)}_{k+p,i} = \alpha^{(k-1)}_{k+p,i} - C^{(k)}_{k+p,i}, \quad p = i+1, i+2, \ldots, n-k; \quad i = 1, 2, \ldots, n-(k+1) \qquad (10)$$

in parallel giving matrix

$$B_{TSk} = \begin{bmatrix} \alpha_{11} & & & & & \\ \alpha'_{21} & & & & & \\ . & \alpha'_{32} \cdot & & 0 & & \\ . & & . & . & & \\ \alpha^{(k)}_{k+1,1} & & . & . & & . \\ \alpha^{(k)}_{k+2,1} & . & & . & . & \\ . & . & . & & . . & \\ . & . & . & . & . & \\ \alpha^{(k)}_{n,1} & . & . \alpha^{(k)}_{n,n-k-1} & \alpha^{(k)}_{n,n-k} & . . \alpha'_{n,n-1} & \alpha_{nn} \end{bmatrix} \qquad (11)$$

GES algorithm totally requires $(n-2)$ diagonal, $(n-2)$ triangular sweeps ($k = 1, 2, \ldots, n-2$, in (7) and (10)) in addition to initial diagonal, initial triangular and final diagonal sweeps (FDS) ($k = n-1$, in (7)) .

3 Performance Characteristics and Discussion

The total number of parallel arithmetic operation steps T_p is obtained as

$$T_p = 3(n-1) \qquad (12)$$

with the maximun number of processors used

$$p_{max} = \frac{n(n-1)}{2}. \tag{13}$$

The speedup $S_p = T_s/T_p$ with respect to the the best sequential algorithm [4] for which $T_s = n^2$ time is obtained as

$$S_p = \frac{n^2}{3(n-1)} \tag{14}$$

and the associated efficiency $E_p = S_p/p_{max}$ is

$$E_p = \frac{2n}{3(n-1)^2}. \tag{15}$$

All efficient parallel algorithms [1,2,3] of triangular matrix inversion run in $O(\log^2 n)$ time using $O(n^3)$ processors and are designed for PRAM system. So we find it appropriate to compare the theoretical performance of GES algorithm with the complexity bounds stated by Sameh and Brent [3] which are the known model reference expressions. We select the same computational model PRAM in the discussions to keep the common comparison base. The speedup and efficiency expressions (denoted by a superscript $*$ to differentiate from those of GES algorithm) for the algorithm of Sameh and Brent [3] over the best sequential time are

$$S_p^* = \frac{2n^2}{(\log^2 n + 3\log n + 6)} \tag{16}$$

and

$$E_p^* = \frac{256}{(\log^2 n + 3\log n + 6)(21n + 60)}. \tag{17}$$

The best comparative study can be done by observing the variation of speedup $R_S = S_p/S_p^*$ and efficiency $R_E = E_p/E_p^*$ ratios with respect to the problem size. These ratios expilicitely are

$$R_S = \frac{log^2 n + 3\log n + 6}{6(n-1)^2} \tag{18}$$

and

$$R_E = \frac{n(21n + 60)(\log^2 n + 3\log n + 6)}{384(n-1)^2}. \tag{19}$$

The efficiency ratio curve (Figure 1) demonstrates a superior behaviour of GES algorithm compared to that of Sameh and Brent for $n > 8$. Even for the worst case (occurs when $n \cong 8$) efficiency of GES exceeds the value of that for Sameh and Brent more than twice. This is the natural result of considerable smaller number of processor needed in GES. As the speedup ratio concerned the GES algrithm is slower than that given by Sameh and Brent and this property is dominant particularly for small size problems.

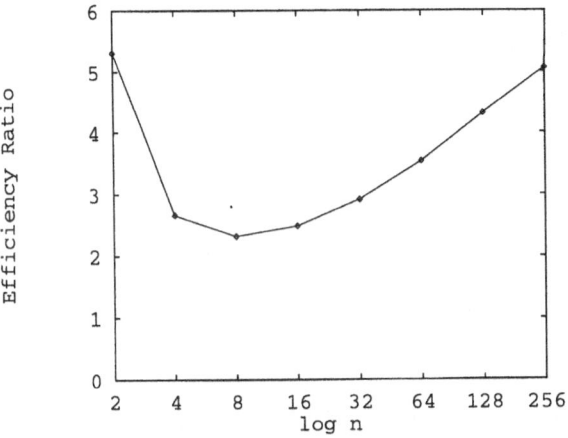

Fig. 1. Efficiency Ratio

The implementation studies of GES on different parallel architectures is open for future work. The structure of the algorithm suggests an easy application on array processors with low (modest) communication cost that meets the PRAM complexity bounds of the problem with a small constant.

References

1. L. Csansky, On the parallel complexity of some computational problems, Ph.D. Dissertation, Univ. of California at Berkeley, 1974.
2. L. Csansky, Fast parallel matrix inversion algorithms, *SIAM J.Comput.,* **5** (1976) 618-623.
3. A. H. Sameh and R. P. Brent, Solving triangular systems on parallel computer, *SIAM J. Numer. Anal.,* **14** (1977) 1101-1113.
4. U. Schendel, *Introduction to Numerical Methods for Parallel Computers* (Ellis Harwood, 1984).
5. A. Schikarski and D. Wagner, Efficient parallel matrix inversion on interconnection networks, *J. Parallel and Distributed Computing,* **34** (1996) 196-201.

Fault Tolerant QR-Decomposition Algorithm and Its Parallel Implementation

Oleg Maslennikow[1], Juri Kaniewski[1],Roman Wyrzykowski[2]

[1] Dept. of Electronics, Technical University of Koszalin,
Partyzantow 17, 75-411 Koszalin, Poland
[2] Dept. of Math. & Comp. Sci, Czestochowa Technical University,
Dabrowskiego 73, 42-200 Czestochowa, Poland

Abstract. A fault tolerant algorithm based on Givens rotations and a modified weighted checksum method is proposed for the QR-decomposition of matrices. The algorithm enables us to correct a single error in each row or column of an input $M \times N$ matrix **A** occurred at any among N steps of the algorithm. This effect is obtained at the cost of $2.5N^2 + O(N)$ multiply-add operations ($M = N$). A parallel version of the proposed algorithm is designed, dedicated for a fixed-size linear processor array with fully local communications and low I/O requirements.

1 Introduction

The high complexity of most of matrix problems [1] implies the necessity of solving them on high performance computers and, in particular, on VLSI processor arrays [3]. Application areas of these computers demand a large degree of reliability of results. while a single failure may render computations useless. Hence fault tolerance should be provided on hardware or/and software levels [4].

The algorithm-based fault tolerance (ABFT) methods [5–9] are very suitable for such systems. In this case, input data are encoded using error detecting or/and correcting codes. An original algorithm is modified to operate on encoded data and produce encoded outputs, from which useful information can be recovered easily. The modified algorithm will take more time in comparison with the original one. This time overhead should not be excessive. An ABFT method called the weighted checksum (WCS) one, especially tailored for matrix algorithms and processor arrays, was proposed in Ref. [5]. However, the original WCS method is little suitable for such algorithms as Gaussian elimination, Choleski, and Faddeev algorithms, etc., since a single transient fault in a module may cause multiple output errors, which can not be located. In Refs. [8,9], we proposed improved ABFT versions of these algorithms.

For such important matrix problems as least squares problems, singular value and eigenvalue decompositions, more complicated algorithms based on the QR-decomposition should be applied [2]. In this paper, we design a fault tolerant version of the QR-decomposition based on Givens rotations and a modified WCS method. The derived algorithm enables correcting a single error in each row or column of an input $M \times N$ matrix $\mathbf{A}$ occurred at any among N steps of the algorithm. This effect is obtained at the cost of $2.5N^2 + O(N)$ multiply-add operations. A parallel version of the algorithm is designed, dedicated for a fixed-size linear processor array with local communications and low I/O requirements.

2 Fault Model and Weighted Checksum Method

Module-level faults are assumed [6] in algorithm-based fault tolerance. A module is allowed to produce arbitrary logical errors under physical failure mechanism. This assumption is quite general since it does not assume any technology-dependent fault model. Without loss of generality, a single module error is assumed in this paper. Communication links are supposed to be fault-free.

In the WCS method [6], redundancy is encoded at the matrix level by augmenting the original matrix with weighted checksums. Since the checksum property is preserved for various matrix operations, these checksums are able to detect and correct errors in the resultant matrix. The complexity of correction process is much smaller than that of the original computation. For example, a WCS encoded data vector $\mathbf{a}$ with the Hamming distance equal to three (which can correct a single error) is expressed as

$$\mathbf{a}^T = [a_1 \quad a_2 \ldots a_N \quad PCS \quad QCS] \tag{1}$$

$$PCS = \mathbf{p}^T [a_1 \quad a_2 \quad \ldots \quad a_N], \quad QCS = \mathbf{q}^T [a_1 \quad a_2 \quad \ldots \quad a_N] \tag{2}$$

Possible choices for encoder vectors $\mathbf{p}$, $\mathbf{q}$ are, for example, [10] :

$$\mathbf{p}^T = \begin{bmatrix} 2^0 & 2^1 & \ldots & 2^{N-1} \end{bmatrix}, \quad \mathbf{q}^T = [1 \quad 2 \quad \ldots \quad N] \tag{3}$$

For the floating-point implementation, numerical properties of single-error correction codes based on different encoder vectors were considered in Refs. [7, 11].

Based on an encoder vector, a matrix $\mathbf{A}$ can be encoded as either a row encoded matrix $\mathbf{A}_R$, a column encoded matrix $\mathbf{A}_C$, or a full encoded matrix $\mathbf{A}_{RC}$ [11]. For example, for the matrix multiplication $\mathbf{A} \, \mathbf{B} = \mathbf{D}$, the column encoded matrix $\mathbf{A}_C$ is exploited [6]. Then choosing the linear weighted vector (3), the equation $\mathbf{A}_C * \mathbf{B} = \mathbf{D}_C$ is computed. To have the possibility of verifying computations and correcting a single error, syndromes S_1 and S_2 for the j-th column of $\mathbf{D}$-matrix should be calculated, where

$$S_1 = \sum_{i=1}^{M} c_{ij} - PCS_j, \quad S_2 = \sum_{i=1}^{M} i * c_{ij} - QCS_j \tag{4}$$

3 Design of the ABFT QR-Decomposition Algorithm

The complexity of the Givens algorithm [2] is determined by $4N^3/3$ multiplications and $2N^3/3$ additions, for a real $N \times N$ matrix $\mathbf{A}$. Based on equivalent matrix transformations, this algorithm preserves the Euclidean norm for columns of $\mathbf{A}$ during computations. This property is important for error detection and enable us to save computations.

In the course of the Givens algorithm, an $M \times N$ input matrix $\mathbf{A} = \mathbf{A}^1 = \{a_{ij}\}$ is recursively modified in K steps to obtain the upper triangular matrix $\mathbf{R} = \mathbf{A}^{K+1}$, where $K = M - 1$ for $M \leq N$, and $K = N$ for $M > N$. The i-th step consists in eliminating elements a_{ji}^i in the i-th column of $\mathbf{A}^i$ by multiplications on rotation matrices $\mathbf{P}_{ji}$, $j = i+1, \ldots, M$, which correspond to rotation coefficients

$$c_{ji} = a_{ii}^{j-1} / \sqrt{(a_{ii}^{j-1})^2 + (a_{ji}^i)^2}, \qquad s_{ji} = a_{ji}^i / \sqrt{(a_{ii}^{j-1})^2 + (a_{ji}^i)^2}$$

Each step of the algorithm includes two phases. The first phase consists in recomputing $M - i$ times the first element a_{ii} of the pivot (i.e. i-th) row, and computing the rotation coefficients. The second phase includes computation of a_{jk}^{i+1}, and resulting elements r_{ik} in the i-th row of $\mathbf{R}$. This phase includes also recomputing $M - i$ times the rest of elements in the pivot row.

Consequently, if during the i-th step, $i = 1, \ldots, K$, an element a_{ii}^j is wrongly calculated, then errors firstly appear in coefficients c_{ji} and s_{ji}, $j = i+1, \ldots, M$, and then in all the elements of $\mathbf{A}^{i+1}$. Moreover, if at the i-th step, any coefficient c_{ji} or s_{ji} is wrongly calculated, then errors firstly appear in all the elements of the pivot row, and then in all the elements of $\mathbf{A}^{i+1}$. All these errors can not be located and corrected by the original WCS method. To remove these drawbacks, the following lemmas are proved. It is assumed that a single transient error may appear at each row or column of $\mathbf{A}^i$ at any step of the algorithm.

Lemma 1. *If at the i-th step, an element a_{jk}^{i+1} $(i < j, i < k)$ is wrongly calculated, then errors will not appear among other elements of $\mathbf{A}^{i+1}$.*

However, if a_{jk}^i is erroneous, then error appears while computing either the element a_{jk}^{j+1} in the pivot row at the j-th step of the algorithm, for $j \leq k$, or values of a_{kk}^j, c_{jk} and s_{jk} $(j = i+1, \ldots, M)$ at the k-th step, for $j > k$. Hence in these cases, we should check and possibly correct elements of the i-th and j-th rows of $\mathbf{A}^i$, each time after their recomputing.

Lemma 2. *Let an element a_{ik}^j of the pivot row $(j = i+1, \ldots, M; k = 1, \ldots, N)$ or an element a_{jk}^{i+1} of a non-pivot row $(k = i+1, \ldots, N)$ was wrongly calculated when executing phase 2 of the i-th step of the Givens algorithm. Then it is possible to correct its value while executing this phase, using the WCS method for the row encoded matrix $\mathbf{A}_R$, where*

$$\mathbf{A}_R = [\mathbf{A} \ \ \mathbf{Ap} \ \ \mathbf{Aq}] = [\mathbf{A} \ \ \mathbf{PCS} \ \ \mathbf{QCS}] \tag{5}$$

$$PCS_j^{i+1} = a_{j,i+1}^{i+1} + a_{j,i+2}^{i+1} + \ldots a_{j,N}^{i+1} = (-s_{ji}a_{i,i+1}^i + c_{ji}a_{j,i+1}^i) + \ldots$$
$$+ (-s_{ji}a_{i,N}^i + c_{ji}a_{j,N}^i) = -s_{ji}(a_{i,i+1}^i + a_{i,N}^i) + c_{ji}(a_{j,i+1}^i + a_{j,N}^i) \tag{6}$$

So before executing phase 2 of the i-th step we should be certain that c_{ji}, s_{ji} and a_{ii}^j were calculated correctly at phase 1 of this step (we assume that the remaining elements were checked and corrected at the previous step). For this aim, the following properties of the Givens algorithm may be used:

– $$(c_{ji})^2 + (s_{ji})^2 = 1 \tag{7}$$

– preserving the Euclidean norm for column of $\mathbf{A}$ during computation

$$\sqrt{(a_{i,i}^i)^2 + (a_{i+1,i}^i)^2 + \ldots + (a_{M,i}^i)^2} = a_{i,i}^M \text{ (where } a_{ii}^M = r_{ii}) \tag{8}$$

The triple time redundancy (TTR) method [4] may also be used for its modest time overhead. In this case, values of c_{ji}, s_{ji} or/and a_{ii}^j are calculated three times.

Hence, for c_{ji} and s_{ji}, the procedure of error detection and correction consists in computing the left part of expression (7), and its recomputing if equality (7) is not fulfilled, taking into account a given tolerance τ [6]. For elements a_{ii}^j, $i = 1, \ldots, K$, this procedure consists in computing the both parts of expression (8), and their recomputing if they are not equal. The correctness of this procedure is based on the assumption that only one transient error may appear at each row or column of $\mathbf{A}^i$ at any step. Moreover, instead of using the correction procedure based on formulae (4), we recompute all elements in the row with an erroneous element detected.

The resulting ABFT Givens algorithm is as follows:

1. The original matrix $\mathbf{A} = \{a_{jk}\}$ is represented as the row encoded matrix $\mathbf{A}_R = \{a_{jk}^1\}$ with $a_{jk}^1 = a_{jk}$, $a_{j,N+1}^1 = PCS_j^1$, for $j = 1, \ldots, M; k = 1, \ldots, N$.
2. For $i = 1, 2, \ldots, K$, stages 3-9 are repeated.
3. The values of $a_{ii}^j = \sqrt{(a_{ii}^{i-1})^2 + (a_{ji}^i)^2}$, are calculated, $j = i + 1, \ldots, M$.
4. The norm $| \mathbf{a}_i |$ for the i-th column of $\mathbf{A}$ is calculated. This stage needs approximately $M - i$ multiply-add operations. The value of $| \mathbf{a}_i |$ is compared with the value of a_{ii}^M. If $a_{ii}^M \neq | \mathbf{a}_i |$, then stages 3,4 are repeated.
5. The coefficients c_{ji} and s_{ji} are computed, and correctness of equation (7) is checked, for $j = i + 1, \ldots, M$. In case of non-equality, stage 5 is repeated.
6. For $j = i + 1, \ldots, M$, stages 7-10 are repeated.
7. The elements a_{ik}^j of the i-th row of $\mathbf{A}^{i+1}$ are computed, $k = 1, \ldots, N + 1$.
8. The value of PCS_i^j is calculated according to (6). This stage needs approximately $N - i$ additions. The obtained value is compared with that of $a_{i,N+1}^j$. In case of the negative answer, stages 7,8 are performed again.
9. The elements a_{jk}^{i+1} in the j-th row of $\mathbf{A}^{i+1}$ are calculated, k = i+1, ...,N+1.
10. The value of PCS_j^{i+1} is computed according to expression (6). This stage also needs approximately $N - i$ additions. The computed value is compared with that of $a_{j,N+1}^{i+1}$. In case of the negative case, stages 9,10 are repeated.

The procedures of error detection and correction increase the complexity of the Givens algorithm on $N^2/2 + O(N)$ multiply-add operations and $N^2 + O(N)$ additions, for $M = N$. Due to increased sizes of the input matrix, the additional overhead of the proposed algorithm is $2N^2 + O(N)$ multiplications and $N^2 + O(N)$ additions ($M = N$). As a result, the complexity of the whole algorithm is increased approximately on $2.5N^2 + O(N)$ multiply-add operations and $2N^2 + O(N)$ additions. At this cost, the proposed algorithm enables us to correct one single transient error occured in each row or column of $\mathbf{A}$ at any among K steps of computations. Consequently, for $M = N$, it is possible to correct up to N^2 single errors when solving the whole problem.

4 Parallel Implementation

The dependence graph $\mathbf{G}_1$ of the proposed algorithm is shown in Fig.1 for $M = 4$, $N = 3$. Nodes of $\mathbf{G}_1$ are located in vertices of the integer lattice $\mathbf{Q} = \{\mathbf{K} =$

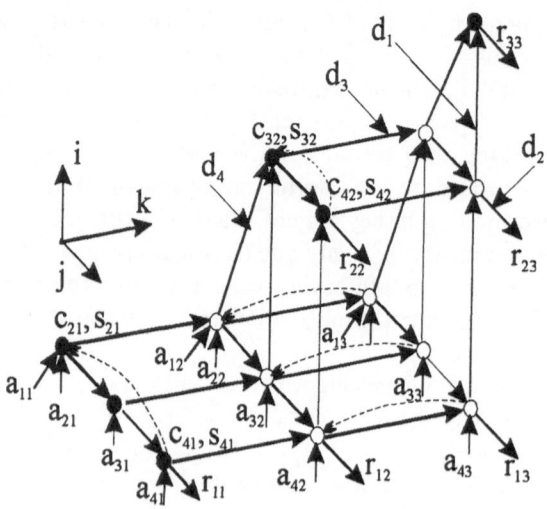

Fig. 1. Graph of the algorithm

(i, j, k) : $1 \leq i \leq K$; $i + 1 \leq j \leq M$; $i \leq k \leq N$}. There are two kind of nodes in G_1. Note that non-local arcs marked with broken lines, and given by vectors $d_5 = (0, i + 1 - M, 0)$, $d_6 = (0, 0, i - N)$ are result of introducing ABFT properties into the original algorithm.

To run the algorithm in parallel on a processor array with local links, all the vectors d_5 are excluded using the TTR technique for computing values of a_{ii}^j, c_{ji} and sji. This needs to execute additionally $6N^2 + O(N)$ multiplications and additions. Then all the non-local vectors d_6 are eliminated by projecting G_1 along k-axis. As a result, a 2-D graph G_2 is derived (see Fig.2).

To run G_2 on a linear array with a fixed number n of processors, G_2 should be decomposed into a set of $s =]N/n[$ subgraphs with the "same" topology and without bidirectional data dependencies. Such a decomposition is done by cutting G_2 by a set of straight lines parallel to j-axis. These subgraphs are then mapped into an array with n processors by projecting each subgraph onto i-axis [12]. The resulting architecture, which is provided with an external RAM module, features a simple scheme of local communications and a small number of I/O channels. The proposed ABFT Givens algorithm is executed on this array in

$$T = \sum_{i=1}^{s} [(N + 3 - n(i - 1)) + (M - n(i - 1))]$$

time steps. For $M = N$, we have

$$T = N^3/n - (N - n)N^2/2 + (2N - n)N^2/(6n)$$

The processor utilization is $E_n = W/(T * n)$, where W is the computational complexity of the proposed algorithm.

Using these formulae, for example, in case of $s = 10$, we obtain

$$E_n = 0.86$$

Note that with increasing in parameter s the value of E_n also increases.

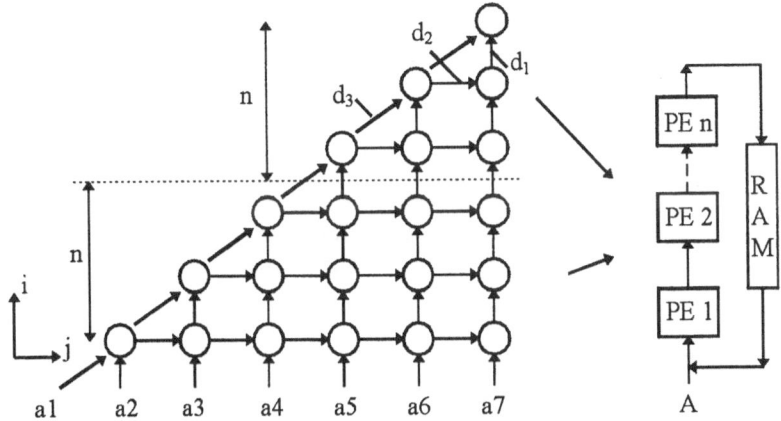

Fig. 2. Fixed-sized linear array

References

1. G.H. Golub, C.F. Van Loan, *Matrix computations*, John Hopkins Univ. Press, Baltimore, Maryland, 1996.
2. Å. Björck, *Numerical methods for least squares problems*, SIAM, Philadelphia, 1996.
3. S.Y. Kung, *VLSI array processors*, Prentice-Hall, Englewood Cliffs, N.J., 1988.
4. S.E. Butner, Triple time redundancy, fault-masking in byte-sliced systems. Tech.Rep. CSL TR-211, Dept. of Elec.Eng., Stanford Univ., CA, 1981.
5. K.H. Huang, J.A. Abraham, Algorithm-based fault tolerance for matrix operations. IEEE Trans. Comput., C-35 (1984) 518-528.
6. V.S. Nair, J.A. Abraham, Real-number codes for fault-tolerant matrix operations on processor arrays. IEEE Trans. Comp., C-39 (1990) 426-435.
7. J.-Y. Han, D.C. Krishna, Linear arithmetic code and its application in fault tolerant systolic arrays, in *Proc. IEEE Southeastcon*, 1989, 1015-1020.
8. J.S. Kaniewski, O.V. Maslennikow, R. Wyrzykowski, Algorithm-based fault tolerant matrix triangularization on VLSI processor arrays, in *Proc. Int. Workshop "Parallel Numerics'95"*, Sorrento, Italy, 1995, 281-295.
9. J.S. Kaniewski, O.V. Maslennikow, R. Wyrzykowski, Algorithm-based fault tolerant solution of linear systems on processor arrays, in *Proc. 7-th Int. Workshop "PARCELLA '96"*, Berlin, Germany, 1996, 165 - 173.
10. D.E. Schimmel, F.T. Luk, A practical real-time SVD machine with multi-level fault tolerance. Proc. SPIE, 698 (1986) 142-148.
11. F.T. Luk, H. Park, An analysis of algorithm-based fault tolerance techniques. Proc. SPIE, 696 (1986) 222-227.
12. R. Wyrzykowski, J. Kanevski, O. Maslennikow, Mapping recursive algorithms into processor arrays, in *Proc. Int. Workshop Parallel Numerics'94*, M. Vajtersic and P. Zinterhof eds., Bratislava, 1994, 169-191.

Parallel Sparse Matrix Computations Using the PINEAPL Library: A Performance Study

Arnold R. Krommer

The Numerical Algorithms Group Ltd,
Wilkinson House, Jordan Hill Road, Oxford, OX2 8DR, UK
arnoldk@nag.co.uk

Abstract. The Numerical Algorithms Group Ltd is currently participating in the European HPCN Fourth Framework project on *Parallel
Industrial NumErical Applications and Portable Libraries* (PINEAPL).
One of the main goals of the project is to increase the suitability of
the existing NAG Parallel Library for dealing with computationally intensive industrial applications by appropriately extending the range of
library routines. Additionally, several industrial applications are being
ported onto parallel computers within the PINEAPL project by replacing sequential code sections with calls to appropriate parallel library
routines.
A substantial part of the library material being developed is concerned
with the solution of PDE problems using parallel sparse linear algebra modules. This talk provides a number of performance results which
demonstrate the efficiency and scalability of core computational routines
– in particular, the iterative solver, the preconditioner and the matrix-
vector multiplication routines. Most of the software described in this
talk has been incorporated into the recently launched Release 1 of the
PINEAPL Library.

1 Introduction

The NAG Parallel Library enables users to take advantage of the increased computing power and memory capacity offered by multiple processors. It provides
parallel subroutines in some of the areas covered by traditional numerical libraries (in particular, the NAG Fortran 77, Fortran 90 and C libraries), such as
dense and sparse linear algebra, optimization, quadrature and random number
generation. Additionally, the NAG Parallel Library supplies support routines for
data distribution, input/output and process management purposes. These support routines shield users from having to deal explicitly with the message-passing
system – which may be MPI or PVM – on which the library is based. Targeted
primarily at distributed memory computers and networks of workstations, the
NAG Parallel Library also performs well on shared memory computers whenever
efficient implementations of MPI or PVM are available. The fundamental design
principles of the existing NAG Parallel Library are described in [7], and information about contents and availability can be found in the library Web page at
http://www.nag.co.uk/numeric/FD.html.

NAG is currently participating in the HPCN Fourth Framework project on *P*arallel *I*ndustrial *N*um*E*rical *A*pplications and *P*ortable *L*ibraries (PINEAPL). One of the main goals of the project is to increase the suitability of the NAG Parallel Library for dealing with a wide range of computationally intensive industrial applications by appropriately extending the range of library routines. In order to demonstrate the power and the efficiency of the resulting Parallel Library, several application codes from the industrial partners in the PINEAPL project are being ported onto parallel and distributed computer systems by replacing sequential code sections with calls to appropriate parallel library routines. Further details on the project and its progress can be found in [1] and in the project Web page at http://www.nag.co.uk/projects/PINEAPL.html.

This paper focuses on the parallel sparse linear algebra modules being developed by NAG.[1] These modules are used within the PINEAPL project in connection with the solution of PDE problems based on finite difference and finite element discretization techniques. Most of the software described in this paper has been incorporated into the recently launched Release 1 of the PINEAPL Library.

The scope of support for parallel PDE solvers provided by the parallel sparse linear algebra modules is outlined in Section 2. Section 3 provides a study of the performance of core computational routines – in particular, the iterative solver, the preconditioner and the matrix-vector multiplication routines. Section 4 summarizes the results and gives an outlook on future work.

2 Scope of Library Routines

A number of routines have been and are being developed and incorporated into the PINEAPL and the NAG Parallel Libraries in order to assist users in performing some of the computational steps (as described in [7]) required to solve PDE problems on parallel computers. The extent of support provided in the PINEAPL Library is – among other things – determined by whether or not there is demand for parallelization of a particular step in at least one of the PINEAPL industrial applications.

The routines in the PINEAPL Library belong to one of the following classes:

Mesh Partitioning Routines which decompose a given computational mesh into a number of sub-meshes in such a way that certain objective functions (measuring, for instance, the number of edge cuts) are optimized. They are based on heuristic, *multi-level* algorithms similar to those described in [4] and [2].

Sparse Linear Algebra Routines which are required for solving systems of linear equations and eigenvalue problems resulting from discretizing PDE problems. These routines can be classified as follows:

---

[1] Other library material being developed by PINEAPL partners includes optimization, Fast Fourier Transform, and Fast Poisson Solver routines.

Iterative Schemes which are the preferred method for solving large-scale linear problems. PINEAPL Library routines are based on *Krylov subspace methods*, including the CG method for symmetric positive-definite problems, the SYMMLQ method for symmetric indefinite problems, as well as the RGMRES, CGS , BICGSTAB(ℓ) and (TF)QMR methods for unsymmetric problems.

Preconditioners which are used to accelerate the convergence of the basic iterative schemes. PINEAPL Library routines employ a range of preconditioners suitable for parallel execution. These include *domain decomposition*-based, specifically *additive* and *multiplicative Schwarz* preconditioners.[2] Additionally, preconditioners based on classical *matrix splittings*, specifically *Jacobi and Gauss-Seidel* splittings, are provided, and *multicolor orderings* of the unknowns are utilized in the latter case to achieve a satisfactory degree of parallelism [3]. Furthermore, the inclusion of parallel incomplete factorization preconditioners based on the approach taken in [5] is planned.

Basic Linear Algebra Routines which, among other things, calculate sparse matrix-vector products.

Black-Box Routines which provide easy-to-use interfaces at the price of reduced flexibility.

Preprocessing Routines which perform matrix transformations and generate auxiliary information required to perform the foregoing parallel sparse matrix operations efficiently.

In-Place Generation Routines which generate the non-zero entries of sparse matrices concurrently: Each processor computes those entries which – according to the given data distribution – have to be stored on it. In-place generation routines are also provided for vectors which are aligned with sparse matrices.

Distribution/Assembly Routines which (re)distribute the non-zero entries of sparse matrices or the elements of vectors, stored on a given processor, to other processors according to given distribution schemes. They also assemble distributed parts of dense vectors on a given processor, etc.

Additionally, all routines are available for both real and complex data.

3 Performance of Library Routines

For performance evaluation purposes, a simple PDE solver module, which uses a number of sparse matrix routines available in Release 1 of the PINEAPL Library, has been developed. See Krommer [6] for a description of the solver module and the numerical problem it solves. Performance data generated by this module is used in this section to demonstrate the efficiency and scalability of core computational routines, such as the iterative solver, the matrix-vector multiplication and the preconditioner routines.

---

[2] The subsystems of equations arising in these preconditioners are solved approximately on each processor, usually based on incomplete LU or Cholesky factorizations.

3.1 Test Settings

All experiments were carried out for a *small*-size problem configuration, $n_1 = n_2 = n_3 = 30$, resulting in a system of $n = 27\,000$ equations, and for a *medium*-size problem configuration, $n_1 = n_2 = n_3 = 60$, resulting in a system of $n = 216\,000$ equations. Most equations, except those corresponding to boundary grid points, contained seven non-zeros coefficients. The low computation cost of the small problem helps to reveal potential inefficiencies – in terms of their communication requirements – of the library routines tested; whereas a comparison of the results for the two different problem sizes makes it possible to assess the scalability of the routines tested for increasing problem size.

All performance tests were carried out on a Fujitsu AP3000 at the Fujitsu European Centre for Information Technology (FECIT). The Fujitsu AP3000 is a distributed-memory parallel system consisting of UltraSPARC processor nodes and a proprietary two-dimensional torus interconnection network. Each of the twelve computing node at FECIT used in the experiments was equipped with a 167 MHz UltraSPARC II processor and 512 Mb of memory.

3.2 Performance of Unpreconditioned Iterative Solvers

Figure 1 shows the speed-ups of unpreconditioned iterative solvers for the two test problems. The iterative schemes tested – BICGSTAB(1), BICGSTAB(2) and CGS – have been found previously to be the most effective in the unsymmetric solver suite. BICGSTAB(2) and CGS can be seen to scale (roughly) equally well, whereas BICGSTAB(1) scales considerably worse (in particular, for the smaller problem). The results are generally better for the larger problem than for the smaller, especially when the maximum number of processors is used. In this case, the improvement in speed-up is between 10 and 20 %, depending on the specific iterative scheme.

3.3 Performance of Matrix-Vector Multiplication

When solving linear systems using iterative methods, a considerable fraction of time is spent performing matrix-vector multiplications. Apart from its use in iterative solvers, matrix-vector multiplication is also a crucial computational kernel in time-stepping methods for PDEs. It is therefore interesting to study the scalability of this algorithmic component separately.

Figure 2 shows the speed-ups of the matrix-vector multiplication routine in the PINEAPL Library. The results demonstrate a degree of scalability of this routine in terms of both increasing number of processors and increasing problem size. In particular, near-linear speed-up is attained for all numbers of processors for the larger problem.

3.4 Performance of Multicolor SSOR Preconditioners

Multicolor SSOR preconditioners are very similar to matrix-vector multiplication in their computational structure and their communication requirements. In

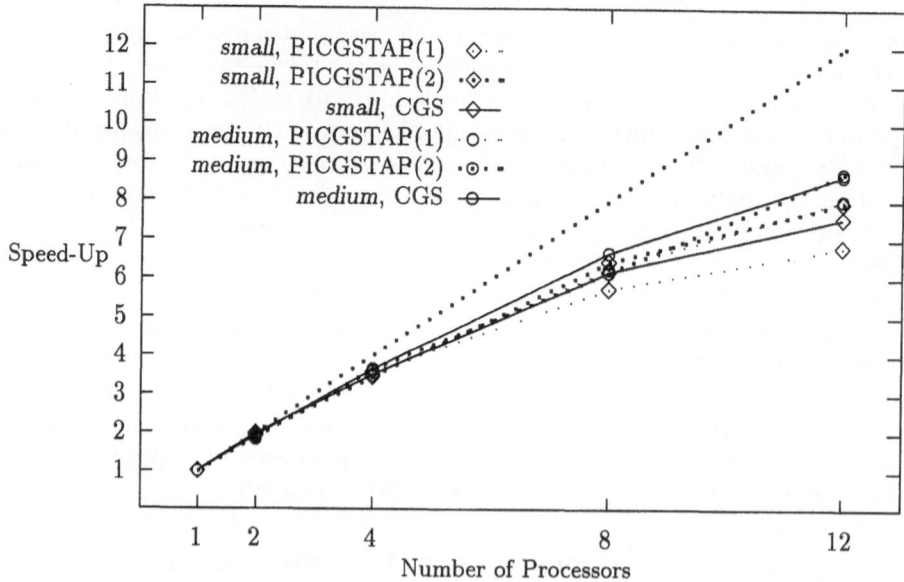

Fig. 1. Speed-Up of Unpreconditioned Iterative Solvers

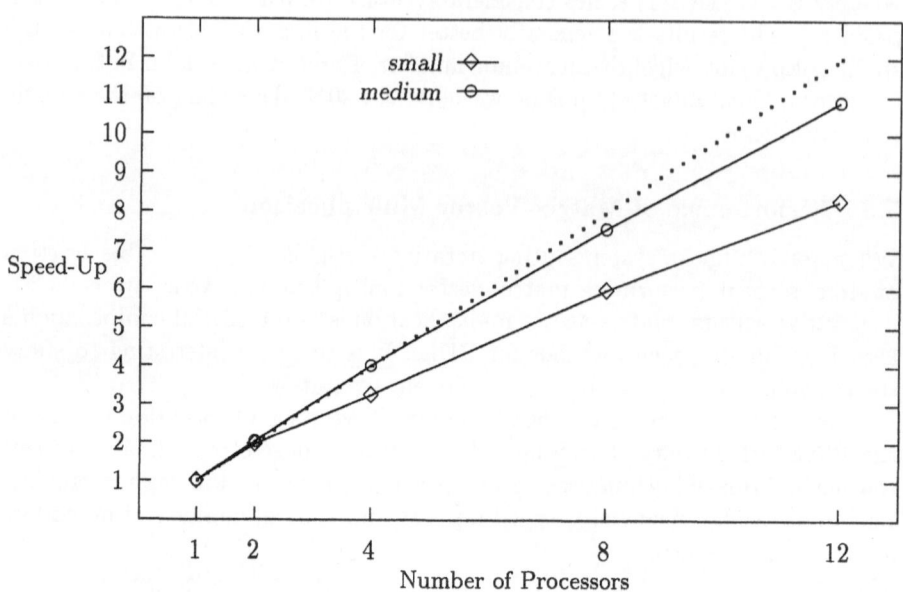

Fig. 2. Speed-Up of Matrix-Vector Multiplication

particular, the two operations are equal in their computation-to-communication ratios. However, the number of messages transferred in multicolor SSOR preconditioners is larger than in matrix-vector multiplication (i. e. the messages are shorter). As a result, one would expect multicolor SSOR preconditioners to exhibit similar, but slightly worse scalability than matrix-vector multiplication. Figure 3 confirms this expectation for configuration with up to eight processors.

Between eight and twelve processors, however, the speed-up curves flatten out significantly. The precise reasons for this uncharacteristic behavior will have to be investigated in future work.

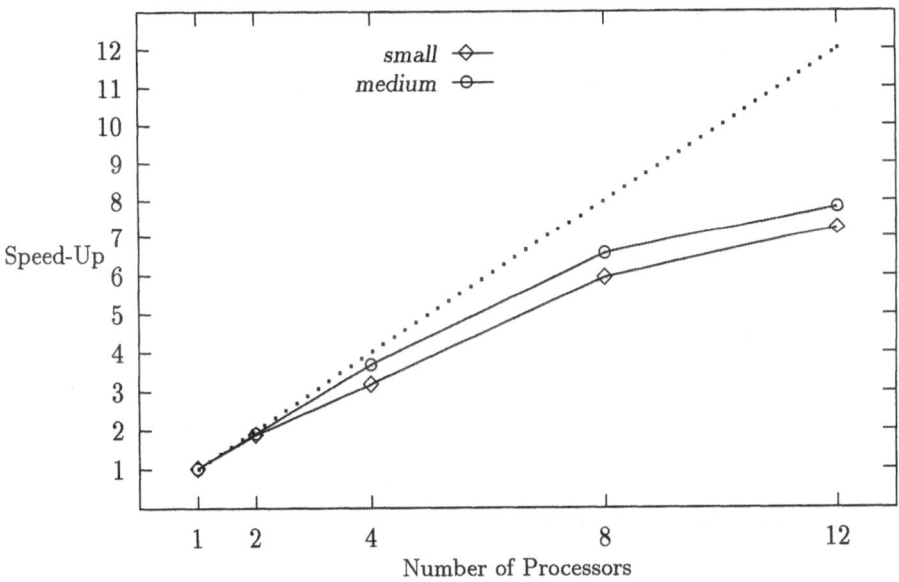

Fig. 3. Speed-Up of Multicolor SSOR Preconditioners

3.5 Performance of Additive Schwarz Preconditioners

The multicolor orderings used in the PINEAPL SSOR preconditioners (see Section 3.4) may vary depending on the number of processors and/or data distribution used. However, experiments have shown that the quality of multicolor SSOR preconditioners is fairly independent of the particular multicolor orderings they are based on: The number of iterations required to achieve convergence does not vary significantly when different multicolor SSOR preconditioners are used in an iterative solver.

With additive Schwarz preconditioners, the situation is very different: The quality of additive Schwarz preconditioners usually decreases when the number of processors – and hence, the number of subdomains – is increased, i. e. the

number of iterations required to achieve converge with a corresponding preconditioned iterative solver increases. Therefore, the performance of additive Schwarz
preconditioners cannot be evaluated separately, but only in conjunction with
their application in iterative solvers.

Figure 4 shows the speed-ups of the BICGSTAB(1) iterative solver in conjunction with several additive Schwarz preconditioners.[3] The additive Schwarz
preconditioners tested differ in the size of the overlap regions between subdomains. In particular, the choice "overlap=0" corresponds to a traditional block
Jacobi preconditioner.

The scalability results are clearly worse than those shown in the previous
sections. In particular, the transition from one to two processors actually results in a performance decrease for the block Jacobi preconditioner. Generally
speaking, the overlapping additive Schwarz preconditioners perform significantly
better than the block Jacobi preconditioner for small numbers of processors; for
larger numbers of processors, the performance gain from using overlap regions
decreases or – for the smaller problem – is completely lost.

In Krommer [6] the *hardware performance*[4] speed-ups corresponding to the
temporal performance[5] speed-ups represented in Figure 4 are investigated. It
turns out that the hardware performance scales significantly better than temporal performance – indeed, the results for the hardware performance are quite
comparable to the results obtained for the unpreconditioned solvers in Figure 1.
This fact demonstrates that (i) the implementation of the additive Schwarz preconditioners is efficient in terms of communication requirements and that (ii) the
main reason for the relatively poor temporal scalability lies in the unsatisfactory
quality of additive Schwarz preconditioners.

4 Conclusion and Outlook

The performance study in Section 3 has established the efficiency and scalability of a number of core computational routines in the sparse linear algebra
modules of Release 1 of the PINEAPL Library. The only main deficiency detected is the poor quality of additive Schwarz preconditioners for two or more
processors when compared to the additive Schwarz preconditioner for a single
processor (=incomplete LU factorization). It is planned to overcome this problem by providing parallel incomplete factorization preconditioners in Release 2
of the PINEAPL Library.

---

[3] The subsystems associated with different domains were approximately solved using
local modified ILU(0) factorizations.

[4] The *hardware performance* of a program is defined as the ratio between the number
of floating-point operations carried out and the run-time of the program.

[5] The *temporal performance* of a program is defined as reciprocal run-time of the
program.

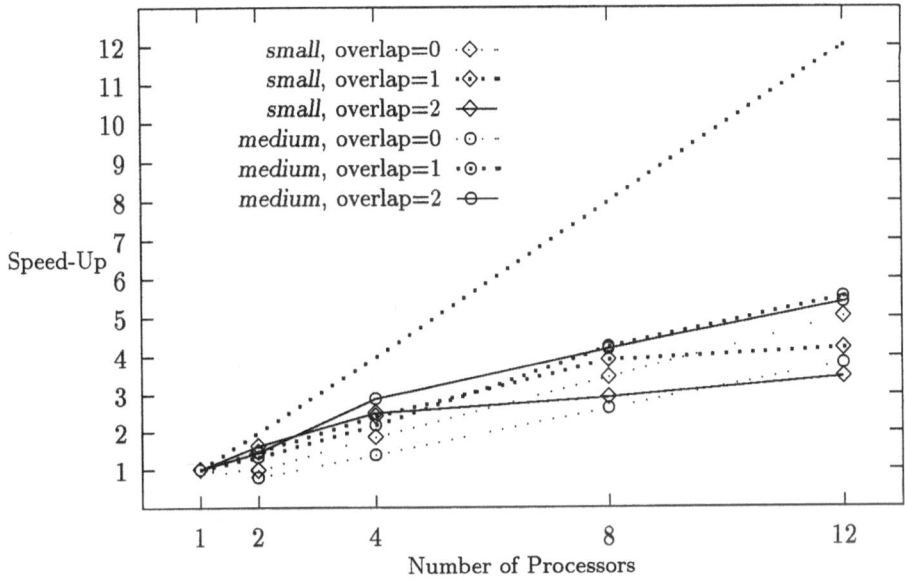

Fig. 4. Speed-Up of PICGSTAB(1) with Additive Schwarz Preconditioners

References

1. M. Derakhshan, S. Hammarling, A. Krommer, *PINEAPL: A European Project on Parallel Industrial Numerical Applications and Portable Libraries*, in "Recent Advances in Parallel Virtual Machine and Message Passing Interface" (M. Bubak et al., Eds.), Vol. 1332 of *LNCS*, Springer-Verlag, Berlin, 1997, pp. 337–342.
2. A. Gupta, *Fast and Effective Algorithms for Graph Partitioning and Sparse Matrix Ordering*, IBM Research Report RC 20496, IBM T.J. Research Center, Yorktown Heights, New York, 1996.
3. M. T. Jones, P. E. Plassman, *Scalable Iterative Solution of Sparse Linear Systems*, Parallel Computing 20 (1994), pp. 753–773.
4. G. Karypis, V. Kumar, *Multilevel k-Way Partitioning Scheme for Irregular Graphs*, Technical Report TR-95-064, Department of Computer Science, University of Minnesota, 1995.
5. G. Karypis, V. Kumar, *Parallel Treshold-Based ILU Factorization*, Technical Report TR-96-061, Department of Computer Science, University of Minnesota, 1996.
6. A. Krommer, *Parallel Sparse Matrix Computations Using the PINEAPL Library: A Performance Study*, Technical Report TR1/98, The Numerical Algorithms Group Ltd, Oxford, UK, 1998.
7. A. Krommer, M. Derakhshan, S. Hammarling, *Solving PDE Problems on Parallel and Distributed Computer Systems Using the NAG Parallel Library*, in "High-Performance Computing and Networking" (B. Hertzberger, P. Sloot, Eds.), Vol. 1225 of *LNCS*, Springer-Verlag, Berlin, 1997, pp. 440–451.

Using a General-Purpose Numerical Library to Parallelize an Industrial Application: Design of High-Performance Lasers[*]

Ida de Bono[1], Daniela di Serafino[1,2] and Eric Ducloux[3]

[1] Center for Research on Parallel Computing and Supercomputers, CPS-CNR, Naples, Italy
[2] The Second University of Naples, Caserta, Italy
[3] Thomson-CSF/LCR, Orsay Cedex, France

Abstract. We describe the development of a parallel version of an industrial application code, used to simulate the propagation of optical beams inside diodes pumped rod lasers. The parallel version makes an intensive use of a 2D FFT routine from a general-purpose parallel numerical library, the PINEAPL Library, developed within an ESPRIT project. We discuss some issues related to the integration of the library routine into the application code and report first performance results.

1 Introduction

The development of high performance lasers is presently a very active field. An industrial challenge is the production of 100W-1KW diodes pumped rod lasers with high compactness, high beam quality and high energy conversion. The numerical simulation of the propagation of optical beams inside these lasers plays a central role in the optimization of the whole design process and leads to a reduction of the development time. On the other hand, the industrial design of rod lasers requires an intensive simulation and an effective use of high-performance computers allows to reduce the so-called "time-to-market". It also gives the possibility of increasing the level of detail in the simulation without increasing the computation time, which is an important requirement in an efficient design process.

In this work we present the main issues concerning the development of a parallel version of an industrial code, used at Thomson-CSF LCR to model the propagation of optical beams in diodes pumped rod lasers. This version, developed for MIMD distributed-memory machines, has been obtained *exploiting software from a general-purpose parallel numerical library*. Numerical software libraries play a significant role in the development of application codes, since they provide building blocks for the solution of computational kernels arising in the modeling of scientific problems. The availability of reliable, accurate and efficient numerical software allows users to develop their application codes without dealing with

---

[*] This work has been supported by the European Commission under the ESPRIT contract 20018 (PINEAPL Project).

details related to the implementation of numerical algorithms, improving at the same time the quality of the computed solutions [2]. In particular, the use of numerical software for High-Performance Computing environments can simplify the porting of application codes to such environments, allowing also the introduction of different mathematical models and numerical techniques, that can better exploit the powerful resources of advanced machines.

This work has been carried out as a part of the PINEAPL (Parallel Industrial NumErical Applications and Portable Libraries) Project, an ESPRIT Project in the area of High-Performance Computing and Networking, with a duration of three years (1996-1998) [3]. The main goal of the project is *to produce a general-purpose library of parallel numerical software suitable for a wide range of computationally intensive industrial applications.* This project is a collaborative effort among academic/research partners (CERFACS (F), CPS-CNR (I), IBM Italia (I), Manchester University (UK) and Math-Tech (DK)) and industrial partners (British Aerospace (UK), Danish Hydraulic Institute (DK), Piaggio Veicoli Europei (I) and Thomson-CSF (F)), under the coordination of NAG (UK). The industrial partners are the end-users of the project and have provided several application codes, that are ported on parallel machines using routines from the PINEAPL Library. A first release of the library is already available and includes numerical routines in the areas of Fast Fourier Transforms (FFTs), Nonlinear Optimization and Sparse Linear Algebra.[1] The parallel version of the rod laser application code, described in this paper, integrates the 2D FFT routine of the PINEAPL Library.

2 Numerical Simulation of Beam Propagation Inside Rod Lasers

The numerical simulation is mainly devoted to describe the evolution of optical beams, which propagate inside the rod laser following a "zig-zag" trajectory (Figure 1). The final goal is to estimate the deformation and the amplification of laser spots, and to find an optimal coupling between laser beams and pumping devices. The complete simulation is performed using three application codes,

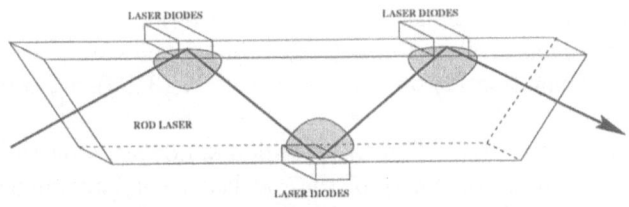

Fig. 1. Schematic picture of a rod laser.

---

[1] For more details see the URL http://www.nag.co.uk/Projects/PINEAPL.html.

modeling the heating of the rod, the propagation of the beams and the transformation due to the reflection on mirrors. The work described here is devoted to the development of a parallel version of the propagation code.

The mathematical model of the propagation of an optical beam is based on the Helmholtz equation:

$$\Delta\Phi + k_0^2 n^2 \Phi = 0 \,,$$

where $\Phi = \Phi(x, y, z)$ is the (polarized) electromagnetic field, n is the index of the medium, $k_0 = 2\pi/\lambda$ is the wave number corresponding to the wavelength λ, and Φ and n have complex values. Let us assume for simplicity that the direction of propagation is z. The fast variation of the field along z is separated from the "modulation", $\Psi(x, y, z)$, leading to an equation of the form:

$$\Delta\Psi + 2i\, k_z \frac{\partial\Psi}{\partial z} + k_0^2(n^2 - n_z^2)\Psi = 0, \tag{1}$$

where k_z is a suitable wave number, $n_z = k_z/n_0$ and $i = \sqrt{-1}$.

The solution of equation (1) is performed using a method in the class of *Beam Propagation Methods* (BPMs) [6], that are widely used in integrated optics. The BPM used here is based on a splitting of equation (1) into two equations, modelling the propagation and the "lens" diffraction, respectively:

$$\Delta_{x,y}\Psi + \frac{\partial^2\Psi}{\partial z^2} + 2i\, k_z \frac{\partial\Psi}{\partial z} = 0, \tag{2}$$

$$2i\, k_z \frac{\partial\Psi}{\partial z} + k_0^2(n^2 - n_z^2)\Psi = 0, \tag{3}$$

where $\Delta_{x,y}$ is the Laplace operator in the (x, y)-plane. The solution is obtained solving in turn equations (2) and (3) along the z-direction, that is repeating the following steps:

- propagation from z to $z + dz/2$ (eq. (2));
- lens diffraction from z to $z + dz$ (eq. (3));
- propagation from $z + dz/2$ to $z + dz$ (eq. (2)).

Each propagation step is carried out performing a transformation from the (x, y)-plane to the 2D frequency space, where the solution is known analytically, and coming back to the (x, y)-plane. The transformations are obtained as direct and inverse Discrete Fourier Transforms (DFTs) on complex data arising from the discretization of equation (2) on a uniform grid. The DFTs are computed using FFT algorithms.

We note that the computational (x, y)-domain depends on the angle of incidence of the beam at the entry of the rod laser. For large angles, the beam can cross the boundaries of the rod; in this case the domain is *unfolded*, i.e. it is reflected with respect to the crossed boundaries, in x- or y-direction or both, and the computation is performed in a domain that is twice or four times the original domain.

3 Integration of the Parallel 2D FFT Routine

Until now, most of the work has been carried out on a simplified version of the propagation code, named PROBLA, which neglects the coupling with thermal effects and some physical details that increase the complexity of the model. This version is considered as a testbed, to evaluate efforts and improvements concerning the integration of the parallel 2D FFT of the PINEAPL Library. The experience gained in working on the simplified code will be exploited to develop a parallel version of the complete industrial code.

PROBLA is written in Fortran 77, apart from some routines for the dynamical allocation of memory, that are written in C. Its structure is sketched in Figure 2. The bulk of computation is in step 5.1, as it is shown by the plots

---

1. *read input parameters;*
2. *initialize the electro-magnetic field;*
3. *compute the mean index of the medium;*
4. *compute the energy of the beam;*
5. *for i = 1, visual_steps*
    5.1 simulate the beam propagation (BPM);
    5.2 rearrange the data to be visualized;
    endfor
6. *reconstruct the real field from the computational field;*
7. *compute the energy of the beam;*
8. *output the electro-magnetic field to be visualized.*

Fig. 2. Structure of the PROBLA code.

of the execution times reported in Section 4. This step implements the BPM method and hence makes an intensive use of 2D FFTs. The BPM is applied between two successive visualization positions, that is positions along the direction of propagation where selected values of the electro-magnetic field are required for a future visualization of the evolution of the beams along that direction.

Taking into account the above considerations, a speedup can be achieved parallelizing the execution of each FFT. Hence, our work has been devoted to the integration of the parallel 2D FFT routine of the the PINEAPL Library into the application code. This routine performs a complex-to-complex FFT on a 2D array of data, distributed in *row-block* form over a 2D logical grid of processors, i.e. assigning approximately the same number of consecutive rows of the array to each processor, moving row by row on the processor grid. In the rod laser application code, such data distribution corresponds to partitioning the (x, y) computational grid along the x-direction, that is into vertical strips. The algorithm implemented in the parallel FFT routine is the so-called *four-step* algorithm, that is: 1D FFTs along the rows of the 2D array, transpose,

1D FFTs along the rows, and transpose. All the communication is performed in the transpose steps, and the second transpose can be avoided when a direct transform followed by an inverse transform must be executed, as it is in the PROBLA code. All the communication is performed using the Basic Linear Algebra Communication Subprograms (BLACS) [4], as in most of the PINEAPL Library routines. More details are given in [1].

An efficient integration of the parallel FFT routine requires the introduction of the row-block distribution in the whole application code, to avoid unnecessary data movements among different stages of the simulation. Therefore, the different steps of PROBLA have been analyzed and modified to use this data distribution. In the BPM step (5.1), the solution in the frequency space and the computation of the lens diffraction are local processes, and hence are performed without any communication among processors, doing only minor changes in the corresponding parts of the application software. A few changes have been introduced to rearrange the data to be visualized (step 5.2), although this step does not require any communication. The initialization of the electro-magnetic field (step 2) is local too, and has been only slightly modified to be run in parallel. Global communications are required in some initial and final steps, such as the computation of the mean index of the medium (step 3) and of the energy of the beam (steps 4 and 7). These steps have been reduced to *global sum* operations, that are executed using *ad hoc* routines from the BLACS. Modifications have been introduced to reconstruct the real field from the computational field (step 6), when the unfolding along the x-direction is performed. In this case, processors holding rows that are symmetric with respect to the central row of the computational domain, exchange data to reconstruct the field in the effective domain. The input/output has not been parallelized and is performed by only one processor, which distribute to the others the input parameters and gathers the data to be printed out. This allows to have the same input/output files as in the sequential code; moreover, it has been observed that generally the input/output does not account for a large percentage of the total execution time (see Section 4).

4   Numerical Experiments

First experiments have been carried out to evaluate the efficiency of the parallel version of the rod laser application code, using two sets of benchmarks of interest for Thomson-CSF. In the following, they are referred to as Benchmark 1 and Benchmark 2. In both benchmark sets, the number of grid points in the y-direction, n_y, is kept fixed, while the number of grid points in the x-direction, n_x, varies. In Benchmark 1 $n_x = 64, 128, 256, 512, 1024$ and $n_y = 64$; in Benchmark 2 $n_x = 64, 128, 256, 512, 1024$ and $n_y = 128$. Moreover, the increment along z is $dz = 10\,\mu m$, the number of visualization steps, i.e. the number of times the BPM step is repeated, is 100, and the domain is unfolded along x, thus the number of grid points in this direction is doubled.

The experiments have been performed on an IBM SP2, at CPS-CNR. This machine has 12 Power2 Super Chip thin nodes (160 MHz), each with 512 MBytes of memory and a 128 Kbytes L1 data cache; the nodes are connected via an SP switch, with a peak bi-directional bandwidth of 110 MBytes/sec. The SP2 has the AIX 4.2.1.operating system; the BLACS 1.1 are implemented on the top of the IBM proprietary version 2.3 of the MPI message-passing library [5]. The Fortran part of the parallel code has been compiled with the XL Fortran compiler, vers. 4.1, and the C part with the XL C compiler, vers. 3.1.4.

Figure 3 shows the execution times, in seconds, of a single BPM step and of the whole PROBLA (including input/output), on 2, 4, 8, 12 processors, for Benchmark 1. Remembering that a BPM step is performed 100 times, we see

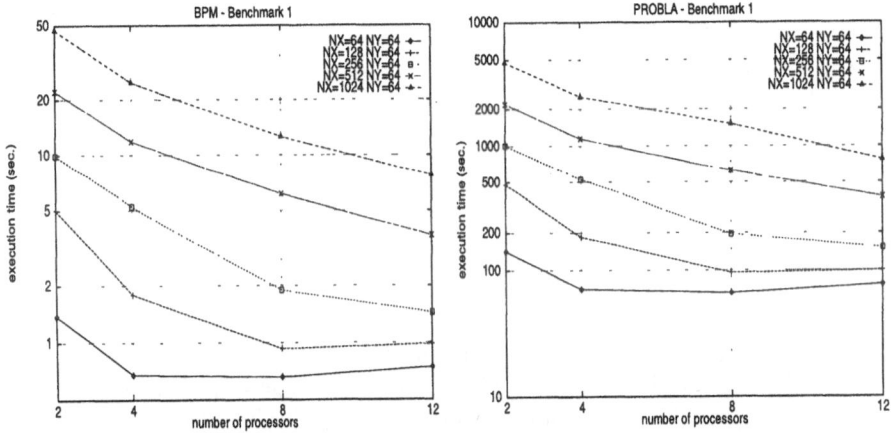

Fig. 3. Execution times of a BPM step and of PROBLA for Benchmark 1.

that it accounts for a very large percentage of the total execution time. This percentage decreases as the number of processors increases; for example, it is more than 98% on 2 processors and reaches the lowest value of 95% on 12 processors. One reason for such reduction is the load inbalance generated by unfolding along the x-direction. In this case, the rearrangement of the data to be visualized (Figure 2, step 5.2) and the reconstruction of the real field from the computational field (Figure 2, step 6) are mainly performed by half the processors, that is by the processors that hold the effective domain, and hence the distribution of the computational work is not balanced. One more reason is the sequential input/output; its time is at most 1.5% of the total execution time on 2 processors, while it reaches 3.3% on 12 processors.

Figure 4 shows the speedup of a BPM step and of PROBLA (including input/output) with respect to the original sequential code, for Benchmark 1.[2]

---

[2] The speedup is here defined as $S_p = T_1/T_p$, where T_1 is the execution time of the original sequential code on one processor and T_p is the execution time of the parallel code on p processors, for the same problem.

The parallel code has been compared with the sequential code to see the effective gain obtained with the parallelization. The speedup lines of a BPM step and of

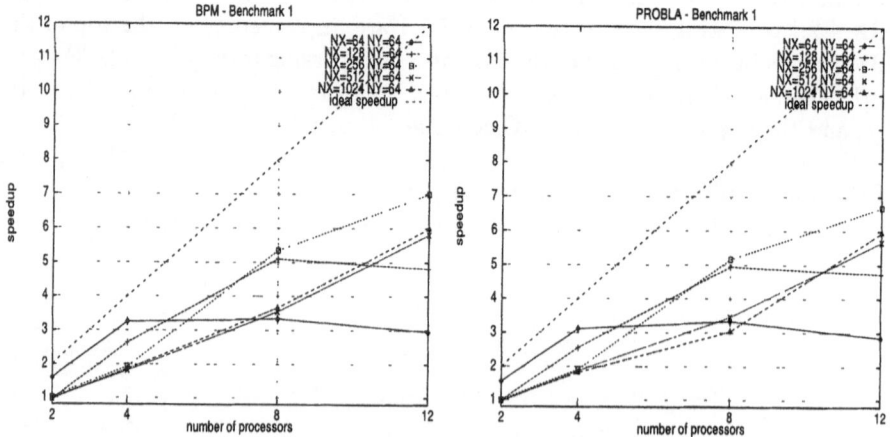

Fig. 4. Speedup of a BPM step and of PROBLA for Benchmark 1.

PROBLA have similar behaviours; the values concerning **PROBLA** are generally slightly lower, according to the previous considerations on the execution times. The speedup plots show that for each number of processors there is a "best problem size" where the parallel code gets the largest speedup. For example, the largest speedup of the BMP step on 4 processors is obtained for $n_x = 64$ and is 3.25, while on 8 processor it is obtained for $n_x = 256$ and is 5.35 (the value of n_y is not considered because it is constant). On the other hand, we can see that there is a "best number of processors" for each problem size, where the efficiency reaches the largest value.[3] For example, for the BPM step with $n_x = 64$ such best number is 4, with an efficiency of 0.78, while for $n_x = 256$ it is 8 with an efficiency of 0.67. Using more processors than the optimal number does not increase the performance, as it is evident for $n_x = 64$ and $n_x = 128$. Experiments on more than 12 processors should be performed to confirm this on problems with larger sizes. The behaviour discussed above is due to the use of the cache memory and to the communications performed in the **parallel global transpose** operation inside the FFT routine. Finally, we note that there is no gain in terms of execution time where only 2 processors are used, except for $n_x = 64$. This is because the sequential FFT used in the original code is generally faster than the parallel FFT on one processor and is comparable with it on 2 processors. This is probably due to the fact that the sequential FFT routine does not perform transpositions of data as the parallel FFT routine; moreover, the sequential FFT has been implemented specifically for radix-2 transforms, that are the transforms

---

[3] The efficiency is defined as S_p/p, i.e. as the speedup divided by the number of processors.

considered in PROBLA, while the parallel FFT has been implemented to perform more general mixed-radix transforms.

The previous considerations apply also to the execution times and the speedup of BPM and PROBLA for Benchmark 2, shown in Figures 5 and 6. Since the

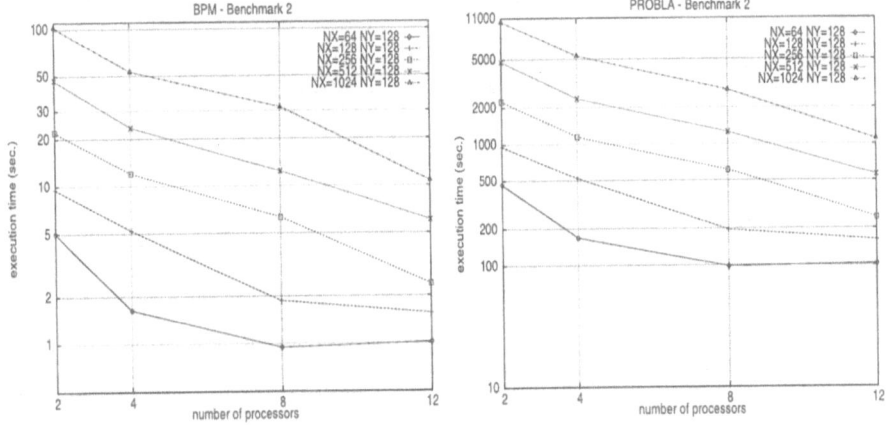

Fig. 5. Execution times of a BPM step and of PROBLA for Benchmark 2.

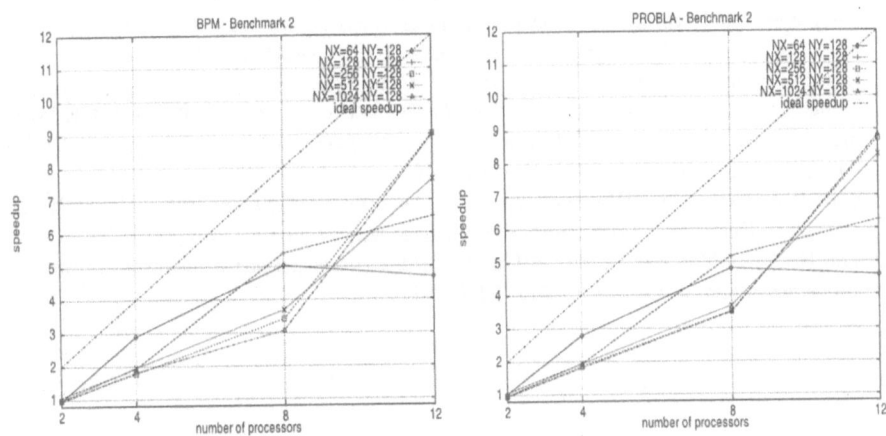

Fig. 6. Speedup of a BPM step and of PROBLA for Benchmark 2.

value of n_y has been doubled with respect to Benchmark 1, the best number of processors after which the performance degrades is generally larger, and is visible only for the cases $n_x = 64$ and $n_x = 128$. Moreover, the speedup values are generally higher than in Benchmark 1; on 12 processors, the BPM step reaches a speedup of 9 and the whole PROBLA code a speedup of 8.

5 Conclusions

A parallel version of an industrial application code has been developed, which integrates the 2D FFT routine of the parallel numerical library developed in the ESPRIT Project PINEAPL. First experiments have been performed to evaluate the performance of the parallel application code, obtaining satisfactory speedup values.

Work is presently devoted to evaluate the scalability of the parallel code. Preliminary tests have been performed on a single BPM step, obtaining encouraging results; values of scaled efficiency[4] between 0.75 and 0.84 have been measured on 4 and 8 processors, using problems of size $n_x \times n_y$, with $n_x/p = 256, 512, 1024$ and $n_y = 128$.

References

1. Carracciuolo, L., de Bono, I., De Cesare, M.L., di Serafino, D., Perla, F.: Development of a Parallel Two-dimensional Mixed-radix FFT Routine. Tech. Rep. TR-97-8, Center for Research on Parallel Computing and Supercomputers (CPS) - CNR, Naples, Italy (1997)
2. di Serafino, D., Maddalena, L., Messina, P., Murli, A.,: Some Perspectives on High-Performance mathematical Software. In A. Murli and G. Toraldo eds., Computational Issues in High Performance Software for Nonlinear Optimization, Kluwer Academic Pub. (to appear)
3. di Serafino, D., Maddalena, L., Murli, A.,: PINEAPL: a European Project to Develop a Parallel Numerical Library for Industrial Applications. In Euro-Par'97 Parallel Processing, C. Lengauer, M. Griebl and S. Gorlatch eds., Lecture Notes in Computer Science **1300** (1997), Springer, 1333–1339.
4. Dongarra, J.J., Whaley, R.C.: A Users' Guide to the BLACS v 1.0. LAPACK Working Note No. 94, Technical Report CS-95-281, Department of Computer Science, University of Tennessee, 107 Ayres Hall, Knoxville, TN (1995)
5. Snir, M., Otto, S.W., Huss-Lederman, S., Walker, D.W., Dongarra, J.J.: MPI: The Complete Reference. MIT Press, Cambridge, MA (1996)
6. Van Roey, J., Van der Donk, J., Lagasse, P.E.: Beam-propagation method: analysis and assessment. J. Opt. Soc. Am. **71** (1981)

---

[4] The scaled efficiency is here defined as $SE_p = T_1(n_x, n_y)/T_p(p \times n_x, n_y)$, where $T_i(r, s)$ is the execution time of the parallel code on i processors, for a problem of size $r \times s$.

Fast Parallel Hermite Normal Form Computation of Matrices over $\mathbb{F}[x]$

Clemens Wagner

Rechenzentrum der Universität Karlsruhe, Germany
clemens@exp-math.uni-essen.de

Abstract. We present an algorithm for computing the Hermite normal form of a polynomial matrix and an unimodular transformation matrix on a distributed computer network. We provide an algorithm for reducing the off-diagonal entries which is a combination of the standard algorithm and the reduce off-diagonal algorithm given by Chou and Collins. This algorithm is parametrised by an integer variable.
We provide a technique for producing small multiplier matrices if the input matrix is not full-rank, and give an upper bound for the degrees of the entries in the multiplier matrix.

1 Introduction

We study the problem of computing the Hermite normal form and an unimodular transforming matrix of a large, rectangular input matrix over $\mathbb{F}[x]$. Our motivation comes from the area of convolutional encoders where we need the transforming matrix to construct the decoder of a given minimal, basic encoder (cf. [6]). Since the degrees of the entries in the transforming matrix determine the size of the decoder's circuit, we are interested in a transforming matrix whose entries have small degrees.

Definition 1. *A matrix* $H \in \text{Mat}_{m \times n}(\mathbb{F}[x])$ *with rank* $r \in \mathbb{N}_0$ *is in* Hermite normal form (HNF) *if*

1. *the first* r *columns of* H *are nonzero, and the last* $n - r$ *columns are zero.*
2. *for* $1 \leq j \leq r$ *let* i_j *be the index of the first nonzero entry in the* j*th column. Then* $1 \leq i_1 < i_2 < \ldots < i_{r-1} < i_r \leq m$.
3. *the pseudo diagonal entries* $H_{i_j,j}$ *for* $1 \leq j \leq r$ *are monic.*
4. *for* $1 \leq j \leq r$ *and* $1 \leq k < j$ *the degree of the off-diagonal entries* $H_{i_j,k}$ *is less than* $\deg(H_{i_j,j})$.

We call $1 \leq i_1 < i_2 < \ldots < i_{r-1} < i_r \leq m$ *the* row indices *of the pseudo diagonal entries of* H*. If only the first two conditions are fulfilled we say* H *is in* echelon form.

Two matrices $A, B \in \text{Mat}_{m \times n}(\mathbb{F}[x])$ are *right equivalent* if there exists an *unimodular* Matrix $V \in \text{GL}_n(\mathbb{F}[x])$ such that $AV = B$. We call V a *transforming matrix* or *multiplier*.

Theorem 1. *For each matrix $A \in \mathrm{Mat}_{m \times n}(\mathbb{F}[x])$ exists an unique, right equivalent matrix $H \in \mathrm{Mat}_{m \times n}(\mathbb{F}[x])$ in Hermite normal form.*

There are many well-known algorithms for computing the Hermite normal form and a multiplier of an integer matrix. *Gaussian elimination based methods*, as in Sims [13, p. 323], eliminate above diagonal entries of a whole row by extended greatest common divisor computations. There are many different specialisations of this general algorithm (e. g. [2], [4], [8]) which can be generalised to arbitrary Euclidean domains [14].

Gaussian elimination has worst-case exponential time and space complexity for matrices over $\mathbb{F}_{2^k}[x]$ and $\mathbb{Q}[x]$ [9]. But it was shown by Wagner [14], [15] that some variants of Gaussian elimination have a better practical performance than algorithms guaranteeing a polynomial time and space complexity.

2 Parallel Implementation

A *parallel computer* $\mathcal{P} := \{\pi_0, \ldots, \pi_{N-1}\}$ consists of N processors with distinct memory and a communication network. Let $A \in \mathrm{Mat}_{m \times n}(\mathbb{F}[x])$, and let

$$\ell(l, k) := \left\lfloor \frac{l}{N} \right\rfloor + \begin{cases} 1 & \text{if } k < (l \bmod N) \\ 0 & \text{otherwise} \end{cases}$$

for each $0 \leq k \leq N - 1$. Each processor π_k on the parallel computer stores a $(\ell(m, k)+1) \times n$ matrix $A^{(k)}$ and a $\ell(n, k) \times n$ matrix $V^{(k)}$ which are submatrices of A and the multiplier $V \in \mathrm{GL}_n(\mathbb{F}[x])$, respectively. The additional row $\ell(m, k)+1$ of $A^{(k)}$ is used for computations. We will refer to this row as the *computation row*.

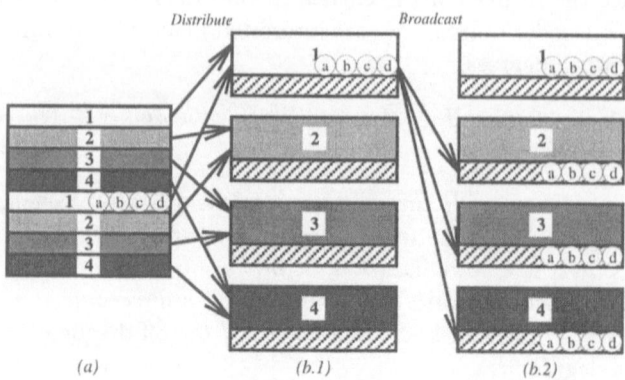

Fig. 1. Distribution of a matrix on a parallel computer (a & b.1), Broadcasting a part of a row (b.1 & b2)

We distribute the input matrix A by storing the ith row of A in row i' of matrix $A^{(k)}$ on processor π_k where $k := (i - 1) \bmod N$ and $i' := \lfloor (i - 1)/N \rfloor + 1$

for $1 \leq i \leq m$. Note: The computation row of every $A^{(k)}$ will not be set here. We construct the identity matrix I_n in the matrices $V^{(k)}$ in a corresponding way.

Figure 1 shows how a matrix with eight rows (a) is distributed on four processors of a parallel computer (b.1). In contrast to Gaussian elimination over fields Gaussian elimination over Euclidean domains needs several steps for eliminating the entries in a row. Over Euclidean domains we have just division with remainder instead of exact division. This is the reason why we distribute rows instead of columns on the parallel computer.

The main HNF algorithm is given by pseudocode in Algorithm 1.

PARALLEL-HNF$(A^{(k)}, V^{(k)}, b)$
1   **input** k *processor index*
         $A^{(k)}$ *π_k-part of a distributed $m \times n$ matrix A*
         $V^{(k)}$ *π_k-part of the distributed $n \times n$ identity matrix V*
         b non-negative integer
2     $r \leftarrow 0$
3   **for** $i \leftarrow 1$ **to** m **do**
4       $g \leftarrow (i-1) \bmod N$
5   **if** $g = k$ **then**
6         $t \leftarrow \lfloor (i-1)/N \rfloor + 1$
7   **broadcast** $(A_{t,r+1}^{(k)}, \ldots, A_{t,n}^{(k)})$ to all other
8   **else**
9         $t \leftarrow \ell(m,k) + 1 \; \triangleright$ *Index of computation row*
10   **receive** $(A_{t,r+1}^{(k)}, \ldots, A_{t,n}^{(k)})$ from π_g
11        $A^{(k)} \leftarrow$ COMPUTE-GCD$(A^{(k)}, V^{(k)}, t, r+1, n)$
12   **if** $A_{t,r+1}^{(k)} \neq 0$ **then**
13          $r \leftarrow r+1$
14          $i_r \leftarrow i$
15     $(A^{(k)}, V^{(k)}) \leftarrow$ PARALLEL-ROD$(A^{(k)}, V^{(k)}, r, b, i_1, \ldots, i_r)$
16   **return** $(A^{(k)}, V^{(k)})$

Algorithm 1. Parallel HNF computation

The matrix $A^{(k)}$ is the *π_k-part* of A which means the rows of A (and the computation row) lying on processor $\pi_k \in \mathcal{P}$. This algorithm is executed on each processor in $\mathcal{P}$ with the appropriate $A^{(k)}, V^{(k)}$ and a non-negative integer b as input. We explain the meaning of b below. In the description of the algorithm we refer to A and V, respectively, if we want to consider the whole matrix consisting of the π_k-parts on each processor.

In Line 4 we compute the index g of the processor which stores row $A_{i,*}$. Processor π_g computes in Line 6 the index t of the row of $A^{(g)}$ which corresponds to row $A_{i,*}$, and broadcasts the entries $A_{i,r+1}, \ldots, A_{i,n}$ to each other processor.

Broadcasting a row to the computation rows of the other processors is illustrated in Figure 1 (b.1) and (b.2). The function COMPUTE-GCD takes two matrices B and C with n columns, an integer $i \geq 1$ and $1 \leq j_1 \leq j_2 \leq n$ as input, where i is a valid row index of B. It produces right equivalent matrices

B' and C' as output where

$$B'_{i,j_1} = \gcd(B_{i,j_1}, \ldots, B_{i,j_2}) \text{ and } B'_{i,j} = 0$$

for $j_1 < j \leq j_2$. This algorithm computes B' from B by *unimodular column operations* (*swapping* of two columns, *multiplying* a column with a unit, and *adding a multiple of one column to another column*) and does the same transformations on C for computing C'. There are several different methods to obtain B' from B (e. g. [1], [3], [7], [14]).

The variable r corresponds to the number of linearly independent rows of A with index less than i. After execution of the for-loop r is the rank of A, $i_1, \ldots, i_r$ are the indices of the pseudo diagonal entries of A, and A is in echelon form.

The technique of computation rows make it easy to reuse already implemented sequential algorithms for the gcd computation. If we use a gcd algorithm whose execution depends only from the entries in row $A_{i,*}$ we need no communication for function COMPUTE-GCD. The algorithms of Blankinship [1], Bradley [3] or Majewski-Havas [12] and the Multiple-Remainder-GCD as in Wagner [14] are examples for that.

The Norm-Driven-Sorting-GCD due to Havas-Majewski-Matthews [8] needs communication for computing column norms of A, and it was shown by examples in Wagner [14], that this algorithm is not useful for parallel HNF computation.

3   Reducing the Off-Diagonal Entries

After leaving the for-loop in Algorithm 1 the current matrix A is in echelon form, and the current V is a multiplier matrix. There are two important algorithms to obtain the Hermite normal form from a matrix in echelon form.

There is a row based (standard), as in [11], [13] and a column based method, due to Chou-Collins [5]. Algorithm 2 is a hybrid technique which is controlled by parameter $b \in \mathbb{N}_0$. If we choose b equal to zero the algorithm does not change the input matrices. For $b = 1$ this algorithm is a parallel variant of the reduction method of Chou-Collins, and if $b + 1 \geq r = \text{rank}(A)$ we have a parallel variant of the standard reduction algorithm.

The function MONIC-MULTIPLIER takes $f \in \mathbb{F}[x]$ as input and gives

$$\begin{cases} \frac{1}{a} & \text{if } f \neq 0 \text{ and } a \text{ is the leading coefficient of } f \\ 1 & \text{if } f = 0 \end{cases}$$

as output. The calls of this function are needed to guarantee point 3. in the definition of the Hermite normal form. For $f, g \in \mathbb{F}[x] \setminus \{0\}$ the expression f/g is equal to (the unique) $q \in \mathbb{F}[x]$ given by $f = g \cdot q + r$ with $\deg(r) < \deg(g)$ where $\deg(0) := -\infty$.

Algorithm 2 parts the input matrix into stripes of width b, and it reduces all off-diagonal entries in each stripe. The *macro* reduction order of the stripes is from right to left, and the *micro* reduction order reduces the entries in the stripes

PARALLEL-ROD$(A^{(k)}, V^{(k)}, r, b, i_1, \ldots, i_r)$

1   **input** k *processor index*
          $A^{(k)}$ π_k-*part of a distributed* $m \times n$ *matrix* A
          $V^{(k)}$ π_k-*part of a distributed* $n \times n$ *matrix* V
          $r, b, i_1, \ldots, i_r$ *non-negative integers*

2   **if** $r > 0$ **and** $b > 0$ **then**

3       $f \leftarrow r$

4   **repeat**

5         $l \leftarrow f - 1$

6         $f \leftarrow \max\{1, f - b\}$

7   **for** $j \leftarrow f + 1$ **to** r **do**

8           $g \leftarrow (i_j - 1) \bmod N$

9           $h \leftarrow \min\{j - 1, l\}$

10   **if** $g = k$ **then**

11            $t \leftarrow \lfloor (i_j - 1)/N \rfloor + 1$

12   **broadcast** $(A_{t,f}^{(k)}, \ldots, A_{t,h}^{(k)}, A_{t,j}^{(k)})$ **to all other**

13   **else**

14            $t \leftarrow \ell(m, k) + 1$ ▷ *Index of computation row*

15   **receive** $(A_{t,f}^{(k)}, \ldots, A_{t,h}^{(k)}, A_{t,j}^{(k)})$ **from** π_g

16   **for** $s \leftarrow f$ **to** h **do**

17            $q \leftarrow A_{t,s}^{(k)} / A_{(k),j}^{t}$

18            $A_{*,s}^{(k)} \leftarrow A_{*,s}^{(k)} - q \cdot A_{*,j}^{(k)}$

19            $V_{*,s}^{(k)} \leftarrow V_{*,s}^{(k)} - q \cdot V_{*,j}^{(k)}$

20   **if** $j \leq l + 1$ **then**

21            $A_{*,j}^{(k)} \leftarrow \text{MONIC-MULTIPLIER}(A_{t,j}^{(k)}) A_{*,j}^{(k)}$

22            $V_{*,j}^{(k)} \leftarrow \text{MONIC-MULTIPLIER}(A_{t,j}^{(k)}) V_{*,j}^{(k)}$

23   **until** $f = 1$

24       $\mu \leftarrow (i_1 - 1) \bmod N$

25   **if** $\mu = \pi_k$ **then**

26         $t \leftarrow \ell(m, k) + 1$

27   **broadcast** $A_{t,1}^{(k)}$ **to all other**

28   **else**

29         $t \leftarrow \lfloor (i - 1)/N \rfloor + 1$

30   **receive** $A_{t,1}^{(k)}$ **from** μ

31       $A_{*,1}^{(k)} \leftarrow \text{MONIC-MULTIPLIER}(A_{t,1}^{(k)}) A_{*,1}^{(k)}$

32       $V_{*,1}^{(k)} \leftarrow \text{MONIC-MULTIPLIER}(A_{t,1}^{(k)}) V_{*,1}^{(k)}$

33   **return** $(A^{(k)}, V^{(k)})$

Algorithm 2. Parallel reduction of off-diagonal entries

from top to bottom. The macro reduction order is similar to the order of Chou-Collins, while the micro reduction order is similar to the standard reduction order.

Figure 2 shows the reduction order of the standard method (a), hybrid with $b = 2$ (b) and for Chou-Collins (c), where the consecutive numbers in the squares describe the order of reduction.

The reduction order of Chou-Collins needs smaller degrees than the standard reduction order, since the method of Chou-Collins uses only reduced column vectors of the input matrix for reducing other columns. On the other hand the standard order needs less communication operations (broadcasts) which we will show in the next section.

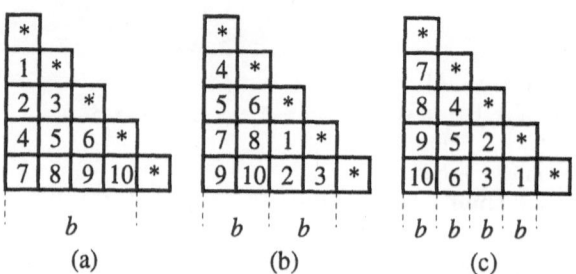

Fig. 2. Reducing of diagonal entries: (a) Standard, (b) hybrid, (c) Chou-Collins

4 Analysis and Performance Examples

Lemma 1. *Given a matrix* $A \in \mathrm{Mat}_{m \times n}(\mathbb{F}[x])$ *with* $r := \mathrm{rank}(A)$ *and* $1 \le b \le r - 1$. *Let* $q := \lfloor r - 2/b \rfloor$. *Computing the Hermite normal form with Algorithm 1 needs at most*

$$m + \frac{q(q+1)}{2}b + r$$

broadcasts if the implementation of COMPUTE-GCD *needs no communication.*

Proof. Algorithm 1 needs m broadcasts at most. The repeat loop in Algorithm 2 is executed $q + 1$ times. In the last run of this loop f is equal to 1 and it does $r - 1$ broadcasts. In the ith run for $1 \le i \le q$ f is equal to $r - ib$ and it does $r - f = ib$ broadcasts. There is one further broadcast after the loop, and we get

$$m + r - 1 + \left(\sum_{i=1}^{q} ib \right) + 1 = m + \frac{q(q+1)}{2}b + r$$

broadcasts. □

For a $m \times n$ matrix with rank r Algorithm 1 needs $m + r$ broadcasts if $b \ge r - 1$. We could reduce the number of broadcasts for $b \ge r - 1$.

Algorithm 3 *We modify Algorithm 1 in the following way.*

– *Replace Line 7 by* **broadcast** $A_{t,*}^{(k)}$ **to all other**

- *Replace Line 10 by* **receive** $A_{t,*}^{(k)}$ **from** π_g
- *Insert the following lines after Line 13:*

$A_{*,r}^{(k)} \leftarrow$ Monic-Multiplier$(A_{t,r}^{(k)})A_{*,r}^{(k)}$

$V_{*,r}^{(k)} \leftarrow$ Monic-Multiplier$(A_{t,r}^{(k)})V_{*,r}^{(k)}$

for $j \leftarrow 1$ **to** $r - 1$ **do**

    $q \leftarrow A_{t,j}^{(k)}/A_{t,r}^{(k)}$

    $A_{*,j}^{(k)} \leftarrow A_{*,j}^{(k)} - qA_{*,r}^{(k)}$

    $V_{*,j}^{(k)} \leftarrow V_{*,j}^{(k)} - qV_{*,r}^{(k)}$

- *Remove Line 15.*

Lemma 2. *Algorithm 3 needs at most m broadcasts for computing the Hermite normal form and a transforming matrix of a $m \times n$ input matrix.*

It seems that Algorithm 3 is faster than Algorithm 1 with $b < r-1$ in practice where r is the rank of the input matrix. We give some running time examples showing that Algorithm 3 needs larger degrees than Algorithm 1 with $b = 1$ for the HNF computation.

We have implemented the algorithms in C/C++ on the IBM SP2 at the GMD in Sankt Augustin. We have used the xlC compiler and the message passing library MPL (both IBM products). We have used the Sorting-GCD algorithm, due to Majewski-Havas [12], for the implementation of the Compute-GCD function, where we used a heap for determining the polynomial with the largest degree in a subvector.

The input matrix is a random 80×80 matrix over $\mathbb{F}_2[x]$, and the degree of the entries are less than or equal to 80. The rank of this matrix is $r = 80$.

Table 1 shows the results of our measurements. We have used 8, 16 and 24 nodes of the SP2. The rows in Table 1 have the following meaning:

<u>b:</u> the value of the parameter b for Algorithm 1. For $b = 80$ we have taken Algorithm 3.

<u>max. degree:</u> The largest degree of all polynomials which occurred during the computation. This value is independent of the number of processors.

<u>$N = x$:</u> Running-time in *minutes:seconds.hundredth seconds* on the SP2 using x processors. We have used the high performance switch and us-protocol for the communication.

<u>t_1/t_b:</u> The fraction of the running time for the b in this column and the running time for $b = 1$.

For $b = 1$ Algorithm 1 runs faster than for every other $b > 1$. The relative running times t_b/t_1 become better if the number of processors become larger (with exceptions for $b \in \{2, 4\}$), since the time for a broadcast depends on the number of processors.

5 Producing Small Multipliers

For every nonzero $A \in \text{Mat}_{m \times n}(\mathbb{F}[x])$ we denote the largest degree of the entries of A as $\|A\| := \max\{\deg(A_{i,j}) \mid 1 \leq i \leq m \wedge 1 \leq j \leq n\}$.

Table 1. Run-times for Algorithm 1 for a 80×80 matrix over $\mathbb{F}_2[x]$ depending on parameter b and the number of processors.

b	1	2	4	8	17	20	40	60	80
max. degree	9965	12571	13129	19306	27227	29009	36802	40989	43486
$N = 8$	2:11.77	2:36.02	3:19.45	4:29.39	6:06.97	6:52.61	8:47.29	9:36.34	11:40.47
t_b/t_1	1.00	1.18	1.51	2.04	2.78	3.13	4.00	4.37	5.32
$N = 16$	1:36.01	1:46.99	2:09.78	2:46.17	3:40.84	4:05.44	5:08.17	5:44.60	7:04.60
t_b/t_1	1.00	1.11	1.35	1.73	2.30	2.56	3.21	3.59	4.42
$N = 24$	1:25.75	1:37.37	1:57.04	2:27.46	3:12.58	3:34.04	4:27.68	4:56.81	5:58.80
t_b/t_1	1.00	1.14	1.36	1.72	2.25	2.50	3.12	3.46	4.18

Lemma 3. *Given matrices $A, H \in \mathrm{Mat}_{m \times n}(\mathbb{F}[x]) \setminus \{0\}$ where $H := \mathrm{HNF}(A)$. A multiplier $V \in \mathrm{GL}_n(\mathbb{F}[x])$ satisfying $AV = H$ is unique if and only if $\mathrm{rank}(A) = n$. In this case $\|V\| \le (n-1)\|A\|$.*

Proof. Let $1 \le i_1 < i_2 < \ldots < i_n \le m$ be the row indices of the pseudo diagonal entries of A, and let $A', H' \in \mathrm{Mat}_n(\mathbb{F}[x])$ be defined as $A'_{j,*} := A'_{i_j,*}$ and $H'_{j,*} := H'_{i_j,*}$ for $1 \le j \le n$, respectively. Note that H' is in Hermite normal form and is right equivalent to A', so $H' = \mathrm{HNF}(A')$. Then $d := \det(A') \ne 0$, and it is easy to prove that

$$\|\mathrm{adj}(A')\| \le (n-1)\|A'\| \le (n-1)\|A\|$$

where $\tilde{A}' \in \mathrm{Mat}_n(\mathbb{F}[x])$ is the *adjoint matrix* of A'. Let $V \in \mathrm{GL}_n(\mathbb{F}[x])$ a transforming matrix with $A'V = H'$. Since

$$dI_n = A'\mathrm{adj}(A') \iff dI_n H' = A'\mathrm{adj}(A')H' \iff H' = A'\left(\frac{\mathrm{adj}(A')H'}{d}\right)$$

and H' as well as $\mathrm{adj}(A')$ are unique V must be unique, and thus

$$\|V\| \le \left\|\frac{\mathrm{adj}(A')H'}{d}\right\| \le (n-1)\|A'\| \le (n-1)\|A\|$$

$AV = H$ because the uniqueness of H. Suppose there exists another $V' \in \mathrm{GL}_n(\mathbb{F}[x])$ then $H' = A'V'$ but this is a contradiction to the uniqueness of the multiplier V for $A'V = H'$. □

In the case $\mathrm{rank}(A) < n$ the largest degree $\|V\|$ of a multiplier matrix V can be arbitrary large. We will modify Gaussian elimination for not full-rank input matrices producing small multipliers. Kaltofen, Krishnamoorthy and Saunders [10] extend the input matrix by the $n \times n$ identity matrix to obtain a full-rank $m + n \times n$ matrix. We will apply this technique to our algorithms.

Algorithm 4 *Input: $A \in \mathrm{Mat}_{m \times n}(\mathbb{F}[x])$, Output: $H := \mathrm{HNF}(A)$ and a multiplier $V \in \mathrm{GL}_n(\mathbb{F}[x])$.*

1. *Set $B := \binom{A}{I_n}$*
2. *Compute $G := \mathrm{HNF}(B)$ without computing a transforming matrix using either Algorithm 1 or 3*
3. *Set H to the first m rows of G and set V to the last n rows of G.*

Algorithm 4 computes the HNF of a full-rank matrix even if the input matrix is not full-rank. Matrix $B := \binom{A}{I_n} \in \mathrm{Mat}_{m+n \times n}(\mathbb{F}[x])$ has an unique multiplier $V \in \mathrm{GL}_n(\mathbb{F}[x])$ satisfying $\|V\| \leq (n-1)\|B\| = (n-1)\|A\|$ for nonzero A which is equal to the returned V. A more accurate analysis leads to the following Lemma.

Lemma 4. *Algorithm 4 computes the Hermite normal form of a matrix $A \in \mathrm{Mat}_{m \times n}(\mathbb{F}[x])$ with rank r and a multiplier $V \in \mathrm{GL}_n(\mathbb{F}[x])$ satisfying*

$$\|V\| \leq \min\{r, n-1\}\|A\|.$$

Proof. If $\mathrm{rank}(A) = n$ we can apply Lemma 3. Otherwise, without loss of generality let $A \neq 0$, and set $B \in \mathrm{Mat}_{m+n \times n}(\mathbb{F}[x])$ to $B := \binom{A}{I_n}$, and proceed with B as with A in the proof of Lemma 3. The full-rank submatrix $B' \in \mathrm{Mat}_n(\mathbb{F}[x])$ of B must contain at least $n - r$ identity vectors, and thus we get $\|\mathrm{adj}(B')\| \leq r\|B'\| \leq r\|A\|$ which completes the proof. $\square$

In Table 2 we give some running time examples for Algorithm 1 and 4.

Table 2. Running times for Algorithm 1 and 4 for $b = 1$.

Algorithm	$m \times n$	$\|A\|$	rank	N	Time	max. degree	$\|V\|$
1	500×500	10	500	16	29:35.26	7763	4987
1	500×500	10	500	32	19:00.96	7763	4987
1	512×512	8	512	16	11:47.61	161	55
1	1000×1000	2	1000	32	34:50.45	3117	1997
1	128×256	32	128	8	4:34.64	9200	9200
4					8:39.56	6395	4092
1	250×260	32	250	16	31:39.07	49400	49400
4					15:32.74	12457	8000

6 Conclusions and Acknowledgements

We have described parallel implementations for Hermite normal form computations and we have studied a technique for computing small multiplier matrices when the input matrix is not full-rank.

The running time examples in Table 1 demonstrate that we can obtain better performance if we allow higher communication costs. We have observed this effect for other examples, too. Algorithm 4 has a good upper bound for the degrees of the computed transforming matrix, and we have seen that this algorithm provides very good transforming matrices in practice.

We have used earlier implemented sequential gcd algorithms for our parallel implementations. This reuse of sequential code where really well supported by the usage of computation rows.

The author is very grateful to the GMD, Sankt Augustin, Germany for offering CPU time on their IBM SP2 installation. Special thanks go to George Havas for his corrections and suggestions.

References

1. W. A. Blankinship. A new version of the Euclidian algorithm. *Amer. Math. Monthly*, 70(9):742–745, 1963.
2. W. A. Blankinship. Matrix triangulation with integer arithmetic. *Comm. ACM*, 9(7):513, 1966.
3. G. H. Bradley. Algorithms and bound for the greatest common divisor of n integers and multipliers. *Comm. ACM*, 13:447, 1970.
4. G. H. Bradley. Algorithms for Hermite and Smith normal matrices and linear Diophantine equations. *Math. Comp.*, 25(116):897–907, 1971.
5. T. W. J. Chou and G. E. Collins. Algorithms for the solution of systems of linear Diophantine equations. *SIAM J. Comput.*, 11(4):687–708, 1982.
6. G. D. Forney. Convolutional codes I: Algebraic structure. *IEEE Trans. Inform. Theory*, IT-16(6):720–738, 1970.
7. G. Havas and B. S. Majewski. Extended gcd calculation. *Congr. Numer.*, 111:104–114, 1995.
8. G. Havas, B. S. Majewski, and K. R. Matthews. Extended gcd algorithms. Technical Report TR0302, The University of Queensland, Brisbane, 1995.
9. G. Havas and C. Wagner. Matrix reduction algorithms for euclidean rings. In *Proc. of the 3rd Asian Symposium Computer Mathematics*, to appear.
10. E. Kaltofen, M. S. Krishnamoorthy, and B. D. Saunders. Parallel algorithms for matrix normal forms. *Linear Algebra Appl.*, 136:189–208, 1990.
11. R. Kannan and A. Bachem. Polynomial algorithms for computing the Smith and Hermite normal forms of an integer matrix. *SIAM J. Comput.*, 8(4):499–507, 1979.
12. B. S. Majewski and G. Havas. A solution to the extended gcd problem. In *ISSAC'95 (Proc. 1995 Internat. Sympos. Symbolic Algebraic Comput.)*, pages 248–253. ACM Press, 1995.
13. C. C. Sims. *Computation with Finitely Presented Groups*. Cambridge University Press, 1994.
14. C. Wagner. *Normalformberechnung von Matrizen über euklidischen Ringen*. PhD thesis, Institut für Experimentelle Mathematik, Universität/GH Essen, 1997. Published by Shaker-Verlag, 52013 Aachen/Germany, 1998.
15. C. Wagner. Hermite normal form computation over Euclidean rings. Interner Bericht 71, Rechenzentrum der Universität Karlsruhe, 1998. submitted for publication.

Optimising Parallel Logic Programming Systems for Scalable Machines

Vítor Santos Costa[1] and Ricardo Bianchini[2]

[1] LIACC and DCC-FCUP, 4150 Porto, Portugal
[2] COPPE Systems Engineering, Federal University of Rio de Janeiro, Brazil

Abstract. Parallel logic programming (PLP) systems have obtained good performance on traditional bus-based shared-memory architectures. However, the scalable multiprocessors being developed today pose new challenges. Our experience with a sophisticated PLP system, Andorra-I, demonstrates that indeed performance suffers greatly on modern architectures. In order to improve performance, we perform a detailed analysis of the cache behaviour of all Andorra-I data structures via execution-driven simulation of a DASH-like multiprocessor. Based on this analysis we optimise the Andorra-I code using 5 different techniques. Our results show that the techniques provide significant performance improvements, leading to the conclusion that PLP systems can and should perform well on modern scalable multiprocessors.

1   Introduction

Parallel computers can improve performance of both numerical and symbolic applications. Logic programs are good examples of symbolic applications that often exhibit large amounts of implicit parallelism and that can greatly benefit from parallel computers. Several PLP systems have been developed so far and have obtained good performance for traditional bus-based shared-memory architectures. However, the scalable multiprocessors being developed today pose new challenges, such as the high latency of memory accesses and the demand for scalability.

The complexity of PLP systems and the large amount of data they process raise the issue of whether PLP systems can obtain good performance on these new parallel architectures. In order to address this issue, we experiment with a sophisticated PLP system, Andorra-I [8], that exploits both dependent and-parallelism and or-parallelism. Andorra-I is a particularly interesting example of PLP system, since most of its data structures are similar to the ones of several other PLP systems. Andorra-I has obtained good performance on the Sequent Symmetry, but our experience with it running on modern multiprocessors demonstrates indeed that scalability suffers greatly on these architectures [7].

This paper addresses the question of whether the poor scalability of Andorra-I is inherent to the complex structure of PLP systems or can be improved through careful analysis and tuning. In order to answer this question, we analyse the cache behaviour of all Andorra-I data areas when applied to several different logic programs. The analysis pinpoints the areas that are responsible for most misses and the main sources of the misses. Based on this analysis we remove the main performance limiting factors in Andorra-I through a small set of optimisations that did not require a redesign of the system. More specifically, we optimise Andorra-I using 5 different techniques: trimming of shared variables, data layout modification, privatisation of shared data structures, lock distribution, and elimination of locking in scheduling.

We present the isolated and combined performance improvements provided by the optimisations on a simulated DASH-like multiprocessor with up to 24 processors. In isolation, shared variable trimming and the modification of the data layout produced the greatest improvements. The improvements achieved when all optimisation techniques are combined are substantial. A few of our programs approach linear speedups as a consequence of our modifications. In fact, for one program the speedup of the modified Andorra-I is a factor of 3 higher than that of the original version of the system on 24 processors. Our main conclusion is then that, even though PLP systems are indeed complex and sometimes irregular, these systems can and should scale well on modern scalable multiprocessors.

2 Methodology

In this section we detail the methodology used in our experiments. The experiments consisted of the simulation of the parallel execution of Andorra-I [8, 10]. The Andorra-I parallel logic programming system employs a very interesting method for exploiting and-parallelism, namely to execute *determinate* goals first and concurrently, where determinate goals are goals that match at most one clause in a program. The Andorra-I system also exploits or-parallelism that arises from the non-determinate goals.

Andorra-I requires access to shared memory for both execution and scheduling of work. In order to study the Andorra-I execution more fully, we divided its shared memory into ten different areas. Andorra-I implements the standard Prolog work areas. The *Code Space* includes the compiled code for every procedure and is read-only during execution of our benchmarks. The *Heap Space* stores structured terms and variables. The *Goal Frame Space* keeps the goals yet to be executed. The *Choicepoint Stack* maintains alternatives for open goals. The *Trail Stack* records any conditional bindings of variables.

Andorra-I also requires several new areas for and/or parallelism. The *Or-Scheduler Data Structures* are used to manage or-parallelism. The data structures for and-parallelism are in the *Worker* area. The *Binding Arrays* are used to implement the SRI model [5] for or-parallelism, by storing conditional bindings. A *Lock Array* was needed in our port to establish a mapping between a shared memory position (such as a variable in the heap) and a lock. Finally, the *Miscellaneous Shared Variables* include the remaining data structures.

To simulate Andorra-I we ported the system to a detailed on-line, execution-driven simulator. The simulator uses the MINT front-end [9], and a back-end

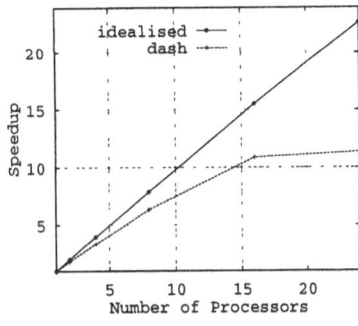

Fig. 1. Speedups for bt-cluster

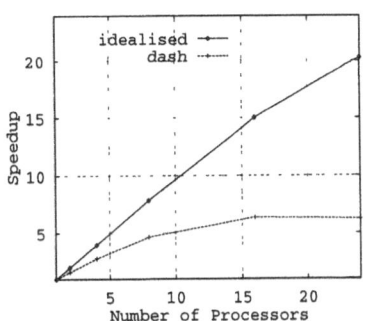

Fig. 2. Speedups for tsp

that simulates the memory and interconnection systems. We simulate a 24-node, DASH-like [4], directly-connected multiprocessor. Each node of the simulated machine contains a single scalar processor, a write buffer, a 128-KB direct-mapped data cache with 64-byte cache blocks, local memory, a full-map directory, and a network interface. We use the DASH write-invalidate protocol with release consistency [3] in order to keep caches coherent. We classify cache misses under this protocol using the algorithm presented in [1].

3   Workload and Original Performance

The benchmarks we used in this work are applications representing predominantly and-parallelism, predominantly or-parallelism, and both and- and or-parallelism. We next discuss application performance for the original Andorra-I (more detailed information on the benchmarks can be found in the extended version of this paper [6] and in Dutra's thesis [2]). Note that our results correspond to the *first* run of an application; results would be somewhat better for other runs.

We use two example *And-parallel* applications, the clustering algorithm for network management from British Telecom, bt-cluster, and a program to calculate approximate solutions to the traveling salesman problem, tsp. To obtain best performance, we rewrote the original applications to make them determinate-only computations.

Figure 1 shows the bt-cluster speedups for the simulated architecture as compared to an idealised shared-memory machine, where data items can always be found in cache. The idealised curve shows that the application has excellent and-parallelism and can achieve almost linear speedups up to twenty four processors. Unfortunately, performance for the DASH-like machine is barely acceptable. Figure 2 shows that the tsp application achieves worse speedups than bt-cluster on a modern multiprocessor. The maximum speedup actually decreases for 24 processors, whereas the ideal machine would achieve a speedup of 20 for 24 processors. Figure 3 illustrates the number and sources of cache misses

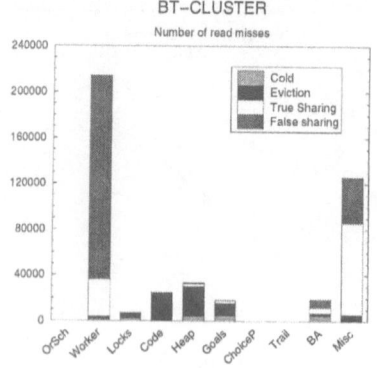

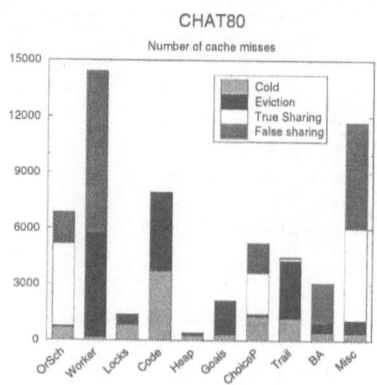

Fig. 3. Misses by data area for bt-cluster

Fig. 4. Misses by data area for chat80

per data area in the bt_cluster application running on 16 processors as a representative example. The Figure shows that the overall miss rate of **bt-cluster** is dominated by true and false sharing misses from the **Worker** and **Misc** areas. This suggests that the system could be much improved by reducing false sharing and studying activity in the **Worker** and **Misc** areas.

We use two *Or-parallel* applications. Our first application, **chat80**, is an example from the well-known natural language question-answering system **chat-80**, written at the University of Edinburgh by Pereira and Warren. The second application, **fp**, is an example query for a knowledge-based system for the automatic generation of floor plans. This application should at least in theory have significant or-parallelism. Figure 5 shows the speedups for the **chat80** application from 1 to 24 processors. These speedups are very similar to those obtained by Andorra-I on the Sequent Symmetry architecture. In contrast, the DASH curve reaches a maximum speedup of 4.2 for 16 processors. Figure 6 shows the speedups for the **fp** application. The theoretical speedup is very good, in fact quite close to linear, in sharp contrast to the actual speedup for the DASH-like machine.

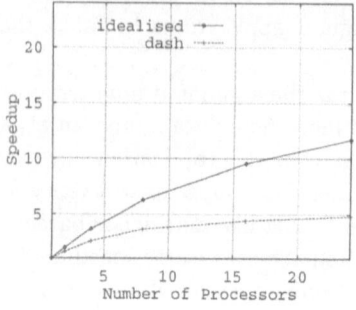

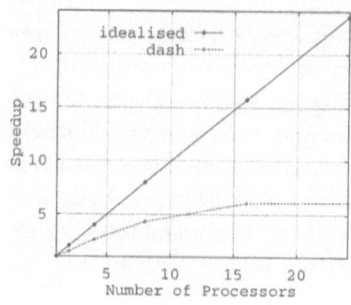

Fig. 5. Speedups for chat80

Fig. 6. Speedups for fp

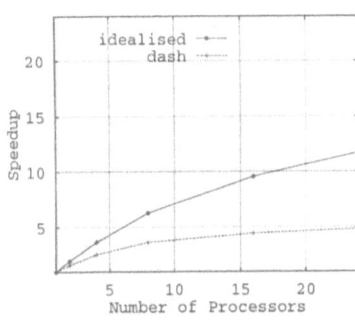

Fig. 7. Speedups for pan2

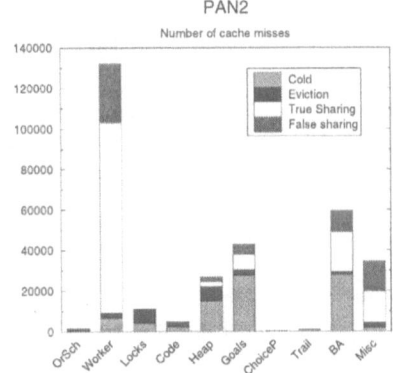

Fig. 8. Misses by data area for pan2

Figure 4 shows the number and source of misses for `chat80` running on 16 processors, again as an example of this type of application. Note that `chat80` does not have enough parallelism to feed 16 processors, suggesting that most sharing misses should result from or-parallel scheduling areas, `OrSch` and `ChoiceP`. Indeed, the figure shows that these areas are responsible for a large number of sharing misses, but the areas causing the most misses are `Worker` and `Misc` as in the and-parallel applications, indicating again that these two areas should be optimised.

As an example of *And/Or* application we used a program to generate naval flight allocations, based on a system developed by Software Sciences and the University of Leeds for the Royal Navy. Figure 7 shows the speedups for `pan2`. The `idealised` curve shows that the application has less parallelism than all other applications; the ideal speedup does not even reach 12 on 24 processors. When run on the DASH simulator, `pan2` exhibits unacceptable speedups for all numbers of processors; speedup starts out at about 1.8 for 2 processors and slowly improves to a maximum of only 4.8 for 24 processors. Figure 8 shows the distribution of cache misses by the different Andorra-I data areas for 16 processors. In this case, the `Worker` area clearly dominates, since the contribution from the `Misc` area is not as significant as in the and-parallel benchmarks. Note that there is more true than false sharing activity in `Worker`. The true sharing probably results from idle processors looking for work.

4 Optimisation Techniques and Performance

The previous analysis suggests that relatively high miss rates may be causing the poor scalability of Andorra-I. It is interesting to note that most misses come from fixed layout areas, such as `Worker` and `Misc`, and not from the execution stacks, as one would assume.

We next discuss how several optimisations can be applied to the system, particularly in order to improve the utilisation of the `Worker` and `Misc` areas.

The first two optimisations were prompted by our simulation-based analysis of caching behaviour, and they are the ones that give the best improvement. The other three were based on our original intuitions regarding the system, and were the ones we would have performed first without simulation data. We studied performance for three applications, bt-cluster, chat80 and pan2. A detailed discussion of our experiments and results can be found in [6]. In the remainder of this section, we simply summarise the impact of each of the techniques studied when applied in isolation.

Variable Trimming. In this technique we investigate the areas that have un-exepected sharing, and try to eliminate this sharing if possible. For two of our applications, the Misc area gave a surprisingly significant contribution to the number of misses. The area is mostly written at the beginning of the execution to set up execution parameters. During execution it is used by the reconfigurer and to keep reduction and failure counters. By investigating each component in the area, we detected that the counters were the major source of misses. As they are only used for research and debugging purposes, we were able to eliminate them from the Andorra-I code.

The results in [6] show that chat80 benefits the most from this optimisation. This is because the failure counter is never updated by and-parallel applications, but often updated by this or-parallel application. The optimisation does not impact the and-parallel benchmarks as much, leading to less than 10% speedup improvements.

Data Layout Modification. All benchmarks but pan2 exhibit a high false sharing rate, showing a need for this technique. On 16 processors, 15% of the misses in pan2 are false sharing misses, whereas in the other applications false sharing causes between 40% (chat80) and 51% (bt-cluster) of all misses. These results suggest that improving false sharing is of paramount importance. According to our detailed analysis of caching behaviour, false sharing misses are concentrated in the Worker, OrSch, ChoiceP and BA areas, besides the Misc area optimised by the previous technique.

The Worker and OrSch data areas are allocated statically. This indicates that we can effectively reduce false sharing. We applied two common techniques to tackle false sharing, *padding* between fields that belonged to different workers or that were logically independent, and *field reordering* to separate fields that were physically close but logically distinct. Although these are well-known techniques, padding required careful analysis, as it increases eviction misses significantly. The field reordering technique was not easily applied either, as the relationships between fields are quite complex.

Padding may lead to serious performance degradation for the dynamic data areas, such as ChoiceP and BA. This restricted our options for layout modification to just field reordering for these areas. The BA area was the target of one final data layout modification, since the analysis of cache behaviour surprised us with a high number of false sharing misses in this area for chat80. Further investigation showed that this was a memory allocation problem. The engines' top of stacks were being *shmalloc'ed* separately and appeared as part of the BA area in the analysis. This increased sharing misses in the area and was especially bad for the or-parallel applications, as different processor's top of stacks would end up in the same cache line. We addressed the problem by moving these pointers into the Worker area, where they logically belong. Our results show that the bt-cluster and chat80 applications benefit the most from this optimisation; speedup improvements can be as significant as 60%. In contrast, the pan2 application achieves improvements of less than 10% from this optimisation.

Privatisation of Shared Variables. This technique reduces the number of shared memory accesses by making local copies in each node of the machine. In the best case, the shared variables are read-only and hence local copies can actually be allocated in private memory. The high number of references to Worker suggested that privatisation could be applied there. In fact, Andorra-I did already use private copies of the variables in Worker and there was little room for improvement. The Locks and Code data areas are the major candidates to privatisation in Andorra-I. The Locks area only includes pointers to the actual locks, is thus read-only during execution, and can be easily privatised. Another area that is also read-only during parallel execution of our benchmarks is Code. Unfortunately, logic programs in general can change the database and, therefore, update Code, making privatisation complex. Our results show that privatisation improves speedups by up to 10% at most and that the impact of this optimisation decreases as the number of processors increases.

Lock Distribution. This technique was considered to reduce contention on accesses to logical variables, and-scheduling, or-scheduling, and stack management. The original implementation used a single array of locks to implement these operations. In the worst case, several workers would contend for the same lock causing contention. To improve scalability, we implemented different lock data structures for different purposes. We expected best results for or-parallel applications, as the optimisation prevents different teams from contending on accesses to logical variables. The cost of this optimisation is that, if the arrays of locks are shared, there will be more expensive remote cache misses. Our results show that the bt-cluster and chat80 applications benefit somewhat from this optimisation, but that the pan2 application already exhibited a significant number of misses in the Locks area and suffers a slowdown.

Elimination of Locking in Scheduling. This technique improves performance in benchmarks with significant and-parallelism by testing whether there is available work, before actually locking the work queue. This modification is equivalent to replacing a test_and_set lock with a test_and_test_and_set lock. This optimisation provides a small speedup improvement for pan2, as it avoids locking when there is no and-work. For bt-cluster the technique does not improve speedups as this application exhibits enough and-work to keep processors busy.

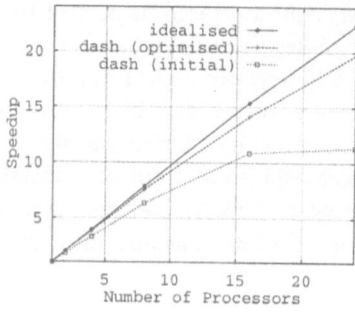

Fig. 9. Speedups for bt-cluster

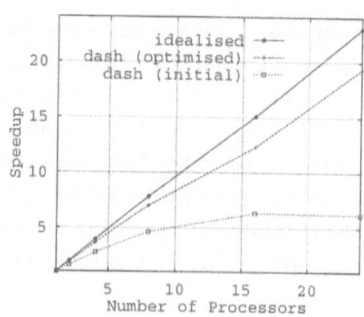

Fig. 10. Speedups for tsp

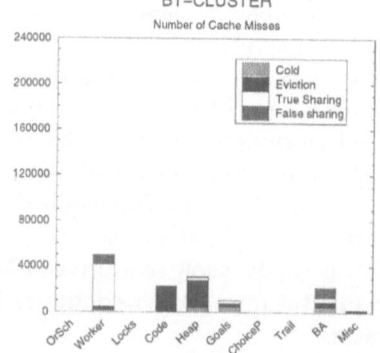

Fig. 11. Misses by data area for bt-cluster

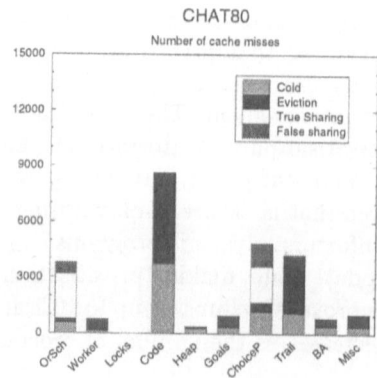

Fig. 12. Misses by data area for chat80

5 Combined Performance of Optimisation Techniques

We next discuss the overall system performance with all optimisations combined. We compare speedups against the idealised and original results. The idealised speedups were recalculated for the new version of Andorra-I, but, as it is shown in the figures, the optimisations did not have any significant impact for the idealised machine. Figures 9 and 10 show the speedups for the two and-parallel applications running on top of the modified Andorra-I system. The maximum speedup for bt-cluster jumped from 12 to 20, whereas the maximum speedup for tsp jumped from 6.3 to 19. This indicates that the realistic machine is now able to exploit the available parallelism more fully. The explanation for the better speedups is a significant decrease in miss rates. For bt-cluster, the new version of Andorra-I exhibits a miss rate of only 0.6% for 16 processors, versus the 1.6% of the previous version. In the case of tsp, the optimisations decreased the miss rate from 3% to 1.2% again on 16 processors.

Figure 11 shows the number and source of misses for bt-cluster on 16 processors. Note that the figure keeps the same Y-axis as in Figure 3 to simplify comparisons against the cache behaviour of the original version of Andorra-I. The figure shows that the number of misses in the `Worker` area was reduced by a factor of 4, while the number of misses in the `Misc` area was reduced by an order of magnitude. The figure also shows that there is still significant true sharing in `Worker`, but false sharing is much less significant. The number of misses from `Misc` is now almost irrelevant.

The or-parallel benchmarks also show remarkable improvements due to the combination of the optimisation techniques we applied. Figure 13 shows the speedups for `chat80` and Figure 14 shows the speedups for `fp`. The maximum speedup for `chat80` almost doubles from one version of the system to the other. Note that speedups for the optimised system still flatten out on 16 processors, but at a much better efficiency. The other benchmark, `fp`, displays our most impressive result. The speedup for 24 processors jumps from 6.2 with the original Andorra-I system to 20 when all our optimisations are applied. This result represents more than a three-fold improvement. Figure 12 shows the distribution of misses for `chat80` with 16 processors. The figure demonstrates that the number of misses in the `Worker` and `Misc` areas was reduced by an order of magnitude. The large number of eviction and cold start misses in the `Code` area remains however. Sharing misses are now concentrated in the `OrSch` and `ChoiceP` areas, as they should.

Figure 15 shows the speedups of the new version of Andorra-I for the `pan2` benchmark. In this case, the improvement resulting from our optimisations was quite small. Figure 16 shows the cache miss distribution for the optimised Andorra-I. The main source of misses was true sharing originating in the `Worker` region. A more detailed analysis proved that these misses originate from lack of work. Workers are searching each other's queues and generating misses. We are investigating more sophisticated scheduling strategies to address this problem.

6 Conclusions and Future Work

Andorra-I is an example of an and/or-parallel system originally designed for traditional bus-based shared-memory architectures. We have demonstrated that

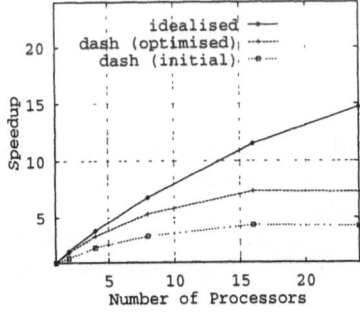

Fig. 13. Speedups for chat80

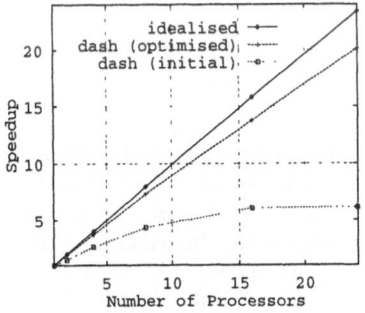

Fig. 14. Speedups for fp

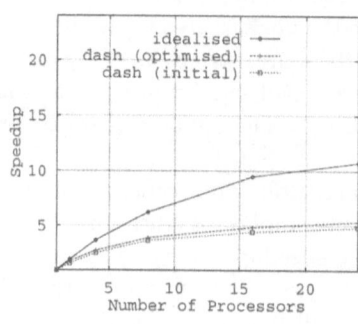

Fig. 15. Speedups for pan2

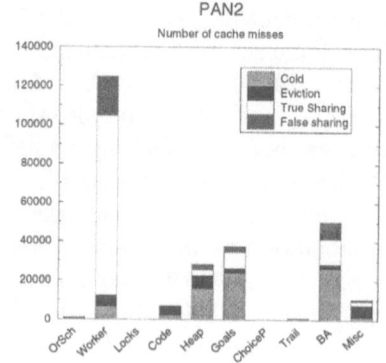

Fig. 16. Misses by data area for pan2

the system can also achieve good performance on scalable shared-memory systems. The key to these results was the extensive data available from detailed simulations of Andorra-I. This information showed that there was no need to restructure the system or its schedulers. Instead, performance could be dramatically improved by focusing on accesses to shared data.

We believe there is potential for improving the performance of PLP systems even further. To prove so will require more radical changes to data structures within Andorra-I itself, as the system was simply not designed for such large numbers of processors. Last, but not least, we are interested in studying performance of other parallel logic programming systems, such as the systems that exploit independent and-parallelism.

Acknowledgements The authors would like to thank Leonidas Kontothanassis and Jack Veenstra for their help with the simulation infrastructure, and Rong Yang, Tony Beaumont, D. H. D. Warren for their work in Andorra-I. This paper results from collaboration work with Inês Dutra, and has benefited from Márcio da Silva's studies. Vítor Santos Costa would like to thank support from the Praxis PROLOPPE and FCT MELODIA projects. Ricardo Bianchini would like to thank the support of the Brazilian CNPq.

References

1. R. Bianchini and L. I. Kontothanassis. Algorithms for categorizing multiprocessor communication under invalidate and update-based coherence protocols. In *Proceedings of the 28th Annual Simulation Symposium*, April 1995.
2. Inês Dutra. *Distributing And- and Or-Work in the Andorra-I Parallel Logic Programming System*. PhD thesis, University of Bristol, Department of Computer Science, February 1995.
3. D. Lenoski, J. Laudon, K. Gharachorloo, A. Gupta, and J. Hennessy. The directory-based cache coherence protocol for the DASH multiprocessor. *Proceedings of the*

17th International Symposium on Computer Architecture, pages 148–159, May 1990.

4. D. Lenoski, J. Laudon, T. Joe, D. Nakahira, L. Stevens, A. Gupta, and J. Hennessy. The dash prototype: Logic overhead and performance. *IEEE Transactions on Parallel and Distributed Systems*, 4(1):41–61, Jan 1993.

5. Ewing Lusk, David H. D. Warren, Seif Haridi, et al. The Aurora Or-parallel Prolog System. *New Generation Computing*, 7(2,3):243–271, 1990.

6. V. Santos Costa and R. Bianchini. Optimising Parallel Logic Programming Systems for Scalable Machines. Technical Report DCC-97-7, DCC - FC & LIACC, UP, October 1997.

7. V. Santos Costa, R. Bianchini, and I. C. Dutra. Evaluating the impact of coherence protocols on parallel logic programming systems. In *Proceedings of the 5th EUROMICRO Workshop on Parallel and Distributed Processing*, pages 376–381, 1997.

8. V. Santos Costa, D. H. D. Warren, and R. Yang. Andorra-I: A Parallel Prolog System that Transparently Exploits both And- and Or-Parallelism. In *Third ACM SIGPLAN PPOPP*, pages 83–93. ACM press, April 1991.

9. J. E. Veenstra and R. J. Fowler. Mint: A front end for efficient simulation of shared-memory multiprocessors. In *Proceedings of MASCOTS '94)*, 1994.

10. Rong Yang, Tony Beaumont, Inês Dutra, Vítor Santos Costa, and David H. D. Warren. Performance of the Compiler-Based Andorra-I System. In *Proceedings of the Tenth International Conference on Logic Programming*, pages 150–166. MIT Press, June 1993.

Experiments with Binding Schemes in LOGFLOW

Zsolt Németh and Péter Kacsuk

MTA Computer and Automation Research Institute
H-1518 Budapest, P.O.Box 63. Hungary
{zsnemeth, kacsuk}@sztaki.hu
http://www.lpds.sztaki.hu

Abstract. The handling of variables is a crucial issue in designing a parallel Prolog system. In the framework of the LOGFLOW project some experiments were made with binding methods. The work is an analysis of possible schemes, a modification of an existing one and a plan for future implementations of LOGFLOW on new architectures. The conditions, principles and results are presented in this paper.

1 Introduction

Logic programming languages offer a high degree of inherent parallelism [4]. The central question in implementing an OR-parallel Prolog system is how the conditional bindings can be handled consistently and effectively.

There are a number of techniques for treating conditional bindings. They can be categorized as stack sharing [1][3][5] [6], stack copying [11] and recomputation methods. All these techniques have a common feature: references (either physical or logical) are allowed to point from an environment to any other environments. However, in a distributed memory system these environments may reside on different processing elements, and dereferencing a variable through several processing elements is a tremendously expensive operation.

LOGFLOW is a distributed-memory implementation of Prolog [8]. Its abstract execution model is the logicflow model based on dataflow principles. Prolog programs are transformed into the so called Dataflow Search Graph (DSG). Nodes in this graph represent Prolog procedures. The execution is governed by token streams. Tokens represent a given state of the machine where the computation is continued.

LOGFLOW is based entirely on the closed binding environment scheme. Thus, tokens holding an environment are allowed to be distributed over the processor space freely, since there are no external references from the token.

The implementation shows some of the drawbacks of the closed binding scheme. There is a long research work in combining the benefits of different binding schemes (e.g. [10][12][14]). The overheads of the closed binding scheme in LOGFLOW were analysed and a hybrid binding method was elaborated where the idea basically relies on the Quasi Closed Environment (QCE [12]) but in fact, the new scheme is an amalgamation of much more binding techniques.

2 The closed binding environment

A closed environment in Conery's definition is a set of frames E so that no pointers or links originating from E dereference to slots in frames that are not members of E [2]. There are two special procedures to maintain the closed state of an environment frame: the (2-stage) unification and the back-unification [7]. The environment closing is a costly operation both in time and in memory consumption. The most important contributors to this overhead are the followings.

Compound terms must be scanned each time when a unification or back- unification takes place. Scanning a compound term means checking each argument of the term whether it is a reference to a variable or not and performing the necessary actions. Since compound terms (structures and lists) are the natural data structures in logic programming languages, they are used extensively.

Structures must not be shared by different environments because their arguments may be changed by bounding variables or by modifying a reference item during the environment closing. If a token is duplicated, all the structures in its environment must be copied for each new environment. This results in an enormous memory consumption and slows down the computation.

3 The proposed solutions

The previously listed overheads may occur because of the very special handling of addresses in the closed binding environment scheme. The proposed solutions try to combine the local addressing scheme of the closed environment with the global addressing scheme of non-closed environments. In [10] the molecules are introduced to avoid copying when all the arguments of a structure became ground. In [14] the so called tagged variable scheme is used which is a variation of hash tables for dataflow machines. In [12] variables are divided into two groups whether they can occur in a structure or not. Variables occurring in a structure are stored as global ones whereas others are stored still according to the local addressing scheme. This distinction does not affect the location of the variables but the way how they can be accessed. In such a way, the most expensive part of the closing procedure, the structure scanning and copying, can be discarded until the task is really migrated between the processors.

4 Combining the closed and non-closed binding schemes in LOGFLOW

The new hybrid binding scheme alloys many ideas from different binding methods. The basis is still a variation of Conery's closed environment. By default, variables are handled in the usual way according to the closed binding environment principles. Variables are made global and thus, are handled differently when they are in the arguments of a structure. The fundamental elements of the new binding scheme are the followings:

1. The special storage of global variables. The variable access and the duplication of environments should be done efficiently. Any well-known global method is a good candidate (e.g. binding arrays, directory trees, etc.); finally

a hashing method was chosen. It is a hash technique on variables rather than on environments as Borgwardt's hash windows [1]. There are comparisons of the two hashing methods in [5].

2. The way how the global variables are detected, created and named according to a global naming scheme. The naming scheme is very similar to that in binding array technique [3].

3. Binding the global variables to a value. Basic considerations were: how to represent unbound variables, where to put newly bound ones.

4. Handling different OR-branches (the question of conditional global variables.) As a result of the decisions made at 3, the duplication of environments at OR-nodes is easy and efficient - there is no need to take extra care of unbound variables.

5. The new binding scheme manipulates within a single PE. Some special procedures should be done when the computation migrates between PEs. The problem shows many common features with the time stamping method.

None of the above mentioned binding techniques can solve all the problems arising in LOGFLOW but their combination could be a promising candidate. Details of the design and implementation principles can be found in [13].

5 Evaluation of the hybrid binding method

The hybrid binding scheme was implemented and tested on programs of different types and sizes. Test results can be summarized as follows:

- the cost of closing procedure within a single processing element (intra- processor overhead) was significantly reduced
- the cost of task migration (inter-processor overhead) became slightly higher due to the procedures involved

Tests showed some further properties of the binding scheme, e.g.:

- the bigger the structures are, the more time can be saved by the new binding method
- when other kinds of optimization are present (e.g. ground term table) the new scheme can add a little to the performance

6 Conclusions and further work

The goal of the revision of the binding method was to improve the performance of a distributed Prolog system. While efforts were made to achieve this goal, test results pointed out an interesting conclusion which can set the direction of the future research work.

The test results suggested that this kind of hybrid binding scheme is an excellent candidate for binding environment on multithreaded architectures like Datarol-II and KUMP/D, where the intra-processor overhead associated with closing procedures is reduced by the binding method whereas, the inter-processor copying overhead is eliminated by the facilities of the architecture. This idea is presented in [9].

There are ongoing research projects on implementing LOGFLOW on a kind of hybrid von Neumann-dataflow multithreaded architecture. It has been proven that the runtime model of the Datarol-II family and that of the LOGFLOW is very similar. Furthermore, the thread scheduling policies and granularities of the two models are quite close to each other. The efficient implementation of LOGFLOW on Datarol-II can be greatly supported by a binding method like the one presented in this paper.

7 Acknowledgments

The work reported in this paper was partially supported by the National Research Grant 'Massively Parallel Implementation of Prolog on Multithreaded Architectures, Particulary on Datarol-II' registered under No. T-022106.

References

1. P. Borgwardt: Parallel Prolog Using Stack Segments on Shared-Memory Multiprocessors. Proceedings of the 1984 Symposium on Logic Programming.
2. J.S. Conery: Binding Environments for Parallel Logic Programs in Non-Shared Memory Multiprocessors. Proceedings of the 1987 Symp. on Logic Programming, 1987.
3. M. Carlsson: Design and Implementation of an OR-Parallel Prolog Engine. SICS Dissertation Series 02.
4. J.Chassin de Kergommeaux, P.Codognet: Parallel Logic Programming Systems. ACM Computing Surveys Vol 26. No 3. Sept 1994
5. A. Ciepielewski, B. Hausman: Performance Evaluation of a Storage Model for OR-Parallel Execution of Logic Programs. SICS 86003 Research Report.
6. B. Hausman, A. Ciepielewski, S. Haridi: OR-Parallel Prolog Made Efficient on Shared Memory Multiprocessors. SICS R87006 Research Report.
7. P. Kacsuk: Distributed Data Driven Prolog Abstract Machine. In: P. Kacsuk, M.J.Wise: Implementations of Distributed Prolog. Wiley, 1992.
8. P. Kacsuk, Zs. Németh and Zs. Puskás: Tools for mapping, Load Balancing and Monitoring in the LOGFLOW Parallel Prolog Project. Parallel Computing Journal, Elsevier, Vol. 22, No. 13, Feb. 1997, pp. 1853-1882
9. P. Kacsuk, M. Amamiya: A Multithreaded Implementation Concept of Prolog for Datarol-II. Proceedings of ISHPC97.
10. L.V. Kalé, B. Ramkumar: The Reduce-Or Process Model for Parallel Logic Programming on Non-Shared Memory Machines. In: P. Kacsuk, M.J.Wise: Implementations of Distributed Prolog. Wiley, 1992.
11. R.Karlsson: A High Performance OR-Parallel Prolog System. SICS Dissertation Series 07, March 1992.
12. H. Kim, J-L. Gaudiot: A Binding Environment for Processing Logic Programs on Large-Scale Parallel Architectures. Technical Report, University of Southern California 1994.
13. Zs. Németh, P. Kacsuk: Analysis and Improvement of the Variable Binding Scheme in LOGFLOW. Workshop on Parallelism and Implementation Technologies for (Constraint) Logic Programming Languages, Port Jefferson, 1997.
14. A.V.S. Sastry, L.M. Patnaik: OR-Parallel Evaluation of Logic Programs on a Multi-Ring Dataflow Machine. New Generation Computing, 10 (1991), pp. 23-53.

Experimental Implementation of Parallel TRAM on Massively Parallel Computer

Kazuhiro Ogata, Hiromichi Hirata, Shigenori Ioroi and Kokichi Futatsugi

JAIST, JAPAN
({ogata, h-hirata, ioroi, kokichi}@jaist.ac.jp)

Abstract. MPTRAM is an abstract machine for parallel rewriting that is intended to be reasonably implemented on massively parallel computers. MPTRAM has been implemented on Cray T3E carrying 128 PEs with MPI. It rewrites terms efficiently in parallel (about 65 times faster than TRAM) if message traffic is little. But we have to reduce the overhead of message passing somehow, e.g. by using the shared globally addressable memory subsystem of Cray T3E, so that MPTRAM can be used to implement algebraic specification languages.

1 Introduction

Algebraic specification languages such as OBJ3 and CafeOBJ allow users to use computers to verify some properties of software systems and observe their dynamic behavior at the early stage of software development. However there are still some problems that have to be solved about algebraic specification languages. One of them is the improvement of their execution (rewriting) speed. That is why we have designed some rewriting abstract machines.

TRAM [5] is an abstract machine for rewriting [4] that may be used to efficiently implement algebraic specification languages. Its rewriting strategy is the user definable E-strategy [2]. Parallel TRAM [7] (PTRAM) is a parallel variant of TRAM that is intended to be implemented on shared-memory multiprocessors. It allows users to control parallel rewriting using the parallel E-strategy that is an extension of the E-strategy and is also a subset of the concurrent E-strategy [3]. Furthermore, we have altered PTRAM so as to reasonably implement it on massively parallel computers. The altered version is called Massively Parallel TRAM (MPTRAM). We have experimentally implemented MPTRAM on Cray T3E [1] carrying 128 processing elements with MPI [8]. In this paper, we describe briefly parallel rewriting with the parallel E-strategy, PTRAM and MPTRAM, and reports some experiments carried out on Cray T3E.

2 Parallel TRAM

Outline of Parallel Rewriting. Parallel TRAM (PTRAM) uses as its rewriting strategy the parallel E-strategy [7], which is an extension of the E-strategy [2] and is also a subset of the concurrent E-strategy [3], so as to control parallel rewriting. The strategy allows each operation have its own parallel local strategy. Parallel local strategies are specified for an n-ary operation by using lists that are basically ones of natural numbers and sets of positive numbers, or to be more exact, that are defined by the following extended BNF notation:

Definition 1 (Syntax of parallel local strategies).

⟨ParallelLocalStrategy⟩ ::= ( ) | (⟨SerialElem⟩* 0)
⟨SerialElem⟩ ::= 0 | ⟨ArgNum⟩ | ⟨ParallelArgs⟩
⟨ArgNum⟩ ::= 1 | 2 | ... | n
⟨ParallelArgs⟩ ::= { ⟨ArgNum⟩+ }

A positive number $x\,(1 \le x \le n)$ in parallel local strategies denotes the xth argument of a term whose top operation (if the term is $f(t_1, \ldots, t_n)$, f is the top operation) is that n-ary operation, and zero stands for the term itself.

The rewriting of terms with the parallel E-strategy is briefly described, which is followed by the operational semantics. A term is rewritten according to the parallel local strategy of its top operation. If the first element of the strategy is some positive number x, the xth argument is rewritten and the rewriting of the original term continues according to the remainder of the strategy. If the first one is zero, the term is first replaced with a new term and then the new one is rewritten according to the parallel local strategy of its top operation if there exists a rewrite rule whose left-hand side matches with the term, or otherwise the rewriting of the original term continues according to the remainder of the strategy. If the first one is a set of positive numbers, all the subterms corresponding to the positive numbers in the set are rewritten in parallel and then the rewriting of the original term continues according to the remainder of the strategy. The following is the operational semantics written in rewrite rules:

Definition 2 (Operational semantics of the parallel E-strategy).

eval(t) → if(eval?(t), t, reduce(t, strategy(t)))
reduce(t, nil) → mark(t)
reduce(t, c(0, l)) → if(redex?(t), eval(contractum(t)), reduce(t, l))
reduce(t, c(x, l)) → reduce(replace(t, x, eval(arg(t, x))), l)
reduce(t, c(m, l)) → reduce(fork(t, m), l)
fork(t, empty) → t
fork(t, p(x, m)) → exit(eval(arg(t, x)), x, fork(t, m))
exit(s, x, t) → replace(t, x, s)

eval rewrites a term according to the parallel local strategy of its top operation by calling reduce if the term has not been rewritten yet. t and s are variables for terms, x for positive numbers (that is why reduce(t, c(x, l)) does not overlap with reduce(t, c(0, l))), l for ⟨ParallelLocalStrategy⟩ and m for ⟨ParallelArgs⟩, and the other symbols are operations. c and nil, and p and empty are constructors for representing ⟨ParallelLocalStrategy⟩ and ⟨ParallelArgs⟩ respectively. exit and if are given ({1 3} 0) and (1 0) as their strategies, and the other operations eager local strategies such as (1 2...0).

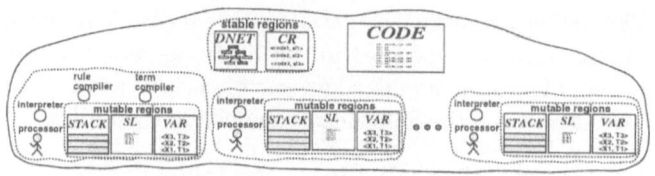

Fig. 1. Rough sketch of PTRAM architecture

PTRAM Architecture. PTRAM is intended to be implemented on shared-memory multiprocessors with a small number of processors. Fig. 1 shows a rough sketch of PTRAM architecture. There are three kinds of processing units (an interpreter, a rule compiler and a term compiler), and six kinds of regions. *DNET* and *CR* are stable regions and are shared with all processors. The three mutable regions *SL*, *STACK* and *VAR* are given to each processor, but *CODE*, on which terms are allocated, is shared with all processors so that terms can be shared with all processors. One of the processors, called the main processor, has also a rule compiler and a term compiler. Given a set of rewrite rules, with the rule compiler the main processor encodes the left-hand sides into a discrimination net that is kind of a decision tree for pattern matching, and translates the right-hand sides into pairs of matching program templates and strategy list templates. The discrimination net is allocated on *DNET*, and the pairs on *CR*. After that, given a term, with the rule compiler the main processor translates it into a pair of a matching program and a strategy list. The matching program is allocated on *CODE*, and the strategy list on the main processor's *SL*. And then, in order to rewrite the original term, the main processor interprets the matching program according to the strategy list with its interpreter, *STACK* and *VAR*, and *DNET*. If a processor meets a situation that some arguments may be evaluated in parallel and there are some available (idle) processors, it asks them to rewrite the arguments independent of it. If there is no idle processor at that time, the processor itself rewrites the arguments.

The strategy list [6] of a term is made from the matching program compiled from the term and the parallel local strategies of the subterms' top operations. It indicates the order in which the term is rewritten. The elements of the strategy list are basically addresses at which the matching programs of the subtemrs and some instructions are stored. There are three instructions for parallel rewriting: fork, join and exit. fork creates a new process for parallel rewriting, join delays its caller process until all of its child processes terminate, and exit terminates its caller process after reporting the termination to the parent process. exit does not have to explicitly return the term that its caller process has rewritten to the parent process because terms are allocated on the shared region *CODE*. Strategy lists may be regarded as process queues. A chunk of consecutive elements of a strategy list represents a process. Each process has two possible states: active and pending. Each process queue contains zero or more pending processes, and zero or one active process that is on the top of the queue if exists.

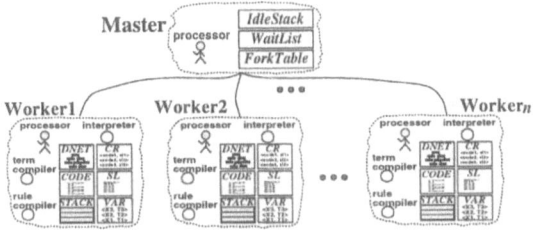

Fig. 2. Rough sketch of MPTRAM architecture

A process whose process queue contains an active process is busy, and otherwise idle. There is no special processor that manages processors, but a table for recording processors' state that is shared with all processors. Each processor itself keeps a record of its own state in the table, and looks at the table so as to find idle processors.

3 Massively Parallel TRAM

We have altered PTRAM to reasonably implement it on massively parallel computers. The altered version is called Massively Parallel TRAM (MPTRAM). Fig. 2 shows a rough sketch of MPTRAM architecture. MPTRAM uses the message passing model as its parallel computational model, and the simple master/worker model as the basic parallel algorithm. The master manages all workers by using the two containers (*IdleStack* and *WaitList*) and the one table (*ForkTable*). Each worker has three possible states: idle, waiting and busy. Idle workers are pushed on the stack *IdleStack*, and waiting workers that are waiting for results from their child processes are added to the linked list *WaitList* because waiting workers may become busy at some time. Workers that are in neither *IdleStack* nor *WaitList* are busy. At first all the workers are idle. Each worker has the three processing units and the six regions so that the amount of messages passed between processors should not explode. That is why each worker can collect garbages in its own *CODE* independent of the others. *ForkTable* will be described later.

Given a set of rewrite rules, the master receives and forwards it to all the workers. Each worker compiles it with its rule compiler in the same way as PTRAM. Given a term afterward, the master receives and forwards it to some worker, e.g. Worker1, and then the worker becomes busy and compiles it with its term compiler like PTRAM. Subsequently, the worker interprets the matching program according to the strategy list with its interpreter so as to rewrite the original term. If a worker meets the fork instruction, it asks the master if there is some available worker. The master searches *IdleStack* and then *WaitList* for an available worker, and changes the available worker to busy and sends its ID to the client worker if exists, or informs the client of no available worker otherwise. The client worker asks the child worker with the ID to rewrite some argument if

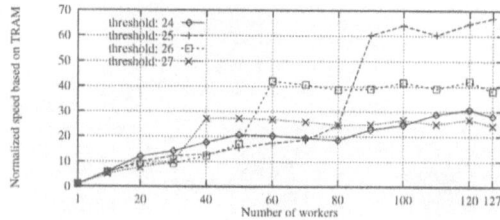

Fig. 3. Experimental results of computing the 34th Fibonacci number

it receives the ID from the master, or otherwise rewrites the argument by itself. If a worker meets the `join` instruction, it basically asks the master whether it can go on, or it must wait for their child workers, in which case it becomes waiting. The master judges this by looking at *ForkTable* in which it keeps a record of how many child workers a worker is waiting for at each join point that is a place where the `join` instruction is referred in the worker's strategy list. If a worker meets the `exit` instruction, it sends the result term to the parent worker, which is either waiting or busy, and notifies the master that it has finished a request rewriting. After the notification, the master decreases the number of how many child workers the parent worker is waiting for at the corresponding join point in *ForkTable*, and makes the parent worker busy and has the parent worker resume if the number becomes zero and there is no other processes above the process corresponding to the join point in the parent worker's strategy list.

Each worker must have some table in which it keeps a record of references to terms that it has asked other workers to rewrite. Since those references become necessary so that a worker can go on rewriting after executing the `join` instruction, the necessary references are put just after each point at which the `join` instruction is referred in the strategy list.

4 Implementation of MPTRAM on T3E with MPI

Implementation. We have experimentally implemented MPTRAM on Cray T3E [1] with MPI [8]. The Cray T3E used carries 128 user processing elements (PEs), each of which has the DECchip 21164 64-bit super-scalar RISC (300MHz DEC Alpha EV5) and 64MB local memory. The PEs are connected by a bidirectional three-dimensional torus system interconnect network. The MPTRAM system uses two intracommunicators called *Master* and *Workers*, and one intercommunicator called *Section* between *Master* and *Workers*. An idle worker waits for a message that informs it that some busy worker is going to ask it to rewrite some term from the master with the intercommunicator *Section*. After it receives such a message, it waits for a message that includes a term from a busy worker with the intracommunicator *Workers*. When a waiting worker waits for some messages from the master or workers that the waiting worker has asked to evaluate some subterms, it is necessary to use the nonblocking receive operation because the waiting worker cannot generally predict the order in which messages will arrive.

Preliminary Experiment. A set of rewrite rules that compute Fibonacci numbers in parallel was used to assess the MPTRAM system on Cray T3E. The rewrite rules used primitive integers (DEC Alpha EV5's integers). They also used a threshold to prevent excess number of processes. If the nth Fibonacci number, where n is less than the threshold, is computed, the rewrite rules compute it sequentially, not in parallel. We had the MPTRAM system compute (rewrite) the 34th Fibonacci number with the rewrite rules in varying the number of workers used from 1 to 127 and also the threshold from 24 to 27. Fig. 3 shows the normalized speed of the MPTRAM system based on TRAM. The result corresponding to each threshold except for 24 has a similar tendency. The rewriting speed drastically increases at some point: 90, 60 and 40 workers if the thresholds are 25, 26 and 27 respectively. This seems to relate to how many processes are created because the fork instruction is executed 88, 54 and 33 times if the thresholds are 25, 26 and 27 respectively. The rewriting speed drastically increases after the number of workers is over the number of forks executed in each case. If the threshold is 24, the fork instruction is executed 143 times that is more than actual PEs. That is why the result has the different tendency from the others.

We used another set of rewrite rules, in which natural numbers were represented á la Peano, to compute the 25th Fibonacci number so as to investigate how the amount of messages passed affected the increase in rewriting speed. The threshold was 14. The speedup was not over twice even with 90 workers while the speedup was approximately nine times if primitive integers were used. Hence we have to reduce the overhead of message passing somehow, e.g. by using the shared globally addressable memory subsystem of Cray T3E, so that MPTRAM can be used to implement efficiently algebraic specification languages.

References

1. Cray T3E: http://www.cray.com/products/systems/crayt3e
2. Futatsugi, K., Goguen, J. A., Jouannaud, J. P. and Meseguer, J.: Principles of OBJ2. Proc. of POPL'85. (1985) 52–66
3. Goguen, J., Kirchner, C. and Meseguer, J.: Concurrent Term Rewriting as a Model of Computation. LNCS 279 (Graph Reduction'86) Springer-Verlag. (1986) 53–93
4. Klop, J. W.: Term Rewriting Systems. In *Handbook of Logic in Computer Science.* **2.** Oxford University Press. (1992) 1–116
5. Ogata, K., Ohhara, K. and Futatsugi, K.: TRAM: An Abstract Machine for Order-Sorted Conditional Term Rewriting Systems. LNCS 1232 (RTA'97) Springer-Verlag. (1997) 335–338
6. Ogata, K. and Futatsugi, K.: Implementation of Term Rewritings with the Evaluation Strategy. LNCS 1292 (PLILP'97) Springer-Verlag. (1997) 225–239
7. Ogata, K., Kondo, M., Ioroi, S. and Futatsugi, K.: Design and Implementation of Parallel TRAM. LNCS 1300 (Euro-Par'97) Springer-Verlag. (1997) 1209–1216
8. Snir, M., Otto, S. W., Lederman, S. H., Walker, D. W. and Dongarra, J.: *MPI: The Complete Reference.* The MIT Press. 1996

Parallel Temporal Tableaux*

R. I. Scott[†], M. D. Fisher[‡] and J. A. Keane[†]

[†] Department of Computation,
UMIST, Manchester UK,
[‡] Department of Computing & Mathematics,
Manchester Metropolitan University, UK.

Abstract. Temporal logic is useful for reasoning about time-varying properties. To achieve this it is important to validate temporal logic specifications. The decision problem for propostional temporal logic is PSPACE complete and thus automated tableau-based theorem provers are both computationally and resource intensive. Here we analyse a standard temporal tableau method using parallelism in an attempt to both reduce the amount of storage needed and improve efficiency. Two shared memory parallel systems are presented which are based on this decomposition of the algorithm but differ in the configuration of parallel processes.

1 Introduction

Temporal logic has been shown [4] to be a powerful tool for reasoning about properties that vary over time. It is important that we are able to test specifications expressed in temporal logic for validity. A decision procedure for determining the validity of a temporal logic formula is given in [6]. As the decision problem for propositional temporal logic (PTL) is PSPACE-complete, and because temporal specifications are typically large, an automated theorem prover based on this algorithm would be computationally intensive and would potentially require a large amount of memory.

The aim of this work is to address ways in which we can minimise the amount of memory required and reduce the computational intensity by utilising parallel systems within temporal tableaux. Wolper's algorithm [6] has two phases in which a graph (representing possible models of the temporal formula) is constructed and, when construction is complete, systematically reduced.

Parallelising Wolper's algorithm has two potential benefits: (1) the reduction of the amount of memory consumed by deleting nodes from the graph during the construction phase; and (2) improving overall efficiency by decreasing the overall time taken.

The paper is structured as follows: in §2 we define PTL; in §3 we describe Wolper's decision procedure for PTL; in §4 we analyse the algorithm in order to identify those operations which can be performed on the graph in parallel;

---

* This work was partially supported by EPSRC grant GR/J48979

from this, in §5, we outline a shared memory parallel program design and show how, by grouping together different operations, the granularity of processes can be refined; in §6 we present results of the parallel temporal tableaux algorithm on a range of parallel systems; conclusions are presented in §7.

2 Propositional Temporal Logic (PTL)

Here we consider a discrete linear future-time temporal logic [1]. The alphabet of this logic is that of classical propositional logic augmented with the unary operators □ ("always"), ◇ ("sometime"), ○ ("next-time") and the binary operators U ("until") and W ("unless"). Informally, $\Box f$ means that f is true in all future states, $\Diamond f$ means that f is true in some future state, $\bigcirc f$ means that f is true in the next state and fUg means that f is true in all future states until g becomes true. fWg is a weaker version of fUg where g is not guaranteed to occur.

An interpretation for PTL is a pair $\langle \sigma, s_0 \rangle$ where σ provides an interpretation for the propositions of the language in each state and s_0 is some initial state. As we are dealing with a linear model of time, each state has a unique successor state. Further discussion of temporal logics can be found in [1]

3 Wolper's Decision Procedure for PTL

Wolper's method for determining the validity of a temporal logic formula is an extension of the semantic tableau decision procedure for propositional logic [1]. As a refutation procedure, it attempts to construct a model for the negation of the formula to be tested. If a model cannot be found, then the negation is unsatisfiable and hence the initial formula is valid. The algorithm works on the observation that a PTL formula can be decomposed into subformulae that are true in the current state or the next state in time. Thus we can attempt to construct a model state by state and test for satisfiability. The algorithm consists of two phases: *Graph Construction*, and *Graph Reduction*.

In the construction phase the logical structure of the formula is abstracted into the form of a graph, whose nodes are labelled with sets of formulae. Models are represented by infinite paths in the graph. The graph is built by applying certain rules to the formulae depending on whether they are disjunctive or conjunctive. Once construction is complete, nodes are deleted according to some reduction rules. If this reduction phase deletes the entire graph then the original formula is valid.

We can classify temporal logic formulae into four types: α formulae (conjunctions); β formulae (disjunctions); next-time formulae (those whose leading connective is ○) and literals (atomic propositions and their negations). *Elementary formulae* are defined to be either literals or next-time formulae. During construction, tableau expansion rules are applied to α and β formulae, these are:

- $\alpha \to \{\{\alpha_1, \alpha_2\}\}$, and
- $\beta \to \{\{\beta_1\}, \{\beta_2\}\}$.

The α and β sub-formulae are given in Table 1.

Table 1. α and β formulae

α	α_1	α_2
$A_1 \wedge A_2$	A_1	A_2
$\square A$	A	$\bigcirc\square A$

β	β_1	β_2
$B_1 \vee B_2$	B_1	B_2
$\Diamond B$	B	$\bigcirc\Diamond B$
$B_1 U B_2$	B_2	$B_1 \wedge \bigcirc(B_1 U B_2)$
$B_1 W B_2$	B_2	$B_1 \wedge \bigcirc(B_1 W B_2)$

We now semi-formally outline the graph construction and graph reduction stages of the algorithm (for a more formal description, see [6]).

3.1 Graph Construction

The digraph $G(N, E, L)$, where N is the set of nodes (which may be marked or unmarked) in the graph, E is the set of edges and L is a function which maps a node n to a set of formulae L_n, is constructed as follows:

1. Start with $N := \{n_1\}$, where n_1 is the initial node and is labelled with the set of formulae $L_{n_1} := \{f\}$ where f, the initial formula, is the negation of the formula to be tested. $E := \emptyset$.
2. While there exists an unmarked node $n \in N$, labelled with the set of formulae L_n, repeatedly apply the following rules:
   (a) if $f \in L_n$ where f is an unmarked non-elementary formula then for all S_i in the tableau expansion rule $(f \to \{S_i\})$, create a new node n' labelled with $L_{n'} := L_n - \{f\} \cup \{S_i\} \cup \{f*\}$ where $f*$ is f marked. $N := N \cup \{n'\}$ and $E := E \cup \{(n, n')\}$
   (b) if L_n contains only elementary and/or marked formulae, create a new node n' where $L_{n'} := \{f \mid \bigcirc f \in L_n\}$ $N := N \cup \{n'\}$ and $E := E \cup \{(n, n')\}$

N.B. when we write $N := N \cup \{n'\}$, we mean that n' is added to N if $\forall n_i \in N.L_{n_i} \neq L_{n'}$, otherwise an edge to the existing node is created. Nodes added by rule 2. (b) are called *prestates* and the parent of a prestate is called a *state*.

3.2 Graph Reduction

The digraph G is reduced by repeatedly applying the following reduction rules:

1. Delete nodes containing *complementary formulae.* $\forall n \in N.\ \forall f \in L_n,\ (\neg f \in L_n \Rightarrow (N = N - \{n\}))$
2. Delete nodes with *no successors.* $\forall n \in N,\ \forall n_i \in N.\ (n, n_i) \notin E \Rightarrow (N = N - \{n\})$
3. Delete prestates containing *unsatisfiable eventualities.*
   (a) $\forall n \in N.\ prestate(n).\ \exists \Diamond f \in L_n \wedge \forall n_i \in N.\ reachablefrom(n_i, n) \wedge f \notin L_{n_i} \Rightarrow (N = N - \{n\})$
   (b) $\forall n \in N.\ prestate(n).\ \exists fUg \in L_n \wedge \forall n_i \in N.\ reachablefrom(n_i, n) \wedge g \notin L_{n_i} \Rightarrow (N = N - \{n\})$

N.B. We say that the formulae $\Diamond g$ and fUg have associated *eventuality g* because both formulas imply that eventually there is a state in which g is satisfied. The predicate $reachablefrom(n_i, n_j)$ is true iff there exists a directed path in the graph from n_j to n_i.

4 Opportunities for Parallelism

To identify possibilities for parallel activity, we divide the sequential algorithm into separate sub-problems. As described above, expansion rules are repeatedly applied to the sets of formulae labelling pre-states. A new node is then added to the graph for each of the new formula sets derived (if no node exists with an identical label) and then a prestate is derived from these states. These prestates can be expanded in parallel as can the two formula sets created upon the application of a β expansion rule. When a set of formulae is fully expanded, a new node is added to the graph only if there is no other node with the same label. Therefore, the graph has to be searched for duplicate labels. Here, two approaches were considered: (1) nodes are added to the graph and a *marker* process can be used to detect duplicate labels; or (2) the graph is searched for a label before any node is added to the graph.

The advantage of the first approach is that it does not restrict the amount of parallel activity, but the graph may become unnecessarily large and the *marker* process could become overloaded. The second approach means that the graph size is kept to a minimum as no speculative work is performed by processes applying tableau expansion rules and adding nodes to the graph. Searching the graph before adding a node can be done in parallel, but synchronisation is required between a process searching the graph and a process adding a new node to the graph. The label of any node added to the graph must be visible to a process concurrently searching the graph for a label to be certain that no two nodes have the same label. This constraint may result in a process having to lock the entire graph while it searches for a particular label.

The graph is reduced by deleting nodes with labels which contain contradictory formulae, with no successors, or whose labels contain unsatisfied eventualities.

The first of these rules can be applied during construction by simply discarding the contradictory formula set, as no global checking of the graph is required.

The other two reduction rules need more consideration. If a process is to delete a node during construction because it has no successors, it must ensure that the node will *never* have any children, and similarly, to delete a node because it contains an unsatisfied eventuality a process must be certain that the eventuality will *never* be satisfied. Coordination is again needed between a process deleting nodes from the graph and a process simultaneously traversing the graph, for instance during search to prevent the actions of the latter from being corrupted. Therefore, it may be more efficient to concurrently mark nodes for deletion, yet actually delete them sequentially at some future point.

As a result we can divide the algorithm into the following operations which may be performed in parallel (these operations are similar to those presented in [3]):

OP1 - applying tableau expansion rules to sets of formulae;
OP2 - searching the graph for duplicate node labels;
OP3 - adding a new node to the graph;
OP4 - identifying nodes which contain contradictory formulae;
OP5 - identifying nodes with no successors;
OP6 - identifying prestates which contain unsatisfied eventualities;
OP7 - deleting nodes identified by OP4, OP5 or OP6.

5 Program Design

The operations described above can be implemented as individual fine-grained processes or grouped together to form processes with a coarser granularity. Below, we describe a medium-grained shared memory program model. This design was then analysed and the configuration of the processes was modified in an attempt to improve efficiency.

5.1 Heterogeneous Processes - Model 1

This model consists of four processes: two *constructors*; a *searcher*; and a *checker*. Each process has read and write access to a shared structure - the graph. The constructors obtain and expand unmarked labels from the graph (OP1). Upon the application of a β rule, two sets of formulae are created, a constructor retains one set and pushes the other onto a shared queue (to which both constructors have access). Sets which contain complementary formulae are simply discarded by the constructor (OP4). Once a formula set has been obtained from the graph (or the shared queue), the constructor does not require further access to the graph and thus the amount of coordination required between constructors is small. Therefore, we allow multiple instances of the constructor process in our model. Fully expanded formula sets are passed to the searcher via *queue 1*.

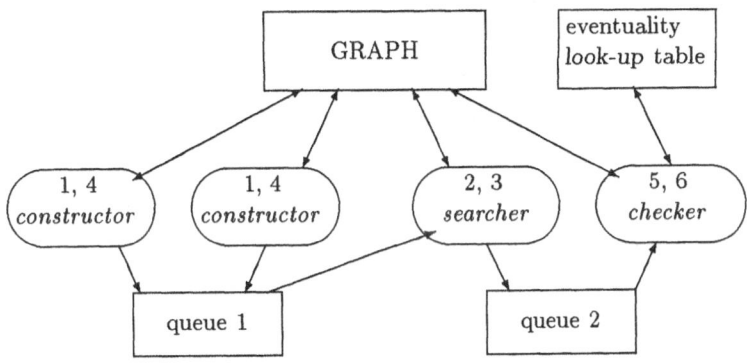

Fig. 1. Program Design with 4 Heterogeneous Processes

The searcher performs OP2 and OP3 and instructs the checker that the graph has been updated via *queue 2*. The checker is responsible for the identification of nodes which can be deleted while the graph is being constructed. The process of identifying nodes that can be deleted is similar to the problem of garbage collection in functional programming languages [5]. The two methods described below for OP5 and OP6 are similar to reference count garbage collection method [5]. To perform OP5 a process has to be sure that a node with no successor nodes can never have any more before marking it for deletion. This is achieved by maintaining a *reference count* for each node. When a node is added to the graph, its *reference count* is set to 1. The node's count is incremented when a β rule is applied during its expansion, and is decremented when a formula set is discarded due to OP4 or when a node or edge is added from it. Thus, if a node's *reference count* is equal to zero and it has no children, a process can be sure that no more successors will be added to the node.

In addition to building up a picture of which eventualities are satisfied at which nodes, a process performing OP6 must also identify when an eventuality can never be satisfied, i.e. when a node cannot reach any more nodes as a result of a node or edge being added to the graph. When a node is added to the graph, this new label has to be checked to see whether it satisfies any of the eventualities that can reach it and when an edge (n_1, n_2) is added to the graph, all the nodes that n_2 can reach must be checked against all the eventualities that can reach n_1. Therefore a process has to identify the subgraph that can be reached by a particular node. The identification of subgraphs is complicated by the existence of loops in the graph and some measure has to be taken to detect this. As the process traverses the graph identifying nodes in a subgraph, it tags each node through which it passes, thus ensuring that it passes through a node once only.

The system must record where eventualities are satisfied and so an *Eventuality Look-Up Table* is maintained which contains information on to which node an eventuality is associated and which nodes satisfy it. In order to delete a node because it is labelled with an unsatisfied eventuality a process must be certain that the node can have no more descendants. This is achieved by maintaining a set, $reachset_n$, of node identifiers for each node n. This set contains the identifiers of the nodes in the subgraph reachable from n which could yet have more successors, i.e. their *reference count* is not zero. These *reachsets* can be maintained during the identification of subgraphs. As both OP5 and OP6 involve the identification of subgraphs, we combine them into one checker process.

The checker is notified of any additions to the graph by the searcher. If a node, n, is added it obtains any eventualities associated with the subgraph that can reach n, updating the relevant *reachsets* as it does so, and checks them against the label of n. If an edge (n_1, n_2) is added it identifies the subgraph, SG_1, that can reach n_1 and the subgraph, SG_2, reachable from n_2, again updating the relevant *reachsets* and checks the labels in SG_2 against any eventualities associated with SG_1.

The actions of a process deleting nodes from the graph (OP7) and those of a process traversing the graph have to be synchronised to ensure that the former process does not corrupt the latter. Locking each node upon access could remove this possibility but could prove unnecessarily restrictive, especially in an example where no nodes can be deleted concurrently. Here the approach taken for OP7 is to suspend the parallel processes and delete nodes sequentially when a predefined number of nodes can be deleted.

5.2   Homogeneous Processes - Model 2

Model 1 was designed to minimise inter-process synchronisation by grouping similar operations into the same process. By statically determining the configuration of the processes in this way it is unlikely that the load across the four processes will be balanced. Neither is model 1 particularly scalable: we can allow more instances of the *constructor* process but we are restricted to single instances of the *seacher* and *checker* processes because of the need for sequential update.

An alternative model has four homogeneous coarse-grained processes which sequentially perform OP1 to OP6. Here OP2, OP3, OP5 and OP6 are only performed once a formula set has been fully expanded and so no process is idle - as long as there are enough sets of formulae needing expansion. We now have more than one process able to update the graph and so some synchronisation is required. Before OP3 is performed we must ensure that the graph contains no node with label similar to the new label. Thus, we have to deal with the possibility that two processes may attempt to add nodes with identical labels at the same time. To facilitate this only one insertion point for new nodes is permitted. When a process has completed OP2, it attempts to lock the last node of a linked list (the point of insertion). If, when the lock is obtained, the locked node is no longer the last node of the linked list, the process knows that more nodes have been added by another process and OP2 is not complete and so the process again searches to the end of the linked-list and attempts to lock the last node. If the locked node is still the last node in the list, then the process can be sure that no more nodes will be added until it releases the lock and thus the graph can never contain two nodes with identical labels. Similarly, OP5 and OP6 need access to the entire graph and so some synchronisation is required - a node is locked when its *reference count* or its *reachset* is updated. OP7 is again performed at some fixed points during construction.

6 Results

The two models were implemented on a Silicon Graphics 4-processor shared memory machine. A sequential version was also implemented for comparison purposes. These programs were tested on four reasonably large formulae which have previously been used to test sequential temporal tableau systems [2]. The sizes of the four formulae are summarised in Table 2.

Table 2. Summary of Test Formulae

Formula	Length/chars	Number of Subformulae
1	876	213
2	422	142
3	116	99
4	418	130

Each program - sequential, model 1 and model 2 - require some preprocessing, this involves converting the formula to post-fix form and building a *formula table* so that a formula can subsequently be referred to simply by an identifier. These preprocessing times are not included in Table 3, containing the timings of the three programs on the four formulae. As an indicator of the nature of the computation involved, it is useful to compare the maximum size of the graph produced in the sequential case, where deletions are carried out once the graph is fully constructed, with that of the parallel case, when nodes are deleted during construction. These are also shown in Table 3.

Table 3. Timings for Each Model & Maximum Graph Size for Sequential and Parallel Models

	TIME			MAX GRAPH SIZE	
Formula	Sequential	Model 1	Model 2	Sequential	Parallel
1	0m41.08s	0m32.97	0m16.57s	1	1
2	28m31.33s	19m27.28s	13m47.13s	197	84
3	0m31.29s	0m49.24s	0m19.87s	257	257
4	3m58.62s	4m51.68s	2m19.32s	134	130

As can be seen in Table 3, Model 2 exhibits some speed-up on all four formulas whereas Model 1 performs better than the sequential implementation on only two occasions. This may be explained by the dependencies between operations, for example OP2 and OP3 rely on OP1, and by the fact that it is difficult to predict the nature of a computation for a given example. Model 1 assigns operations to processors statically, whereas Model 2 is more dynamic, operations 2, 3, 5 and 6 are performed *on demand*. Model 2 does not exhibit ideal speed-up because of the amount of inter-process synchronisation required when accessing the graph and because of the extra work needed to delete nodes concurrently.

7 Conclusions and Further Work

Automated reasoning methods for temporal logics consume a large amount of memory and can be slow. This restricts the size of temporal theorem that can be validated. Our aim, in parallelising the decision procedure for these logics, was two-fold: to reduce the time taken to prove a theorem; and to reduce the size of graph produced in proving a theorem.

The algorithm was divided into seven concurrent operations. Two parallel models were designed and implemented along with a sequential algorithm. The heterogeneous model proved to be slower than the sequential algorithm for two of the four formulas tested, whereas the second homogeneous model exhibited satisfactory speed-ups for all four formulas. This can be attributed to the more dynamic nature of the second model.

We have also compared the maximum size of the graph produced for the sequential and parallel cases. The sequential algorithm deletes nodes from the graph only when construction is complete whereas the parallel models attempt to delete as many nodes as possible during construction. A considerable amount of deletion was possible for formula 2 (over 50%), but very little parallel deletion was achieved for the other formulas. If the amount of concurrent reduction possible could be predicted by some pre-analysis, efficiency may be improved by deciding not to test for concurrent reduction if it were deemed unlikely. Unfortunately, formulas 2 and 4 are similar in length and structure and yet formula 2 allows much more reduction.

The parallel temporal logic theorem prover outlined above achieves reasonable speed-up on all the temporal formulas tested. We have also shown that significant parallel graph reduction is possible for some but not all formulas. Thus, for some cases, a parallel theorem prover may not be able to handle larger formulae by concurrently reducing the graph.

References

1. M. Ben-Ari. Mathematical Logic for Computer Science. Prentice Hall, 1993.
2. G. D. Gough. Decision Procedures for Temporal Logic. Technical Report UMCS-89-10-1, Department of Comp. Sci., Univ. of Manchester, 1984.
3. R. Johnson. A BlackBoard Approach to Parallel Temporal Tableaux. In *Proc. of Artificial Intelligence, Methodologies, Systems, and Applications*, World Scientific, 1994.

4. Z. Manna and A. Pneuli. Verification of Concurrent Programs: The Temporal Framework. In R.S. Boyer and J.S. Moore (Eds.), *The Correctness Problem in Computer Science*, Academic Press, 1982.

5. R. Plasmeijer and M. van Eekelen. Functional Programming and Parallel Graph Rewriting. Addison-Wesley, 1993.

6. P. Wolper. Temporal Logic Can Be More Expressive. *Information and Control*, 56, p72-99, 1983.

Workshop 10+17+21+22
Theory and Algorithms for Parallel Computation

Bill McColl and David Walker

Co-chairmen

Theory and Algorithms for Parallel Computation

Bridging models such as BSP and LogP are playing an increasingly important role in scalable parallel computing. They provide a convenient architecture-independent framework for the design, analysis, implementation and comparison of parallel algorithms and data structures.

The Theory and Algorithms sessions at EuroPar '98 will cover a number of important issues concerning models of parallel computation. The topics covered will include: The BSP model and its relationship to LogP, BSP algorithms for PC clusters, distributed shared memory models, computational complexity of parallel algorithms, BSP algorithms for discrete event simulation, cost modelling of shared data types, systolic algorithms, cache optimisation, optimal scheduling of task graphs.

Parallel I/O

In recent years it has become increasingly apparent that I/O presents significant challenges to the goal of making parallel applications truly portable, while atthe same time maintaining good performance. The need for a parallel I/O standard has become particularly acute as more and more large-scale scientific applications have migrated to parallel systems. Many of these applications produce hundreds or gigabytes of output which would result in a serious bottleneck if handled sequentially. Parallel I/O is also of central importance in very large-scale problems requiring out-of-core solution, and is closely related to the subject of distributed file systems. Hardware vendors have developed customized parallel I/O subsystems for their particular platforms, but at most this only addresses portability within a vendor's family of systems. Recently, the MPI Forum and published a standard specification for parallel I/O called MPI-IO. This has not yet been widely implemented, and much research remains to be done before it can be adequately assessed. The Scalable I/O Initiative has also addressed related issues. Nevertheless, parallel I/O remains a wide open field, and a fertile research area.

The parallel I/O session of EuroPar'98 contains three papers that address different aspects of parallel I/O. The first, by Jensch, Lüling, and Sensen, is of interest to anyone who has ever been frustrated by the slow response time of a popular web site. They consider the problem of how the average response time can be speeded up by mapping the site's files over multiple disks in such a way that the overall throughout from the disks to the outside world is maximised. This is done by assigning files to disks in such a way that collisions, i.e., quasi-simultaneous accesses to files on the same disk, are minimised. A graph theoretic approach is used to solve the optimisation problem, and found to be effective when compared with a random assignment strategy. The second paper, by Schikuta, Fuerle, and Wanek, describes ViPIOS, the Vienna Parallel I/O System. ViPIOS uses a client-server approach to parallel I/O which seeks to achieve maximum I/O bandwidth by reacting dynamically to the behavior of the application. Applications are the clients whose I/O requests are serviced by one or more ViPIOS servers. ViPIOS can be used either as an independent sub-system, or as a runtime library. The last paper by Dickens and Thakur describes different approaches to two-phase I/O, and presents performance results for the Intel Paragon. The two-phase approach is a technique often used for performing parallel I/O in which processors coordinate their I/O requests so that a few requests for large contiguous blocks of data are made, rather than a larger number of requests for fragmented data. This reduces latency and improves I/O performance. One phase combines I/O requests, and the other phase actually performs the I/O.

The three papers presented in this session give some idea of the diversity of ideas currently being pursued in the search for a solution to the parallel I/O problem. For those interested in finding out more about the subject the parallel I/O archive at Dartmouth University is an excellent place to begin: http://www.cs.dartmouth.edu/pario/.

BSP, LogP, and Oblivious Programs

Jörn Eisenbiegler   Welf Löwe   Wolf Zimmermann

Institut für Programmstrukturen und Datenorganisation, Universität Karlsruhe,
76128 Karlsruhe, Germany,
{eisen|loewe|zimmer}@ipd.info.uni-karlsruhe.de

Abstract. We compare the BSP and the LogP model from a practical point of view. Using compilation instead of interpretation improves the (best known) simulations of BSP programs on LogP machines by a factor of $O(\log P)$ for oblivious programs. We show that the runtime decreases for classes of oblivious BSP programs if they are compiled into LogP programs instead of executed directly using a BSP runtime library. Measurements support the statements above.

1   Introduction

Parallel programming suffers from the lack of a uniform and commonly accepted machine model providing abstract programming of parallel machines and describing their costs adequately. Two candidates, the BSP model (Valiant [9]) and the LogP model (Culler et al. [4]), have been considered in an increasing number of papers. The comparison of the two models in [2] determines the delays for a simulation of LogP programs on the BSP machine and vice versa. For our observations we make two additional assumptions: we only consider oblivious programs and message passing architectures. A program is *oblivious* if source and destination processors of communications are statically determined[1]. The target machines are processor-memory-nodes connected by a communication network. We explicitly exclude shared memory machines. Virtual shared memory architectures are covered since they implicitly require communication via the interconnection network for remote memory operations. We only mention send and receive communications and deliberately ignore remote store and load operations.

The first part of our paper shows that oblivious BSP programs can be compiled to the LogP machine. We further show, that compilation reduces the delay of the simulation of oblivious BSP programs on the LogP machine by a factor of $O(\log(P))$ compared to the result in [2], i.e., there is asymptotically no delay for the compiled LogP program compared to the BSP program. Even better, it turns out that the compiled LogP program could outperform a direct execution of the BSP program on the same architecture. To sharpen this observation, we consider three classes of oblivious programs: first we study Multiple-Program-Multiple-Data (MPMD) solutions. Those programs are in general hard to partition into

---

[1] Many algorithms, especially those in scientific computing are oblivious, e.g. Matrix Multiplication, Discrete Simulation, Fast Fourier Transform, etc.

supersteps, i.e. in global phases of receive-, compute-, and send-operations. As an example, we discuss the optimal broadcast problem. The second ánd third classes of problems allow Single-Program-Multiple-Data (SPMD) solutions and there is a natural partition into supersteps. They differ in the data dependencies between the phases: in the second class there are sparse dependencies between the phases in third class these dependencies are dense. We call a data dependency sparse iff the BSP communication of each superstep last longer than the corresponding communication in the compiled LogP program. As representatives of the second class of problems, we discuss a numeric wave simulation. The fast Fourier transform is a representative of the third class. Measurement of the parameters and run time results for an optimal broadcast, the simulation and the FFT support our theoretical results.

2 The Machine Models

This section describes the two machine models a little more in detail. In order to distinguish between the parameters of both models, we use capital letters for the parameters of the LogP model and small letters for the parameters of the BSP model.

2.1 The LogP Model

The LogP model assumes a finite number P of processors with local memory, which are connected by a data network. It abstracts from the network topology, presuming that the position of the processor in the network has no effect on communication costs. Each processor has its own clock, synchronization and communication is done via message passing. All send and receive operations are initiated by the processor which sends or receives, respectively. From the programmers point of view, the network has no direct connection to the local memory. All communication is done via the processor.

In the LogP model, communication costs are determined by the parameters L, O, and G. Sending a message costs time O (overhead) on the processor. The time the network connection of this processor is busy with sending the message into the network is bound by G (gap). A processor can not send or receive two messages within time G, but if a processor returns from a send or receive routine, the difference between gap and overhead can be used for computation. The time between the end of sending a message and the start of receiving this message is defined as latency L. There are most $\lceil L/G \rceil$ messages in transit from any or to any processor at any time, otherwise, the communication stalls. We only consider programs satisfying this capacity constraint. If the sending processor is still busy with sending the last bytes of a message while the receiving processor is already busy with receiving, the send and the receive overhead for this message overlap. In this case the latency is negative. This happens on many systems especially for long messages or if the communication protocol is too complicated. L, O, and G have been determined for quite a number of machines; all works confirmed runtime predictions based on the parameters by measurements. In contrast to [1] and [5], we assume the LogP parameters to be constants (as proposed in the early LogP works [4]). This assumption is admissible if the message size does not vary in a single program.

2.2 The BSP Model

The BSP (bulk synchronous parallel) machine was defined by Valiant [9]. We refer to its modification by McColl in [8,6]. Like the LogP model, the BSP model assumes a finite number of processors P with local memory, local clock, and a network connection to an arbitrary network. It also abstracts from the network topology. In contrast to the LogP model, the BSP machine can explicitly (barrier) synchronize all processors. The synchronization barriers subdivides the calculation into *supersteps*. All send operations in a superstep i are guaranteed to be completed before superstep $i+1$. In the BSP model as invented by Valiant, processor communicate via remote memory access. For oblivious programs we may focus on a message based communication: Consider the remote memory accesses at one superstep. If processor π_i reads from processor π_j, then π_i should send a request to π_j and π_j sends its answer for the general case. However, for oblivious algorithms, it is already known that π_i reads from the memory of π_j. Thus, the request can be saved in the case of oblivious algorithms: it is sufficient to send the result of the read request from π_j to π_i. A write of processor π_i to the memory of processor π_j, is equivalent to sending a message from processor π_i to π_j containing the memory address and the value to be written.

The cost model uses two parameters: the time for the barrier synchronization l, and the reciprocal of the network bandwidth g. With theses parameters, the time for one superstep is bounded by $l + 2h \cdot g + w$, where h is the maximal number of messages sent or received by one processors and w is the maximal computation time needed by one processor in this superstep. A BSP machine is able to route a $\lceil l/g \rceil$-relation in a superstep which is a capacity constraint analogous to the LogP model. The total computation time is the sum of the time for all supersteps. Like for the LogP model, we assume the parameters to be constant.

3 BSP vs. LogP for Oblivious Algorithms

In this section, we discuss the compilation of oblivious BSP programs to the LogP machine. First, we discuss how the communication between subsequent supersteps can be mapped onto the LogP machine without exceeding the LogP capacity constraints. Second, we define the actual compilation and prove execution time bounds for the compiled LogP programs. Third, we compare the direct execution of a BSP program on a target machine with the execution of the (compiled) LogP program. Therefore, we conclude this section by determining lower bounds for the BSP and LogP parameters, respectively, for the same target machine.

3.1 Communication of a Superstep on the LogP Machine

For simplicity, we assume that a h-relation is implemented, i.e. each processor sends and receives exactly h messages. It is well known that a "pipelined" communication can be computed using edge coloring on a bipartite graph (U, V, E) where U and V are the set of processors and $(u, v) \in E$, iff u communicates with v. Each color, represented by an integer $j \in \{0, \ldots, k - 1\}$ where k is the number of required colors, defines set of non-conflicting communications that can be started simultaneously. A $send(v)$ on processor u is scheduled at time $j \cdot \max(O, G)$ and $recv(u)$ on processor v at time $L + O + j \cdot \max(O, G)$.

Since we consider oblivious BSP algorithms, the edge coloring of the communication graph for each superstep (and therefore the communication phase itself) can be computed prior to execution of the BSP algorithm. Thus, the time for edge coloring ($\mathcal{O}(|E| \log(|V| + |U|)$ due to [3]) can be ignored when considering the execution time of the BSP algorithm.

It is easy to see that the schedule obtained from the above algorithm does not violate the capacity LogP constraints if $L \geq (h - 1) \cdot \max(O, G)$. If follows

Lemma 1. *If* $L \geq (h - 1) \max(O, G)$, *then every fixed h-relation can be implemented on the LogP machine such that its execution time is* $L + 2O + (h - 1) \max(O, G)$.

If $L < (h - 1) \max(O, G)$ the scheduling algorithm must be modified to avoid stalling. First we discuss the simplified model where each channel is allowed to contain an arbitrary number of messages. In this case, the first receive operation is performed after the last send operation. The same approach as above then yields execution time $2O + 2(h-1) \max(O, G)$ since the first receive operation can be performed at time $O + (h-1) \max(O, G)$ instead of $O + L < (h-1) \max(O, G)$. If the number of messages is bounded, the message is received greedily, i.e. as soon as possible after a send operation on the processor is finished at the time when the message arrives. This does not increase the overall execution time of a communication phase since no new gaps are introduced. Thus, the following lemma holds:

Lemma 2. *If* $L < (h - 1) \max(O, G)$, *then every fixed h-relation can be implemented on the LogP machine such that its execution time is* $2O + 2(h - 1) \max(O, G)$.

3.2 Execution Time Bounds for the Compiled LogP Programs

The actual compilation of an oblivious BSP algorithms to the LogP machine is now straightforward: Each BSP processor corresponds one to one to a LogP processor. Beginning with the first superstep we map the tasks of BSP processor to a corresponding LogP processor in the same order. Communication is mapped as described in the previous subsection. Since a processor can proceed its execution when it received all its messages of the preceding superstep, there is no need for a barrier synchronization. Together with Lemmas 1 and 2, this observation leads to the

Theorem 1 (Simulation of BSP on LogP). *Every superstep of an oblivious BSP algorithm with work w and h remote memory accesses can be implemented on the LogP Machine in time $w + 2O + (h-1)\max(O,G) + \max(L, (h-1)\max(O,G))$.*

If we choose $g = \max(O,G)$ and $l = \max(L, (h-1)\max(O,G)) + 2O - \max(G - O, 0)$ then, the execution time of a BSP algorithm in the BSP model and the compiled BSP algorithms in the LogP model is the same. Especially, the bound for the simulation in [2] is improved by a factor of $\log P$ for oblivious BSP programs.

3.3 Interpretation vs. Compilation

For the comparision of a direct execution of a BSP program with the compiled LogP program, the parameters for the BSP machine and LogP machine, respectively, cannot be chosen arbitrarily. They are determined by the target architecture, for comparable runtime predictions we must choose the smallest admissible values for the respective model.

A superstep implementing a h-relation costs in the BSP-model $w + l + h \cdot g$. According to Theorem 1, it can be executed in time $w + 2O + (h-1)\max(O,G) + \max(L, (h-1)\max(O,G))$ on the LogP machine. The speedup is the ratio of these two numbers. Easy calculations prove the following

Corollary 1 (Interpreting vs. Compiling). *Let M be a parallel computer with BSP parameters $l, g,$ and P and LogP parameters $L, O, G,$ and P. Let A an oblivious BSP algorithm where each superstep executes at most h remote memory accesses. If $l \geq O + \max(L, (h-1)\max(O,G))$ and $g \geq \max(O,G)$ then the execution time of the compiled BSP algorithm on the LogP model is not slower than the execution time of the BSP algorithm on the BSP model. If one of these inequalities is strict, then the execution time of the compiled BSP algorithm is faster.*

3.4 Lower Bounds for the Parameters

Let g^* and l^* (resp. $\max^*(O,G)$ and L^*) be the smallest values of the BSP parameters (resp. LogP parameters) admissible on a certain architecture. Our simulation implies that $g^* = \mathcal{O}(\max^*(O,G))$ and $l^* = \mathcal{O}(L^*)$. Together with the simulation result of [2] that proves a constant delay in simulation of LogP algorithms on the BSP machine, we obtain

Theorem 2 (BSP and LogP parameters). *For the smallest values of the BSP parameters g^* and l^* (resp. LogP parameters $\max^*(O,G)$ and L^*) achievable on any given topology, it holds that*

$$g^* = \Theta(\max(O^*, G^*)) \quad \text{and} \quad l^* = \Theta(L^*),$$

provided that both machines run oblivious programs.

So far we only considered the worst case behavior of the two models. They are equivalent for oblivious programs in the sense that bi-simulations are possible with constant delay. We now consider the time for implementing a barrier synchronization and a packed router on topology networks.

Theorem 3. *Let d be the diameter of a point-to-point network with P processors. Let dg be the degree of the processors of the network. Each processor may route up to dg messages in a single time step but it receives and sends only one message at each time step[2]. On any topology, the time l* of its BSP model is*

$$l^* = \Omega(\max(d(P), \log P)).$$

Proof. A lower bound for synchronization is broadcasting. Hence, $d(P)$ is obviously a lower bound for l^*. Assume an optimal broadcast tree [7] could be embedded optimally on the given topology. Its depth for P processors is $\Theta(\log P)$, a message could be broadcasted in time $\Theta(\log P)$. Hence, synchronization takes at least time $\Omega(\log P)$.

Remark 1. Actually $\Omega(P^{1/3})$ is a physical lower bound for l^* and L^* under the assumption that the processors must be layouted (in 3-dimensions) and signals have a run duration. Then the minimal average distance is $\Omega(P^{1/3})$. Due to the small constant factors of this bound, we may abstract from the layout and model signal delay by discrete hops from one processor to its neighbors.

For many packet routing problems (specific h-relations) and topologies, the latency is considerable smaller than l^* which could lead to a speed up of the LogP programs compared to oblivious BSP programs. This hypothesis is addressed by the next section.

4   Subclasses of Oblivious Programs

In general, there is a higher effort in designing LogP programs instead of BSP programs. E.g., the proof that the BSP capacity constraints are guaranteed reduces to simply counting the number of messages sent in a superstep. Due to the asynchronous execution model, the same guarantee is not easy to prove for the LogP machine. BSP programs cannot deadlock, LogP programs can. Hence, for all considered classes, we will have to evaluate whether:

- the additional effort in programming directly a LogP program pays out in efficiency, or
- compilation of BSP programs to LogP programs speeds up the program's execution, or
- such a compilation cannot guarantee a speedup.

For our practical runtime comparisons, we determine the parameters of the two machine models for the IBM RS6000/SP with 16 processors[3].

---

[2] This behavior lead to the LogP model where only receiving and sending a message takes processor time while routing is done by the communication network.

[3] The IBM RS6000/SP is a IBM SP, but uses other processors. Therefore, we compiled the BSP tools without processor specific optimizations.

Table 1. LogP vs. BSP parameters (IBM RS6000/SP), message size < 16 Bytes.

BSP	LogP
$l = 502\mu s$	$L = 17.1\mu s$
—	$O = 9.0\mu s$
$g = 30.1\mu s$	$G = 9.8\mu s$

4.1 The Parameters

We use the native message passing library for the LogP model and the Oxford BSP tools[4] for the BSP model. The results are shown in Table 1. For all measurements in this and the following sections we used compiler option -O3 -qstrict and for the BSP library option -flibrary-level 2.

4.2 MPMD-Programs

If an efficient parallel solution for a problem is hard to partition in supersteps, the BSP model is not appropriate. For those programs the LogP model seems preferable. Due to its asynchronous execution model, it avoids unnecessary dependencies of processors by synchronization. However, if the problem is low level, as our example broadcast is, the solution may be hidden in a library. It is not hard to extend the BSP model by a set of such library functions. Their implementation could be tuned using the LogP model.

Optimal Broadcast: A basic algorithm on distributed memory machines is the optimal broadcast of data from one processor to all others, which is used in many applications. For the LogP model, an optimal solution is given by Karp at al. in [7]. Each processor that received the item immediately initiates a repeated sending to processors which have not received the item until there is no such processors. We achieve the optimal BSP broadcast for our machine if the first processor sends messages to all others in one superstep. After synchronization the other processors receive their message. This is optimal for our target machine, since all other implementations would need at least two synchronization steps, which alone cost more than the algorithm given above. The measured runtime of broadcast for LogP and BSP on our machine can be seen in Table 2.

Remark 2. In the measurements for the BSP parameters, we used taged messages where the tag identifies the sender. For an optimal broadcast, this identification of the sender is not necessary. This explains the difference between estimation and measurements. In contrast to the LogP model, it is not possible on the IBM RS6000/SP to receive a message and send a it immediately without a gap. The LogP estimation ignore this property and are therefore too small.

However, even if we assume $\max(O, G) = g$, $L = l$ then the optimal LogP broadcast is a lower bound for the optimal BSP broadcast. The latter requires

---

[4] see http://www.BSP–worldwide.org

Table 2. Predictions vs. runtimes of broadcast, wave simulation, and FFT.

	BSP		LogP	
	measurement	estimation	measurement	estimation
broadcast	822 μs	980 μs	101 μs	99.6 μs
simulation (n=1,000)	6.84 s	6.35 s	1.50 s	1.56 s
simulation (n=100,000)	20.6 s	19.4 s	14.9 s	14.24 s
FFT (n=1024)	3.16 ms	3.51 ms	2.65 ms	2.67 ms
FFT (n=16384)	34.8 ms	35.6 ms	39.3 ms	35.2 ms

global synchronization barriers between subsequent send and receive operations. These synchronizations lead to delays on other processors if the LogP broadcast tree is not balanced by chance. For each send and receive pair *all* processors have to be synchronized instead of two.

4.3 SPMD-Programs with Sparse Dependencies

Assume that a BSP process sends at most h messages to the subsequent supersteps. The dependencies in the BSP program are *sparse* for M if the for LogP and BSP parameters for M it holds that

$$l + 2h \cdot g > 2O + (h-1)\max(O, G) + \max(L, (h-1)\max(O, G)) \qquad (1)$$

If a BSP programs communicates at most an h-relation from one superstep to the next, these communication costs are bounded by the right hand side of inequation (1) in the compiled LogP program (due to Theorem 1). Then we expect a speed up if the BSP program is compiled instead of directly executed.

However, if the problem size n is large compared to P, computation costs could dominate communication costs. Hence, if the problem scales arbitrarily, the speed-up gained by compilation could approach zero for increasing n.

Wave Simulation: For simulating a one-dimensional wave, a new value for every simulated point is recalculated in every time step according to the current value of this point(y_0), its two neighbors (y_{-1}, y_{+1}), and the value of this point one time step before (y_0'). This update is performed by the function

$$\Phi(y_0, y_{-1}, y_{+1}, y_o') = 2 \cdot y_0 - y_o' + \Delta_t^2/\Delta_y \cdot 2 \cdot (y_{+1} - 2 * y_0 + y_{-1}).$$

Since recalculation of one node needs the values of direct neighbors only, it is optimal to distribute the data balanced and block by block on the processors. Figure 1 sketches two successive computational steps and the required communication. The two programs for the LogP and BSP model only differ in the communication phase of each computation step. In both models, the processes communicate with their neighbors, i.e, the communication phase must route a 2-relation. The BSP additionally performs a synchronization of all processors. For our target machine, the data dependencies are sparse i.e., each communications phase is faster on the LogP machine than on the BSP, cf Table 2.

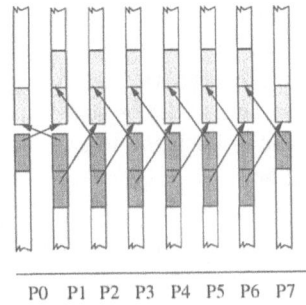

PO P1 P2 P3 P4 P5 P6 P7

Fig. 1. A communication phase of the Wave Simulation.

4.4 SPMD-Programs with Dense Dependencies

We call data dependencies of a BSP program *dense* for a target machine if they are not sparse. Obviously for those programs compilation does not pay as the following example show.

Fast Fourier Transform: We consider an one-dimensional parallel FFT and assume the input vector v to be of size $n = 2^k, k \in \mathbb{N}$. Furthermore, let $n \geq 2 \times P$. Then the following algorithm requires only one communication phase: v is initially distribute block-by-block and we may perform the first $\log(n/P)$ operations locally. Then v is redistributed in a cyclic way, which requires an all-to-all communication. The remaining operations can be executed also locally. The computation is balanced and a barrier synchronization is done implicitly on the LogP machine with the communication. Hence, we expect no noteworthy differences in the runtimes. The measurements in Table 2 confirm this hypothesis.

Remark 3. For the redistribution of data from block-wise to cyclic, we used a LogP–library routine, which gathers data by memcopy calls with variable data length. For the BSP, such a routine does not exist and was therefore implemented by hand. This implementation allowed the compiler to use more efficient copying routines and explains the difference of the LogP runtime to its estimation and to the BSP runtime.

5 Conclusions

We show how to compile of oblivious BSP algorithms to LogP machines. This approach improves the best known simulation delay of BSP programs on the LogP machine [2] by a factor of $O(\log(P))$. It turns out, that the both models are asymptotically equivalent for oblivious programs. We identified a subclass of oblivious programs that are potentially more efficient if directly designed for the LogP machine. Due to a more comfortable programming, other programs are preferably designed for the BSP machine. However, among those we could identify another subclass that are more efficient if compiled to the LogP machine instead of executed directly. Others are not. Our measurements determine the parameters for a LogP and a BSP abstraction of a IBM RS6000/SP with 16 processors. They compare broadcast, wave simulation, and FFT programs as representative of the described subclasses of oblivious programs. Predictions and measurements of the programs in both models confirm our observations.

Further work could combine the best parts of both worlds. We could use the same classification to identify parts of BSP programs that are executed directly, namely the non-oblivious parts and those which do not profit from compilation, and compile the other parts. Low level programs from the first class, like the broadcast, could be designed and tuned for the LogP machine and inserted to BSP programs as black boxes.

References

1. Albert Alexandrov, Mihai F. Ionescu, Klaus E. Schauser, and Chris Scheiman. LogGP: Incorporating long messages into the LogP model. In *7th Annual ACM Symposium on Parallel Algorithms and Architectures*, pages 95–105, 1995.
2. Gianfranco Bilardi, Kieran T. Herley, Andrea Pietracaprina, Geppino Pucci, and Paul Spirakis. Bsp vs logp. In *SPAA'96: 8th Annual ACM Symposium on Parallel Algorithms and Architectures*, pages 25–32. ACM, acm press, June 1996.
3. R. Cole and J. Hopcroft. On edge coloring bipartite graphs. *SIAM Journal on Computing*, 11(3):540–546, 1982.
4. D. Culler, R. Karp, D. Patterson, A. Sahay, K. E. Schauser, E. Santos, R. Subramonian, and T. von Eicken. LogP: Towards a realistic model of parallel computation. In *4th ACM SIGPLAN Symposium on Principles and Practice of Parallel Programming (PPOPP 93)*, pages 235–261, 1993.
5. Jörn Eisenbiegler, Welf Löwe, and Andreas Wehrenpfennig. On the optimization by redundancy using an extended LogP model. In *International Conference on Advances in Parallel and Distributed Computing (APDC'97)*, pages 149–155. IEEE Computer Society Press, March 1997.
6. Jonathan M. D. Hill, Paul I. Crumpton, and David A. Burgess. Theory, practice, and a tool for BSP performance prediction. In Luc Bougé, Pierre Fraigniaud, Anne Mignotte, and Yves Robert, editors, *Euro-Par'96 Parallel Processing*, number 1123 in Lecture Notes in Computer Science, pages 697–705. Springer, August 1996.
7. R.M. Karp, A. Sahay, E.E. Santos, and K.E. Schauser. Optimal broadcast and summation in the logp model. *ACM-Symposium on Parallel Algorithms and Architectures*, 1993.
8. W. F. McColl. Scalable computing. In Jan van Leeuwen, editor, *Computer Science Today*, number 1000 in Lecture Notes in Computer Science, pages 46–61. Springer, 1995.
9. Leslie G. Valiant. A bridging model for parallel computation. *Communications of the ACM*, 33(8), August 1990.

Parallel Computation on Interval Graphs Using PC Clusters: Algorithms and Experiments

A. Ferreira[1], I. Guérin Lassous[2], K. Marcus[3] and A. Rau-Chaplin[4]

[1] CNRS, INRIA, Projet SLOOP, BP 93, 06902 Sophia Antipolis, France.
ferreira@sophia.inria.fr.
[2] LIAFA – Université Paris 7, Case 7014, 2, place Jussieu, F-75251 Paris Cedex 05.
guerin@liafa.jussieu.fr.
[3] Eurecom, 2229, route des Cretes - BP 193, 06901 Sophia Antipolis cedex - France.
marcus@eurecom.fr.
[4] Faculty of Computer Science, Dalhousie University, P.O. Box 1000, Halifax, NS,
Canada B3J 2X4. arc@cs.dal.ca.

Abstract. The use of PC clusters interconnected by high performance
local networks is one of the major current trends in parallel/distributed
computing. We give coarse-grained, BSP-like, parallel algorithms to solve
many problems arising in the context of interval graphs, namely connected components, maximum weighted clique, BFS and DFS trees, minimum interval covering, maximum independent set and minimum dominating set. All of the described p-processor parallel algorithms require
only constant or $O(\log p)$ number of communication rounds and are efficient in practice, as demonstrated by our experimental results obtained
on a Fast Ethernet based PC cluster.

1 Introduction

The use of PC clusters interconnected by high performance local networks with
raw throughput close to 1Gb/s and latency smaller than $10\mu s$ is one of the
major current trends in parallel/distributed computing. The local networks are
either realized with off-the-shelf hardware (e.g. Myrinet and Fast Ethernet), or
application-driven devices, in which case additional functionalities are built-in,
mainly at the memory access level. Such cluster-based machines (called henceforth PCC's) typically utilize some flavour of Unix and any number of widely
available software packages that support multi-threading, collective communication, automatic load-balance, and others. Note that such packages typically
simplify the programmers task by both providing new functionality and by promoting a view of the cluster as a single virtual machine. Clusters based on
off-the-shelf hardware can yield effective parallel systems for a fraction of the
price of machines using special purpose hardware. This kind of progress may
thus be the key to a much wider acceptance of parallel computing, that has been
postponed so far, perhaps primarily due to issues of cost and complexity.

Although a great deal of effort has been undertaken on system-level and programming environment issues as described above, little attention has been paid
to methodologies for the design of algorithms for this kind of parallel systems.

Despite the availability of a large number of built-in and/or highly optimized procedures, algorithms are still designed at the machine level and claims to portability lay only on the fact that they are implemented using communication libraries such as PVM or MPI.

In this paper we show that theoretical (BSP-like) coarse-grained models are well adapted to PCC's. In particular, algorithms designed for such models are portable and their theoretical and practical performance are closely related. Furthermore, they allow a reduction on the costs associated with software development since the main design paradigm is the use of existing sequential algorithms and communication sub-routines, usually provided with the systems.

Our approach will be to study a class of problems from the start of the algorithm design task until the implementation of the algorithms on a PCC. The class of problems will be those arising on a family of intervals on the real line which can model a number of applications in scheduling, circuit design, traffic control, genetics, and others [16].

Previous Work

This class of problems has been studied extensively in the parallel setting and many work-optimal fine-grained PRAM algorithms have been described in the literature [3, 14, 15, 16]. Their sequential complexity is $\Theta(n \log n)$ in all cases.

Whereas fine-grained PRAM algorithms are likely to be efficient on fine-grained shared memory architectures, it is common knowledge that they tend to be impractical on PCC's due to their failure to exploit locality. Therefore, there has been a recent growth of interest in coarse-grained computational models [4, 5, 18] and the design of coarse-grained algorithms [5, 6, 8, 10, 13].

The BSP model, described by Valiant [18], uses slackness in the number of processors and memory mapping via hash functions to hide communication latency and provide for the efficient execution of fine grained PRAM algorithms on coarse-grained hardware. Culler et. al. introduced the LogP model which, using Valiant's BSP model as a starting point, focuses on the technological trend from fine grained parallel machines towards coarse-grained systems and advocates portable parallel algorithm design [4]. Other coarse grained models focus more on utilizing local computation and minimizing global operations. These include the Coarse-Grained Multicomputer ($CGM(n,p)$) model used in this paper [5], where n is the input size and p the number of processing elements. In this mixed sequential/parallel setting, there are three important measures of any coarse-grained algorithm, namely, the amount of *local computation* required, the number and type of *global communication phases* required and the *scalability* of the algorithm, that is, the range of values for the ratio $\frac{n}{p}$ for which the algorithm is efficient and applicable. We refer to [5, 9, 10] for more details on this model.

Recently, Cáceres et al. [2] showed that many problems in general graphs, such as list ranking, connected components and others, can be solved in $O(\log p)$ communication rounds in BSP and CGM. However, unlike general graphs, interval graphs can be more easily partitioned and treated in the distributed memory setting. Since each interval is given by its two extreme points, they can be sorted by left and/or right endpoints and distributed according to this ordering. This partitioning allows us to design less complex parallel algorithms; moreover, the derived algorithms are easier to implement and faster both in theory and in practice.

The following result will be used in the remaining to achieve a constant number of communication rounds in the solution of many problems.

Theorem 1. [13] *Given a set S of n items stored $O(n/p)$ per processor on a $CGM(n,p)$, $n/p \geq p$, sorting S takes a constant number of communication rounds.* □

The algorithms proposed for the CGM are independent of the communication network. Moreover, it was proved that the main collective communication operations can be implemented by a constant number of calls to global sort ([5]). Hence, by Theorem 1, these operations take a constant number of communication rounds. However, in practice these operations will be implemented through built-in, optimized system-level routines. In the remainder, let $T_S(n,p)$ denote the time complexity of a global sort in the CGM.

Our Work

We describe constant communication round coarse-grained parallel algorithms to solve a set of the standard problems arising in the context of interval graphs [16], namely connected components [3], maximum weighted clique [15] and breadth-first-search (BFS) and depth-first-search (DFS) trees [14]. We also propose $O(\log p)$ communication round algorithms for optimization problems as minimum interval covering, maximum independent set [15] and minimum dominating set [17].

In order to demonstrate the practicability of our approach, we implemented three of the above algorithms on a PCC interconnected by a Fast Ethernet backbone. Because of the paradigms used, the programs were easy to develop and are quite portable. The results presented in this paper show that high performance can be achieved with off-the-shelf PCC's along with the right model for algorithm design. Interestingly, super-linear speedups were observed in some cases due to memory swapping effects. Using multiple processors allows us to effectively utilize more RAM and therefore allows computation on data sets that are simply too large to be effectively processed on single processor machines.

In Section 2 the required basic operations are described. Then, in Section 3, chosen problems in interval family model are presented, and solutions are proposed using the basic operations from Section 2. In Section 4, we describe experiments on a Fast Ethernet based PCC. We close the paper with some conclusions and directions for further research.

2 Basic Operations

In the CGM model, any parallel prefix (suffix) associative function to be performed in an array of elements can be done in $O(1)$ communication steps, since each processor can compute locally the function, and then with a total exchange all the processors get to know the partial result of all the other processors and can compute the final result for each element in the array.

We will also use the pointer-jump operation, to identify the elements in a linked list. This operation can be easily done in $O(\log p)$ communication steps, at each step each processor keeps track of the pointers of its elements.

2.1 Interval Operations

In the following algorithms two functions will be widely used, the *Ominright* and *Omaxright* [1]. Given an interval I, *Omaxright*(I) (*Ominright*(I)) denotes, among all intervals that intersect I, the one whose right endpoint is the furthest right (left). The formal definition is the following.

$$Omaxright(I_i) = \begin{cases} I_j, & \text{if } b_j = \max\{b_k | a_k \le b_i < b_k\} \\ nil, & \text{otherwise.} \end{cases}$$

The function *Omaxright* can be computed with time complexity $O(T_S(n,p))$, as follows.

1. Sort the left endpoints of the interval in ascending order as $a'_1, a'_2, \ldots, a'_n$.
2. Compute the prefix maxima of the corresponding sequence $b'_1, b'_2, \ldots, b'_n$ of right endpoints and let the result be $b''_1, b''_2, \ldots, b''_n$. ($b''_k = \max_{1 \le i \le k}\{b'_i\}$.)
3. For every i ($1 \le i \le n$) compute the rank $r(i)$ of b_i with respect to $a'_1, a'_2, \ldots, a'_n$
4. For every i ($1 \le i \le n$), set *Omaxright*$(I_i) = I_j$, such that $b_j = b''_{r(i)}$ and $b_i \ne b''_{r(i)}$; otherwise set *Omaxright*$(I_i) = nil$.

We define also the parameter First$(\mathcal{I})$ as the segment I which "ends first", that is, whose right endpoint is the furthest left:

$$\text{First}(\mathcal{I}) = I_j, \text{ with } b_j = \min\{b_i | 1 \le i \le n\}.$$

To compute it, we need only to compute the minimum of the sequence of right endpoints of intervals in the family $\mathcal{I}$.

Finally, we will use the function next$(I) : \mathcal{I} \to \mathcal{I}$ defined as

$$\text{next}(I_i) = \begin{cases} I_j, & \text{if } b_j = \min\{b_k | b_i < a_k\}, \\ nil, & \text{otherwise.} \end{cases}$$

That is, *next*(I_i) is the interval that ends farthest to the left among all the intervals beginning after the end of I_i. To compute next(I_i), $1 \le i \le n$, we use the same algorithm used for *Omaxright*(I_i), with a new step 2.

1. Sort the left endpoints of the interval in ascending order as $a'_1, a'_2, \ldots, a'_n$.

2. Compute the suffix minima of the corresponding sequence $b_1', b_2', \ldots, b_n'$ of right endpoints and let the result be $b_1'', b_2'', \ldots, b_n''$.
   ($b_k'' = \min_{k \leq i \leq n}\{b_i'\}$.)
3. For every i ($1 \leq i \leq n$) compute the rank $r(i)$ of b_i with respect to $a_1', a_2', \ldots, a_n'$
4. For every i ($1 \leq i \leq n$), set $Next(I_i) = I_j$, such that $b_j = b_{r(i)}''$ and $b_i \neq b_{r(i)}''$; otherwise set $Next(I_i) = nil$.

It is easy to see that the above procedure implements the definition of $next(I_i)$, with the same complexity as for computing $Omaxright(I_i)$.

3 Interval Graph Problems and Algorithms

Formally, given a set n of intervals $\mathcal{I} = \{I_1, I_2, \ldots, I_n\}$ on a line, the corresponding *interval graph* $G = (V, E)$ has the set of nodes $V = \{v_1, \ldots, v_n\}$, and there is an edge in E between nodes v_i, v_j if and only if $I_i \cap I_j \neq \emptyset$.

In this section, solutions for some important problems in interval graphs are proposed for the CGM model. Some of these algorithms use techniques derived from their corresponding PRAM algorithms while others require different methods, e.g. to compute the connected components, as shown below.

3.1 Maximum Weighted Clique

A *clique* is a set of nodes that are mutually adjacent. In the *maximum weighted clique problem* for an interval graph, we want to know the maximum weight of such a set, given weights $p(I_i) \geq 0$ on the intervals, and identify a maximum weighted clique by marking its nodes. The CGM algorithm is as follows:

1. Sort the endpoints of the segments such that each processor receives $2n/p$ endpoints.
2. Assign to each endpoint c_i a weight w_i defined by
$$w_i = \begin{cases} p(I_j), & \text{if } c_i = a_j, \text{ for some } 1 \leq j \leq n, \\ -p(I_j), & \text{if } c_i = b_j, \text{ for some } 1 \leq j \leq n, \end{cases}$$
3. Compute the prefix sum of the resulting weighted sequence - the maximum obtained is the cardinality of a maximum clique; let $d_1, \ldots, d_{2n}$ denote the resulting sequence.
4. Consider the sequence $e_1, \ldots, e_{2n}$ obtained by replacing every d_j corresponding to a right endpoint of an interval with -1 and compute the rightmost maximum of the resulting sequence; this occurs at a_k.
5. Broadcast a_k. Every interval I_u such that $a_u \leq a_k < b_u$ is marked to be in the final maximum weighted clique.

Due to space limitations, the correctness and the complexity of the algorithm can be found in [7].

Theorem 2. *The maximum weighted clique problem in an interval graph of size n can be solved on a CGM(n,p) in $O(T_S(n,p) + n/p)$ time, with a constant number of communication rounds.* □

3.2 Connected Components

The *connected components* of a graph G are the maximal connected subgraphs of G. The *connected components problem* consists of assigning to each node the label of the connected component that contains it. For the $CGM(n,p)$ we have the following algorithm:

1. Sort the intervals by left endpoints distributing n/p elements to each processor.
2. Each processor P_i computes the connected components for the subgraph corresponding to its n/p intervals, giving labels to the components and associating the labels to the nodes.
3. Each processor detects the farthest right segment amongst its n/p intervals - tail t_i - and broadcasts it (with its label) to all other processors.
4. Each processor checks if any of the tails intersects its components, and updates its local labels using in each case the smallest such new label.
5. Each processor P_i records the pair $(t_i, $ new label$)$ and sends it to processor P_0.
6. Processor P_0 performs a connected components algorithm on the tails and updates the tail labels using the smallest such new labels and sends
   the tails and their new labels to all processors.
7. Each processor updates the labeling accordingly.

Due to space limitations, the correctness and the complexity of the algorithm can be found in [7].

Theorem 3. *The connected components problem in interval graphs can be solved on a $CGM(n,p)$ in $O(T_S(n,p) + n/p)$ time, with a constant number of communication rounds.* □

3.3 BFS and DFS Tree

The problem of finding a Breadth First Search Tree in an interval graph reduces to the problem of computing the function *Omaxright* described earlier. The tree given by the edges $(I_i, Omaxright(I_i))$ is a BFS tree [14]. And the tree formed by the edges $(I_i, Ominright(I_i))$ is a DFS tree [14]. The algorithm is the following:

1. Compute $Omaxright(I_i)$, for $1 \leq i \leq n$.
2. Let $father(I_i) = Omaxright(I_i)$.
3. The edges $(I_i, father(I_i))$ form a BFS tree.

With the appropriate modifications, this algorithm may be used to find a DFS tree. The obtained BFS and DFS trees have their roots in the segments ending farthest to the right in each connected component. With respect to its complexity, the algorithm takes a constant number of communication steps and requires a total running time of $O(T_S(n,p) + n/p)$.

Theorem 4. *Given an interval graph* G, *BFS and DFS trees can be found using a* $CGM(n, p)$ *in* $O(T_S(n, p) + n/p)$ *time, with a constant number of communication rounds.*   □

3.4 Minimum Interval Covering

Given a family $\mathcal{I}$ of intervals and a special interval $J = (J_a, J_b)$, the problem of the minimum interval covering is to find a subset $\mathcal{J} \subseteq \mathcal{I}$ such that $J \subseteq \cup(\mathcal{J})$, and $|\mathcal{J}|$ is minimum; i.e., to find the minimum number of intervals in $\mathcal{I}$ needed to cover J. To solve this problem we may only consider the intervals $I_i = (a_i, b_i) \in (\mathcal{I})$ such that $b_i \geq J_a$ and $a_i \leq J_b$. Let $\mathcal{I}_J$ be the family of the intervals in $\mathcal{I}$ satisfying this condition.

An algorithm to solve this problem is as follows:

1. Compute $Omaxright(I)$, $I \in \mathcal{I}_J$.
2. Find the interval I_{init} such that $b_{\text{init}} = \max\{b_k | a_k \leq J_a\}$.
3. Mark I_{init} and all the intervals in the path given by $Omaxright$ pointers beginning at I_{init}.

Due to space limitations, the correctness and the complexity of the algorithm can be found in [7].

Theorem 5. *The minimum interval covering problem in interval graphs can be solved using a* $CGM(n, p)$ *in* $O(T_S(n, p) + \log p)$ *time, with* $O(\log p)$ *communication rounds.*   □

3.5 Maximum Independent Set and Minimum Dominating Set

The first problem consists of finding a largest set of mutually non-overlapping intervals in the family $\mathcal{I}$, called the *maximum independent set*. The second problem consists of finding a *minimum dominating set*, i.e., a minimum set of intervals which are adjacent to all remaining intervals in the family $\mathcal{I}$. To solve these problems, we simply show a coarse-grained implementation of the algorithms proposed in [17]. In fact, it can be shown that both problems can be solved by building a linked list from First($\mathcal{I}$):

MAXIMUM INDEPENDENT SET:

1. Compute First($\mathcal{I}$)
2. Compute next(I_i), for i, $1 \leq i \leq n$
3. Let father(I_i) = next(I_i), for i, $1 \leq i \leq n$
4. Using the pointer-jump operation, mark all the intervals in the linked list given by *father* and beginning at First($\mathcal{I}$)

MINIMUM DOMINATING SET:

1. Compute First($\mathcal{I}$)
2. Compute Omaxright(I_i), for $1 \le i \le n$
3. Compute next(I_i), for $1 \le i \le n$
4. Let father(I_i) = Omaxright(next(I_i)), for $1 \le i \le n$
5. Using the pointer-jump operation, mark all the intervals in the linked list given by *father* and beginning at Omaxright(First($\mathcal{I}$))

The number of communication rounds in each of the algorithms is $O(\log p)$, giving us a total time complexity of $O(T_S(n,p) + \log p)$ and $O(\log p)$ communication rounds. Their correctness stems from the arguments in [17].

Theorem 6. *The maximum independent set and the minimum dominating set problems in interval graphs can be solved using a $CGM(n,p)$ in $O(T_S(n,p)+\log p)$ time, with $O(\log p)$ communication rounds.* □

4 Experimental Results

This section describes the implementations of three of the algorithms presented previously. Our aim here is to demonstrate that these algorithms are not only theoretically efficient but that they lead to simple fast codes in practice. They were implemented on a Fast Ethernet-PCC platform which consists of a set of 12 Pentium Pro 200 Mhz processors each with 64M of RAM that are linked by a 100Mb/s Fast Ethernet network. The processors run the Linux Operating System and the programs are written in C utilizing the *PVM* communication library [11] for all interprocessor communications.

Since many of the algorithms rely on sorting, the choice of the sorting method was critical. In the following we first present the implemented sort and its performance before describing the implementation and performance of our algorithms.

4.1 Global Sort

The sorting algorithm implemented is described in [12]. The algorithm requires a constant number of communication steps and its single drawback is that data may not be equally distributed at the end of the sort. Nevertheless, a partial sum procedure and a routing can be used to redistribute the data with a constant number of communication rounds so that each processor stores $\frac{n}{p}$ data in its memory.

Figure 1 shows the execution time for the global sort on an array of integers with the data redistributed in comparison to the sequential performance of quicksort (from the standard C library). The results shown are the average of ten execution times over ten different inputs generated randomly. The abscissa represents n the size of the input array and the ordinate the execution time in seconds.

For less than 7,000,000 integers, the achieved speedup is about 2.5 for four processors and 6 for twelve processors. Beyond the size of 7,000,000 integers, the

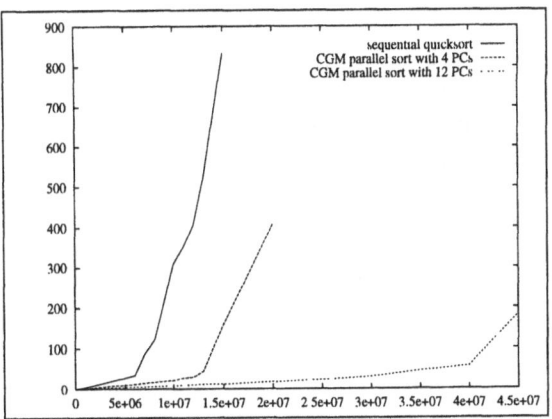

Fig. 1. Sorting on the Fast Ethernet-PCC.

memory swapping effects increase significantly the execution time on a single processor and super-linear speedup is obtained, with 10 for four processors and 35 for twelve processors. Sorting 40,000,000 integers takes less than one minute with twelve processors.

4.2 Maximum Weighted Clique

The algorithm requires a constant number of communication rounds and $O(\frac{n}{p}log\frac{n}{p})$ local operations. Note that only the local computations involved in the sort require $O(\frac{n}{p}log\frac{n}{p})$ operations, whereas all the other steps require only $O(\frac{n}{p})$ operations.

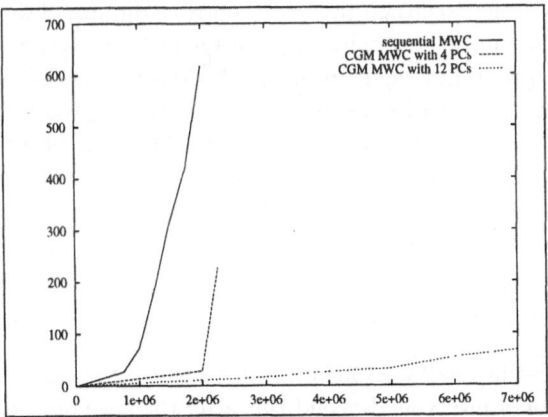

Fig. 2. Maximum Weighted Clique on the Fast Ethernet-PCC.

Figure 2 presents the execution time when the number of intervals increases.

With a graph having less than 1,000,000 intervals, the speedup is 2.5 with four processors, whereas it is equal to 7 for twelve processors. Beyond 1,000,000 intervals, the speedup is 10 for four processors and 35 for twelve processors, these superlinear timings being due to memory swapping effects. Again note that with this algorithm larger data sets than in the sequential case can be handled in a reasonable time.

4.3   Connected Components

As in the maximum weighted clique algorithm above, here also only the sort requires $O(\frac{n}{p}log\frac{n}{p})$ local operations, all the other steps being linear in $\frac{n}{p}$.

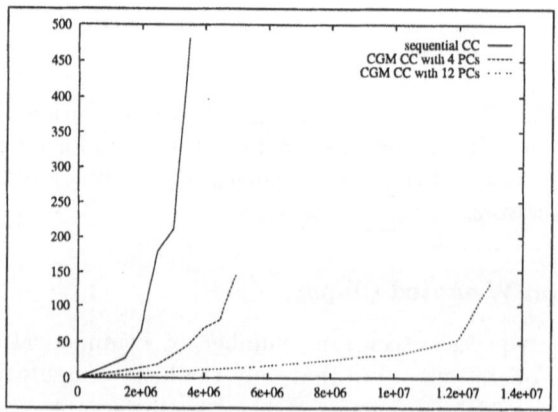

Fig. 3. Connected Components on the Fast Ethernet-PCC.

Figure 3 shows the execution time in seconds as the size of the input increases. The achieved speedup is approximately 2.5 for four processors and 7 for twelve processors for a graph having at most 2,000,000 intervals. With more intervals, the speedup is 8 for four processors and 30 for twelve processors. Also observe that with one processor, at most 2 million of data can be processed whereas twelve processors can process 12 million of data reasonably. Beyond 12 million data items the execution time increases more steeply due to memory swapping effects, even with twelve processors.

4.4   BFS Tree

The achieved speedup is 2 for four processors and 6 for twelve processors with at most 2 million of intervals. Beyond this size, the speedup becomes 7 for four processors and 20 for twelve processors. The measured times are slower than those obtained for the previous algorithms, because two steps of the function Omaxright require $O(\frac{n}{p}log\frac{n}{p})$ operations, whereas for the previous problems only one step required this number of local operations.

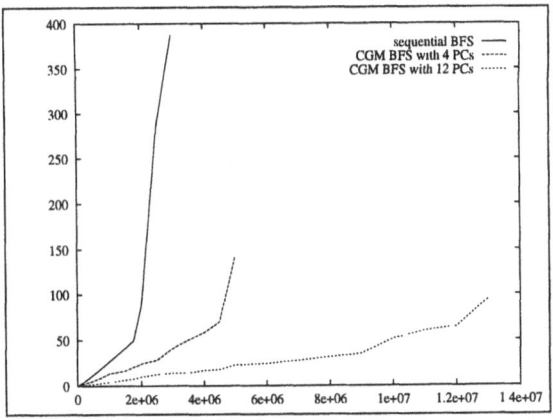

Fig. 4. BFS Tree on the Fast Ethernet-PCC.

5   Conclusion

In this paper we have shown how to solve many important problems on interval graphs using a coarse-grained parallel computer such as a cluster of PC's. The proposed algorithms were shown to be theoretically efficient, easy to implement and fast in practice. We believe this can largely be attributed to the use of the CGM model which accounts for distributed memory effects, mixes sequential and parallel coding, and encourages the use of a constant or very small number of communication rounds.

Note that the use of the CGM model, which was primarily developed for algorithm design in the context of interconnection networks, has led to efficient implementations even in the context of a bus-based network like Ethernet. We speculate that this is due to several factors including: 1) the model focuses on sending a small number of large messages rather than a large number of small ones 2) it relies on standard, and typically well optimized, communications operations and 3) it focuses on reducing the number of communication rounds and therefore the number on interdependencies between rounds. Of course at some point such bus-based networks always become saturated and more attention must be paid to bandwidth and broadcast conflict concerns, particularly as one scales up. We are currently exploring how such concerns can best be dealt with within the context of a CGM-like model.

References

1. A.A. Bertossi and M.A. Bonuccelli. Some Parallel Algorithms on Interval Graphs. *Discrete Applied Mathematics*, 16:101–111, 1987.
2. E. Caceres, F. Dehne, A. Ferreira, P. Flocchini, I. Rieping, A. Roncato, N. Santoro, and S. Song. Efficient parallel graph algorithms for coarse grained multicomputers and BSP. In *Proc. of ICALP'97*, pages 131–143. Lecture Notes in Computer Science. Springer-Verlag, 1997.

3. R. Cole and U. Vishkin. The accelerated centroid decomposition technique for optimal tree evaluation in logarithmic time. *Algorithmica*, 3:329–346, 1988.

4. D.E. Culler, R.M. Karp, D.A. Patterson, A. Sahay, K.E. Shacuser, E. Santos, R. Subramonian, and T. von Eicken. LogP: Towards a realistic model of parallel computation. In *Proc. 4th ACM SIGPLAN Symp. on Princ. and Practice of Parallel Programming*, pages 1–12, 1993.

5. F. Dehne, A. Fabri, and A. Rau-Chaplin. Scalable parallel geometric algorithms for coarse grained multicomputers. In *Proc. 9th ACM Symp. on Computational Geometry*, pages 298–307, 1993.

6. M. Diallo, A. Ferreira, and A. Rau-Chaplin. Communication-efficient deterministic parallel algorithms for planar point location and 2d Voronoi diagram. In *Proceedings of the 15th Symposium on Theoretical Aspects of Computer Science – STACS'98*, Lecture Notes in Computer Science, Paris, France, February 1998. Springer Verlag.

7. A. Ferreira, I. Guerin-Lassous, K. Marcus, and A. Rau-Chaplin. Parallel computation of interval graphs on PC clusters: Algorithms and experiments. RR LIAFA 97-30, University of Paris 7, http://www.liafa.jussieu.fr/~guerin/biblio.html, 1997.

8. A. Ferreira, C. Kenyon, A. Rau-Chaplin, and S. Ubéda. d-dimensional range search on multicomputers. *Algorithmica*, in press. Special Issue on Coarse Grained Algorithms.

9. A. Ferreira and M. Morvan. Models for parallel algorithm design: An introduction. In A. Migdalas, P. Pardalos, and S. Storoy, editors, *Parallel Computing in Optimization*, pages 1–26. Kluwer Academic Publisher, Boston (USA), 1997.

10. A. Ferreira, A. Rau-Chaplin, and S. Ubéda. Scalable 2d convex hull and triangulation for coarse grained multicomputers. In *Proc. of the 6th IEEE Symposium on Parallel and Distributed Processing, San Antonio, USA*, pages 561–569. IEEE Press, October 1995.

11. A. Geist, A. Beguelin, J. Dongarra, W. Jiang, R Manchek, and V. Sunderman. *PVM: Parallel Virtual Machine - A Users' Guide and Tutorial for Networked Parallel Computing*, 1994.

12. A.V Gerbessiotis and L.G Valiant. Direct bulk-synchronous parallel algorithms. *Journal of Parallel and Distributed Computing*, pages 251–267, 1994.

13. M.T. Goodrich. Communication-efficient parallel sorting. In *Proc. of 28th Symp. on Theory of Computing*, 1996.

14. S.K. Kim. Optimal Parallel Algorithms on Sorted Intervals. In *Proc. 27th Annual Allerton Conference Communication, Control and Computing*, volume 1, pages 766–775, 1990.

15. A. Moitra and R. Johnson. PT-Optimal Algorithms for Interval Graphs. In *Proc. 26th Annual Allerton Conference Communication, Control and Computing*, volume 1, pages 274–282, 1988.

16. S. Olariu. Parallel graph algorithms. In A. Zomaya, editor, *Handbook of Parallel and Distributed Computing*, pages 355–403. McGraw-Hill, 1996.

17. S. Olariu, J.L. Schwing, and J. Zhang. Optimal Parallel Algorithms for Problems Modelled by a Family of Intervals. *IEEE Transactions on Parallel and Distributed Systems*, 3(3):364–374, 1992.

18. L.G. Valiant. A bridging model for parallel computation. *Communications of the ACM*, 33:103–111, 1990.

Adaptable Distributed Shared Memory: A Formal Definition[*]

Jordi Bataller and José M. Bernabéu-Aubán

Dept. de Sistemes Informatics i Computació, Universitat Politècnica de València

Abstract. In this paper we introduce a new model of consistency for distributed shared memory called Mume. It offers only the essentials to be considered as a shared memory system. This allows an efficient implementation on a message passing system, and due to this, it can be used to emulate other memory models.

1 Introduction

Distributed shared memory (DSM) is a paradigm for programming parallel and distributed systems. It offers the agents of the system a shared space address that they use to communicate each other. The main problem of a DSM implementation is performance, specially when it is implemented on top of a message passing system. This, of course, depends on the consistency the DSM system offers. Many systems have been proposed, each one supporting a different level of consistency. Unfortunately, experience has shown that none is well suited for the whole range of problems. This is also true for different implementations of the same memory model, since no single one is able to efficiently support all of the possible data access patterns.

In this paper, we introduce a novel model called Mume. Following the direction of the main techniques to improve shared memory systems, Mume tries to give the programmer means to tell the system how to carry out the work more efficiently. Mume is a low level layer close to the level of the message passing interface. Mume interface only offers the strictly necessary requirements to be considered as shared memory thus allowing an efficient implementation. The interface also includes three types of synchronization primitives, namely, total ordering, causal ordering and mutual exclusion. Because the Mume interface is low enough, it can be used either to develop final applications or to build new primitives and use them in turn to develop final programs. This latter possibility allows, for example, to emulate the main memory models proposed to date. This paper describes Mume from a formal point of view.

---

[*] This work has been supported in part by research grants TIC93-0304 and TIC96-0729

2 Formal Definition of Mume

The formalism we use was presented in [3]. It is a simple extension of the I/O automata model, [5]. We define a memory model by giving the set of allowable executions for it. An execution is an ordered sequence of actions.

The basic orders we consider for an execution α are Process Order ($<_{PO}^{\alpha}$), Writes-to ($\mapsto$) and Causal Relation ($<_{CR}^{\alpha}$). Process Order is the order in which a processor issues actions. Writes-to is the order defined between a write and the read that gets the value left by the write. Causal Relation is the transitive closure of the sum of PO and $\mapsto$. We also consider the Mutual Exclusion ($<_{ME}^{\alpha}$) order for executions produced by models that include a synchronization primitive. Obviously, the order is defined only on the synchronized actions and they are ordered as they appear in the execution. An execution is admissible for a model if it can be rearranged in a way that the new sequence respects a given ordering and it is *serial*. An execution is serial if every read gets the value written by the immediately preceding write on the same variable. We express this concept by saying that a given execution is $<_{?}^{\alpha}$-serializable, this is, serializable respecting the ordering $<_{?}^{\alpha}$.

In the case of Mume, the actions or primitives which compose executions are classified into ordinary ones (*read* and *write*) and synchronizations (*sync*). Mume considers that there exists a set of independent memory deposits (*caches*). Caches apply ordinary actions as they arrive, if they are not synchronized. Ordinary actions must be labelled to indicate to which cache they should be applied. Of course, read actions can only operate at one cache. Formally, the ordinary actions are: $write(i, x, v)[A]$ and $read(i, x, v)[a]$, where $i \in \mathcal{P}$ (set of processes), $x \in \mathcal{V}_D$ (set of data variables), v is the value being read or written and $A \subseteq C, a \in C$ (set of caches).

Because Mume ordinary actions are independent of each other by default, it is necessary to synchronize them to set the order in which they are applied to caches. The synchronizations are the following ones: $TOsync(i, s_{TO}, t)$, $CRsync(i, s_{CR}, t)$ are $MEsync(i, s_{ME}, t)$, where $i \in \mathcal{P}$, $s_{TO} \in \mathcal{V}_{S_{TO}}$, $s_{CR} \in \mathcal{V}_{S_{CR}}$, $s_{ME} \in \mathcal{V}_{S_{ME}}$ (sets of synchronization variables) and $t \in S = \{acq, rel\}$ (synchronization type : acquire or release). The actions issued by a process between an acquire and its following release are the ones that are synchronized. We say that they are *X-synchronized* being X the type of synchronization: TO, CR or ME. Actions of different processors are related if they are synchronized by synchronizations acting on the same variable. We say that they are *X-related*, being X TO, CR or ME. The meaning of each type of synchronization is the following.

TOsync ensures that actions TO-related are performed in the same total order to any cache. It also ensures that if two actions of the same process are TO-related, then, Process Order is respected. *CRsync* maintains causal consistency: it ensures that actions CR-related are applied to any cache in an order consistent with the causal dependences among the CR-related actions. *MEsync* is the strictest synchronization. It behaves like a TO-synchronization and, in addition, it guarantees mutual exclusion. Furthermore, when a process obtains the exclusion it is sure that it will see all the writes made by all processes which held the exclusion in the past.

Formally, the Mume memory model is defined by requiring its executions to satisfy the properties **TOsync, CRsync,** and **MEsync.** Those properties are defined as follows:

- **TOsync:** an execution α respects $TOsync$ iff $\forall c \in C$ α is $(<^{\alpha}_{TOsync} |c)$-serializable.
- **CRsync:** an execution α respects $CRsync$ iff $\forall c \in C$ α is $(<^{\alpha}_{CRsync} |c)$-serializable.
- **MEsync:** an execution α respects $MEsync$ iff $\forall c \in C$, α is $(<^{\alpha}_{MEsync} |(c, sync))$-serializable.

In this definition, $<^{\alpha}_{Xsync}$ is the ordering that is defined considering only the X-related actions of an execution. $<_? |c$ denotes the restriction of $<_?$ to c.

3 Emulation of Known DSM Models

The primitives of Mume can be used to emulate other memory models including Sequential consistency, PRAM, Causal consistency, Processor consistency, Cache consistency, Slow consistency, Local consistency as well as synchronized models such as Release consistency. The references for these models can be found in [3]. We want to point out that each model can be emulated in different ways. For instance, any model can be trivially emulated by using a single cache for all the processes. But as many systems are implemented this way, we have chosen to give emulations in which one memory deposit is used per process, writes act on every cache, and reads act on the "local" cache. Now, as an example, we explain how Mume can emulate some of the cited models. The proofs can be found in [2].

Sequential. The executions of this model satisfy that they are $<_{PO}$-serializable. To emulate it, we have to TO-relate every action. Formally:

- $Mume = (\mathcal{P}, \mathcal{V}_D, C = \{m_i\}_{i \in \mathcal{P}}, \mathcal{V}_{S_{TO}} = \{o\}, \mathcal{V}_{S_{CR}} = \{\}, \mathcal{V}_{S_{ME}} = \{\})$.
- Labelling of ordinary actions: $\forall i \in \mathcal{P} : read(i, \cdot, \cdot)[m_i] \ write(i, \cdot, \cdot)[\{m_j\}_{j \in \mathcal{P}}]$.
- The first and last action of every process are $TOsync(i, o, acq)$ and $TOsync(i, o, rel)$.

PRAM. The executions of this model satisfy that, for all process i, they are $<_{PO} |i$-serializable.

To emulate the PRAM model, actions of the same process must be TO-related to achieve PO and to ensure that writes of that process are performed in the same order at each cache. Formally:

- $Mume = (\mathcal{P}, \mathcal{V}_D, C = \{m_i\}_{i \in \mathcal{P}}, \mathcal{V}_{S_{TO}} = \{o_i\}_{i \in \mathcal{P}}, \mathcal{V}_{S_{CR}} = \{\}, \mathcal{V}_{S_{ME}} = \{\})$.
- Labelling of ordinary actions: $\forall i \in \mathcal{P} : read(i, \cdot, \cdot)[m_i] \ write(i, \cdot, \cdot)[\{m_j\}_{j \in \mathcal{P}}]$.
- The first and last action of every process are $TOsync(i, o_i, acq)$ and $TOsync(i, o_i, rel)$.

Causal. The executions of this model satisfy that, for all process i,they are $<_{CR}|i$-serializable.

In order to emulate this model, every action of each process must be CR-related. Formally:

- $Mume = (\mathcal{P}, \mathcal{V_D}, \mathcal{C} = \{m_i\}_{i \in \mathcal{P}}, \mathcal{V}_{\mathcal{STO}} = \{\}, \mathcal{V}_{\mathcal{SCR}} = \{r\}, \mathcal{V}_{\mathcal{SME}} = \{\})$.
- Labelling of ordinary actions: $\forall i \in \mathcal{P}: read(i, \cdot, \cdot)[m_i] \; write(i, \cdot, \cdot)[\{m_i\}_{i \in \mathcal{P}}]$.
- The first and last action of every process are $CRsync(i, r, acq)$ and $CRsync(i, r, rel)$.

Cache. The executions of this model satisfy that, for all variable x,they are $<_{PO}|x$-serializable. It can be emulated by TO-relating actions operating on the same data variable Formally:

- $Mume = (\mathcal{P}, \mathcal{V_D}, \mathcal{C} = \{m_i\}_{i \in \mathcal{P}}, \mathcal{V}_{\mathcal{STO}} = \{o_v\}_{v \in \mathcal{V_D}}, \mathcal{V}_{\mathcal{SCR}} = \{\}, \mathcal{V}_{\mathcal{SME}} = \{\}$.
- Labelling of ordinary actions: $\forall i \in \mathcal{P}: read(i, \cdot, \cdot)[m_i] \; write(i, \cdot, \cdot)[\{m_i\}_{i \in \mathcal{P}}]$.
- Each action is surrounded by the pair $TOsync(i, o_v, acq)$ and $TOsync(i, o_v, rel)$ if it operates on the data variable v.

Processor. Every execution of this model satisfies that, for any pair of processes i and j, the execution can be $<_{PO}|i$-serialized and $<_{PO}|j$-serialized in α_i and α_j respectively, and for all variable x $\alpha_i|x = \alpha_j|x$.

The Processor model can be emulated using one cache per process. This model can be emulated by TO-relating actions of each process to achieve PO. Write actions are TO-related when they operate on the same data variable. Formally:

- $Mume = (\mathcal{P}, \mathcal{V_D}, \mathcal{C} = \{m_i\}_{i \in \mathcal{P}}, \mathcal{V}_{\mathcal{STO}} = \{s_v\}_{v \in \mathcal{V_D}}) \cup \{o_i\}_{i \in \mathcal{P}}, \mathcal{V}_{\mathcal{SCR}} = \{\}, \mathcal{V}_{\mathcal{SME}} = \{\}$.
- Labelling of ordinary actions: $\forall i \in \mathcal{P}: read(i, \cdot, \cdot)[m_i] \; write(i, \cdot, \cdot)[\{m_i\}_{i \in \mathcal{P}}]$.
- The first and last action of every process are $TOsync(i, o_i, acq)$ and $TOsync(i, o_i, rel)$. Each write is surrounded by the pair $TOsync(i, s_v, acq)$ and $TOsync(i, s_v, rel)$ if it operates on the data variable v.

4 Related Work and Concluding Remarks

In this paper we have defined a new memory model called Mume. Mume ordinary operations are close to the message passing interface since the programmer has to declare to which caches they have to be applied. In spite of this, ordinary operations still possess memory semantics because a write can be hidden by a subsequent one, this is, not all written values have to be read. Apart of ordinary operations, Mume also have three types of synchronizations in order to relate the ordinary ones.

One of the choices followed to improve the performance of shared memory systems has been to give the programmer means to tell the system useful information so the system can know which is the best way to do things or when actions should be performed. An example of this are synchronized models such as Release consistency. These models use synchronizations not only to achieve mutual exclusion but also to tell the system when to update the data. Another case, [4], is to allow the programmer to classify the variables of his programs according to the expected access pattern of each one. This permits the system to apply the best strategy to maintain the consistency of each variable.

Compared to other synchronized models, Mume has two synchronizations more. They help to relate ordinary actions without the need of using 'mutual exclusion. Furthermore, because the ordinary actions of Mume are marked with the target cache, the programmer has more means to influence the performance than in other systems. For example, some systems have to use complex protocols (*write-shared protocols*) in order to eliminate the problem of false sharing. The overhead introduced is significant and may reduce the gains dramatically. In Mume, the programmer can eliminate false sharing by fine tuning concrete and particular points of his program, carefully choosing the caches to where ordinary actions are applied. But it is not mandatory to use Mume at the lowest level. Just because it is low level, libraries can be developed in order to emulate different memory models or to support different access patterns to variables. If libraries are used, the programmer still have the choice of fine tuning a program. The "high level" program can be translated to the level of the native primitives where such tuning can be done.

Because we are concerned with formal issues, we have presented here the formal definition of Mume. You can find a more elaborate version of this work along with the correctness proofs of the emulations of other memory models in [2]. You can also see [1] to find a wider discussion on the practical issues of Mume.

We are currently developing an implementation of Mume for a loosely coupled system in order to compare it with other memory systems and to evaluate its expected benefits.

References

1. J. Bataller and J. Bernabeu. Constructive and adaptable distributed shared memory. In *Proc. of the 3rd Int'l Workshop on High-Level Parallel Programming Models and Supportive Environments*, pages 15–22, March 1998.
2. J. Bataller and J. Bernabeu-Auban. An adaptable dsm model: a formal description. Technical Report II-DSIC-17/97, Dep. de Sistemes Informatics i Computacio (DSIC), Universitat Politecnica de Valencia (Spain), 1997.
3. J. Bataller and J. Bernabeu-Auban. Synchronized DSM models. In *Europar'97. Third Intl. Conf. on Parallel Processing*, Lecture notes on computer science, pages 468–475. Springer-Verlag, August 1997.
4. J. B. Carter, J. K. Bennett, and W. Zwaenepoel. Techniques for reducing consistency-related communication in distributed shared memory systems. *ACM Transactions on Computer Systems*, 13(3):205–243, August 1995.
5. N. Lynch. I/O Automata: A model for discrete event system. Technical Report MIT/LCS/TM-351, Laboratory for Computer Science, Massachusetts Institute of Tecnology, March 1988.

Parameterized Parallel Complexity

Marco Cesati[1] and Miriam Di Ianni[2]

[1] Dip. di Informatica, Sistemi e Produzione, Univ. "Tor Vergata", via di Tor Vergata, I-00133 Roma, Italy.
cesati@uniroma2.it
[2] Istituto di Elettronica, Università di Perugia, via G. Duranti 1/A, I-06123 Perugia, Italy.
diianni@istel.ing.unipg.it

Abstract. We introduce a framework to study the parallel complexity of parameterized problems, and we propose some analogs of NC.

1 Introduction

The theory of NP-completeness [8] is a theoretical framework to explain the apparent asymptotical intractability of many problems. Yet, while many natural problems are intractable in the limit, the way by which they arrive at the intractable behaviour can vary considerably. For instance, deciding if the nodes of a graph can be properly colored by k colors is NP-complete even for a constant $k \geq 3$ [8], while there exists an algorithm deciding if the edges of a n nodes graph can be covered by k nodes in time $\mathcal{O}(n)$ for any fixed value of k. The parameterized complexity setting [6,7] has been introduced in order to overcome the intrinsic inability of the standard NP-completeness model to give insight into this variety of behaviours.

In this paper we investigate the issue of which problems do admit efficient fixed parameter parallel algorithms. A first attempt to formalize the concept of efficiently fixed parameter parallelizable problems has been pursued by Bodlaender, Downey and Fellows: in a one-page abstract [3] they suggested the introduction of the class PNC as the parameterized analogue of NC. However, neither theoretical results nor applications to concrete natural problems were presented. We now want to give a deeper insight to such concept. According to the degree of efficiency we are interested in, several kinds of efficient parallelization for parameterized problems can be considered. In section 2, after reviewing some basic concepts of the parameterized complexity theory, we define the two classes of efficiently parallelizable parameterized problems, PNC and FPP, and we study their relationship with the class of sequentially tractable parameterized problems (FPT). We also present a non trivial tool for proving FPP-membership of parameterized graph problems based on the concept of treewidth and on the results in [1, 4, 2]. In section 3 we study the relationship between NC, PNC, and FPP. In section 4 we give two alternative characterizations of both FPP and

PNC, and we use them to prove the PNC-completeness of two parameterized structural problems.

2   The Classes PNC and FPP

Let Σ be a finite alphabet. A *parameterized problem* is a set $L \subseteq \Sigma^* \times \Sigma^*$. Tipically, the second component represents a parameter $k \in \mathbf{N}$. The kth slice of the problem is defined as $L_k = \{ x \in \Sigma^* : \langle x, k \rangle \in L \}$. The class FPT of *fixed parameter tractable problems* contains all parameterized problems that have a solving algorithm with running time bounded by $f(k) |x|^\alpha$, where $\langle x, k \rangle$ is the instance of the problem, k is the parameter, f is an arbitrary function and α is a constant independent of x and k.

From now on, with the term "parallel algorithm" we always refer to a PRAM algorithm. The first class of efficiently parallelizable parameterized problems, PNC, has been defined in [3]. PNC (*parameterized analog of* NC) contains all parameterized problems which have a parallel solving algorithm with at most $g(k) |x|^\beta$ processors and running time bounded by $f(k)(\log |x|)^{h(k)}$, where $\langle x, k \rangle$ is the instance of the problem, k is the parameter, f, g and h are arbitrary functions, and β is a constant independent of x and k. The definition of PNC is coherent with the definition of FPT; indeed we can prove that PNC is a subset of FPT.

A drawback with the definition of PNC is the exponent in the logarithm which bounds the running time. We thus introduce the class of *fixed-parameter parallelizable problems* FPP that contains all parameterized problems having a parallel solving algorithm with at most $g(k) |x|^\beta$ processors and running time bounded by $f(k)(\log |x|)^\alpha$, where $\langle x, k \rangle$ is the instance of the problem, k is the parameter, f and g are arbitrary functions, and α and β are constants independent of x and k. Observe that, by definition, FPP $\subseteq$ PNC.

Several parameterized parallel problems can be proved to be included in FPP by directly showing a PRAM algorithm for them. Among them we recall parameterized VERTEX COVER and MAX LEAF SPANNING TREE. However, we can now present a different way for proving FPP-membership, based on the concept of treewidth. *Treewidth* was introduced by Robertson and Seymour [9], and it has proved to be a useful tool in the design of graph algorithms. Bodlaender and Hagerup [4] showed that there exists an optimal parallel algorithm on a EREW PRAM using $\mathcal{O}((\log n)^2)$ time and $\mathcal{O}(n)$ space which is able to construct a minimum-width tree decomposition of G or correctly decide that $tw(G) > k$. Therefore parameterized TREEWIDTH belongs to FPP.

The concept of *treewidth* turns out to be an useful tool for proving FPP-membership. In [1] it is shown that many problems that are (likely) intractable for general graphs are in NC when restricted to graphs of bounded treewidth. In particular, the authors prove that some graph properties (*MS* and *EMS properties*) are verifiable in NC for graphs of bounded treewidth. A careful study of the proofs in [1] reveals that such properties are also verifiable in FPP under the same hypothesis, since any optimal tree decomposition of width k can be transformed in $\mathcal{O}(\log n)$ time and $\mathcal{O}(n)$ operations on a EREW PRAM into a binary tree decomposition of depth $\mathcal{O}(\log n)$ and width at most $3k + 2$ [4]. Therefore: *all problems involving MS or EMS properties are in* FPP *when restricted to graphs of parameterized treewidth*. Thus, for any (E)MS property P, the following BOUNDED TREEWIDTH GRAPHS VERIFYING P problem belongs to FPP: given a graph G, is G satisfying P and such that $tw(G) \leq k$ (k being the parameter)? Lists of graph problems defined over (E)MS properties can be found in [2].

In a few cases it is possible to prove FPP-membership even without restrictions on the treewidth since some properties directly imply a bound on the treewidth. Therefore several natural problems (like FEEDBACK VERTEX SET) are in FPP, because (i) yes-instances have bounded treewidth, and (ii) the corresponding property is (E)MS.

3 Relationship Between FPP, PNC, and NC

Trivially, the slices of every parameterized problem belonging to FPP or PNC are included in NC. It is now worthwhile to verify whether the converse is true, that is, whether the parameterized versions of all problems whose slices belong to NC are included in FPP or PNC. Consider the CLIQUE problem. Each slice CLIQUE$_k$ is trivially in NC. Indeed, CLIQUE$_k$ can be easily decided by a parallel algorithm that uses $\mathcal{O}(n^k)$ processors and requires constant time. Since CLIQUE is likely not in FPT [7], it is also likely not in PNC. Thus, slice-membership to NC is not a sufficient condition for membership to PNC. Therefore, the classes FPP and PNC are somewhat "orthogonal" to NC, as much as the definition of FPT is orthogonal to the one of P. Nevertheless, there is a strong relation between P-completeness and FPT-completeness, where the completeness for FPT is defined with respect to parameterized reductions preserving membership to FPP or PNC. Indeed, consider the WEIGHTED CIRCUIT VALUE problem WCV: its instance consists of the description of a logical decision circuit C and of an input word w; the parameter k is the Hamming weight of w, and the question is whether C accepts w. We have proved the following.

Lemma 1. *WCV is in* FPT, *while each slice* WCV$_k$ *is P-complete.*

Previous lemma has an important consequence. Let L, L' be parameterized problems. We say that L FPP-reduces (PNC-reduces) to L' if there are an FPP-algorithm (PNC-algorithm) Φ and a function f such that $\Phi(x, k)$ computes $\langle x', k' = f(k) \rangle$ and $\langle x, k \rangle \in L$ if and only if $\langle x', k' \rangle \in L'$. It is not hard to prove the following.

Corollary 1. *Let $\leq_\Phi$ be either the* FPP-*reduction or the* PNC-*reduction. If there exists an* FPT-*hard problem L with respect to $\leq_\Phi$ such that all its slices L_k are included in* NC, *then* P=NC.

Corollary 1 implies that every FPT-hard parameterized problem must have at least one slice which is not in NC unless P = NC. Thus, the notion of FPT-completeness does not add anything to the classical notion of P-completeness to characterize hardly parallelizable parameterized problems. In particular, to distinguish between parallel intractability in the two contexts, we must find an hardly parallelizable parameterized problem with slices in NC. This is the aim of next section. However, we can hope to characterize "FPP's intractability" by proving the PNC-completeness of some problem (in the hypothesis FPP $\neq$ PNC), but we have no way to characterize "PNC's intractability".

4 Alternative Characterizations of FPP and PNC

NC is primarily defined as the class of problems that can be decided by uniform families of polynomial size, polylogarithmic depth circuits. Successively, it has been proved that uniform families of circuits and PRAM algorithms are polynomially related. In this paper, we have taken the inverse approach, by initially defining the classes FPP and PNC in terms of PRAM algorithms. In this section, we extend the relation between PRAM algorithms and uniform families of circuits to the parameterized setting. In this case, *uniform* means that the circuit C_n^k, able to decide instances of size n when the value of the parameter is k, can be derived in space $f(k)(\log n)^\alpha$, for some function f and constant α. It is possible to prove the following theorem by essentially using the same proof technique in [10].

Theorem 1. *A uniform circuit family of size $s(n,k)$ and depth $d(n,k)$ can be simulated by a PRAM-PRIORITY algorithm using $\mathcal{O}(s(n,k))$ active processors and running in time $\mathcal{O}(d(n,k))$.*

Conversely, there are a constant c and a polynomial p such that, for any FPP (PNC) algorithm Φ operating in time $T(n,k)$ with processor bound $P(n,k)$, there exist a constant d_Φ and, for any pair $\langle n,k \rangle$, a circuit C_n^k of size at most $d_\Phi\, p(T(n,k), P(n,k), n)$ and depth $c\, T(n,k)$ realizing the same input-output behaviour of Φ on inputs of size n. Furthermore, the family $\{C_n^k\}$ is uniform.

The previous theorem allows us to show the first PNC-complete problem. It is denoted as BOUNDED SIZE-BOUNDED DEPTH-CIRCUIT VALUE PROBLEM (in short, BS-BD-CVP) and is defined as follows: given a constant α, three functions f, g and h, a boolean circuit C with n input lines, an input vector x and an integral parameter k, decide if the circuit C (having size $g(k)\, n^\alpha$ and depth $h(k)(\log n)^{f(k)}$) accepts x. We have proved the following.

Corollary 2. BS-BD-CVP *is PNC-complete with respect to FPP-reductions.*

A second characterization of the classes of parameterized parallel complexity is based on random access alternating Turing machines. There is a strong relation between the time required by a random access alternating Turing machine and the depth of a simulating circuit, and between the space required by a random access alternating Turing machine and the size of a simulating circuit [5]. Such relation also holds in the parameterized setting. For the sake of brevity, we do not state the corresponding theorem. We only want to remark that such a characterization allows us to define another PNC-complete problem,

Random Access Alternating Turing Machine Computation (in short, RA-ATMC): given a random access alternating Turing machine AT, an input word x and an integral parameter k, does $AT(x)$ halts within $\mathcal{O}(f(k)(\log|x|)^k)$ steps with $\text{SPACE}^*(n) \in \mathcal{O}(g(k)\log n)$?

Corollary 3. RA-ATMC *is* PNC-*complete with respect to* FPP-*reductions.*

References

1. Stefan Arnborg, Jens Lagergren, and Detlef Seese. Easy problems for tree-decomposable graphs. *Journal of Algorithms*, 12, 308–340, 1991.
2. Hans L. Bodlaender. A partial k-arboretum of graphs with bounded treewidth. Technical Report, Department of Computer Science, Utrecht University, 1995.
3. Hans L. Bodlaender, Rodney G. Downey, and Michael R. Fellows. Applications of parameterized complexity to problems of parallel and distributed computation, 1994. Unpublished extended abstract.
4. Hans L. Bodlaender and Torben Hagerup. Parallel algorithms with optimal speedup for bounded treewidth. Technical Report UU-CS-1995-25, Department of Computer Science, Utrecht University, 1995.
5. A. K. Chandra, D. C. Kozen, and L. J. Stockmeyer. Alternation. *J. of the ACM*, 28, 114–133, 1981.
6. Rodney G. Downey and Michael R. Fellows. Parameterized computational feasibility. *Feasible Mathematics II*, 219–244. Birkhäuser, Boston, 1994.
7. Rodney G. Downey and Michael R. Fellows. Fixed-parameter tractability and completeness II: On completeness for $W[1]$. *TCS*, 141, 109–131, 1995.
8. Michael R. Garey and Davis S. Johnson. *Computers and Intractability — A Guide to the Theory of NP-Completeness.* W. H. Freeman and Co., New York, 1979.
9. Neil Robertson and Paul D. Seymour. Graph Minors. II. Algorithmic aspects of tree-width. *Journal of Algorithms*, 7, 309–322, 1986.
10. Larry Stockmeyer and Uzi Vishkin. Simulation of parallel random access machines by circuits. *SIAM J. Comput.*, 13(2), 409–422, 1984.

Asynchronous (Time-Warp) Versus Synchronous (Event-Horizon) Simulation Time Advance in BSP

Mauricio Marín

Programming Research Group, Computing Laboratory, University of Oxford
`mmarin@comlab.ox.ac.uk`

Abstract. This paper compares the very fundamental concepts behind two approaches to optimistic parallel discrete-event simulation on BSP computers [10]. We refer to (**i**) *asynchronous* simulation time advance as it is realised in the BSP implementation of Time Warp [3], and (**ii**) *synchronous* time advance as it is realised in the BSP implementation of Breathing Time Buckets [8]. Our results suggest that *asynchronous* time advance can potentially lead to more efficient and scalable simulations in BSP.

1 Introduction

As BSP is a (bulk) synchronous model of parallel computing [10], the obvious method of parallel discrete-event simulation [2, 7] on BSP computers seems to be synchronisation protocols based on *synchronous* time advance, such as Breathing Time Buckets [8]. In this paper we show that using synchronous time advance on a BSP computer might not be a good idea for two reasons. Firstly, synchronous simulation methods force the execution of periodical parallel min-reductions during which no interesting advance in simulation time can be made. These operations take some number of BSP *supersteps* to complete [10], and the simulation is stopped during their execution. The cumulative cost of these operations is relevant since the total number of min-reductions is equal to the total number of supersteps used to process the simulation events. Secondly and more importantly, synchronous methods impose barriers in simulation time which tend to reduce the rate of time advance per superstep of the simulation. This increases the total number of supersteps dedicated to parallel event-processing. Equivalently, time barriers tend to decrease the number of events that are processed in each superstep.

In BSP, supersteps are delimited by the barrier synchronisation of all the processors. In turn this operation takes some amount of real time to complete and it can become comparatively expensive in simulation models where the amount of computation involved in the processing of events is small (e.g., queuing networks). As many real life simulations can easily demand the execution of hundreds of thousands of event-processing supersteps, a significant reduction of the total number of supersteps can also cause a significant reduction of the total running time of the simulation. In addition, the cost of the remaining barrier synchronisation of processors can be amortised by the processing of a comparatively larger number of events per superstep.

On the other hand, approaches based on *asynchronous* time advance do not stop the simulation to perform periodical min-reductions, and simulation time barriers are not required (cf., [6]). For symmetric work-loads and excluding min-reduction supersteps, we have observed that as the size of the simulation model scales up, the rate of simulation time advance per superstep decreases comparatively faster using synchronous time advance. In particular, synchronous time advance requires on the average $O(\sqrt{P}/\ln \ln P)$ times more event-processing supersteps than asynchronous time advance, with P being the total number of logical processes (LPs). Moreover, for any simulation model, we show that the number of supersteps for the asynchronous approach is at most the supersteps required by the synchronous approach, and in the end the total number of supersteps required by the asynchronous approach is optimal for any model. In addition, for the same symmetric work-loads, we have observed that load balance tends to be better under asynchronous time advance for moderate level of aggregation of LPs onto processors (here load balance refers to balance in computation and communication as it is understood in BSP), whereas load balance is optimal in both approaches when the aggregation of LPs is large enough (domain of current practical simulations).

Obviously the nature of simulation work-loads is completely irregular and it is hard (perhaps impossible) to draw general conclusions about the overall running time achieved by protocols based on the two approaches analysed in this paper. We contribute with the analysis of an important performance indicator, namely total number of supersteps. In our view, the results presented in this paper suggest that the domain of simulation models and current BSP machines in which protocols based on synchronous time advance can be efficient should be expected to be limited. As to the development of general purpose BSP simulation environments, we conjecture that protocols based on asynchronous time advance offer better opportunities to achieve scalable performance.

2 Two Approaches to Time Advance

Approaches based on the event-horizon concept [9] like Breathing Time Buckets [8], advance in simulation time in a *synchronous* manner by consuming supersteps as in figure 1.a. Note that here we do not consider the additional supersteps required for min-reductions which compute a new global event-horizon on each cycle. That is, each cycle of the pseudo-code shown in this figure should be followed by a min-reduction. Approaches like Time Warp [3], on the other hand, advance in simulation time in an *asynchronous* manner [6] by consuming supersteps as in figure 1.b. Note that only silent min-reductions are needed in these approaches since values such as *global virtual time* (GVT) [3] are calculated without stopping the simulation. That is, these min-reductions are carried out using the same supersteps used to simulate events. We call these two styles of superstep advance as SYNC and ASYNC respectively.

Lemma 1. *The total number of supersteps required by ASYNC is optimal.*

Proof: Dependencies among events form trees. If the occurrence of event e generates event e_1, then e_1 is a child of e. If e_1 is scheduled to occur in a different processor and e takes place at superstep s, then e_1 must be processed at least in the superstep $s + 1$. But the simulation of e_1 may be delayed up to some superstep $s' > s + 1$ if earlier events take place in e_1's processor at superstep s'. In this case, each descendant of e_1 may only take place at superstep $\geq s'$ and if an e_1's decendant takes place in another processor, it may occur at least in superstep $s' + 1$. In this sense we say that every event has an initial "energy" which indicates the minimum superstep at which it can take place.

Figure 1.b helps us to realise that given any set of events to be processed during the whole simulation, the superstep counters are only updated with the *initial energy* of chronological events that cannot be processed in earlier supersteps. This rule is inductively applied in each processor from the first to the last superstep. In this way the simulation, as mapped onto the processors, is completed in the minimum number of supersteps. □

Lemma 2. *The total number of supersteps required by ASYNC is at most the supersteps of SYNC.*

Proof: A subset E_H of all simulation events are the events that mark the event-horizon times. These events are not necessarily causally related and they may be generated in different processors. Also note that these events are the least timestamped messages buffered during the respective SYNC supersteps. The size $|E_H|$ of this subset is the total number of supersteps required by SYNC since by construction these events cannot occur in the same superstep.

Consider the simulation with ASYNC. The first chronological event in E_H is necessarily the first event processed by one of the processors in its second superstep. However, the remaining processors may advance farther in time during their first superstep since they are not barrier synchronised by event-horizon times. This implies that more than one horizon event may be processed in the second superstep of ASYNC. Similar argument applies to the following supersteps so that, in general, SYNC is not optimal in supersteps. Both approaches require identical number of supersteps when, for example, each event in E_H is causally related so that they must be simulated sequentially. □

Lemma 3. *Under assumption of unlimited memory, Time Warp as realised in BSP can approximate the supersteps of ASYNC within $\log P$ supersteps.*

Proof: The lemma follows trivially by letting Time Warp process every available event (with time within the simulation period), correct erroneous computations accordingly, and start a new GVT calculation in each superstep. □

Note that processing every available event per superstep may lead to large roll-back overheads. However, near-optimal supersteps can be achieved at low overheads by limiting the number of events processed in each superstep [6].

Generate N initial pending events;
$T_Z := \infty$; [event horizon time]
$S_Z \leftarrow \Phi$; [buffer]
loop
    if TimeNextEvent() $> T_Z$ then
        SStep := SStep $+ 1$;
        Schedule(S_Z);
        $T_Z := \infty$;
        $S_Z \leftarrow \Phi$;
    endif
    e := NextEvent();
    $e.t := e.t +$ TimeIncrement();
    $p := e.p$; [e occurs in processor p]
    $e.p :=$ SelectProcessor();
    if $e.p \neq p$ then
        $S_Z \leftarrow S_Z \cup \{e\}$;
        $T_Z :=$ MinTime(S_Z);
    else
        Schedule(e);
    endif
endloop

(a) SYNC

Generate N initial pending events;
[$e.s$ indicates the minimal superstep at
  which the event e may take place in
  processor $e.p$.]
loop
    e := NextEvent();
    $p := e.p$; [e occurs in processor p]
    if $e.s >$ SStep[p] then
        SStep[p] := $e.s$;
    endif
    $e.t := e.t +$ TimeIncrement();
    $e.p :=$ SelectProcessor();
    if $p = e.p$ then
        $e.s :=$ SStep[p];
    else
        $e.s :=$ SStep[p] $+ 1$;
    endif
    Schedule(e);
endloop

The total number of supersteps is the
maximum of the P values in array $SStep$.

(b) ASYNC

Fig. 1. Sequential programs describing the rate of superstep advance in two approaches to parallel simulation. This program, called the hold-model [11], simulates work-loads of systems such as queuing networks. *Schedule*() stores events in the set of pending events. *NextEvent*() retrieves from this set the event with the least time. *TimeIncrement*() returns random time values. *SelectProcessor*() returns a number between 0 and $P - 1$ selected uniformly at random. The variable/array *SStep* maintains the current number of supersteps of the *simulated* BSP machine.

3   Average Case Analysis

It is not difficult to see that the number of supersteps per unit simulation time, denoted by S_p, required by SYNC and ASYNC is the optimal, $S_p = 1$, for fully connected communication topology when the *TimeIncrement* function of figures 1 returns 1. However, when this function returns random values, say exponentially distributed, the S_p values increase noticeably as shown in the following analysis (details in [5]). We assume one LP per processor.

Suppose that the initial event-list has N pending events e with timestamps $e.t$. Let X be a continuous random variable with p.d.f. $f(x)$ and c.d.f. $F(x)$. A hold operation consists of (i) retrieving the event e^* with the least timestamp from the event-list, (ii) creating an event e with timestamp $e.t = e^*.t + X$, and (iii) storing e in the event-list. e^* is discarded so that N remains constant throughout the whole sequence of hold operations. It is known [11] that after executing a long sequence of hold operations the probability distribution of the times $e.t$ approaches an steady state distribution with p.d.f. $g(y) = (1 - F(y))/\mu$ where $\mu = E[X]$. We represent the $e.t$ values with the random variable Y. The values of Y are measured *relative* to the timestamp of the last event e^* retrieved by the last hold operation. Let $G(y)$ be the c.d.f. of Y.

SYNC: We use the above defined $f(x)$ and $g(x)$ probability density functions to calculate SYNC's S_p, say S_p^s, as follows. Let A be the minimum of the N random variables Y. Since $\text{Prob}[A > x] = \overline{G}(x)^N$ the c.d.f. of A is $M_A(x) = 1 - \overline{G}(x)^N$. Let B be the minimum of the N random variables resulting from the sum of X and Y. The variable B represents the time of the event-horizon. The c.d.f. of $X + Y$, say $F_2(x)$, can be calculated using $F_2(x) = \int_0^x F(x - t)g(t)dt$ or $F_2(x) = \int_0^x G(x - t)f(t)dt$. The c.d.f. of B is then $M_B(x) = 1 - \overline{F_2}(x)^N$. Note that $E[B] - E[A]$ is the average time advance per superstep. Thus if the simulation ends at time $T \gg 1$, then it will require an average of $T/(E[B] - E[A])$ supersteps to complete. Therefore $S_p^s = 1/(E[B] - E[A])$.

Let us consider the negative exponential distribution with mean $\mu = 1$ for time increments X. In this case $f(x) = g(x) = e^{-x}$, therefore $E[A]$ is given by

$$E[A] = \int_0^\infty \overline{M}_A(t) \, dt = \int_0^\infty e^{-Nt} \, dt = \frac{1}{N},$$

and $E[B]$ is

$$E[B] = \int_0^\infty (e^{-t} + t e^{-t})^N \, dt = \sum_{i=0}^N \binom{N}{i} \int_0^\infty t^i e^{-Nt} \, dt = \frac{1}{N} \sum_{i=0}^N \binom{N}{i} i! \left(\frac{1}{N}\right)^i.$$

Using the approximation given in [4] (pp. 112-117) we obtain

$$E[B] \approx \frac{5}{4} \frac{1}{\sqrt{N}} + \frac{3}{4} \frac{1}{N},$$

so that

$$S_p^s \approx \frac{4}{5} \sqrt{N}.$$

Let us now consider SYNC's event-efficiency E_f, say E_f^s, which is defined as the ratio of the optimal number of events to the average *maximum* number of events processed in each processor per superstep. This is a measure of load balance in computation and communication for the studied symmetric workload. On average, a total of m events take place in each superstep. Given a sensible distribution of LPs onto the P processors, by Valiant's theorem [10] we learn that the average maximum number of events per superstep should tend to the optimal m/P as m scales up. Thus for m large enough the efficiency is close to 1. For exponential distribution, $m \approx \frac{5}{4}\sqrt{N}$ [5,9]. Define $D = N/P$. Regression analysis on simulation data from the program in figure 1.a (validated with numerical evaluation of the expected maximum of P binomial random variables) produces (asymptotically) for exponential distribution [5],

$$E_f^s = \frac{D^{1/4}}{P^{1/4} \ln P + D^{1/4}}.$$

ASYNC: In this case there is no global synchronisation in simulation time. In a given superstep each LP p_i simulates events with time less than the minimum event time of all new events (messages) arriving to p_i from other LPs by the next superstep. For large number of LPs (which justifies parallel simulation itself), it is reasonable to assume that a significant amount of LPs will work with its own statistically independent local event-horizon. The average instance of this local event-horizon being the average minimum of D independent random variables $Z = X + Y$. Thus, from the previous results for SYNC, these comments lead to a first view of ASYNC's S_p, say $S_p^a \approx O(\sqrt{D})$ for fixed number of LPs P. That is, we conjecture that for $P \gg 1$ and $D \gg 1$, S_p^a asymptotically tends to the S_p value of a SYNC simulation with just D events. In the following we provide evidence supporting this claim.

The LPs advance the simulation in a generally different amount of time in each superstep. Note that the *average* time advance per superstep is by definition $1/S_p^a$. Let us consider a lower bound T_p for this time advance. Consider all the order statistics associated with the set of N random variables $Z = X + Y$. Thus $T_p = $ Average among the first P values $E[Z_i]$, where the lower bound T_p comes from the assumption that the first P order statistics of Z are all located in a different LP. However, the actual average cannot be much larger than T_p since any difference $Z_i - Z_k$ increases very slowly with N. For example, for exponential distribution, the maximum Z_N increases only logarithmically with N [1]. In this case we conjecture that the true average time advance per superstep is $\approx 1.25/\sqrt{D}$. Let us then compare T_p with this quantity.

Unfortunately calculating T_p is mathematically intractable. Thus we performed the following experiment. For large N, say $N = 10^4$, we generated $I = 10^3$ different instances of a set of N random values $X + Y$, with X and Y being exponentially distributed. For each instance i we calculated the partial sums $\text{Sum}[i, P] = \frac{1}{P} \sum_{k=1}^{P} Z_k$ with $1 \leq P \leq N$, so that T_p for a particular D is given by $T_p = \frac{1}{I} \sum_{i=1}^{I} \text{Sum}[i, \frac{N}{D}]$. The results show that in the range $D \geq 10$, T_p behaves like $O(1/\sqrt{D})$. Assuming $T_p = \beta (P/N)^\alpha$ least squares regression on the points $(\log T_p, \log P)$ produced $\alpha = 0.47738$ and $\beta = 0.01194$ which should be compared with the conjectured $\alpha = 0.5$ and $\beta = 0.0125$ resulting from the curve $1.25/\sqrt{D}$.

We now consider the effect of P in S_p^a when D is fixed. Let χ_k be the sum of $k \geq 1$ random variables with p.d.f. $f(x)$ so that they represent the time increments of events forming a thread of causally related events. Then, for any positive difference between LP time advances, say $\Delta = T_j - T_i$, there is some non-zero probability that $T_j < T_i + \epsilon_i + \chi_k < T_j + \epsilon_j$ for some $\epsilon_i, \epsilon_j \geq 0$, such that $T_i + \epsilon_i + \chi_k$ is the time of the next event e_j in LP p_j after superstep s. This event e_j is processed in some superstep s' with $s + 1 < s' \leq s + k + 1$. An increase in P implies an increase in the diversity of T_i values and thereby an increase in the probability of events of type "e_j". Thus, for fixed D, S_p^a may increase in about k supersteps with P. For exponential distribution the k values are Poisson. The average k for the maximum time interval $\Delta = Z_N - Z_1$ is bounded from above by $Z_N = O(\ln N)$. So a conservative upper bound for S_p^a is $O(\ln P)$. However, similar experiment to the above described tells us $Z_P < 1 \ln \ln P$ in the range $D \geq 10$. This leads to

$$S_p^a \approx \ln \ln P \sqrt{D}.$$

This expression was validated with simulation results obtained with the program in figure 1.b. Regression analysis on data from the same program produces (asymptotically) for exponential distribution [5],

$$E_f^a = \frac{D^{1/4}}{(\ln \ln P)^{1/2} \ln P + D^{1/4}} \left(\frac{1}{\ln P \ln D} \right).$$

Comments: Note that the ratio S_p^s/S_p^a behaves in practice as $O(\sqrt{P})$ for large systems. For fixed P and large D values, SYNC's efficiency is better than ASYNC's efficiency. For large P's, $P^{1/4}$ goes to ∞ faster than $(\ln \ln P)^{1/2} \ln P$. Thus for large P and small D (practical simulation models) the efficiency of ASYNC is better than SYNC efficiency. However, the expression for E_f^a was obtained considering just one LP per processor. As more LPs are put on the processors, E_f^a improves noticeably. This is so because the variance of the cumulative sum of co-resident LP time advances is reduced as the number of LPs per processor increases. Consequently, the variance of the number of events simulated in each processor decreases [5]. Conversely, E_f^s is insensitive to the relation $D_{lp} D = m/P$ with D_{lp} being the number of LPs per processor. Another important point here is that we can always move ASYNC towards SYNC by imposing upper limits to the number of events processed in each processor and superstep (this filters peaks). If these limits are sufficiently small, ASYNC degenerates to SYNC [5].

To compare the two approaches under more practical grounds we performed experiments with programs similar to those shown in figure 1. First note that the work-load generated by these programs is equivalent to a fully connected queuing network where each node (LP) contains infinite servers. The service times are given by the time increments of events. In our experiments we considered a more realistic queuing network: one non-preemptive server per node, exponential service times, fully connected topology, and unlimited queue capacities with moderate number of jobs flowing throughout the network (closed system). Figure 2 shows the results. The comparison is made in terms of $R_S = S_p^s/S_p^a$ and $R_E = E_f^s/E_f^a$. The plotted data clearly show that ASYNC outperforms SYNC in a wide range of parameters (the minimal value of R_S in figure 2.a is 1.9). The data indicate that R_S increases as $\sqrt{P}$ and R_E decreases when N and P scale up simultaneously. In other words, the results show that ASYNC has better scalability than SYNC. In addition, the level of aggregation of LPs onto processors contribute to further increase R_S and decrease R_E. In particular, for fixed P, the ratio R_S increases as $\sqrt{D_{lp}}$ in figure 2.a.

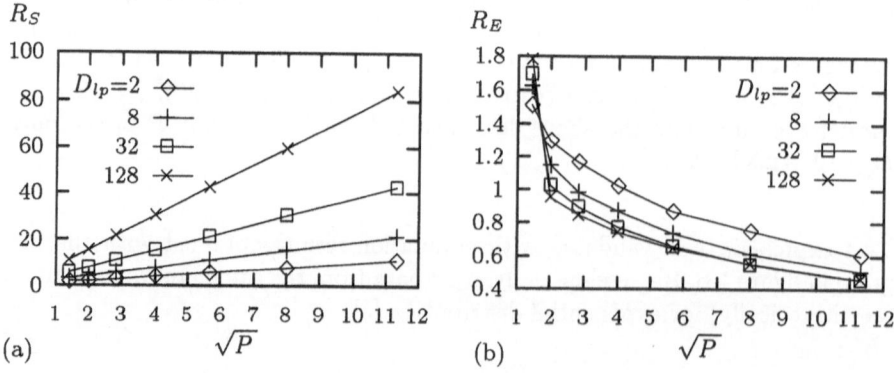

Fig. 2. Fully-connected non-preemptive single-server queuing network with exponential service times. Figure a: $R_S=$ supersteps-SYNC/supersteps-ASYNC. Figure b: $R_E=$ AvgMaxEv-SYNC/AvgMaxEv-ASYNC where AvgMaxEv is the average maximum number of events per superstep. $D_{lp}=$ number of LPs per processor, and $P=$ number of processors. Initially $D = 8$ "jobs" are scheduled in each LP (server).

References

1. R. Felderman and L. Kleinrock. "An upper bound on the improvement of asynchronous versus synchronous distributed processing". In *SCS Multiconference on Distributed Simulation V.22*, pages 131–136, Jan. 1990.
2. R.M. Fujimoto. "Parallel discrete event simulation". *Comm. ACM*, 33(10):30–53, Oct. 1990.
3. D.R. Jefferson. "Virtual Time". *ACM Trans. Prog. Lang. and Syst.*, 7(3):404–425, July 1985.
4. D. E. Knuth. *"The Art of Computer Programming, Vol. 1*, Fundamental algorithms"*. Addison-Wesley, Reading, Mass., 1973.
5. M. Marín. "Simulation time advance in BSP". Technical Report Oxford University, May 1998. http://www.comlab.ox.ac.uk/oucl/groups/bsp/.
6. M. Marín. "Time Warp On BSP Computers". Technical Report Oxford University, Feb. 1998. To appear in 12th European Simulation Multiconference.

7. D.M. Nicol and R. Fujimoto. "Parallel simulation today". *Annals of Operations Research*, 53:249–285, 1994.

8. J.S. Steinman. "SPEEDES: A multiple-synchronization environment for parallel discrete event simulation". *International Journal in Computer Simulation*, 2(3):251–286, 1992.

9. J.S. Steinman. "Discrete-event simulation and the event-horizon". In *8th Workshop on Parallel and Distributed Simulation (PADS'94)*, pages 39–49, 1994.

10. L.G. Valiant. "A bridging model for parallel computation". *Comm. ACM*, 33:103–111, Aug. 1990.

11. J.G. Vaucher. "On the distribution of event times for the notices in a simulation event list". *INFOR*, 15(2):171–182, May 1977.

Scalable Sharing Methods
Can Support a Simple Performance Model

Jonathan Nash

Scalable Systems and Algorithms Group
School of Computer Studies, The University of Leeds
Leeds LS2 9JT, West Yorkshire, UK

Abstract. The Bulk Synchronous Parallelism (BSP) model provides a simple and elegant cost model, as a result of using supersteps to develop parallel software. This paper demonstrates how the cost model can be preserved when developing software for irregular problems, which typically require dynamic load balancing and introduce runtime task dependencies. The solution introduces shared data types within a superstep, which support weakened forms of shared data consistency for scalable performance. An example of a priority queue to support a solution of the travelling salesman problem is given, with predicted and observed performance results provided for 256 processors of a Cray T3D MPP.

1 Introduction

The Bulk Synchronous Parallelism (BSP) model [7, 12] provides a simple and elegant cost model, as a result of using supersteps to develop parallel software. A key feature is the independent execution of processors in generating remote accesses, allowing a superstep cost to be characterised by an $h - relation$ [7]. Specifically, given a machine with network performance g, barrier cost L and computational performance s, a superstep can be costed as $gh + sw + L$, where h and w are the maximum usage of each resource by any processor.

The BSP model seems less suited to the efficient support of irregular problems, which require dynamic load balancing and introduce runtime task dependencies [9]. This paper describes work on supporting irregular problems with scalable high performance, while preserving the BSP-style cost model (based on earlier experiences with the WPRAM model [10]). The key idea has been to develop scalable data structures which support dynamic sharing patterns [8, 9, 3], using highly concurrent implementations provided by the use of weakened forms of shared data consistency [4, 1]. This highly concurrent behaviour approximates the superstep operation, allowing the BSP-style cost model to be preserved.

Section 2 describes the use of abstractions for sharing in parallel systems. Section 3 uses the example of a priority queue to describe the techniques to support scalable high performance. Section 4 describes the extension of the BSP cost model to characterise the priority queue performance, with Section 5 providing predicted and observed performance, both for the queue and its use in the travelling salesman problem, for 256 processors of a Cray T3D. Section 6 provides a summary and points to some current work.

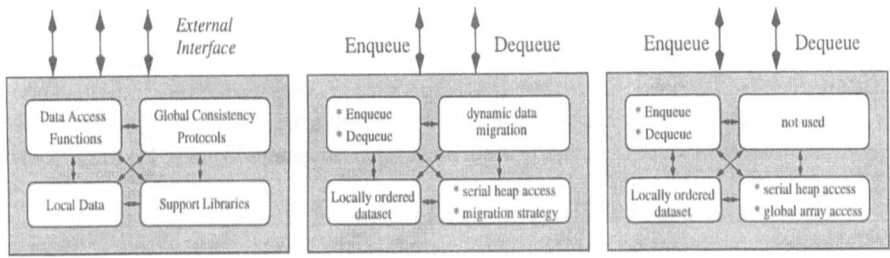

Fig. 1. (a) Generic SADT; (b) PriQueue#1 SADT; (c) PriQueue#2 SADT

2 Abstractions for Sharing in Parallel Systems

The programming framework is based on the idea of shared abstract data types (SADTs) [5]. An abstract data type (ADT) encapsulates local data, and provides methods to change the state of the data (and return some result). An SADT augments an ADT to include concurrent method invocation, and must specify when changes in state are visible to the methods. SADTs support the key sharing patterns in a parallel system [1, 2, 6], allowing the encapsulation of possibly complex concurrency issues. This well defined layer of abstraction allows for optimised implementations on specific platforms, code portability and re-use.

The diagrams in Figure 1 point to the four main features of an SADT, both in generic terms, and related specifically to the support of two alternative priority queue (PriQueue) SADT implementation strategies. A more detailed description can be found in [3].

Local data: Each processor maintains a local state. This data might be replicated or partitioned across the processors, depending on the access patterns which the SADT generates. The PriQueue maintains a partitioned segment of the data set on each processor.

Data access functions: In order to support the external interface methods, the internal SADT data will have to be accessed. The first PriQueue implementation adopts a bulk synchronous approach in which the Enqueue and Dequeue operations will access local data items, with dynamic data migration occurring within regular superstep periods [3]. The second implementation supports a shared data structure across the processors, and so Enqueue and Dequeue operations may result in the access of remote data.

Global consistency: This is typically only explicitly defined if the data access functions, defined above, make only local data accesses. In the case of the first PriQueue, the consistency is specified by the load balancing superstep frequency and the number of data items migrated. The second implementation does not contain any explicit consistency routines, since a shared data structure is used to directly store and retrieve the data items.

Support libraries: Both the data access functions and the global consistency protocols will typically need to make use of serial and parallel libraries.

The first PriQueue requires libraries to support a serial heap for local data access, and various migration strategies are supported for the load balancing superstep. The second implementation also makes use of the heap library, and also routines to support the access of a shared array data structure, so that data items on other processors may be accessed [3] (see Section 3.3).

SADTs are utilised within a bulk synchronous superstep, to support dynamic data sharing patterns in a portable and scalable manner:

```
Enqueue sequence #1; Barrier
Enqueue sequence #2; Barrier
Dequeue operations
```

If a FIFO Queue SADT is being accessed [8], with the FIFO ordering defined at the barrier points (and an arbitrary ordering between barriers) then the Dequeue operations would initially retrieve items from the first Enqueue sequence. If the SADT was the PriQueue defined earlier, the first barrier could be removed, since this will not effect the ordering of the data items (an explicit priority ordering is maintained). The second barrier guarantees that all Enqueue operations have completed, and the following Dequeues will begin to access the globally highest priority items. This barrier could also be removed if this strict guarantee is not required, allowing processors to access the currently highest priority tasks (or if the creation of the data items is depicted by runtime task dependencies [3]).

While a serialised implementation of the PriQueue might be practical for a small-scale parallel machine, highly parallel computers consisting of tens or hundreds of processors require high degrees of concurrency, if performance is to scale. The next section describes how this has been achieved for the PriQueue, based around the second PriQueue implementation strategy described above.

3 Techniques for Supporting Scalable Performance

The example of the travelling salesman problem (TSP) is given here, to illustrate the use of the PriQueue SADT (full details of the solution are presented in [3]). As shown in Figure 2(a), the solution of the TSP makes use of the PriQueue to store and retrieve new tasks, representing tours, which are dynamically created within a tree-based computation. The priority is related to the length of the (partially generated) tour. An Accumulator SADT is also used to note the current best tour, to which new tours can be compared (for reasons of brevity, further details of this SADT are excluded). A lower level lightweight Lock SADT is also used to support the operation of the PriQueue and Accumulator [3].

The next sections describe the use of weak data consistency, bulk synchronous parallelism [1] and fine-grain parallelism to support the PriQueue with scalable performance. Specific implementation details are provided for the Cray T3D.

---

[1] The algorithm which solves the TSP in fact only uses a single superstep

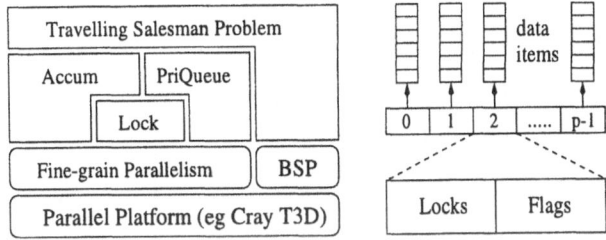

Fig. 2. (a) Structure of the TSP solution; (b) PriQueue implementation

Table 1. The SHMEM operations

Bulk synchronous operations	Action
Get (buf, x, Buf, i)	$buf[0..x-1] = Buf_i[0..x-1];$
Put (buf, x, Buf, i)	$Buf_i[0..x-1] = buf[0..x-1];$
Quiet $()$	suspend upon completion of Put(s)
Barrier $()$	synchronise all processors
Fine-grain operations	**Action**
w = Swap (c, W, i)	$w = W_i;\ W_i = c;$
w = Inc (W, i)	$w = W_i;\ W_i = W_i + 1;$
Wait (w, c)	suspend until location $w \neq c$

3.1 Weak Data Consistency

A key characteristic for scalable performance was to use weakened data consistency [4, 1]. Sequential consistency can provide a global priority ordering on the PriQueue elements, but typically through the use of a serialising lock. This becomes the bottleneck in large systems, limiting performance. The implementation here removes this lock by removing the strict ordering guarantee. Each processor $0..p-1$ holds a locally ordered data segment, as shown in Figure 2(b). These are accessed cyclically by the processors when storing and retrieving data, distributing the priorities approximately evenly. A processor is then only guaranteed to remove one of the p highest priorities (Section 5 will show that this does not have any substantial effect on the operation of the algorithm). This approach provides for a highly concurrent implementation (see below), allowing scalable performance characteristics as the number of processors grow.

3.2 Bulk Synchronous Parallelism

For the implementation study on the Cray T3D, the SHMEM library of operations has been used, as given in Table 1. In the table, V_i denotes a variable V which is accessed on processor i, in contrast to v, which is a local variable. SHMEM allows the direct access of remote memory locations through the specification of a processor-address pair, supporting high bandwidth and low latency operations. The BSP programming model can be supported through the use of the *Put* and *Get* operations, which write to and read from remote memories

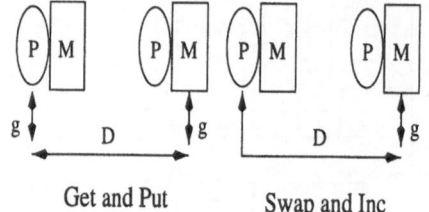

Operation	Cost
Swap/Inc	$D + 2g$
Get	$D + 4gx$
Put	$2gx$
Quiet	D
Wait	–
Barrier	L

Cost	Value
g	0.04-0.06
D	0.75-1.27
L	2.00
s	0.0533
t	0.0664

Fig. 3. (a) Modelling *SHMEM*; (b) SHMEM costing; (c) Cost values

respectively, and a *Barrier* operation. *Put* supports the pipelining of multiple requests, to amortize the network latency cost (and allow overlapping of communication with subsequent local computation), with *Quiet* being used to suspend until these complete. The *Put* and *Get* operations can be used to maintain a priority ordering on a given remote data segment of the PriQueue, through a simple binary search method.

3.3 Fine-grain Parallelism

The PriQueue concurrency can be exploited by operations which support fine-grain parallelism. Looking back to Figure 2(b), it was mentioned that the local data segments are accessed cyclically by the processors. The SHMEM library provides an atomic increment operation, *Inc*, which can be used to coordinate the global access of these segments, as shown in Table 1. The result in w can be taken modulus p to obtain the required PriQueue data segment.

Figure 2(b) shows that each data segment also contains a number of locks and local flags, which are used to control the access to the data segments [3]. The implementation of these locks, through a lower level Lock SADT [3], can use the SHMEM atomic *Swap* operation, in which a word c is exchanged with the current contents of W on processor i. This operation allows for the construction of a scalable linked list [8]. In addition, *Wait* can be used to support fine-grain point-to-point synchronisation methods [10], which are typically used to coordinate the access of the shared linked lists [8]. The next section will show how the specific hardware support for these operations allows for the construction of high performance software.

4   A BSP-style Performance Model

This section describes how a model of the performance of parallel applications can be constructed from a simple BSP-like cost model, as described in the introduction. The key to preserving this costing approach has been the scalable implementation of the PriQueue SADT, which preserves the independent super-step behaviour of the processors.

Table 2. PriQueue SADT costing

Operation	Cost
Enqueue (max,tot)	Inc(max,tot) +
	Get(1,4*max,4*max)+Put(10,max,max) +
	882*max*s + max*Add_Heap(HEAP_SIZE) +
	2 { Acquire(max,tot/p) + Release(max,tot/p) }
Dequeue (max,tot)	Inc(max,tot) +
	Get(1,max,max)+Get(9,3*max,3*max)+Put(1,max,max) +
	1107*max*s + max*Sub_Heap(HEAP_SIZE) +
	2 { Acquire(max,tot/p) + Release(max,tot/p) }

4.1 Characterising Machine Performance

As described in the introduction, a superstep cost can be modelled as $gh+sw+L$, where g, s and L cost the machine operations (network access, local computation and barrier synchronisation respectively), with h and w specifying their maximum usage by any one processor within the superstep. In order to effectively model the operations for fine-grain parallelism, two additional machine costs, D and t, are required. The cost D measures the round-trip network latency, and is incurred when there exists a data dependency within a superstep (for example, when accessing the results of an atomic increment). The cost t represents the time to transfer words within the local memory, and is present due to the fact that irregular applications typically access complex local data structures. So the cost of a superstep can now be represented by $gh + sw + tn + Dc + L$, where n and c again reflect the maximum resource usage.

4.2 Characterising the Performance of the SHMEM Library

Figure 3(a) shows how the SHMEM operations which access the network are costed. For the Swap and Inc operations, performance is measured by the latency term D and the network access cost g. For the Put and Get operations, the time g is incurred at both the sending and receiving ends. Figure 3(b) provides a summary of all the costs. Put allows the pipelining of multiple remote accesses (totaling x words), with the processor suspending using Quiet, or Barrier, until they complete. Put and Get requests are split into packets of 4 words, with an acknowledgement packet returning for a Put, and the data returned for a Get. However, the effective network bandwidth is halved for the Get (hence the factor of 4 in the table), since each packet must return before the next packet can be sent. Figure 3(c) gives each of the cost parameters for the Cray T3D. The terms of g and D for each operation were derived from network performance results under randomised and deterministic traffic patterns.

4.3 Characterising the Performance of the PriQueue and TSP

As described in Section 4.1, the performance at the machine level can be modelled using the terms $gh + sw + tn + Dc + L$, where h, w, n and c represent the

Table 3. Costs for maintaining an ordered heap segment

Operation	Cost
Add_Heap (n)	Put(18,n,n) + 2*Get(9,n,n)
Sub_Heap (n)	Put(18,n,n) + 3*Get(9,n,n)

Table 4. Lock SADT costing

Operation	Cost
Acquire (max,tot)	Swap(max,tot) + Swap(max,max) + 37*max*s
Release (max,tot)	Put(1,tot,tot) + 36*tot*s

maximum usage of each associated machine resource. When characterising the performance of parallel software, two terms are used to model the workload at a given level of abstraction. The term *tot* measures the total usage of a given resource across all processors, whereas *max* measures the maximum resource usage by any given processor. For example, the performance of the solution to the travelling salesman problem given in [5] can be described as:

Enqueue (max,tot) + Dequeue (max,tot) +

Accum_Read (max,tot) + Accum_Update (u) + Compute (max)

In this case, *tot* and *max* refer to the number of generated tours. The costs *Enqueue* and *Dequeue* are for the PriQueue, and *Accum_Read* and *Accum_Update* the Accumulator (the u term has been found by experiment to be a small constant number). A *Compute* term measures the local work required for a given number of tours. In this example, SADT access occurs within a single superstep. Processors independently produce and consume tours through the access of the PriQueue. The only exception is when all items have been temporarily removed from the PriQueue and new items are being enqueued (causing any idle processors to momentarily suspend execution), and at the termination stage. The former case typically only occurs for a short time at the start of execution, when the first few PriQueue items are being generated. Thus, the independent superstep behaviour is preserved at the application level. The highly concurrent behaviour of the SADTs preserves this state of affairs (see below and Section 3.3), allowing the simple BSP cost model to be employed.

Table 2 gives the costing of the *Enqueue* and *Dequeue* operations. These include access to the *Inc*, *Put* and *Get* SHMEM operations. The *Inc* cost represents a potential source of contention, including a term of *tot*, reflecting its use to coordinate the access of the local data segments which make up the PriQueue, as described in Section 3.3. The PriQueue costing also includes the need to maintain a local priority ordering within each local data segment after an *Enqueue* or *Dequeue* occurs, using *Add_Heap* and *Sub_Heap* respectively. These costs are parameterised by the size of the local segment (2^{HEAP_SIZE} items). Also included are costs to *Acquire* and *Release* a lock at a local segment. The term of *tot/p* in these costs reflects the scalable distributed implementation, in which the local data segments are accessed cyclically by the processors.

Table 5. Parameterised SHMEM costs

Operation	Cost
Swap/Inc (max,tot)	$max * D + tot * 2 * g$
Get (x,max,tot)	$max * D + tot * 4 * g * x$
Put (x,max,tot)	$max * D + tot * 2 * g * x$

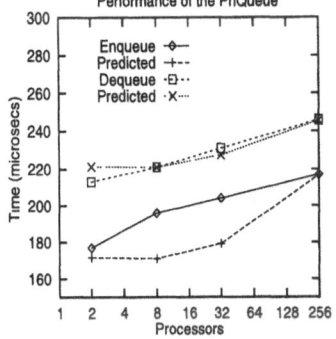 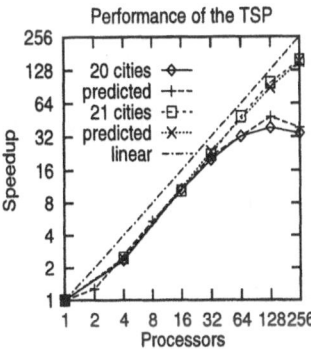

Fig. 4. (a) PriQueue performance (b) TSP performance

Tables 3 and 4 give the costs for maintaining the priority ordering of the data segments and the Lock SADT. The ordering uses a binary search to exchange data items within the segment, resulting in the $HEAP\_SIZE$ term. The Lock SADT implementation is too involved to describe here, but further information can be found in [3]. Finally, Table 5 gives the cost of the SHMEM operations, when parameterised by *max* and *tot*. It can be seen, for example, that when a *Swap* operation is concurrently performed on a given shared word, the *max* term gives the usage of the network latency resource D (since $max * D$ gives the total time that any processor incurs the latency resource), and the *tot* term relates to the usage of the network resource g (since $tot * g$ provides the total contention at the memory module holding the word being accessed).

5 Predicted and Observed Performance

The performance of the PriQueue is given in Figure 4(a), in which all processors continuously generate *Enqueue* and *Dequeue* requests. It can be seen that the predicted and observed performance agree closely. Figure 4(b) shows the TSP performance for 20 and 21 city problems. Speedups of over 150 on 256 processors for a 21 city problem are typical, reducing the time from 53 seconds on one processor to around $\frac{1}{3}$ second on 256 processors. It can be seen that the predicted times are again close, using the performance models derived for the SADTs. The values for *tot* and *max*, described in Section 4.3, were taken from actual runs of the TSP solution, on a given number of processors.

Figure 5(a) shows the number of tasks generated for both problems. The total tasks, mean tasks (ie tot/p) and maximum tasks dealt with by any processor are

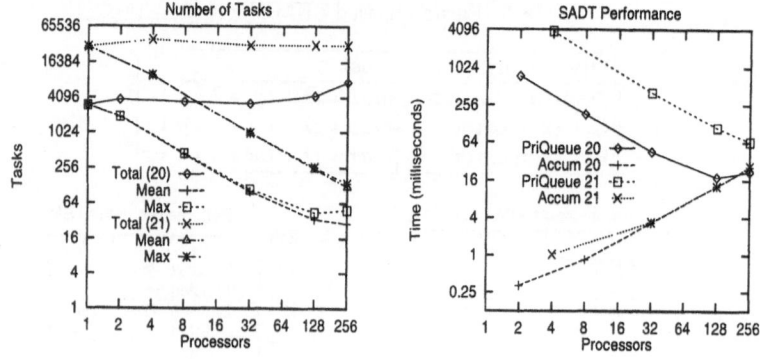

Fig. 5. (a) Number of generated tasks (b) SADT overheads

given. The total number of tasks stays quite constant as the processors vary. Obviously, a larger sample set is required if specific conclusions are to be drawn on the change in task numbers. The maximum number of tasks is very close to the mean value, even for large numbers of processors, demonstrating the good dynamic load balancing properties of the PriQueue.

Figure 5(b) presents the SADT overheads for both problem sizes. The 20 city problem overhead is chiefly due to the rising cost of Accumulator updates. The 21 city problem overheads increase slowly from 23% for 4 processors up to 37% for 256 processors. The PriQueue overheads essentially decreases linearly with the number of processors. This is due to the even load balance of tasks, together with the highly concurrent implementation of the PriQueue providing scalable high performance. The Accumulator overheads show an approximately linear increase, due to its serial implementation updating replicated copies on each processor in turn. This is not a serious limitation for the TSP, since the update frequency is very low, but for larger number of processors a tree-based solution may offer improved performance.

6 Summary and Current Work

This paper has described an approach to the development of scalable high performance parallel software for irregular problems, and how this can lead to the support of a simple analytical costing, derived from the BSP model. Further work is required to describe the assumptions under which the cost model can be used (in particular, the degree of concurrency within the sharing methods), and to expand the number of machines to demonstrate the generality of the approach. However, results for the Cray T3D have demonstrated both the scalable and high level of performance which can be achieved.

Current work is centering around the support of an adaptive unstructured 3D mesh SADT, for solving computational fluid dynamic problems, from existing MPI codes [11]. Each processor maintains a local state which consists of both partitioned internal mesh elements and replicated shared halo elements, with global consistency points defining when these halos are updated.

Acknowledgements

Thanks to Professor Peter Dew and the members of the TallShiP and Pallas research projects for their helpful comments and the use of the serial code for the travelling salesman problem.

References

1. C. Clemencon, B. Mukherjee and K. Schwan, *Distributed Shared Abstractions (DSA) on Multiprocessors*, IEEE Transactions on Software Engineering, vol 22(2), pp 132-152, February 1996.
2. J. Darlington and H. W. To, *Building Parallel Applications Without Programming*, Abstract Machine Models for Highly Parallel Computers, (eds J.R. Davy and P.M. Dew) Oxford University Press, pp 140-154, 1995.
3. P. M. Dew and J. M. Nash, *The High Performance Solution of Irregular Problems*, MPPM'97: Massively Parallel Programming Models Workshop, Royal Society of Arts, London, November 1997 (to be published in IEEE Press).
4. K. Gharachorloo, S. V. Adve, A. Gupta and J. H. Hennessy, *Programming for Different Memory Consistency Models*, Journal of Parallel and Distributed Computing, vol 15, pp 399-407, 1992.
5. D. M. Goodeve and S. A. Dobson, *Programming with Shared Data Abstractions*, Irregular'97, Paderborn, Germany, June 1997.
6. L. V. Kale and A. B. Sinha, *Information sharing mechanisms in parallel programs*, Proceedings of the 8th International Parallel Processing Symposium, pp 461-468, April 1994.
7. W. F. McColl, *An Architecture Independent Programming Model For Scalable Parallel Computing*, Portability and Performance for Parallel Processing, J. Ferrante and A. J. G. Hey eds, John Wiley and Sons, 1993.
8. J. M. Nash, P. M. Dew and M. E. Dyer, *A Scalable Concurrent Queue on a Message Passing Machine*, The Computer Journal 39(6), pp 483-495, 1996.
9. J. M. Nash, P. M. Dew, J. R. Davy and M. E. Dyer, *Scalable Dynamic Load Balancing using a Highly Concurrent Shared Data Type*, The 2nd European School of Computer Science: Parallel Programming Environments for High Performance Computing, pp 123-128, April 1996.
10. J. M. Nash, P. M. Dew, J. R. Davy and M. E. Dyer, *Implementation Issues Relating to the WPRAM Model for Scalable Computing*, Euro-Par'96, Lyon, France, pp 319-326, 1996.
11. P. M. Selwood, M. Berzins and P. M. Dew, *3D Parallel Mesh Adaptivity: Data-Structures and Algorithms*, Proceedings of the 8th SIAM Conference on Parallel Processing for Scientific Computing, SIAM, 1997.
12. L. G. Valiant, *A Bridging Model for Parallel Computation*, Communications of the ACM 33, pp 103-111, 1990.

Long Operand Arithmetic on Instruction Systolic Computer Architectures and Its Application in RSA Cryptography

Bertil Schmidt[1], Manfred Schimmler[2], and Heiko Schröder[3]

[1] Lehrstuhl für Informatik I, RWTH Aachen,
52056 Aachen, Germany,
bes@i1.informatik.rwth-aachen.de
[2] Inst. f. Datenverarbeitungsanlagen, TU Braunschweig,
38106 Braunschweig, Germany,
Schimmler@ida.ing.tu-bs.de
[3] Department of Computer Studies, Loughborough University,
Loughborough, LE11 3TU, England,
H.Schroder@lboro.ac.uk

Abstract. Instruction systolic arrays have been developed in order to combine the speed and simplicity of systolic arrays with the flexibility of MIMD parallel computer systems. Instruction systolic arrays are available as square arrays of small RISC processors capable of performing integer and floating point arithmetic. In this paper we show, that the systolic control flow can be used for an efficient implementation of arithmetic operations on long operands, e.g. 1024 bits. The demand for long operand arithmetic arises in the field of cryptography. It is shown how the new arithmetic leads to a high-speed implementation for RSA encryption and decryption.

1   Introduction

Instruction systolic arrays (**ISAs**) provide a programmable high performance hardware for specific computationally intensive applications [5]. Typically, such an array is connected to a sequential host, thus operating like a coprocessor which solves only the computationally intensive tasks within a global application. The ISA model is a mesh connected processor grid, which combines the advantages of special purpose systolic arrays with the flexible programmability of general purpose machines [3].

In this paper we illustrate how the capabilities of ISAs are exploited to derive efficient parallel algorithms of addition, subtraction, multiplication and division of long operands. The demand for long operand arithmetic arises in the field of cryptography, e.g. RSA encryption and decryption. Their implementations on Systola 1024 show that the concept of the ISA is very suitable for long operand arithmetic and results in significant run time savings.

The ISA concept is explained in detail in Section 2. Section 3 gives an overview over the architecture of Systola 1024. It is documented how the ISA has been integrated on an low cost add-on board for commercial PCs. The new ISA algorithms for long operand arithmetic are explained in Sections 4 to 6. The implementation of RSA based on these arithmetic routines is given in Section 7. Section 8 discusses its performance and concludes the paper.

2 Principle of the ISA

The basic architecture of the ISA is a quadratic $n \times n$ array of identical processors, each connected to its four direct neighbours by data wires. The array is synchronized by a global clock. The processors are controlled by instructions, row selectors and column selectors. The instructions are input in the upper left corner of the processor array, and from there they move step by step in horizontal and vertical direction through the array. This guarantees that within each diagonal of the array the same instruction is active during each clock cycle. In clock cycle $k + 1$ processor $(i + 1, j)$ and $(i, j + 1)$ execute the instruction that has been executed by processor (i, j) in clock cycle k.

The selectors also move systolically through the array: row-selectors horizontally from left to right, column-selectors vertically from top to bottom (Fig. 1). The selectors mask the execution of the instructions within the processors, i.e. an instruction is executed if and only if both selector bits, currently in that processor, are equal to one. This construct leads to a very flexible structure which creates the possibility of very efficient solutions for a large variety of applications.

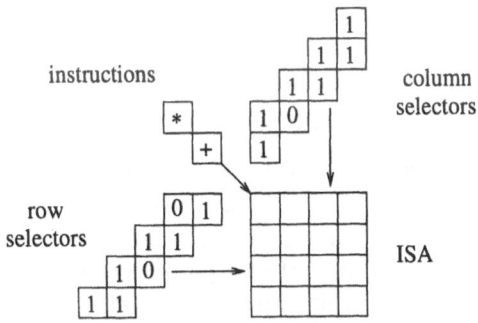

Fig. 1. Control flow in an ISA

Every processor has read and write access to its own memory. Besides that, it has a designated communication register (**C-register**) that can also be read by the four neighbour processors. Within each clock phase reading access is always performed before writing access. Thus, two adjacent processors can exchange data within a single clock cycle in which both processors overwrite the contents of their own C-register with the contents of the C-register of their neighbour. This convention avoids read/write conflicts and also creates the possibility to broadcast information across a whole row or column with one single instruction:

Row broadcast: Each processor reads the value from its left neighbour. Since the execution of this operation is pipelined along the row, the same value is propagated from one C-register to the next, until it finally arrives at the rightmost processor. Note that the row broadcast requires only a single instruction.

The **period** of ISA programs is the number of their instructions, which is $2n - 2$ clock cycles less than their execution time. The period describes the minimal time from the first input of an instruction of this program to the first input of an instruction of the next program. In the following the period of ISA programs is used to specify their time-complexity. This is appropriate because they will be used as subroutines of much larger ISA programs in Section 7.

3 Architecture of Systola 1024

The ISATEC Systola 1024 parallel computer is a low cost add-on board for standard PCs. The ISA on the board is a 4×4 array of processor chips. Each chip contains 64 processors, arranged as an 8×8 square. This provides 1024 processors on the board. In order to exploit the computation capabilities of this unit, it is necessary to provide data and control information at an extremely high speed. Therefore, a cascaded memory concept, consisting of interface processors and board RAM, is implemented on board that forms a fast input and output environment for the parallel processing unit (see Fig. 2).

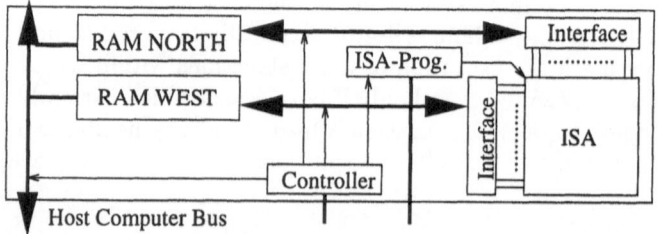

Fig. 2. Data paths in the parallel computer Systola 1024

4 Addition and Substraction of Long Operands

The difficulty of partitioning an addition in an array where only neighbours can talk to each other is the carry propagation. If it meets a sum value consisting of ones only, then an incoming carry produces a new carry in the most significant digit. In the worst case this can hold for all blocks of the partitioned addition, such that the carry-bit of the first block propagates through all other blocks. Therefore, we use an **accumulation technique** based on the generate and propagate signals of a *carry-look-ahead-adder*.

If every processor knows, whether it has to propagate an incoming carry from the left to the right, it can set a flag (e.g. the zero-flag) to one if and only if it propagates an incoming carry. Furthermore every processor can store the carry generated by itself in its C-register. With the accumulation operation

(*) if zeroflag then C:=C[WEST]

all carry-bits travel to the processor one position left of their destinations. This is again because of the skewed instruction execution: Suppose, processors $(i, j-1)$, (i, j), and $(i, j+1)$ have set their zero-flag to one (in order to propagate a carry) and processor $(i, j-2)$ has generated a carry-bit one. The instruction (*) is executed in processor $(i, j-1)$ which replaces its carry-bit (0) with that of processor $(i, j-2)$. In the next clock cycle the same instruction is executed in processor (i, j), where again the own carry (0) is replaced by the value of the left neighbour (which has been set to one in the previous instruction cycle). In the following cycle, processor $(i, j+1)$ replaces its carry with the one from the left neighbour, etc. With the final assignment C:=C[WEST] all carries are at the place where they are needed. Note that the carry propagation requires only two instructions, although the carry-bits can travel over a distance of up to $n-1$ processors. Of course, this technique only works if no processor propagating a carry has generated a carry-bit on its own. But this is obviously impossible.

The operands to be added are distributed over the rows of the ISA. If the length of the operands is 512/1024 bits, every processor in a row of 32 processors gets 16/32 bits. The LSBs are stored in the leftmost and the MSBs in the rightmost processor of the row. Every processor has 16 bits of the operand A in register RA and 16 bits of B in register RB. The operations

```
RSUM:=RA + RB; C:=carry;
```

store the sum of RA and RB in RSUM and the carry-bit in C. Now every processor must find out whether it has to propagate an incoming carry. This is the case if and only if the value of RSUM is consisting of ones only. Thus, we can complete the long addition as pointed out above with

```
if (RSUM + 1=0) then C:=C[WEST]; RSUM:=C[WEST] + RSUM;
```

Table 1. Addition of two 24-bit numbers in a row with 6 processors

Processor	1	2	3	4	5	6	
A	0101	1100	1010	0010	1101	1000	Every pr. gets 4 bits of A and B. LSB/MSB
B	0001	0011	0101	1010	1010	0100	in leftmost/rightmost bit in pr. 1/6.
SUM	0100	1111	1111	1001	0000	1100	SUM:=A + B
C	1	0	0	0	1	0	C-register stores the carry-bit
Zeroflag	0	1	1	0	0	0	if (SUM+1=0) then zeroflag:=1 else :=0
C	1	1	1	0	1	0	C after prop.: if zeroflag then C:=C[WEST]
SUM	0100	0000	0000	0101	0000	0010	SUM:=SUM + C[WEST]

The **subtraction** works in principle in the same way as the addition, using RDIFF:=RA-RB, the propagation condition if (RDIFF=0) and the carry-bit subtraction RDIFF:=RDIFF-C[WEST].

Due to the systolic control flow the n-bit addition/subtraction is possible using $O(n)$ ISA processors with constant period. The complete 512/1024-bit addition/subtraction in one Systola 1024 processor-row requires only 5/8 instructions. However, having 32 processor-rows, 32 different additions/subtractions can be calculated in parallel in the same time.

5 Multiplication of Long Operands

The idea behind the multiplication on the ISA is related to the *school method for integer multiplication*: One operand is cut into pieces and all pieces are multiplied with the other operand in parallel. The results are then shifted with respect to each other in an appropriate way and added to form the final result.

The ISA-program for the $m \times n$-bit ($m = 16 \cdot p$, $n = 16 \cdot q$) multiplication in each processor-row is explained below. Every processor stores 16 bits of the first operand in register RA, the least significant 16 bits in the leftmost and the most significant 16 bits in the rightmost processor. The complete second operand is stored in $RB_0, \ldots, RB_{q-1}$ in each processor. At the end the most significant m bits of the result will be distributed over the rows in RP_q. The least significant n bits will be stored in the first processor of each row in $RP_0, \ldots, RP_{q-1}$.

```
RP0..RPq := RA · RB0..RBq-1 // compute 16 × 16 · q-bit multiplication
C:=RP0                      // load least significant result in C-register
for i:=1 to q do            // add RP0..RPq along processor-row in q steps:
begin                          add 16 bit more significant word in C[EAST]
  RPi-1:=C                     and RPi. Sum is stored in C and RPi
  C:=C[EAST]+RPi+carry         This addition is repeated up to the most
end                            significant word
RPq:=C                      // store most significant result
C:=carry
if RPq+1=0 then C:=C[WEST] // propagate the carry-bit of the last addition
RPq:=C[WEST]+RPq              as described in Section 4
```

The implementation on Systola 1024 requires 223/507 instructions for one single $512 \times 512/1024 \times 1024$-bit multiplication in one row of 32 processors.

6   Division of Long Operands

Division can be efficiently reduced to multiplication and subtraction by using the *Newton-Raphson-method*. For a value B the reciprocal value is computed by an iteration that converges rapidly towards $\frac{1}{B}$. To achieve a fast convergence ($2m$ bits precision in m iteration steps), $0.5 \leq B < 1$ must hold for B. The iteration is given by the equations

$$x_0 := 1 \quad , \quad x_{i+1} := (2 - B \cdot x_i) \cdot x_i \tag{1}$$

Obviously, any division $Q = \frac{A}{B}$ can be computed by multiplying A with $\frac{1}{B}$. For division on the ISA we assume B to be in the range between 0.5 and 1. If this is not the case, B is initially shifted in the ISA such that the most significant one is the MSB of the rightmost word of B.

In order to understand the choice of intermediate operand lengths in our division method it is useful to consider the behavior of the convergence in the Newton-Raphson-algorithm. Let x_i differ from the precise value of $\frac{1}{B}$ by some ε. Then the following holds for x_{i+1}:

$$x_{i+1} = (2 - B \cdot x_i) \cdot x_i = (2 - B \cdot (\tfrac{1}{B} - \varepsilon)) \cdot (\tfrac{1}{B} - \varepsilon) = \tfrac{1}{B} - B \cdot \varepsilon^2 \tag{2}$$

Since $B < 1$, this means that the precision of x_{i+1} (i. e. number of correct leading digits) is at least double the precision of x_i. This means that we can restrict the operand lengths to 2^i bits for B and x_{i-1} while computing the ith iteration x_i. E.g. the first four iterations can be performed in one processor, since the required precision of the operands is ≤ 16. Only the last iteration step needs full precision of m bits. The subtractions and multiplications are performed efficiently on the ISA as explained above.

7   Parallel Implementation of RSA Encryption

For obtaining a high-speed implementation of RSA encryption a fast *modular exponentiation* ($M^e \bmod N$) is necessary. To achieve a sufficient degree of security the operand-lengths have to be relatively large (currently ranging from

512 up to 2048 bits). Modular exponentiation can be accomplished by iterating *modular multiplications* using the *square-and-multiply algorithm* [2].

The difficulty in implementing an efficient modular multiplication ($A \cdot B$ mod N) is the modular reduction step. In the case of RSA encryption the modulus N is known in advance to each modular multiplication. Thus, the precomputation of $\frac{1}{N}$ requires only a negligible amount of work. Now, the division x div N can be replaced by the more efficient multiplication $x \cdot \frac{1}{N}$. Of course, we cannot produce $\frac{1}{N}$ precisely. We instead produce a number μ such that $x \cdot \mu$ is close enough to x div N; i. e. the result needs only a small correction by at most 4 subtractions with N. Thus, the modular multiplication for RSA can be implemented efficiently on the ISA by three multiplications and at most four subtractions.

One single n-bit modular multiplication for $n = 512/1024$ requires $581/1709$ instructions in one processor- row of Systola 1024.

8   Performance Evaluation and Conclusions

The performance of the parallel implementation is compared with an optimized sequential implementation on a PC (see Table 2). The current ISA-prototype Systola 1024 is a machine based on technology far from being state-of-the-art. Extrapolating to technology used for processors such as the Pentium II 266 MHz would lead to a speedup of the ISA by a factor of at least 80. Thus resulting in encryption/decryption speeds of more than **500 Kbit/s** for 1024-bit RSA (without taking advantage of the *Chinese Remainder Theorem*). Such performance can be expected from the already announced next generation Systola board **Systola 4096**.

Table 2. Performance of RSA encryption

	Systola 1024	Pentium II 266	Speedup
512-bit RSA with full length exponent	128 KBit/s	18 KBit/s	7
1024-bit RSA with full length exponent	56 KBit/s	6 KBit/s	9

We have presented ideas for fast implementations of addition, subtraction, multiplication and division on operands that are too long to be handled within one processor. We take advantage of the ability of the ISA to implement accumulation operations extremely efficiently. We have used the results for finding a high-speed instruction systolic implementation of RSA encryption and decryption. This leads to efficient software solutions with respect to performance and hardware cost. We also used the long operand arithmetic routines for the implementation of a prime number generator for RSA keys on Systola 1024[1].

References

1. Hahnel, T.: The Rabin-Miller Prime Number Test on Systola 1024 on the Background of Cryptography. Master Thesis, University of Karlsruhe (1998)
2. Knuth, D.E.: The Art of Computer Programming: Seminumerical Algorithms. Volume 2, Reading, Addison-Wesley, second edition (1981)
3. Kunde, M., et al.: The Instruction Systolic Array and its Relation to other Models of Parallel Computers. Parallel Computing **7** (1988) 25–39

4. Rivest, R.L., Shamir, A., Adleman, L.: A method for obtaining digital signatures and public key cryptosystems. Comm. of the ACM **21** (1978) 120–126
5. Schmidt, B., Schimmler, H., Schröder, H.: Morphological Hough Transform on the Instruction Systolic Array. Euro-Par'97, LNCS 1300, Springer (1997) 798–806

Hardware Cache Optimization for Parallel Multimedia Applications

C. Kulkarni[1], F. Catthoor[1,2] and H. De Man[1,2]

[1] IMEC, Kapeldreef 75, B-3001 Leuven, Belgium
[2] Professor at the Katholieke Universiteit Leuven

Abstract. In this paper, we present a methodology to improve hardware cache utilization by program transformations so as to achieve lower power requirements for real-time multimedia applications. Our methodology is targeted towards embedded parallel multimedia and DSP processors. This methodology takes into account many program parameters like the locality of data, size of data structures, access structures of large array variables, regularity of loop nests and the size and type of cache with the objective of improving cache performance for lower power. Experiments on real life demonstrators illustrate the fact that our methodology is able to achieve significant gain in power requirements while meeting all other system constraints. We also present some results about software controlled caches and give a comparison between both the types of caches and an insight about where the largest gains lie.

1 Introduction

Parallel machines were mainly, if not exclusively, being used in scientific communities until recently. Lately, the rapid growth of real-time multimedia processing (RMP) applications have brought new challenges in terms of the required processing (computing) power, power requirements and a host of other issues. For these type of applications, especially video, the processing power of traditional uniprocessors is no longer sufficient, which has lead to the introduction of small- and medium-scale parallelism in this field too, but then mostly oriented towards single chip systems for cost reasons. Today, many parallel video and multimedia processors are emerging ([20] and its references), increasing the importance of parallelization techniques.

However, the cost functions to be used in these new application fields are no longer purely performance based. Power is also a crucial factor, and has to be optimized for a given throughput. RMP applications are usually memory intensive and most of the power consumption is due to the memory accesses for data transfers i.e. in the memory hierarchy [22]. Power management and reduction is becoming a major issue in such applications [3].

Many of the current algorithm standards (including the ones which are still under development like MPEG-x and object-oriented coders [2]) are based on a hierarchy of subsystems. The main data-dependencies and irregularities are situated in the higher layers. However, from there several submodules are called,

924

such as motion estimation/compensation, wavelet schemes, DCT etc. These lend themselves for extensive compile-time analysis enabling more efficient use of caches[1], as will be shown in the paper. This is in contrast with much of the previous work in a more "general" application context. The performance of these systems is related to the critical path in the algorithm (requires only local analysis, as shown with bold line in figure 1 where segments represent loop nests), whereas power requirements are related to the global data flow over conditional and/or parallel paths (as shown with the other lines in figure 1). Also this is a significant difference with the previous work (see below). These observations are the motivations for us to come up with an optimization strategy to utilize especially the cache hierarchy efficiently for low power.

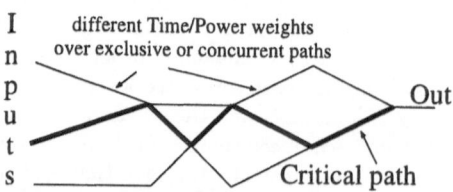

Fig. 1. Illustration of the difference between speed oriented critical path optimization which operates on the worst-case path in the global condition tree (with mutually exclusive branches) and power optimization which acts on the global flow.

Our target architecture is based on (parallel) programmable video or multi-media processors (TMS320C6x, TriMedia, etc.) as this is a current trend to use in RMP applications [7] (see section 2).

Most of the work related to efficient utilization of caches has been directed towards optimization of the throughput by means of (local) loop nest transformations to improve locality [1, 15] and loop blocking transformation to improve cache utilization [14]. Some work [17] has been reported on the data organization for improved cache performance in embedded processors but also they do not take into account a power oriented model. Instead they try to reduce cache conflicts in order to increase the throughput. Similar work regarding reduction of cache conflicts and improvement of register allocation has been reported in [10,15]. Architectural techniques to reduce the overhead in cache related power are available in [23]. Recently a study has been presented in [8] which reports on the effect with respect to power requirements on the type of cache used in an architecture. But this work also does not address the basic issue of improving the cache utilization by means of program transformations.

Program transformations for reducing memory related power and area overhead are discussed in [5, 6]. These transformations are part of our ATOMIUM [16] global data transfer and storage oriented approach and have focussed upto now on uni-processors and customized memory organizations. Program transfor-

---

[1] in this paper unless specifically mentioned cache generally means a hardware controlled cache

mations for improving cache usage for software controlled caches are discussed in [12]. Some other complementary aspects of our work on system level memory management for parallel processor mapping have been discussed in [4].

The paper is organised as follows. In section 2, the correlation between cache and power is discussed. Section 3 gives a formalized methodology to explore system level optimization for efficient cache usage. This is followed by discussion of two real life demonstrators and their results in section 4. Section 5 presents conclusions from the above study and proposes some future directions.

2   Power Models for Hardware Controlled Caches

The relation between the extent of data caching and power is explored in this section. Consider the simple but representative architecture model that we are using in this paper as shown in figure 2 (extensions to this model are feasible e.g. as in [12]). The power model used in our work comprises a power function which is dependent upon access frequency, size of memory, number of read/write ports and the number of bits that can be accessed in every (data) fetch operation.

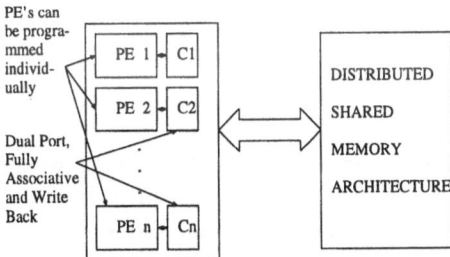

Fig. 2. The target architecture used, where each processing element (PE) has its own cache hierarchy (C_i) without prefetching or branch prediction.

Let P be the total power, N be the total number of accesses to a memory (can be SRAM or DRAM or any other type of memory)[2] and S be the size of the memory. The total power is then given by:

$$P = N \times F(S) \tag{1}$$

where F is a polynomial function with the different coefficients representing a vendor specific energy model per access. F is completely technology dependent.

For an architecture with no cache at all i.e. only a CPU and (off chip) main memory, power can be approximated as below:

$$P_{total} = P_{main}^{hi} = N_{main} \times F(S_{main}) \tag{2}$$

---

[2] Power consumption varies for different types of SRAM or DRAM depending on the individual power models (vendor and architecture specific)

Let us introduce a level one cache as in figure 2, then we observe that the total number of accesses is distributed into number of cache accesses (N_{cache}) and number of main memory accesses (N_{main}). The total power is now given by:

$$P_{total} = P_{main}^{lo} + P_{cache} \tag{3}$$

$$P_{main}^{lo} = (N_{mc} + N_{main}) \times F(S_{main}) \tag{4}$$

$$P_{cache} = (N_{mc} + N_{cache}) \times F(S_{cache}) \tag{5}$$

Here the term N_{mc} represents the amount of traffic between the two levels of memory. It is a useful indicator of the power overhead due to various characteristics like block size, associativity, replacement policy etc. for an architecture with hierarchy.

Let the factor of gain for an on-chip cache access compared to that of an off-chip main memory access be α. Here α is technology dependent. Then we have:

$$F(S_{cache}) = \alpha \times F(S_{main}) \quad where \quad \alpha < 1. \tag{6}$$

The constants in this formula is instantiated for a proprietary vendor model. In the literature several more detailed cache power models have been reported recently (e.g. [11]) but we do not need these because we only require relative comparisons.

3 Methodology

The steps comprising our methodology in this work are shown in figure 3. We assume that all the parallelization related steps (i.e data and/or task level partitioning) are done before applying this methodology. We only focus on the cache related issues here since other issues in our work have been addressed in [4, 5].

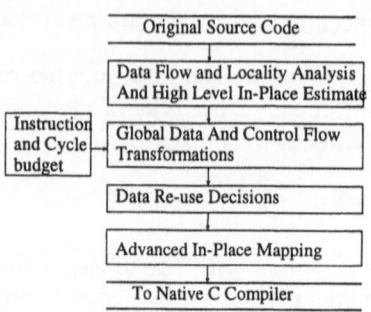

Fig. 3. Proposed methodology

The first step, Data Flow and Locality Analysis, is necessary to identify the parts and data structures in the program where there is the most potential gain or where the performance bottlenecks are located. Next, a high-level estimation is applied of the potential in-place optimization for data [6]. In step 3, Loop and Control flow transformations are applied [4, 5] but over a much more global scope than typically done in compilers [24, 13], so as to improve the regularity and locality for power reduction. The constraints are the given cycle budget and the code size limitations and data dependencies. In the Data Re-use Decisions step, different data re-use trees [18] are explored and the best hierarchy mapping (from the power point of view) is obtained for a particular application.

Finally, during Advanced In-place Mapping and Caching, both inter-signal and detailed intra-signal in-place mapping [6] are applied to obtain better cache utilization.

4 Experimental Results

In this section we illustrate the validity and effectiveness of the methodology on two real-life applications, namely a Quad-tree Structured DPCM video compression algorithm and a voice coder algorithm, using a DLX machine simulator and the Dinero III cache simulator [9]. It is important to emphasize that the results of the hardware controlled cache for both the test vehicles are based on relatively small test data[3] due to the small address space available in the DLX simulator. But we also observe that relative gains for larger frame sizes will almost be similar for hardware controlled caches. Hence comparing relative gains for hardware and software controlled caches is justified.

4.1 QSDPCM Application

The Quadtree Structured Difference Pulse Code Modulation (QSDPCM) [21] technique is an interframe adaptive coding technique for video signals. The algorithm consists of two basic parts: the first part is motion estimation and the second is a wavelet-like quadtree mean decomposition of the motion compensated frame-to-frame difference signal. A global view of the QSDPCM coding algorithm is given in figure 4(a). It is worth noting that the algorithm spans 20 pages of complex C code.

It should be stressed that the functional partitioning in the QSDPCM algorithm was not based on ad hoc algorithmic issues but based on our methodology proposed in [4] where the best possible one was selected for power. As motivated above, a small cache size of 64 words is chosen for all the functions of the QSDPCM algorithm throughout this paper (i.e. in figure 2, C1=C2=...=Cn=64 words). The results of applying the above methodology for a hardware controlled cache are given in table 1. Here we give the power requirement of the most memory intensive parts of the QSDPCM algorithm for different levels of optimization.

We observe that the initial specification consumes a relatively large amount of power. The gains due to the different steps of the methodology are quite clear

---

[3] This means that e.g. for QSDPCM the frame size for software controlled cache was the real frame of 288 × 528, whereas for hardware controlled caching this was scaled down to 48 × 64

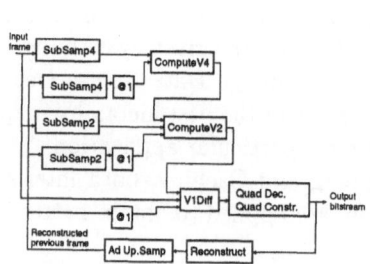
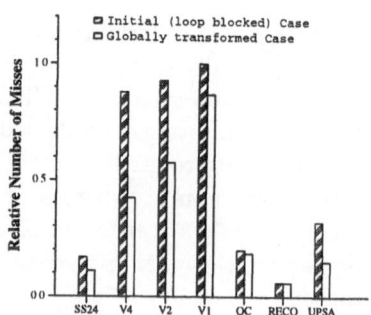

Fig. 4. (a) The QSDPCM algorithm (b) Gain in number of misses between the globally optimized version and the initial loop blocked version for QSDPCM algorithm

from the columns 3,4 and 5. The functions V4 and V2 have almost similar gains due to the different steps in the methodology. This indicates that apart from gains due to improving locality by means of global transformations and data re-use (column 3), there are equal gains due to only applying the aggressive in-place mapping and caching step (column 4). The various sources of power consumption are described below. In general we observe that the gain in power for the above methodology is quite significant for hardware controlled caches.

Table 1. Normalized power requirements of the QSDPCM algorithm for hardware controlled cache

Function	Initial (no transf)	Modified (globally transformed)	Initial, loop blocked and cached	In-placed caching on globally transformed
Subsamp2/4	0.57	0.218	0.117	0.0282
V4	0.89	0.398	0.344	0.0753
V2	1.00	0.381	0.334	0.0721
V1	1.51	0.597	0.452	0.1161
QuadConstr.	0.28	0.140	0.073	0.0230

Table 2 gives the power consumption for the various functions of QSDPCM algorithm for a software controlled cache. Note that initially the major power consumption is in functions V4,V2,V1 and QuadConstr. These are due to multiple reads from the frame memories as well as large number of intermediate signals. After the global transformations, the power consumption is reduced close to the absolute minimum for the given algorithm. All other sources of power consumption are reduced (eliminated) to a negligible amount. We observe a gain (on average) of a factor 9 globally. This illustrates the huge impact of global transformations combined with more aggressive cache exploitation at compile time, even on a fully predefined processor architecture.

Figure 4(b) shows the number of misses for each function in the QSDPCM algorithm for initial and optimized versions. The initial version has been optimized for caching with loop blocking, whereas the entire methodology presented

Table 2. Normalized power requirements of the QSDPCM algorithm for software controlled cache

Function	Initial (no transf)	Modified (globally transformed)	Initial,In-placed locally transformed and cached	In-placed caching on globally transformed
Subsamp2/4	0.21	0.092	0.095	0.0359
V4	0.75	0.002	0.374	0.0010
V2	1.90	0.005	0.492	0.0024
V1	3.29	0.208	0.644	0.0836
QuadConstr.	0.71	0.18	0.149	0.074

in section 3 is applied for the optimized version. We observe a gain of (on average) 26%. This shows that our methodology is not only beneficial for power but also for performance (speed). For the software controlled case, an even larger gain of about 40% was measured however.

It is quite clear that the gain due to software controlled cache is relatively large compared to that of the hardware controlled cache. The reasons for this lie in the two major parameters that govern a hardware controlled cache. First, if the updating policy of the cache is not taken into account in the code at compile time, a large amount of power is consumed in the write-backs (copy-backs). So a write through policy consumes more power than a write back one. The second issue is the block size. In general, increasing block size reduces the miss rate. But nevertheless it also increases cache pollution[4] in case of algorithms with less than average spatial locality. It can be concluded that more software control over the cache has a very positive impact on power as well as for speed. This is an important consideration for the cache designers of (embedded) multi-media processors.

4.2 Voice Coder Application

We have used a Linear Predictive Coding (LPC) vocoder which provides an analysis-synthesis approach for speech signals [19]. In our algorithm, speech characteristics (energy, transfer function and pitch) are extracted from a frame of 240 consecutive samples of speech. Consecutive frames share 60 overlapping samples. All information to reproduce the speech signal with an acceptable quality is coded in 54 bits. The general characteristics of the algorithm are presented in figure 5(a). It is worth noting that the steps LPC analysis and Pitch detection use autocorrelation to obtain the best matching predictor values and pitch (using a Hamming window). The autocorrelation function has irregular access structures due to algebraic manipulations, and in-place mapping performs better on irregular access structures than just loop blocking [12].

Figure 5(b) shows the power requirements of the initial version (with loop blocking) and the in-place mapped version. Here only the in-place mapping and

---

[4] Cache pollution is a phenomenon wherein data which are not accessed (or demanded) are brought in due to larger block sizes

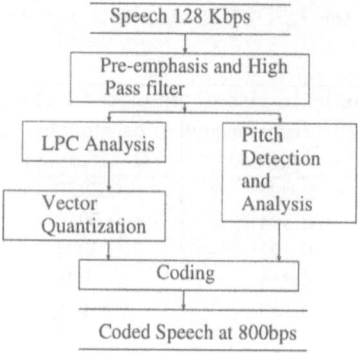
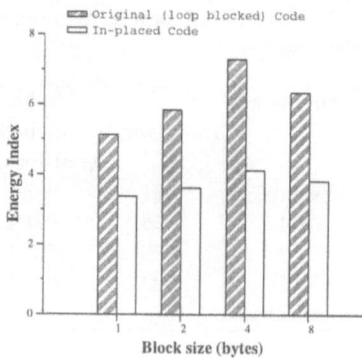

Fig. 5. (a) The different parts of the voice coder algorithm and the associated global dependencies (b)Gain in terms of energy for a voice coder autocorrelation function for different block sizes on DLX simulator

caching steps of section 3 are used so as to demonstrate that for irregular access structures it acts much better from the point of view of cache utilization than just loop blocking. We have performed this experiments for varying block sizes and have observed that our approach indeed performs much better for all the block sizes. A gain of a factor 2 is observed.

5   Conclusion

The main contributions of this paper are: (1) a formalized methodology for power-oriented source-level transformations to improve cache utilization in a hardware controlled cache context (2) the effectiveness of this methodology is demonstrated on two real-life test vehicles: the QSDPCM video compression algorithm and a voice coder (3) insight is provided into the differences in power and speed gains between hardware and software controlled caches, with consequences for low-power cache design in embedded multi-media applications.

References

1. J.Anderson, S.Amarasinghe, M.Lam, "Data and computation transformations for multiprocessors", in *5th ACM SIGPLAN Symposium on Principles and Practice of Parallel Programming*, pp.39-50, August 1995.
2. V.Bhaskaran and K.Konstantinides, "Image and video compression standards: Algorithms and Architectures", *Kluwer Academic publishers*, Boston, MA, 1995.
3. R.W.Broderson, "The network computer and its future", *Proc. IEEE Int. Solid-State Circuits Conf.*,San Francisco, CA, pp.32-36, Feb.1997.
4. K.Danckaert, F.Catthoor and H.De Man, "System-level memory management for weakly parallel image processing", *In proc. EUROPAR-96*, Lecture notes in computer science series, vol. 1124, Lyon, Aug 1996.

5. E.De Greef, F.Catthoor, H.De Man, "Program transformation strategies for reduced power and memory size in pseudo-regular multimedia applications", accepted for publication in *IEEE Trans. on Circuits and Systems for Video Technology*, 1998.

6. E.De Greef, F.Catthoor, H.De Man, "Memory Size Reduction through Storage Order Optimization for Embedded Parallel Multimedia Applications", *Intnl. Parallel Proc. Symp.(IPPS)* in Proc. Workshop on "Parallel Processing and Multimedia", Geneva, Switzerland, pp.84-98, April 1997.

7. T.Halfhill and J.Montgomery, "Chip fashion: multi-media chips", *Byte Magazine*, pp.171-178, Nov 1995.

8. P.Hicks, M.Walnock and R.M.Owens, "Analysis of power consumption in memory hierarchies", *In Proc. Int'l symposium on low power electronics and design*, pp.239-242, Monterey, CA, Aug 1997.

9. Marck Hill, "Dinero III cache simulator", Online document available via http://www.cs.wisc.edu/ markhill, 1989.

10. M.Jimenez, J.M.Liaberia, A.Fernandez and E.Morancho, "A unified transformation technique for multilevel blocking", *In proc. EUROPAR-96*, Lecture notes in computer science series, vol. 1124, Lyon, Aug 1996.

11. Milind Kamble and Kanad Ghose, "Analytical Energy Dissipation Models for Low Power Caches", *In Proc. Int'l symposium on low power electronics and design*, pp.143-148, Monterey, Ca., Aug 1997.

12. C.Kulkarni, F.Catthoor and H.De Man, "Code transformations for low power caching in embedded multimedia processors", *Proc. of Intnl. Parallel Processing Symposium (IPPS)*, pp.292-297, Orlando, FL, April 1998.

13. D.Kulkarni and M.Stumm, "Linear loop transformations in optimizing compilers for parallel machines", *The Australian computer journal*, pp.41-50, May 1995.

14. M.Lam, E.Rothberg and M.Wolf, " The cache performance and optimizations of blocked algorithms", *In Proc. ASPLOS-IV*, pp.63-74, Santa Clara, CA, 1991.

15. N.Manjikian and T.Abdelrahman, "Array data layout for reduction of cache conflicts", *In Proc. of 8th Int'l conference on parallel and distributed computing systems*, September, 1995.

16. L.Nachtergaele, F.Catthoor, F.Balasa, F.Franssen, E.De Greef, H.Samsom and H.De Man, "Optimisation of memory organisation and hierarchy for decreased size and power in video and image processing systems", *Proc. Intnl. Workshop on Memory Technology, Design and Testing*, San Jose CA, pp.82-87, Aug. 1995.

17. P.R.Panda, N.D.Dutt and A.Nicolau, " Memory data organization for improved cache performance in embedded processor applications", *In Proc. ISSS-96*, pp.90-95, La Jolla, CA, Nov 1996.

18. J.P. Diguet, S. Wuytack, F.Catthoor and H.De Man, "Formalized methodology for data reuse exploration in hierarchical memory mappings", *In Proc. Int'l symposium on low power electronics and design*, pp.30-36, Monterey, CA, Aug 1997.

19. L.R.Rabiner and R.W.Schafer, "Digital signal processing of speech signals", *Prentice hall Int'l Inc.*, Englewood cliffs, NJ, 1988.

20. K.Ronner, J.Kneip and P.Pirsch, "A highly parallel single chip video processor", *Proc. of the 3rd International Workshop, Algorithms and Parallel VLSI architectures III*, Elsevier, 1995.

21. P.Strobach, "QSDPCM – A New Technique in Scene Adaptive Coding," *Proc. 4th Eur. Signal Processing Conf.*, EUSIPCO-88, Grenoble, France, Elsevier Publ., Amsterdam, pp.1141–1144, Sep. 1988.

22. V.Tiwari, S.Malik and A.Wolfe, "Instruction level power analysis and optimization of software", *Journal of VLSI signal processing systems*, vol. 13, pp.223-238, 1996.

23. Uming Ko, P.T.Balsara and Ashmini K.Nanda, "Energy optimization of multi-level processor cache architectures", *In Proc. Int'l symposium on low power electronics and design*, pp.45-49, Dana Point, CA, 1995.
24. M.E.Wolf and M.Lam, "A loop transformation theory and an algorithm to maximize parallelism", *IEEE Trans. on parallel and distributed systems*, pp.452-471, Oct 1991.

Parallel Solutions of Simple Indexed Recurrence Equations

Yosi Ben-Asher[1] and Gady Haber[2]

[1] Dep. of Math. and CS. Haifa University 31905 Haifa, Israel,
yosi@mathcs.haifa.ac.il
[2] IBM Science and Technology, Haifa, Israel,
haber@haifasc3.vnet.ibm.com

Abstract. We consider a type of recurrence equations called "Simple Indexed Recurrences" (SIR) wherein ordinary recurrences of the form $X[i] = op_i(X[i-1], X[i])$ $(i = 1 \ldots n)$ are extended to $X[g(i)] = op_i(X[f(i)], X[g(i)])$, such that op_i is an associative binary operation, $f, g : \{1 \ldots n\} \mapsto \{1 \ldots m\}$ and g is distinct.[1] This extends our capabilties for parallelizing loops of the form: for $i = 1$ to n { $X[i] = op_i(X[i-1], X[i])$ } to the form: for $i = 1$ to n { $X[g(i)] = op_i(X[f(i)], X[g(i)])$ }. An efficient solution is presented for the special case where we know how to compute the inverse of op_i operator. The algorithm requires $O(\log n)$ steps with $O(n/\log n)$ processors. Furthermore, we present a practical and a more improved version of the non-optimal algorithm for SIR presented in [1] which uses repeated iterations of pointer jumping. A sequence of experiments was performed to test the effect of synchronous and asynchronous message-passing executions of the algorithm for $p << n$ processors. This algorithm computes the final values of $X[]$ in $\frac{n}{p} \cdot \log p$ steps and $n \cdot \log p$ work, with p processors. The experiments show that pointer jumping requires $O(n)$ work in most practical cases of SIR loops, thus forming a more practical solution.

1 Introduction

Ordinary recurrence equations of the form $X_i = op_i(X_{i-1}, X_i)$ $i = 1 \ldots n$ can be generalized to what we refer to as *Indexed Recurrence* (IR) equations. In IR equations, general indexing functions of the form $X_{g(i)} = op_i(X_{f(i)}, X_{g(i)})$ replace the $i, i-1$ indexing of the ordinary recurrences. Efficient parallel solutions to such equations can be used to parallelize sequential loops of the form:

for $i = 1$ to n { $X[g(i)] = op_i(X[f(i)], X[g(i)])$ }

if the following conditions are met:

1. $op_i(x, y)$ is any segment of code that is equivalent to a binary associative operator and may be dependent on i.
2. the index functions $f, g : \{1..n\} \mapsto \{1..m\}$ do not include references to elements of the $X[]$ array.

---

[1] This paper is a continuation of the work on IR equations presented in [1].

3. the index function g is distinct, i.e., for every indexes i, j if $i = j$ then $g(i) = g(j)$.

In practice, such solutions can be used to parallelize sequential loops within given source code segments.

2 An Efficient *SIR* Algorithm for the Case Where the Inverse of op_i is Known

In this section we describe an efficient parallel algorithm for *SIR* loops, providing we know how to compute the inverse operation op_i^{-1}. The algorithm requires $O(\log n)$ steps with $O(n/\log n)$ processors.

The first step in the algorithm is to translate the *SIR* loop into its upside down tree representation T, in which every vertex i $(1 \leq i \leq n)$ represents the $g(i)$ index and every edge $e = \{i, j\}$ represents the dependency $op_i(X_{g(j)}, X_{g(i)})$ where $g(j) = f(i)$ and $j < i$. For example consider the following upside-down tree representation of a *SIR* loop:

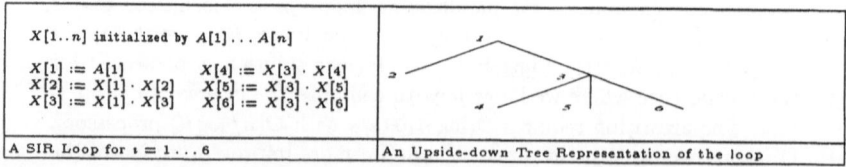

| A SIR Loop for $i = 1 \ldots 6$ | An Upside-down Tree Representation of the loop |

We then apply the *Euler Tour* circuit on T in order to compute all the prefixes of the vertices in T. An Euler Tour in a graph is a path that traverses each edge exactly once and returns to its starting point. The Euler tour circuit which operates on a tree assumes that the tree is directed and that it is represented by an Adjacency List. In the directed tree version T_D of T, each undirected edge $\{i, j\}$ in T has two directed copies - one for (i, j) and one for (j, i). For example, consider the directed version of the tree for the above example, and its Adjacency List representation:

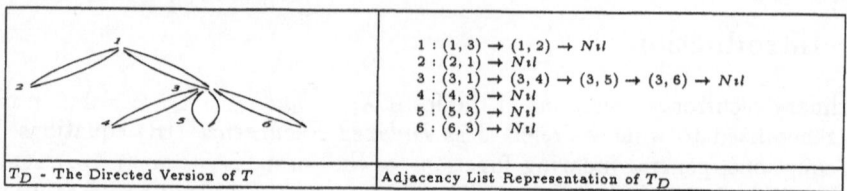

| T_D - The Directed Version of T | Adjacency List Representation of T_D |

Our main concern when coming to construct the adjacency list of T_D, is finding the list of all the children of each vertex i in T_D efficiently; i.e., finding all vertices $j < i$ which satisfy that $g(j) = f(i)$. Sorting all the pairs $< f(1), g(1) >, \ldots, < f(n), g(n) >$ using $f(i)$ as the sorting key, will automatically group together all vertices which have the same parent. The sorting process can be done efficiently using the Integer Sort algorithm [3] which operates on keys in the range of $[1..n]$.

After the adjacency list of T_D is formed, we can construct the Euler tour of T_D in a single step according to the following rule:

$$EtourLink(v, u) = \begin{cases} Next(u, v) & \text{if } Next(u, v) \neq Nil \\ AdjList(u) & \text{Otherwise} \end{cases}$$

by allocating a processor to each pair of edges $(v, u), (u, v)$ in T_D.

Once the Euler tour of T_D is constructed, in the form of a linked list, we can compute the prefixes of all the vertices of T_D, using the efficient parallel prefix algorithm on the linked list [2], after initializing all the edges of the tree with the following weights:

- The weight of each forward edge $< f(i), g(i) >$ in T_D is initialized by $A_{g(i)}$, where A is the array of initial values of the X array of SIR.
- The weight of each backward edge $< g(i), f(i) >$ is initialized by the inverse value $A_{g(i)}^{-1}$.

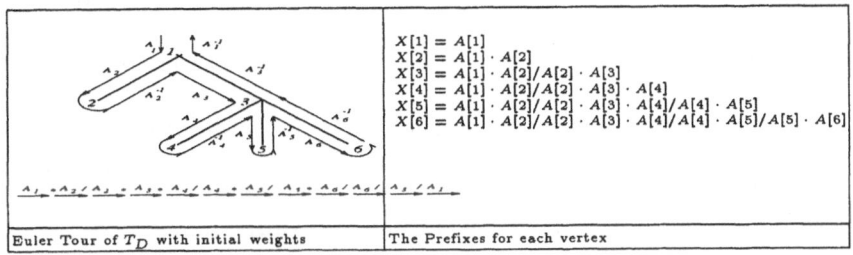

	X[1] = A[1]
	X[2] = A[1] · A[2]
	X[3] = A[1] · A[2]/A[2] · A[3]
	X[4] = A[1] · A[2]/A[2] · A[3] · A[4]
	X[5] = A[1] · A[2]/A[2] · A[3] · A[4]/A[4] · A[5]
	X[6] = A[1] · A[2]/A[2] · A[3] · A[4]/A[4] · A[5]/A[5] · A[6]

Euler Tour of T_D with initial weights	The Prefixes for each vertex

3 Parallel Solution for SIR by Using Pointer Jumping

Consider for example the SIR problem $A'[2i] := A'[i+1] \cdot A'[2i]$; (for $i = 1, 2..n$) where $g(i) = 2i$ and $f(i) = i+1$. The final values of each element $A'[i]$ is a product of a varying number of items. We refer to this sequence of multiplications of every element i in $A'[]$ as the **trace** of $A'[i]$. In general the trace of each element $A'[g(i)]$ satisfies that for all $i = 1 \ldots n$: $A'[g(i)] = A[f(j_k)] \oplus \ldots \oplus A[f(j_1)] \oplus A[g(i)]$ such that:

- $j_1 = i$.
- for $t = 2 \ldots k$ the indices j_t satisfy that $j_t < j_{t-1}$ and $g(j_t) = f(j_{t-1})$.
- j_k is the last index for which $g(j_t) = f(j_{t-1})$, i.e., there is no $1 \leq j_{k+1} < j_k$ such that $g(j_{k+1}) = f(j_k)$.

The above rule suggests a simple method for computing $A'[g(i)]$ in parallel. Let $A^{-t}[g(i)]$ denote the sub-trace with $t + 1$ rightmost elements in the trace of $A'[g(i)]$, i.e., $A^{-t}[g(i)] = A[f(j_{k-t})] \oplus \ldots \oplus A[f(i)] \oplus A[g(i)]$, and now consider the **concatenation** (or **multiplication**) of two "successive" sub-traces: $A^{-(t_1+t_2)}[g(i)] = A^{-t_1}[g(j)] \oplus A^{-t_2}[g(i)]$ where $g(j) = f(j_{k-t_2})$ and $A[f(j_{k-t_2})]$ is the last element in $A^{-t_2}[g(i)]$.

The proposed algorithm (as shown in [1]) is a simple greedy algorithm that keeps iterating until all traces are completed. In each iteration all possible concatenations of successive sub-traces are computed in parallel where the concatenation operation of sub-traces $A^{-t_1}[g(j)], A^{-t_2}[g(j)]$ can be implemented as follows:

1. The value of a sub-trace $A^{-t}[g(i)]$ is stored in its array element $A[g(i)]$.
2. A pointer $Next[g(i)]$ points to the sub-trace $A^{-t_1}[g(j)]$ to be concatenated to $A^{-t_2}[g(i)]$ (to form $A^{-(t_1+t_2)}[g(i)]$). Hence, $A[Next[g(i)]]$ contains the value of the sub-trace $A^{-t_1}[g(j)]$.

Following is the code of an improved and a more practical version of the algorithm for which we ran our experiments with $p << n$ processors. The new algorithm improves the above algorithm in the following points:

- It works with compressed array of size n (the number of loop iterations) instead of the original array size, which is of size $m > n$, by mapping every $A'[g(i)]$ element to an auxiliary array $X[i]$.
- The index function $f(i)$ is transformed to a decreasing function which never refers to future indexes of $X[]$. This reduces the total number of iterations required by the algorithm from an order of $\log n$ to $\log p$, improving the execution time to $\frac{n}{p} \log p$. Suppose each processor computes $\frac{n}{p}$ traces and uses a sequence of passes wherein it calculates the traces sequentially one after another. Since $f()$ always points backwards to earlier elements in the input array, then after each step i we already have the final values of the first 2^i elements, requiring a total of $\log p$ steps in order to complete the entire array of size n. The same argument however, does not hold for the case where $f()$ can refer to future elements in the array.

```
Input: Initialized Arrays A[1..m], g[1..m], f[1..m], ⊕.     (where ⊕ is a binary associative operator)

Output: Array X[1..n] where X[i] will hold the value of A[g[i]]
    at the end of the IR loop.

Initialize Auxiliary Arrays X[1..n], G[1..m], Next[1..n] to zero.

forall i ∈ {1..n} do in parallel
    G[g[i]] := i; (The auxiliary array G represents the inverse of g, i.e. G[i] ≡ g⁻¹(i))
end parallel-for

forall i ∈ {1..n} do in parallel
    u := G[f[i]]; (The iteration where A'[f(i)] was modified by the loop).
    if (u ≥ i or u = 0) { (The trace of A'[g(i)] is of size two.)
        X[i] := A[f[i]] ⊕ A[g[i]]; (i.e., A'[g(i)] = A[f(i)] ⊕ A[g(i)].)
        Next[i] := 0;
    } else {
        if (f[u] ≥ f[i] or f[u] = 0) { (The trace of A'[f(i)] is of size three.)
            X[i] := A[f[u]] ⊕ A[f[i]] ⊕ A[g[i]];
            Next[i] := 0;
        } else { (The trace of A'[g(i)] is of size greater than three.)
            X[i] := A[f[i]] ⊕ A[g[i]];
            Next[i] := f[u];
        }
    }
end parallel-for
```

Initialization Stage

```
for t = 1 to log n do (The algorithm performs, at most, log n iterations)
    forall i ∈ {(2^(t−1) + 1)..n} do in parallel
        if (Next[i] > 0) then { (Apply the concatenation operation followed by the pointer updating)
            X[i] := X[Next[i]] ⊕ X[i];
            Next[i] := Next[Next[i]];
        }
```

The code for the iteration phases

4 Experimental Results for Message Passing Sync and Async SIR

For the message-passing version of SIR we assume that the initial values $A[1..m]$ are partitioned between the processors, with each processor holding $\frac{m}{p}$ elements.

Thus the first step is for each processor to fetch the $O(\frac{n}{p})$ elements of $A[]$ that it will use during the initialization stage. Each processor will sort the requested elements according to their processor destination, and then fetch them using p "big-messages" (i.e., using message packaging). Two variants of this algorithm have been tested:

Sync SIR: where all processors compute one iteration of the main loop synchronously.

Async SIR(k): where each processor performs l iterations of the main-loop before passing the control to the next processor. The number l is chosen randomly (every time) from the range $1 \ldots k$, where k is the parameter of the algorithm. Thus, by choosing different values of k we can change the amount of asynchronous deviation of the execution.

We simulated different numbers of processors, $p = 4, 8, 16, 32, 64, 128, 256, 512, 1024$ with $n = 500,000$. The simulation of the algorithm was made on a sequential machine, since this allowed us an exact and simple measure of the total size of generated messages (communication), the work (in units of C instructions) and the execution time (also in units of C instructions). All diagrams contain an artificial curve (e.g., $f(p) = \frac{n}{p} \cdot \log n$) for comparison. We first chose to test a SIR loop where $g(i) = i$ and $f(i) = i - 1$, as this maximizes the length of each trace and consequently the expected work. For $Sync\ SIR$, we already know the results: execution time approximately $\frac{n}{p} \cdot \log p$, work $n \cdot \log p$, total of $p \cdot \log p$ big messages and communication (total number of data items that have been sent) of $p \cdot \log p$. The results of the execution time in fig. 1 verify this computation and show that increasing k cost of $Async\ SIR(k)$ can reduce the speedups significantly. Another observation regarding $Async\ SIR(k)$ is that for a relatively small number of processors the impact of k is much larger than with a large number of processors. According to fig. 1, the difference between the various $Async\ SIR(k)$ results, becomes constant (independent on p) for values greater than 32. This is because of the relatively small number of $A[]'s$ elements that are allocated to each processor $(\frac{n}{p} \approx p)$. The same results are obtained for the work of the algorithms as described in fig. 1. The communication and number of big messages is exactly $p \cdot \log p$, as expected. These experiments were repeated for a random setting $f(i) = random(1 \ldots n)$. In general, we hoped to reach optimal performances of execution time $\frac{n}{p}$ and work $O(n)$. The execution time has improved, and $sync\ SIR$ (see fig. 2) is now between $\frac{n}{p}$ and $\frac{n}{p} \cdot \log p$. In addition, the effect of k in $Async\ SIR(k)$ on the execution time is reduced, and for $k = 1, 5, \sqrt{p}$ we get an execution time that is less than $\frac{n}{p} \cdot \log n$. In particular, we can see that the work (fig. 2) behaves like $O(n)$ rather than $n \log p$. The communication (fig. 2) is below n. Unlike the case $f(i) = i$, the effect of k in $Async\ SIR(k)$ on the communication is negligible (all the results for $k = 1, 5, \sqrt{p}$ are the same). In addition, asynchronous execution seems to improve the communication compared to $sync\ SIR$. An asynchronous execution might, for example, cause the middle processor to advance its $Next[i]$ pointers before other processors started to work. This might reduce the communication that would have occured between

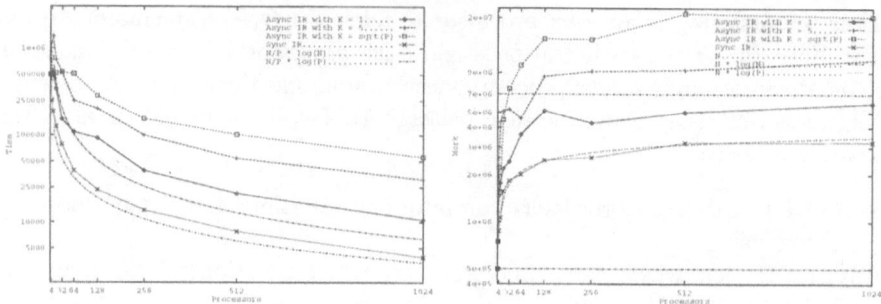

Fig. 1. *Execution time and Work for* $f(i) = i - 1$.

these processors if the middle processor hadn't advanced its pointers. The total number of messages was not affected by the random setting and (as is described in fig. 2) is around $p \log p$.

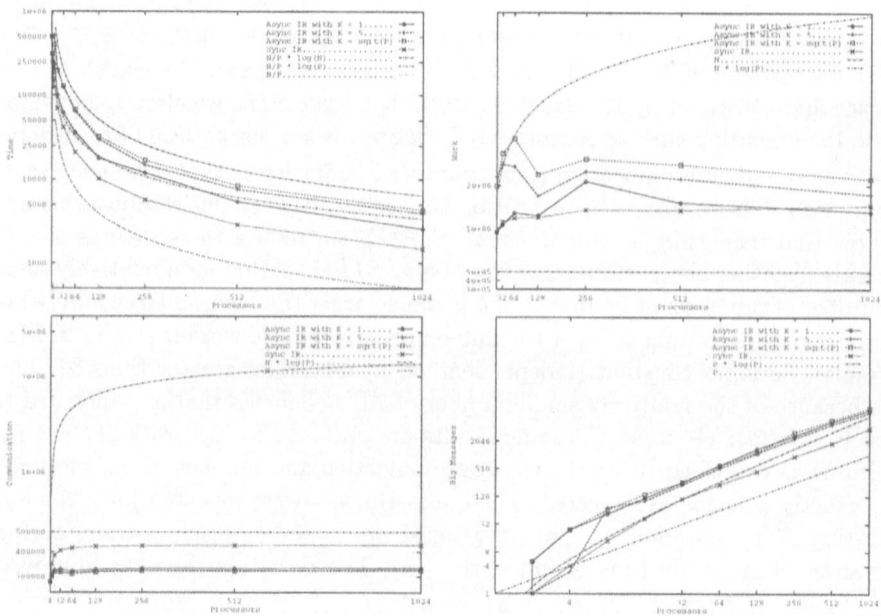

Fig. 2. *Execution time, Work, Comm. and big-messages for* $f(i) = random(1..n)$.

References

1. Y. Ben-Asher and G. Haber. Parallel solutions of indexed recurrence equations. In *In Proceedings of the IPPS'97 conference, Geneva*, 1997.

2. L. Rudolph Kruskal C. P. and M. Snir. The power of parallel prefix. *IEEE Transactions on Computers*, 34(C):965–968, 1985.
3. S. Rajasekaran, S. Sen. On parallel integer sorting. Technical Report To appear in ACTA INFORMATICA, Department of Computer Science, Duke University, 1987.

Scheduling Fork Graphs under LogP with an Unbounded Number of Processors

Iskander Kort and Denis Trystram

LMC-IMAG
BP53 Domaine Universitaire
38041 Grenoble Cedex 9, France
{kort,trystram}@imag.fr

Abstract This paper deals with the problem of scheduling a specific precedence task graph , namely the *Fork* graph, under the LogP model. LogP is a computational model more sophisticated than the usual ones which was introduced to be closer to actual machines.
We present a scheduling algorithm for this kind of graphs. Our algorithm is optimal under some assumptions especially when the messages have the same size and when the gap is equal to the overhead.

1 Motivation

The last decade was characterized by the huge development of many kinds of parallel computing systems. It is well-known today that a universal computational model can not unify all these varieties. PRAM is probably the most important theoretical computational model. It was introduced for shared-memory parallel computers. The main drawback of PRAM is that it does not allow to take into account the communications through an interconnection network in a distributed-memory machine. Practical PRAM implementations have often bad performances. Many attempts to define standard computational models have been proposed. More realistic models such as BSP and LogP [3] appeared recently. They incorporate some critical parameters related to communications.

The LogP model is getting more and more popular. A LogP machine is described by four parameters L, o, g and P. Parameter L is the interconnection network latency. Parameter o is the overhead on processors due to local management of communications. Parameter g represents the minimum duration between two consecutive communication events of the same type. Parameter P corresponds to the number of processors. Moreover, the model assumes that at most $\lceil \frac{L}{g} \rceil$ messages from (resp. to) a processor may be in transit at any time.

We present in this paper an optimal scheduling algorithm under the LogP model for a specific task graph, namely the *Fork* graph. We assume that the number of processors is unbounded.

In the remainder of this section we give some definitions and notations concerning *Fork* graphs, then we briefly describe some related work. In Sect. 2, the scheduling algorithm is presented.

1.1 About the Fork Graph

A *Fork* graph is a tree of height one. It consists of a root task denoted by T_0 preceeding n leaf tasks $T_1, \ldots, T_n$. The root sends, after its completion, a message to each leaf task. In the remainder of this paper, symbol F will refer to a *Fork* graph with n leaves. In addition, we denote by w_i the execution time of task T_i, $i \in \{0, \ldots, n\}$. The processors are denoted by $p_i, i = 0, 1, \ldots$. We assume, without loss of generality, that task T_0 is always scheduled on processor p_0. Let S be a schedule of F, then $df_i(S)$ denotes the completion time of processor p_i in S where $i \in \{0, 1, \ldots\}$. Furthermore, $df(S)$ denotes the *makespan* (length of S) and df^* the length of an optimal schedule of F.

The contents of the messages sent by T_0 must be considered when dealing with scheduling under LogP. Indeed, assume that T_0 sends the same data to some tasks T_i and T_j. Then assigning these tasks to the same processor (say $p_i, i \neq 0$) saves a communication. Two extreme situations are generally considered. In the first one, it is assumed that T_0 sends the same data to all the leaves. This is called *a common data semantics*. In the second situation, the messages sent by T_0 are assumed to be pairwise different. This is called *an independent data semantics* [4].

1.2 Related Work

The problem of scheduling tree structures with communication delays and an unbounded number of processors has received much interest recently. The major part of the available results focused on the extended Rayward-Smith model. Chrétienne has proposed a polynomial-time algorithm for scheduling Fork graphs with arbitrary communication and computation delays [1]. He has also showed that finding an optimal schedule for a tree with a height of at least two is NP-Hard [2]. More recently, some results about scheduling trees under LogP have been presented. In [6], Verriet has proved that finding optimal *Fork* schedules is NP-Hard when a common data semantics is considered. In [5], the authors have presented a polynomial-time algorithm that determines optimal linear schedules for inverse trees under some assumptions.

2 An Optimal Scheduling Algorithm

We present in this section an optimal scheduling algorithm of F when the number of processors is unbounded $(P \geq n)$. Furthermore , we assume that:

- The messages sent by the root have the same size.
- The gap is equal to the overhead: $g = o$. This is the case for systems where the communication software is a bottleneck.
- An independent data semantics is considered.
- $w_i \geq w_{i+1}, \forall i \in \{1, \ldots, n-1\}$.

We start by presenting some dominant properties related to *Fork* schedules.

Lemma 1. *Each of the following properties is dominant:*

π_1 *Each leaf task T_i such that $w_i \leq o$ is assigned to p_0.*

π_2 *Processor p_0 executes T_0 at first, then it sends the messages to tasks T_i which are assigned to other processors. These messages are sent according to decreasing values of w_i. Finally, p_0 executes the tasks that were assigned to it in an arbitrary order.*

π_3 *Each processor $p_i (i \neq 0)$ executes at most one task, as early as possible (just after receiving its message).*

Every schedule that satisfies properties $\pi_1 - \pi_3$ will be called a dominant schedule. Such schedules are completely determined when the subset A^* of the leaves that should be assigned to p_0 is known. We propose in the sequel an algorithm (called CLUSTERFORK) that computes this subset. Initially, each task is assigned to a distinct processor. The algorithm manages two variables b and B which are a lower bound and an upper bound on df^* respectively. More specifically, at any time we have $b \leq df^* < B$. These bounds are refined as the algorithm proceeds. CLUSTERFORK explores the tasks from T_1 to T_n. For each task $T_i, i \in \{1, \ldots, n\}$, one of the following situations may be encountered. If the completion time of T_i in the current schedule is not less than B then T_i is assigned to p_0. If df_i is not greater than b, then the algorithm does not assign this task to p_0. Finally, if df_i is in the range $]b, B[$, then the algorithm checks whether there is a schedule S of F such that $df(S) < df_i$. If such a schedule exists, then T_i is assigned to p_0 and B is set to df_i, otherwise T_i is not assigned to p_0 and b is set to df_i.

Theorem 1. *Let A^* be a subset produced by CLUSTERFORK, then any dominant schedule S^* associated with A^* is optimal.*

Finally, it is easy to see that CLUSTERFORK has a computational complexity of $O(n^2)$.

3 Concluding Remarks

In this paper we presented an optimal polynomial time scheduling algorithm for *Fork* graphs under the LogP model and using an unbounded number of processors. This problem was solved under some assumptions which hold for a current parallel machine namely the IBM-SP. We remark that a slight modification in the problem parameters (for instance when the messages are the same) leads to an NP-hardness result. Indeed, in this latter case, scheduling at most one task on each processor p_i, $i \neq 0$ is no more a dominant property and gathering some tasks on the same processor may lead to better schedules.

References

1. P. Chrétienne. Task Scheduling over Distributed Memory Machines. In North Holland, editor, *International Workshop on Parallel and Distributed Algorithms*, 1989.

Algorithm 1 CLUSTERFORK

Begin
{Initially: $df_i = w_0 + (i+1) * o + L + w_i, \forall 1 \leq i \leq n$}
$b := 0$; {lower bound on df^*}
$B := MAXVAL$; {upper bound on df^*}
for $(i := 1$ **to** $n)$ **do**
  if $(df_i \geq B)$ **then**
    $x_i := 1$; {T_i is assigned to p_0}
    $UPDATE\_DF(i)$;
  else if $(df_i > b)$ **then**
    {Look for a schedule such that $df < df_i$}
    $j := i$; $k := 0$; $v := df_0$;
    while $(j \leq n)$ **do**
      if $(df_j - k * o \geq df_i)$ **then**
        {The assignment of T_j to p_0 is necessary}
        if $(v + w_j - o \geq df_i)$ **then**
          $j := n + 1$;
        else
          $v := v + w_j - o$; $k := k + 1$;
      $j := j + 1$;
    if $(j = n + 2)$ **then**
      {The looked for schedule does not exist}
      $x_i := 0$; $b := df_i$;
    else {$df^* < df_i$}
      $x_i := 1$; $B := df_i$;
      $UPDATE\_DF(i)$; {Update processors completion times}
  else {$df_i \leq b$}
    $x_i := 0$;
End CLUSTERFORK

2. P. Chrétienne. Complexity of Tree-scheduling with Interprocessor Communication Delays. Technical Report 90.5, MASI , Pierre and Marie Curie University, Paris, 1990.

3. D. E. Culler et al. LogP: A practical model of parallel computation. *Communications of the ACM*, 39(11):78–85, November 1996.

4. L. Finta and Z. Liu. Complexity of Task Graph Scheduling with Fixed Communication Capacity. *International Journal of Foundations of Computer Science*, 8(1):43–66, 1997.

5. W. Löwe, M. Middendorf, and W. Zimmermann. Scheduling Inverse Trees under the Communication Model of the LogP-Machine. *Theoretical Computer Science*, 1997. to appear.

6. J. Verriet. Scheduling Tree-structured Programs in the Logp Model. Technical Report UU-CS-1997-18, Department of Computer Science, Utrecht University, 1997.

A Data Layout Strategy for Parallel Web Servers*

Jörg Jensch, Reinhard Lüling, and Norbert Sensen

Department of Mathematics and Computer Science
University of Paderborn, Germany
{jaglay, rl, sensen}@uni-paderborn.de

Abstract. In this paper a new mechanism for mapping data items onto the storage devices of a parallel web server is presented. The method is based on careful observation of the effects that limit the performance of parallel web servers, and by studying the access patterns for these servers.

On the basis of these observations, a graph theoretic concept is developed, and partitioning algorithms are used to allocate the data items. The resulting strategy is investigated and compared to other methods using experiments based on typical access patterns from web servers that are in daily use.

1 Introduction

Because of the exponential growth in terms of number of users of the World Wide Web, the traffic on popular web sites increases dramatically. To resolve capacity requirements on the server systems hosting popular web sites, one way to strengthen the capability of a server system is to use parallel server systems that contain a number of disks attached to different processors which are connected by a single bus or a scalable communication network. The same solution can be applied for very large web sites which cannot be stored on one disk but use a number of disks connected to a single processor system.

The problem that shows up if a number of disks are used to store the overall amount of data is to balance the number of requests issued to the disks as evenly as possible in order to achieve the largest overall throughput from the disks. The requests that have to be served by one disk are strictly dependent on the data layout of the server system, i.e. the mapping of the files stored on the server onto the different storage subsystems (disks). The investigation of a new concept for the data layout of parallel web servers is the focal point of this paper.

The problem is of major relevance for the performance of web servers and can be assumed to become even more important if the current technological trends concerning the bandwidth offered by storage devices, processors and wide area

* This work was partly supported by the MWF Project "Die Virtuelle Wissensfabrik", the EU project SICMA, and the DFG Sonderforschungsbereich 1511 "Massive Parallelität: Algorithmen, Entwurfsmethoden, Anwendungen".

communication networks are regarded: Whereas the performance of processors and external communication networks (ATM, Gigabit Ethernet) has been dramatically increased over the last years, the read/write performance of storage devices shows only little increase over the last years. Thus, the performance of a server system can only be increased by the use of a larger number of disks that delivers data elements in parallel to the external communication network.

Different approaches can be found in the literature in order to increase the performance of web servers: NCSA [7] and SWEB [4] have built a multi-workstation HTTP server based on round-robin domain name resolution(DNS) to assign requests to a set of workstations. In [5] this DNS-strategy is extended, the time-to-live (TTL) periods are used to distribute the load more evenly. In the past different researchers tried to identify the workload behavior with emphasis on the development of a caching strategy. NCSA developers analyzed user access patterns for system configurations to characterize the WWW traffic in terms of request count, request data volume, and requested resources [9]. Arlitt and Williamsion [3] studied workloads of different Internet Web servers. Their emphasis placed on finding universal invariants characterizing Web servers workloads. Bestavros et. al [2] tried to characterize the degree of temporal and spatial locality in typical Web server reference streams. A general survey on file allocation is given in [13]. In [9] the problem of declustering is solved by transforming the problem onto a MAX-CUT problem.

The data layout method that is presented in this paper is based on the following idea: The goal of the layout is to minimize the number of items that resides on the same disk and are requested often simultaneously. So we examine the access pattern of the past and place those items on different disks. Since the items which are often simultaneously requested are roughly the same at following days, this strategy leads to a good data layout.

The model of our parallel web server is presented in section 2. In section 3 we show that typical access patterns remain constant over some time, and collisions are typical patterns of a web server. In section 4 the data layout strategy is presented and the strategy is studied in detail, compared with other methods and limits of the strategy are shown. The paper finishes on some concluding remarks in section 5.

2 Model

This section presents the model of the parallel web server that is used to describe the data layout algorithm in the next sections. Thus, in our model we concentrate on the aspects that are important for the presentation in the rest of the paper.

A parallel web server is built by the following entities: A number of processing modules that are connected by some kind of network or bus architecture, a number of communication devices that connect the processing modules to the external clients accessing the server and requesting information, and a number of storage devices (disks) that are connected to the processing modules.

The parallel web server stores data items (files) on the disks and works as follows:

- The server is able to accept one or more requests arriving via the communication devices from the external clients per time step.
- The server forwards a request to the disk holding the requested item. We assume here, that each item is only stored once on the whole disk pool.
- Every disk can accept only one request per time step. If more than one request is sent to a disk per time step, these requests queue up.
- The processing time for each request on the disk is constant and takes one time unit, so one disk can process only one request in one time unit.
- All disks are independent, so that the server can process a maximum of n requests per time step if n is the number of disks.

In case that two requests are forwarded to the same disk in one time step, one request can be served and the other request has to wait for one time unit. This conflict resolution can be done arbitrarily, i.e. we do not assume a specific protocol here. In case that two (or more) requests access the same storage device (disk) we say, that these requests are *colliding*, i.e. these requests are in *collision*. We say that two requests collide if they arrive on the web server within a time interval of time δ. We have chosen δ to be one second.

The aim of our work is now to develop a data layout strategy in order to map the data items in a way onto the disks, that the requests that arrive at the server lead to a minimal number of collisions and therefore to a minimal latency in answering the data requests issued by the external clients.

3 Monitoring the access to web servers

In order to increase the overall performance of a parallel web server it is not important to balance the overall number of requests issued to the disks as evenly as possible, but to avoid that a larger number of requests are submitted to a single disk. This means that in each small time interval the load has to be distributed as evenly as possible minimizing the latency time for a request this way. Thus, the aim is to minimize the number of collisions on the disks.

In order to get an impression of the collisions that occur we have taken the log files from the web server of the University of Paderborn (www.uni-paderborn.de) for November and December 1997. We built all tuples (i, j) of files i and j stored on the web server in Paderborn and measured the number of collisions, i.e. the number of hits that we made on files i and j in a time interval δ. Figure 1 shows the number of collisions for all pairs of files for two consecutive days sorted by the collisions of the first day. The x-axis lists all tuples (i, j) of files according to the collisions on the first day, and the y-axis lists the number of collisions for each pair of files. Now it is interesting to observe that the collisions are very similar on the two consecutive days.

So if we develop an algorithm, that minimizes the collisions for a typical access pattern of the web server monitored at one day, these collisions will also

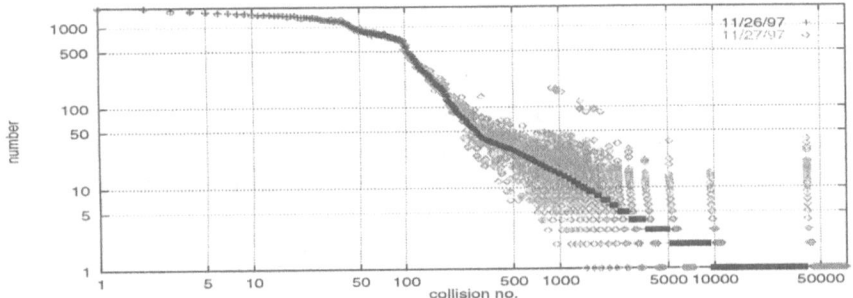

Fig. 1. Distribution of collisions

be eliminated on the next day. Thus, we can take the similarity of access patterns into account for the construction of our data layout strategy. This is the basic observation and the foundation of our data layout principle.

4 Data layout strategies

As described above the data layout strategy aims at minimizing the number of collisions for a typical access pattern. The mapping that is computed in this way can be assumed to also avoid a large number of collisions in the future as the distribution of collisions is very similar on consecutive days.

In the following we will firstly explain the algorithm used to compute the data layout, and secondly evaluate the performance of this algorithm in detail.

4.1 Algorithm

Throughout the rest of the paper we define F to be the set of data items (files) and $\{(t_1, f_1), (t_2, f_2), \ldots, (t_m, f_m), \ldots\}$ be an access pattern for a web server with t_i being the time when the request for data item (file) $f_i \in F$ arrives on the server. Then $c(i, j) = | \{\{(t, i), (t', j)\} \mid | t - t' | \leq \delta\} |$ is defined as the number of collisions of files i and j for the given access pattern.

Our strategy is to distribute the objects stored on the web server onto the given disks in such a way so that collisions are minimized. For a given access pattern this leads to the following algorithmic problem:

given: A set of data items F, the access pattern, and a given number of storage devices n.

question: Determine mapping π, $\pi = \min_{l:F \to \{1,\ldots,n\}} \sum_{i,j \in F, l(i)=l(j)} c(i, j)$.

It is easy to map this problem to the MAX-CUT-problem. The MAX-CUT-problem is defined as follows:

given: A graph $G = (V, E)$, weights $w(e) \in I\!N$ and a number of partitions n.

question: Determine a partition of V into n subsets, such that the sum of the weights for the edges having the endpoints in different subsets is maximal.

The mapping is done in a way that the data items (files) are the nodes of the graph that has to be partitioned, and the number of collisions $c(i, j)$ determines the weight of edge $\{i, j\}$. The MAX-CUT problem is known to be NP-hard [6]. However, there are good polynomial approximation algorithms which deliver good solutions. In our experiments we make use of the PARTY-Library [12] containing an efficient implementation of an extension from the partitioning algorithm described in [8].

4.2 Collision Resolving if access pattern are known in advance

In a first step we examine the gain of our algorithm if the access pattern and therefore the collisions are known in advance. Therefor we take the access statistics of one day and determine the number of collisions of each pair of data items. On the base of these statistics, we build the graph, partition the graph, and look how many collisions have remained and how many have been resolved.

In the following we compare the results of our algorithm with the random mapping strategy for a number of access patterns. Each access patterns exactly represents all requests that were issued to the server during one day. We compare the number of collisions induced by the access pattern (k_a) with the number of remaining collisions that occur when applying the mapping algorithm described above (k_r). The factor f describes the ratio between the number of remaining collisions for the random mapping (which is $\frac{k_a}{n}$) and the mapping that is determined by the algorithm (k_r).

Table 1. Results for a number of access patterns, each representing one day

| day | nodes | edges | k_a | results, $n = 8$ | | |
				k_r	$\frac{k_r}{k_a}[\%]$	f
sun 11/23/97	5533	15148	75334	3152	4.18	2.99
mon 11/24/97	8877	48136	239365	12100	5.06	2.50
tue 11/25/97	7932	43720	228870	11367	4.97	2.52
wed 11/26/97	8206	41172	215825	10800	5.00	2.50
thu 11/27/97	7464	44656	231364	12021	5.20	2.40
fri 11/28/97	6976	30919	174120	8671	4.98	2.51
sat 11/29/97	5065	9059	37702	1305	3.46	3.61
sun 11/30/97	4798	8657	42544	1579	3.71	3.37

In Table 1 the statistics and results of the partition of one week are shown. The table shows that all collisions up to 4 - 5 percent can be resolved and that the algorithm has about 2 to 3 times the performance of the random mapping.

4.3 Realistic optimization

In this section we examine how many collisions can be resolved if we use the access-statistics of the past. This approach can only be successful if the collisions of successive days have some similarity. We already examined this similarity in section 3.

Table 2 presents the average results for a number of days where the mapping was determined by the access pattern of day $i - 1$ and this mapping was used for collision avoiding of day i. The table shows the factor f_{pd} comparing the performance of the random mapping method with the algorithm presented above in respect to the number of disks n. It also shows the percentage of remaining collisions $\frac{k_{r,pd}}{k_a}[\%]$ that could not be resolved. It is shown that the number of remaining collisions with this mapping is clearly lower than with a random mapping. The advantage of our mapping increases with the number of available disks.

Table 2. Comparison of random placement and algorithm using access pattern of the previous day and algorithm using access pattern of the same day

n	$\frac{k_{r,pd}}{k_a}[\%]$	f_{pd}	$\frac{k_{r,sd}}{k_a}[\%]$	f_{sd}	$\frac{\frac{k_a}{n}-k_{r,pd}}{\frac{k_a}{n}-k_{r,sd}}$
2	45.2	1.11	43.1	1.16	0.70
3	27.4	1.22	24.9	1.34	0.70
4	18.7	1.34	16.3	1.54	0.72
5	13.7	1.47	11.3	1.78	0.72
6	10.5	1.61	8.2	2.05	0.73
7	8.2	1.76	6.0	2.41	0.73
8	6.7	1.87	4.7	2.73	0.74

The table also shows the loss which occurs from the fact that the access pattern of successive days are not identical. To see this the according percentage of remaining collisions $\frac{k_{r,sd}}{k_a}[\%]$ and the factor f_{sd} are given which could be achieved if the access pattern of each day would be known in advance.

The value of the term $\frac{\frac{k_a}{n}-k_{r,pd}}{\frac{k_a}{n}-k_{r,sd}}$ shows the relative performance difference of the random mapping and the data layout determined by the algorithm presented above, for the case that the algorithm knows the exact access pattern or only knows the access pattern of the day before. The results show that the performance of our method decreases by about 30 percent if the data layout is computed on the basis of the access pattern from day $i - 1$ instead of day i if the access pattern of day i is applied. This loss seems to be nearly independent from the number of disks but becomes smaller for larger number of disks.

In general the results show, that the data layout that is based on the access pattern of a previous day leads to a large reduction of the collisions later on.

4.4 Using a number of access patterns to determine the data layout

Up to now we only used the access pattern of one day to compute the data layout. Here we will present results for the case that a number of access patterns

Table 3. Results of using larger access patterns for determination of data layout

days	2 partitions			8 partitions		
	avg	max	min	avg	max	min
1	44.65	44.99	43.95	6.58	7.08	5.96
2	44.49	44.92	43.60	6.40	6.90	5.80
3	44.35	44.91	43.26	6.33	6.66	5.71
4	44.38	44.85	43.43	6.35	6.93	5.78
5	44.48	45.21	43.47	6.28	6.83	5.63
6	44.48	45.15	43.66	6.26	6.85	5.64
7	44.48	45.24	43.51	6.30	6.88	5.53
8	44.42	44.94	43.73	6.26	6.88	5.68
9	44.46	45.37	43.63	6.24	6.86	5.58
10	44.39	45.22	43.32	6.25	6.71	5.58
11	44.46	45.45	43.36	6.27	6.80	5.63
12	44.38	45.07	43.40	6.28	6.75	5.64
13	44.40	45.30	43.47	6.33	6.83	5.73
14	44.37	44.91	43.43	6.29	7.00	5.67

were used to determined this layout. All collisions from these patterns were taken into account to determine the data layout.

The results in Table 3 were determined using m previous days of a fixed date i and determining the data layout using all the collisions that occur in these m previous days. The experiments were made for different start dates i, so average, maximum and minimum could be studied. The table presents the percentage of the collisions that could not resolved for the access pattern of date i using the different data layouts. It is shown, that the best results were achieved with a data layout that is determined using only a few days of access patterns. A larger number of access patterns does not increase the overall performance of the data layout method.

4.5 Update frequency for data layout

Using the results of the previous sections we know that the data layout method presented here provides reasonable performance improvements to randomization strategies when it considers the access pattern. The results also show that a typical access pattern which is used to determine the data layout should contain the access information of only a few days. No better results could be made with more information. The question is now, if the data layout is determined on a day i taking the access pattern of day $i - 1$, $i - 2$, ... $i - x$ into account (small x) how will the performance be on a day $i + d$, thus d days in the future. This will answer the question how often the data layout has to be updated to achieve the best results.

The following experiment was done to answer this question: A data layout was determined taking the access pattern of data i into account. Then the access patterns of the days $i + d$, was applied to this data layout and the number of

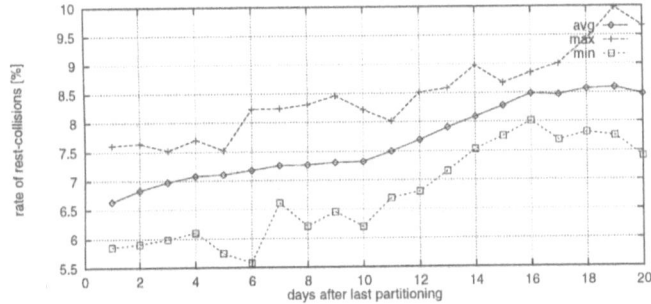

Fig. 2. Influence of time since last update of data layout on performance

collisions that had not been resolved was measured. For different values (x-axis) the results are shown in Figure 2. As the results were gained for different starting days i, average values, maximum and minimum could be determined for every d. The results show, that the number of unresolved collisions is steadily increasing, thus the data layout should be updated as often as possible (taking the last access pattern into account) to achieve the best results.

5   Conclusions

This paper presents a new strategy for allocating data items onto the storage devices of a parallel or sequential web server that contains a number of disks to store this data. The method is based on the observation that requests that often arise in parallel on the web server should be forwarded to different storage devices. Using this observation a graph-theoretic formulation is developed and the problem is reduced to a MAX-CUT problem that is solved using heuristics.

The results show that the performance improvements of the strategy are considerable compared to randomization strategies that are usually used and that the performance improvement scales up with the number of disks used in the parallel server. The various investigations presented in the paper show that it makes no sense to observe the access to the parallel web server for a long time to determine a typical access pattern which then builds the basis for the determination of a data layout and that the data layout has to be updated from time to time as for a fixed data layout the number of resolved collisions decreases steadily.

Future work will focus on the investigation of an extended method that determines the number of duplicates for data items considering different restrictions (in terms of disk capacity, or number of copies) into account.

References

1. M. Adler, S. Chakrabarti, M. Mitzenmacher, L. Rasmussen: Parallel Randomized Load Balancing. *Proc. 27th Annual ACM Symp. on Theory of Computing (STOC '95)*, 1995.
2. Virgilio Almeida, Azer Bestavros, Mark Crovella and Adriana de Oliveira: Characterizing Reference Locality in the WWW. *Proceedings of PDIS'96: The IEEE Conference on Parallel and Distributed Information Systems*, December 1996.
3. Martin Arlitt and Carey Williamson: Web server workload characterization: The search for invariants. *Proceedings of ACM SIGMETRICS'96*,1996.
4. D. Andresen, T. Yang, V. Holmedahl and O. Ibarra: SWEB: Towards a Scalable WWW Server on MultiComputers, *Proceedings of the 10th International Parallel Processing Symposium (IPPS'96)*, April 1996.
5. M. Colajanni and P.S. Yu: Adaptive TTL schemes for Load Balancing of Distributed Web Servers, *ACM SIGMETRICS Perfermance Evaluation Review*, 1997, volume 25,2
6. R. M. Karp: Reducibility among combinatorial problems, in R. E. Miller and J. W. Thatcher, *Complexity of Computer Computations*, pages 85–103, 1972
7. Eric Dean Katz, Michelle Butler and Robert McGrath: A scalable HTTP server: The NCSA prototype. *Computer Networks and ISDN Systems*, volume 27, pages 155–164, 1994.
8. B.W. Kernighan and S.Lin. An effective heristic procedure for partitioning graphs. *The Bell Systems Technical Journal*, pages 192–308, Feb 1970.
9. Kwan, T.T., R.E. McGrath and D.A. Reed: NCSA World Wide Web Server: Design and Performance. IEEE Computer, November 1995, 28(11), pages 68-74.
10. Y.H. Liu and P. Dantzig and C.E. Wu and J. Challenger and L.M. Ni, A distributed {Web} server and its performance analysis on multiple platforms. *Proceedings of the 16th International Conference on Distributed Computing Systems*, IEEE, 1996
11. D.R. Liu and S. Shekhar. Partitioning Similarity Graphs: A Framework for Declustering Problems, *Information Systems*, 1996, volume 21,6, pages 475–496
12. R. Preis and R. Diekmann. The PARTY Partitioning – Library User Guide. *Technical Report tr-rsfb-96-024*, University of Paderborn.
13. Benjamin W. Wah., File placement on distributed computer systems. *Computer Magazine of the Computer Group News of the IEEE Computer Group Society*, 17(1), January 1984.

ViPIOS: The Vienna Parallel Input/Output System*

Erich Schikuta, Thomas Fuerle and Helmut Wanek

Institute for Applied Computer Science and Information Systems
Department of Data Engineering, University of Vienna,
Rathausstr. 19/4, A-1010 Vienna, Austria
schiki@ifs.univie.ac.at

Abstract. In this paper we present the Vienna Parallel Input Output System (ViPIOS), a novel approach to enhance the I/O performance of high performance applications. It is a client-server based tool combining capabilities found in parallel I/O runtime libraries and parallel file systems.

1 Introduction

In the last few years the applications in high performance computing (Grand Challenges [1]) shifted from being CPU-bound to be I/O-bound. Performance can not be scaled up by increasing the number of CPUs any more, but by increasing the bandwidth of the I/O subsystem. This situation is commonly known as the I/O bottleneck in high performance computing ([5])

In reaction all leading hardware vendors of multiprocessor systems provided powerful concurrent I/O subsystems. In accordance researchers focused on the design of appropriate programming tools to take advantage of the available hardware resources.

1.1 The ViPIOS Approach

Conventionally two different directions in developing programming support are distinguished: Runtime libraries for high-performance languages (e.g. Passion [6]) and parallel file systems, (e.g. IBM Vesta [4]).

We see a solution to the parallel I/O problem in a combination of both approaches, which results in a dedicated, smart, concurrently executing runtime system, gathering all available information of the application process both during the compilation process and the runtime execution. Initially it can provide the optimal fitting data access profile for the application and may then react to the execution behavior dynamically, allowing to reach optimal performance by aiming for maximum I/O bandwidth.

* This work was carried out as part of the research project "Language, Compiler, and Advanced Data Structure Support for Parallel I/O Operations" supported by the Austrian Science Foundation (FWF Grant P11006-MAT)

This approach led to the design and development of the Vienna Input Output System, ViPIOS ([2,3]).

ViPIOS is an I/O runtime system, which provides efficient access to persistent files, by optimizing the data layout on the disks and allowing parallel read/write operations. ViPIOS is targeted as a supporting I/O module for high performance languages (e.g. HPF).

2 System Architecture

The basic idea to solve the I/O bottleneck in ViPIOS is *de-coupling*. The disk access operations are de-coupled from the application and performed by an independent I/O subsystem, ViPIOS. This leads to the situation that an application just sends general I/O requests to ViPIOS, which performs the actual disk accesses in turn. This idea is caught by figure 1.

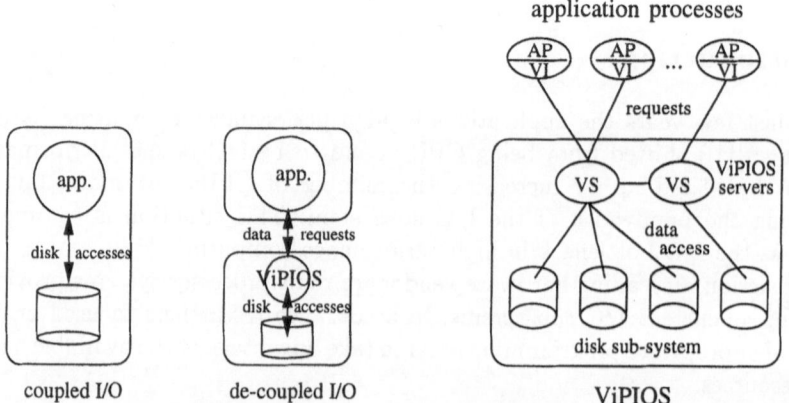

Fig. 1. Disk access de-coupling   **Fig. 2.** ViPIOS system architecture

Thus ViPIOS's system architecture is built upon a set of cooperating server processes, which accomplish the requests of the application client processes. Each application process AP is linked by the ViPIOS interface VI to the ViPIOS servers VS (see figure 2).

The server processes run independently on all or a number of dedicated processing nodes on the underlying MPP. It is also possible that an application client and a server share the same processor.

Generally each application process is assigned exactly one ViPIOS server (which is called the *buddy server* to the application), but one ViPIOS server can serve a number of application processes, i.e. there exists a one-to-many relationship between the application and the servers (see figure 3). The other ViPIOS servers are called *foe server* to the application.

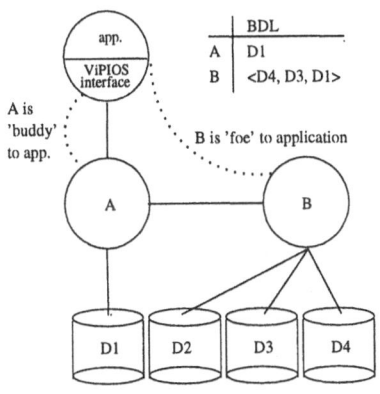

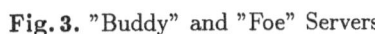

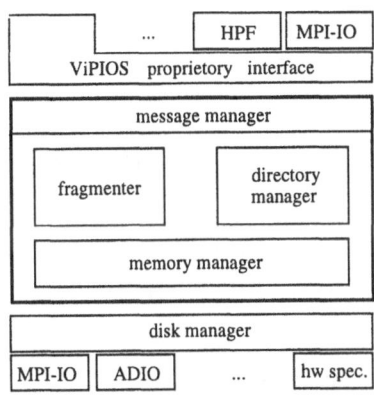

Fig. 3. "Buddy" and "Foe" Servers

Fig. 4. ViPIOS server architecture

2.1 ViPIOS Server

A ViPIOS server process consists of several functional units as depicted by figure 4.

Basically we differentiate between 3 layers:

- The *Interface layer* provides the connection to the "outside world" (i.e. applications, programmers, compilers, etc.). Different interfaces are supported by *interface modules* to allow flexibility and extendibility. Until now we implemented an HPF interface module (aiming for the VFC, the HPF derivative of Vienna FORTRAN) a (basic) MPI-IO interface module, and the specific ViPIOS interface which is also the interface for the specialized modules.
- The *Kernel layer* is responsible for all server specific tasks.
- The *Disk Manager layer* provides the access to the available and supported disk sub-systems. This layer too is modularized to allow extensibility and to simplify the porting of the system. At the moment ADIO [7], MPI-IO, and Unix style file systems are supported.

The ViPIOS kernel layer is built up of four cooperating functional units:

- The *Message manager* is responsible for the external (to the applications) and internal (to other ViPIOS servers) communication.
- The *Fragmenter* can be seen as "ViPIOS's brain". It represents a smart data administration tool, which models different distribution strategies and makes decisions on the effective data layout, administration, and ViPIOS actions.
- The *Directory Manager* stores the meta information of the data. We designed 3 different modes of operation, centralized (one dedicated ViPIOS directory server), replicated (all servers store the whole directory information), and localized (each server knows the directory information of the data it is storing only) management. Until now only localized management is implemented.
- The *Memory Manager* is responsible for prefetching, caching and buffer management.

Requests are issued by an application via a call to one of the functions of the ViPIOS interface, which in turn translates this call into a request message which is sent to the buddy server.

The *local directory* of the buddy server holds all the information necessary to map a client's request to the physical files on the disks. The *fragmenter* uses this information to decompose (*fragment*) a request into sub-requests which can be resolved locally and sub-requests which have to be communicated to other ViPIOS-servers (foe servers). The *I/O subsystem* actually performs the necessary disk accesses and the transmission of data to/from the AP.

2.2   System Modes

ViPIOS can be used in 3 different system modes, as

- runtime library,
- dependent system, or
- independent system.

These modes are depicted by figure 5.

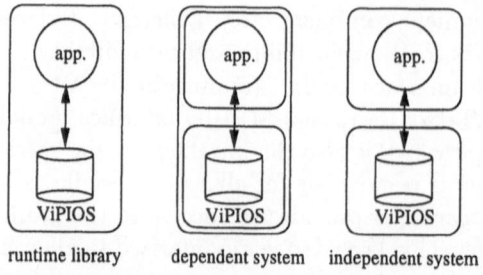

Fig. 5. ViPIOS system modes

Runtime Library. Application programs can be linked with a ViPIOS runtime module, which performs all disk I/O requests of the program. In this case ViPIOS is not running on independent servers, but as part of the application. The ViPIOS interface is therefore not only calling the requested data action, but also performing it itself. This mode provides only restricted functionality due to the missing independent I/O system. Parallelism can only be expressed by the application (i.e. the programmer).

Dependent System. In this case ViPIOS is running as an independent module in parallel to the application, but is started together with the application. This

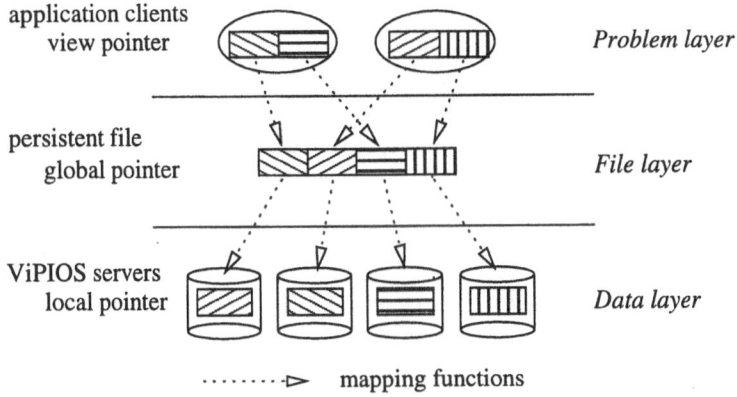

Fig. 6. ViPIOS data abstraction

is inflicted by the MPI[1] specific characteristic that cooperating processes have to be started together in the same communication world. Processes of different worlds can not communicate until now. This mode allows smart parallel data administration but objects a preceeding preparation phase.

Independent System. This is the mode of choice to achieve highest possible I/O bandwidth by exploiting all available data administration possibilities. In this case ViPIOS is running similar to a parallel file system or a database server waiting for applications to connect via the ViPIOS interface. This connection is realized by a proprietary communication layer bypassing MPI. We implemented two different approaches, one by using PVM, the other by patching MPI. A third promising approach is just evaluated by employing PVMPI, a possibly uprising standard under development for coupling MPI worlds by PVM layers.

3 Data Abstraction in ViPIOS

ViPIOS provides a data independent view of the stored data to the application processes.

Three independent layers in the ViPIOS architecture can be distinguished, which are represented by file pointer types in ViPIOS.

- Problem layer. Defines the problem specific data distribution among the cooperating parallel processes (View file pointer).
- File layer. Provides a composed view of the persistently stored data in the system (Global file pointer).
- Data layer. Defines the physical data distribution among the available disks (Local file pointer).

---

[1] The MPI standard is the underlying message passing tool of ViPIOS to ensure portability

Thus data independence in ViPIOS separates these layers conceptually from each other, providing mapping functions between these layers. This allows *logical data independence* between the problem and the file layer, and *physical data independence* between the file and data layer.

This concept is depicted in figure 6 showing a cyclic data distribution.

4 Conclusions and Future Work

In this paper we presented the Vienna Parallel Input Output System (ViPIOS), a novel approach to parallel I/O based on a client server concept which combines the advantages of existing parallel file systems and parallel I/O libraries. We described the underlying design principles of our approach and gave an in-depth presentation of the developed system.

References

1. A Report by the Committee on Physical, Math., and Eng. Sciences Federal Coordinating Council for Science, Eng. and Technology. *High-Performance Computing and Communications, Grand Challenges 1993 Report*, pages 41 – 64. Committee on Physical, Math., and Eng. Sciences Federal Coordinating Council for Science, Eng. and Technology, Washington D.C., October 1993.
2. Peter Brezany, Thomas A. Mueck, and Erich Schikuta. Language, compiler and parallel database support for I/O intensive applications. In *Proceedings of the International Conference on High Performance Computing and Networking*, volume 919 of *Lecture Notes in Computer Science*, pages 14–20, Milan, Italy, May 1995. Springer-Verlag. also available as Technical Report of the Inst. f. Software Technology and Parallel Systems, University of Vienna, TR95-8, 1995.
3. Peter Brezany, Thomas A. Mueck, and Erich Schikuta. A software architecture for massively parallel input-output. In *Third International Workshop PARA'96 (Applied Parallel Computing - Industrial Computation and Optimization)*, volume 1186 of *Lecture Notes in Computer Science*, pages 85–96, Lyngby, Denmark, August 1996. Springer-Verlag. Also available as Technical Report of the Inst. f. Angewandte Informatik u. Informationssysteme, University of Vienna, TR 96202.
4. Peter F. Corbett and Dror G. Feitelson. The Vesta parallel file system. *ACM Transactions on Computer Systems*, 14(3):225–264, August 1996.
5. Juan Miguel del Rosario and Alok Choudhary. High performance I/O for parallel computers: Problems and prospects. *IEEE Computer*, 27(3):59–68, March 1994.
6. Rajeev Thakur, Alok Choudhary, Rajesh Bordawekar, Sachin More, and Sivaramakrishna Kuditipudi. Passion: Optimized I/O for parallel applications. *IEEE Computer*, 29(6):70–78, June 1996.
7. Rajeev Thakur, William Gropp, and Ewing Lusk. An abstract-device interface for implementing portable parallel-I/O interfaces. In *Proceedings of the Sixth Symposium on the Frontiers of Massively Parallel Computation*, pages 180–187, October 1996.

A Performance Study of Two-Phase I/O

Phillip M. Dickens

[1] Department of Computer Science
Illinois Institute of Technology
[2] Rajeev Thakur
Mathematics and Computer Science Division
Argonne National Laboratory

Abstract. Massively parallel computers are increasingly being used to solve large, I/O intensive applications in many different fields. For such applications, the I/O subsystem represents a significant obstacle in the way of achieving good performance. While massively parallel architectures do, in general, provide parallel I/O hardware, this alone is not sufficient to guarantee good performance. The problem is that in many applications each processor initiates many small I/O requests rather than fewer larger requests, resulting in significant performance penalties due to the high latency associated with I/O. However, it is often the case that *in the aggregate* the I/O requests are significantly fewer and larger. Two-phase I/O is a technique that captures and exploits this aggregate information to recombine I/O requests such that fewer and larger requests are generated, reducing latency and improving performance. In this paper, we describe our efforts to obtain high performance using two-phase I/O. In particular, we describe our first implementation which produced a sustained bandwidth of 78 MBytes per second, and discuss the steps taken to increase this bandwidth to 420 MBytes per second.

1   Introduction

Massively parallel computers are increasingly used to solve large, I/O-intensive applications in several different disciplines. However, in many such applications the I/O subsystem performs poorly, and represents a significant obstacle to achieving good performance. The problem is generally not with the hardware; many parallel I/O subsystems offer excellent performance. Rather, the problem arises from other factors, primarily the I/O patterns exhibited by many parallel scientific applications [1,5] In particular, each processor tends to make a large number of small I/O requests, incurring the high cost of I/O on each such request.

   The technique of *collective I/O* has been developed to better utilize the parallel I/O subsystem [2,7,8]. In this approach, the processors exchange information about their individual I/O requests to develop a picture of the *aggregate* I/O request. Based on this global knowledge, I/O requests are combined and submitted in their proper order, making a much more efficient use of the I/O subsystem.

Two significant implementation techniques for collective I/O are two-phase I/O [2, 7] and disk-directed I/O [4, 6]. In the two-phase approach, the application processors collectively determine and carry out the optimized approach. In this paper, we deal only with the two-phase approach.

Consider a collective read operation. If the data is distributed across the processors in a way that conforms to the way it is stored on disk, each processor can read its local array in one large I/O request. This distribution is termed the *conforming* distribution, and represents the optimal I/O performance. Assume the array is *not* distributed across the processors in a conforming manner. The processors can still perform the read operation *assuming* the conforming distribution, and then use interprocessor communication to redistribute the data to the desired distribution. Since interprocessor communication is orders of magnitude faster than I/O calls, it is possible to obtain performance that approaches that of the conforming distribution.

The question then arises as to what is the best approach to implement the two-phase I/O algorithm. While much has been published regarding the performance gains using this technique, relatively little has been written about specific implementation issues, and how these issues affect performance. In this paper, we focus on the implementation issues that arise when implementing a two-phase I/O algorithm. We start by describing a very simple implementation, and go through a sequence of steps where we explore different optimizations to this basic implementation. At each step, we discuss the modification to the algorithm and discuss its impact on performance.

2 Experimental Design

We performed all experiments on the Intel Paragon located at the California Institute of Technology. This Paragon has 381 compute nodes and an I/O subsystem with 64 SCSI I/O processors, each of which controls a 4GB seagate drive.

We define the *maximum* I/O performance as each processor writing its portion of the file assuming the conforming distribution. We define the *naive* approach as each processor performing its own independent file access without any global strategy, i.e. no collective I/O. We assume an SPMD computation, where each processor operates on the portion of the global array that is located in its local memory. We study the costs of writing a two-dimensional array to disk for various two-phase I/O implementations. Our metric of interest is the percentage of the maximum bandwidth achieved by each approach. The application does nothing except make repeated calls to the two-phase I/O routine. Our experiments involved a two dimensional array of integers with dimensions 4096 X 4096 (for a total file size of 64 megabytes). We used MPI for all communications. To conserve space, we present one graph which maps the performance of each of the various approaches.

3   Experimental Results

The initial implementation of the two-phase I/O algorithm is quite simple. First, the processors exchange information related to their individual I/O requirements to determine the collective I/O requirement. Next, each processor goes through a series of sending and receiving messages to perturb the data into the conforming distribution. When a processor receives a portion of its data it performs a simple byte to byte copy into its write buffer. When a processor sends a portion of its data it performs a byte for byte copy from its local array into the send buffer. After all data has been exchanged, each processor performs its write operation in one large request. This initial implementation uses both blocking sends and blocking receives.

The cost associated with the naive implementation (no collective I/O) is the latency incurred from issuing many small I/O requests. There are four primary costs associated with the simple implementation of two-phase I/O. First is the extra buffer space required for the write buffer and the communication buffers. Second is the cost of copying data into the communication buffers, copying data from the communication buffers to the write buffer, and copying data from the local array into the write buffers. Third is interprocessor communication and fourth is the actual writing of the data to the disk.

The results for the non-collective I/O implementation is depicted in the curve labeled the *naive approach* of Figure 1. The initial implementation of the two-phase I/O algorithm is depicted in the curve labeled *Step1*. As noted, the graph depicts the percentage of the maximum bandwidth achieved by each approach given 16, 64 and 256 processors. With 16 processors, each processor must allocate a four megabyte write buffer and the messages passed between the processors are one megabyte. With these values, the time to perform the many small I/O operations is virtually the same as the extra costs associated with the two-phase approach. When we move to 64 processors however, each processor allocates a much smaller write buffer (one megabyte), and the messages between the processors are much smaller (131072 bytes). Thus its relative performance is improved. With the naive approach however the additional processors are all issuing many small I/O requests resulting in significant contention for the IOPs and communication network. With 64 processors, this very simple implementation of the two-phase I/O algorithm improves performance by a factor of 3. With 256 processors, it improves performance by a factor 40.

3.1·   Step2: Reducing Copy Costs

As noted above, two-phase I/O requires a significant amount of copying. For this reason, we would expect modifications that reduced the cost of copying data would have a significant impact on performance.

The next step then is to change all of the copy routines to use the **memcpy()** library call whenever possible. This optimization, labeled as *Step2* in Figure 1, has a tremendous impact on performance. As can be seen, with the optimized copying routine this implementation of the two-phase I/O algorithm outperforms both the naive approach and the initial implementation. The improvement in performance over the initial implementation varies between a factor of 1.7 and a factor of 15. The reason for such a tremendous difference is that **memcpy()** uses a block move instruction, and requires only four assembly language instructions per *block* of memory. Performing a byte for byte copy requires 20 assembly language instructions *per byte*.

3.2 Step3: Asynchronous Communication

The next step we investigated was performing asynchronous rather than synchronous communication. In this implementation, a processor first posts all of its sends (using **MPI_Isend**) and then posts all of its receives (using **MPI_Irecv**). After posting its communications, the processor copies any of its own data into the write buffer. The processor then waits for all of the messages it needs to receive, and copies the data into the write buffer as they arrive. It then waits for all of the send operations to complete and performs the write.

The trade-offs in this implementation are quite interesting. In previous versions of the algorithm, a processor would alternate between sending a message and waiting to receive a message, blocking until a *specific* message arrives before posting its next send. The advantage is that the processor frees the send and receive buffers immediately, thus maintaining only one communication buffer at any time. With reduced buffering requirements the cost of paging is decreased.

There are two advantages to using asynchronous communication. First, the underlying system can complete *any* message that is ready without having to wait for a *particular* message. Second, it can perform the copying between its local array and its write buffer with the asynchronous communication going on in the background. The primary drawback of this approach is that each asynchronous request requires its own communication buffer. Thus the cost of buffering is considerably higher reducing performance.

The results are shown in the curve labeled *Step 3* in Figure 1. With 16 processors, the asynchronous approach further improves performance by approximately 20%. With 64 processors however, this improvement over the previous implementation is reduced to approximately 5%. There is no noticeable improvement with 256 processors. The reason for the decrease in performance with 64 and 256 processors is that as the number of processors increase, the time required to perform the actual write to disk begins to dominate the costs of the algorithm. Thus using asynchronous communication is less important than it is with a smaller number of processors.

3.3 Step4: Reversing the Order of the Asynchronous Communication

The next approach is to reverse the order of the asynchronous sends and receives. Thus a processor first posts all of its asynchronous receives and then posts its asynchronous sends. The idea behind this optimization is that MPI communications are generally much faster if the receive has been posted (and thus a buffer in which to receive the message has been provided) before the message is sent [3]. Thus pre-allocating all of the receive buffers before the corresponding sends are initiated should improve the interprocessor communication costs. There is of course the same issue discussed above: pre-allocating all of the communication buffers can result in increased paging activity.

The results are given in the curve labeled *Step4* in Figure 1. With 16 processors, reversing the order of sends and receives provides a 15% improvement over the previous approach. Again this improvement in performance decreases as the number of processors is increased. With 64 processors there is an improvement of approximately 5%, and the improvement disappears with 256 processors. This is again due to the fact that the write time begins to dominate the cost of the algorithm as the number of processors increases.

3.4 Step5: Combining Synchronous with Asynchronous Communication

The final optimization we pursued was to combine asynchronous receives with synchronous sends. The idea behind this optimization is that posting all of the receive buffers will improve communication costs, and releasing the send buffer after each use will reduce the buffering costs. The results are shown in the curve labeled *step5*. This implementation results in performance gains of up to 10%.

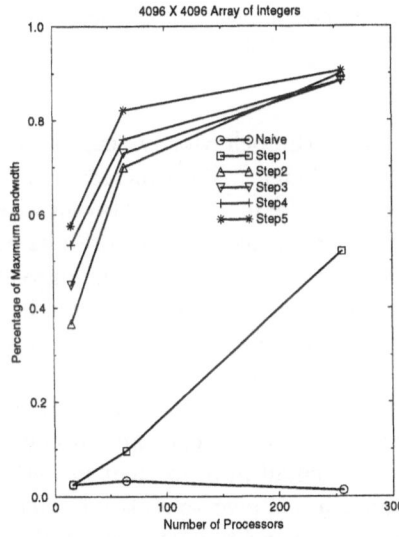

Fig. 1. Improvement in performance as a function of the implementation and the number of processors.

3.5 Scalability of Two-Phase I/O

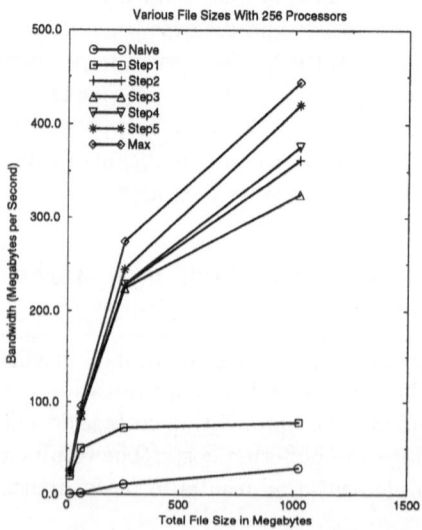

Fig. 2. The bandwidth achieved by each approach as the file size is increased and the number of processors is a constant 256.

For completion, we examine the behavior of these two-phase I/O implementations as the amount of data being written increases and the number of processors remains a constant 256. We looked at total file sizes of 16 megabytes, 64 megabytes, 256 megabytes and 1 gigabyte. The results are shown in Figure 2.

In this figure, the bandwidth achieved by each approach is given as a function of the total file size. For comparison, the maximum bandwidth is also shown. As can be seen, neither the naive approach nor the first two-phase implementation scale well. It is interesting to note that step3, where all communication is asynchronous and the sends are posted before the receives, performs more poorly in the limit than does step2 where all communication is synchronous. Also in the limit there is a small improvement in performance between step2 and step4 and both approaches appear to scale relatively well. The combination of asynchronous receives and synchronous sends scales very well, and, in the limit, approaches the optimal performance.

It is important to note that in the limit the initial implementation of two-phase I/O resulted in a bandwidth of 78 megabytes per second. The final implementation resulted in a bandwidth of 420 megabytes per second.

4 Conclusions

In this paper, we investigated the impact on performance of various implementation techniques for two-phase I/O, and outlined the steps we followed to obtain high performance. We began with a very simple implementation of two-phase I/O that provided a bandwidth of 78 megabytes per second, and ended with an optimized implementation that provided 420 megabytes per second. During the course of this analysis, we provided a good look at the trade-offs involved in the implementation of a two-phase I/O algorithm. Current research is aimed at extending these results to other parallel architectures such as the IBM SP2.

References

1. Crandall, P., Aydt, R., Chien, A. and D. Reed Input-Output Characteristics of Scalable Parallel Applications, In *Proceedings of Supercomputing '95*, ACM press, December 1995.
2. DelRosario, J., Bordawekar, R. and Alok Choudhary. Improved parallel I/O via a two-phase run-time access strategy. In *Proceedings of the IPPS '93 Workshop on Input/Output in Parallel Computer Systems* pages 56-70, Newport Beach, CA, 1993.
3. Gropp, W., Lusk, E. and A. Skjellum. Using MPI. Portable Parallel Programming with the Message-Passing Interface. The MIT Press, Cambridge, Massachusetts. 1996.
4. Kotz, D. Disk-directed I/O for MIMD multiprocessors. *ACM Transactions on Computer Systems* 15(1):41-74, February 1997.
5. Kotz, D. and N. Nieuwejaar. Dynamic file-access characteristics of a production parallel scientific workload. In *Supercomputing '94* pages 640-649, November 1994.
6. Kotz, D. Expanding the potential for disk-directed I/O. In *Proceedings of the 1995 IEEE Symposium on Parallel and Distributed Processing*. Pages 490 - 495, IEEE Computer Society Press.
7. Thakur, R. and A. Choudhary. An Extended Two-Phase Method for Accessing Sections of Out-of-Core Arrays. *Scientific Programming* 5(4):301-317, Winter 1996.
8. Thakur, R., Choudhary, A., More, S and S. Kuditipudi. Passion: Optimized I/O for parallel applications. *IEEE Computer*, 29(6):70-78, June 1996.

Workshop 13+14
Architectures and Networks

Kieran Herley and David Snelling

Co-chairmen

This workshop embraces two broad themes: routing and communication networks and parallel computer architecture. The first theme is devoted to communication in parallel computers and covers all aspects of the theory and practice of the design, analysis and evaluation of interconnection networks and their routing algorithms. The continued vitality of this area is reflected by the strength and quality of the papers presented in Sessions 1 and 3 (Routing Networks and Algorithms I and II) of this workshop. Although recent years have seen computer architecture undergoing a period of contraction and a tendency toward uniformity in the form of shared memory systems composed of super scalar processors, there is nonetheless no lack of innovation in the field. The computer architecture theme, represented by the papers of Sessions 2 and 4 (Computer Architecture I and II), highlights not only innovation within the converging area of super scalar systems, but the continued presence of novel ideas as well.

A total of 32 submissions were received (19 on the theme of routing and communication networks and 13 on the computer architecture theme) of which a total of 15 were accepted by the workshop committee for presentation at the conference.

Session 1 is devoted to the issue of routing algorithms for unstructured or loosely structured communication on various networks. The paper by S. R. Donaldson, J. M. D. Hill and D. B. Skillicorn entitled "Predictable Communication on Unpredictable Networks: Implementing BSP over TCP/IP" addresses the problem of supporting the well-known BSP model on networks of workstations running TCP/IP. Techniques that offer high throughput with low variance are presented and evaluated experimentally with encouraging results. N. Agrawal and C. P. Ravikumar consider novel techniques for efficient wormhole routing where deadlock is handled not by avoidance but by detection and recovery in their paper "Adaptive Routing Based on Deadlock Recovery". An adaptive routing technique is presented that ensures that deadlocks are infrequent and two schemes are outlined for deadlock recovery are described and evaluated. M. Ould-Khaoua's paper "On the Optimal Topology for Multicomputers with Fully-Adaptive Wormhole Routing: Torus or Hypercube?" revisits the hypercube versus torus comparison for wormhole routing networks. Whereas previous studies, generally based on nonadaptive routing, indicated the superiority of torus topologies, the paper suggests that under certain conditions the hypercube may offer superior performance when adaptive techniques are employed. The fundamental h-relation routing problem is the focus of the paper "Constant Thinning Protocol for Routing in Complete Networks" by A. Kautonen, V. Leppännen and M. Penttonen. They present and evaluate a randomized algorithm for this problem on the OCPC model with encouraging results.

T. Grün and M. A. Hillebrand's paper, "NAS Integer Sort on Multi-Threaded Shared Memory Machines", proposes novel techniques for improving performance in the already novel environment of multi-threaded computer architecture. They use both the commercial Tera MTA and the experimental SB-PRAM machines as platforms for these techniques. In "Analysing a Multistreamed Superscalar Speculative Instruction Fetch Mechanism", R. R. dos Santos and P. O. A. Navaux address the possibility of a more conservative step toward multi-threading within the more traditional super scalar environment. Their primary target is the performance loss due to instruction fetch latency resulting from miss predicted branches. Not surprisingly we find detailed consideration of implementation costs critical in the design of special purpose processor arrays. In "Design of Processor Arrays for Real-time Applications", D. Fimmel and R. Merker study various algorithms for optimizing the layout of uniform processes on processor arrays.

Session 3 is largely devoted to routing algorithms for relatively structured communication such as gossiping or all-to-all scattering. The paper "Interval Routing and Layered Cross Product: Compact Routing Schemes for Butterflies, Mesh of Trees and Fat Trees" by T. Calamoneri and M. Di Ianni show how shortest-path routing information may be maintained in a compact (space-efficient) way for a class of interconnections that includes the butterfly, the mesh of trees and certain fat-trees. The results rely on an elegant exploitation of interval routing techniques. The gossiping problem involves simultaneously broadcasting a set of packets where each node is the origin of a distinct packet and must receive a copy of the packet that originates at every other node. This problem is the subject of both the paper "Gossiping Large Packets on Full-Port Tori" by U. Meyer and J. F. Sibeyn and "Time-Optimal Gossip in Noncombining 2-D Tori with Constant Buffers" by M. Šoch and P. Tvrdík. The former presents a near-optimal algorithm for two- and higher-dimensional tori that are time-independent (each node repeats the same simple routing action repeatedly for the duration of the algorithm's execution), whereas the latter presents a time-optimal algorithm for two-dimensional tori that requires only a constant amount of buffer-space per node. The paper "Divide-and-Conquer Algorithms on Two-Dimensional Meshes" by M. Valero-García, A. González, L. Díaz de Cerio and D. Royo explores techniques for supporting divide-and-conquer computations on the two-dimensional mesh. The approach provides efficient solutions to this problem under both the wormhole and store-and-forward routing regime. The all-to-all-scatter problem involves routing a set of distinct packets, one for each source-destination node-pair, where each packet is to routed from its source to its destination. The paper "Average Distance and All-to-All Scatter in Kautz Networks" by P. Salinger and P. Tvrdík describes an asymptotically optimal algorithm for this problem for the Kautz network.

Modern ccNUMA systems represent the merger of easy to build distributed memory systems and easy to program SMP systems. In their paper, "Reactive Proxies: a Flexible Protocol Extension to Reduce ccNUMA Node Controller Contention", S. A. M. Talbot and P. H. J. Kelly propose and evaluate a technique for distributing the contention that arises naturally in ccNUMA systems when processes share a data structure. The approach is novel in that the degree of distribution is driven by the degree of contention and that well behaved applications do not suffer because of the protocol. The flip side of performance is always reliability. T. Skeie presents techniques for "Handling Multiple Faults in Wormhole Mesh Networks". The infrastructure for large distributed systems needs to provide both performance and reliability. The presence of multiple paths between nodes on a mesh provides both. In this case, fault tolerance is achieved without dramatic losses in performance. In another instance of novel approaches in modern computer architecture, "Shared Control Supporting Control Parallelism using a SIMD-like Architecture", N. Abu-Ghazaleh and P. Wilsey provide mechanisms for supporting control parallelism in SIMD systems. The aim is to provide flexibility in conditional and similar computation that does not require global redundant computation necessary in traditional SIMD systems.

The members of the workshop committees wish to extend their sincere thanks to those who acted as reviewers during the selection process for their invaluable help during that period and to all of those who by their submissions have contributed to the success of this workshop.

Predictable Communication on Unpredictable Networks: Implementing BSP over TCP/IP

Stephen R. Donaldson[1], Jonathan M.D. Hill[1], and David B. Skillicorn[2]

[1] Oxford University Computing Laboratory, UK.
[2] CISC, Queen's University, Canada

Abstract. The BSP cost model measures the cost of communication using a single architectural parameter, g, which measures permeability of the network to continuous traffic. Architectures, typically networks of workstations, pose particular problems for high-performance communication because it is hard to achieve high throughput, and even harder to do so predictably. Yet both of these are required for BSP to be effective. We present a technique for controlling applied communication load that achieves both. Traffic is presented to the communication network at a rate chosen to maximise throughput and minimise its variance. Performance improvements as large as a factor of two over MPI can be achieved.

1 Introduction

The BSP (Bulk Synchronous Parallel) model [10,8] views a parallel machine as a set of processor-memory pairs, with a global communication network and a mechanism for synchronising all processors. A BSP calculation consists of a sequence of *supersteps*. Each superstep involves all of the processors and consists of three phases: (1) processor-memory pairs perform a number of computations on data held locally at the start of a superstep; (2) processors communicate data into other processor's memories; and (3) all processors synchronise.

The BSP cost model treats communication as an aggregate operation of the entire executing architecture, and models the cost of delivery using a single architectural parameter, the permeability, g. This parameter can be intuitively understood as defining the time taken for a processor to communicate a single word to a remote processor, *in the steady state where all processors are simultaneously communicating*. The value of the g parameter will depend upon: (1) the bisection bandwidth of the communication network topology; (2) the protocols used to interface with and within the communication network; (3) buffer management by both the processors and the communication network; and (4) the routing strategy used in the communication network.

However, the BSP runtime system also makes an important contribution to the performance by acting to improve the effective value of g by the way it uses the architectural facilities. For example, [6] shows how orders of magnitude improvements in g can be obtained, for architectures using point-to-point connections, by packing messages before transmission, and by altering the order of transmission to avoid contention at receivers.

In this paper, we address the problem raised by shared-media networks and protocols such as TCP/IP, where there is far greater potential to waste bandwidth. For example, if two processors try to send more or less simultaneously, collision in the ether means that neither succeeds, and transmission capacity is permanently lost. The problem is compounded because it is hard for each processor to learn anything of the global state of the network. Nevertheless, as we shall show, significant performance improvements are possible.

We describe techniques, specific to our implementation of *BSPlib* [5], that ensure that the variation in g is minimised for programs running over bus-based Ethernet networks. Compared to alternative communications libraries such as Argonne's implementation of MPI [2], these techniques have an absolute performance improvement over MPI in terms of the mean communication throughput, but also have a considerably smaller standard deviation. Good performance over such networks is of practical importance because networks of workstations are increasingly used as practical parallel computers.

2 Minimising g in Bus-Based Ethernet Networks

Ethernet (IEEE 802.3) is a bus-based protocol in which the media access protocol, 1-persistent CSMA/CD (Carrier Sense Multiple Access with Collision Detection) proceeds as follows. A station wishing to send a frame listens to the medium for transmission activity by another station. If no activity is sensed, the station begins transmission and continues to listen on the channel for a collision. After twice the propagation delay, 2τ, of the medium, no collision can occur, as all stations sensing the medium would detect that it is in use and will not send data. However, a collision may occur during the 2τ window. On detection, the transmitting station broadcasts a jamming signal onto the network to ensure that all stations are notified of the collision. The station recovers from a collision by using a *binary exponential back-off* algorithm that re-attempts the transmission after $t \times 2\tau$, where t is a random variable chosen uniformly from the interval $[0, 2^k]$ (where k is the number of collisions this attempted transmission has experienced). For Ethernet, the protocol allows k to reach ten, and then allows another six attempts at $k = 10$ (see, for example, King [7]).

Analysis of this protocol (Tasaka [9]) shows that $S \to 0$ as $G \to \infty$ (where S is the rate of successful transmissions and G the rate at which messages are presented for delivery), whereas for a p-processor BSP computer one would expect that $S \to B$ as $G \to \infty$, from which one could conclude that $g = p/B$; where B is a measure of the bandwidth.

In the case of BSP computation over Ethernet, the effect of the exponential backoff is exaggerated (larger delays for the same amount of traffic) because the access to the medium is often synchronised by the previous barrier synchronisation and the subsequent computation phase. For perfectly-balanced

972

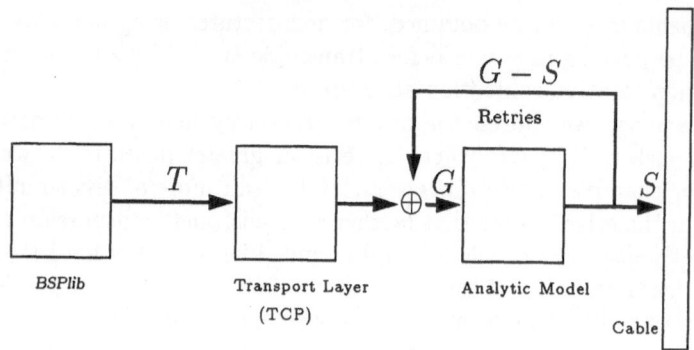

Fig. 1. Schematic of aggregated protocol layers and associated applied loads

computations and barrier synchronisations, all processors attempt to get their first message onto the Ethernet at the same time. All fail and back off. In the first phase of the exponential back-off algorithm, each of p processors choose a uniformly-distributed wait period in the interval $[0, 4\tau]$. Thus the expected number of processors attempting the retransmits in the interval $[0, 2\tau]$ is $p/2$, making secondary collisions very likely. If the processors are not perfectly balanced, and a processor gains access to the medium after a short contention period, then that process will hold the medium for the transmission of the packet, which will take just over $1000\mu s$ for 10Mbps Ethernet. With high probability, many of the other processors will be synchronised by this successful transmission due to the 1-persistence of this protocol. The remaining processors will then contend as in the perfectly-balanced scenario.

In terms of the performance model, this corresponds to a high applied load, G, albeit for a short interval of time. If S were (at least) linear in G then this burstiness of the applied load would not be detrimental to the throughput and would average out.

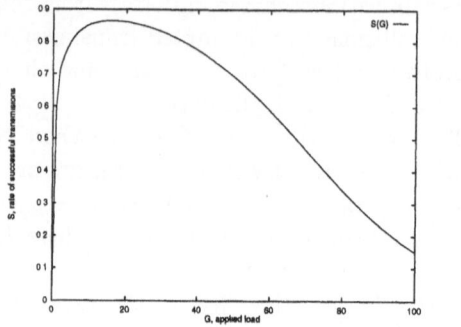

Fig. 2. Plot of applied load (G) against successful transmissions (S)

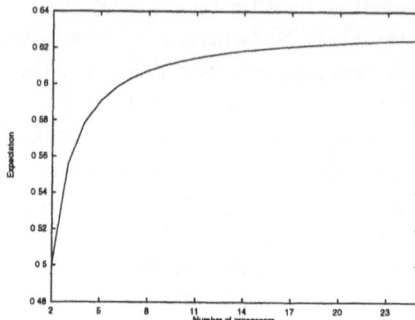

Fig. 3. Contention expectation for a particular slot as a function of p

Fortunately, the BSP model allows assumptions to be made at the global level based on local data presented for transmission. At the end of a superstep and before any user data is communicated, *BSPlib* performs a reduction, in which all processors determine the amount of communication that each processor intends sending. From this, the number of processors involved in the communication is determined. For the rest of the communication the number of processors involved and the amount of data is used to regulate (at the transport level) the rate at which data is presented for transmission on the Ethernet. By using BSD Socket options (TCP_NDELAY), the data presented at the transport layer are delivered immediately to the MAC layer (ignoring the depth of the protocol stack and the availability of a suitable window size). Thus, by pacing the transport layer, pacing can be achieved at the MAC or link layer. This has the effect of removing burstiness from the applied load.

Most performance analyses give a model of random-access broadcast networks which provide an analytic, often approximate, result for the successful traffic, S, in terms of the offered load, G. Hammond and O'Reilly [3] present a model for slotted 1-persistent CSMA/CD in which the successful traffic, S, (or efficiency achieved) can be determined in terms of the offered load G, the end-to-end propagation delay, τ (bounded by $25.65\mu s$, i.e., the time taken for a signal to propagate $2500m$ of cable and 4 repeaters), and E, the frame transmit time (which for 10Mbps Ethernet with a maximum frame size of 1500 bytes is approximately $1200\mu s$). Figure 2 shows the predicted rate of successful traffic against applied load, assuming that that jamming time is equal to τ.

Since both S, the rate of successful transmissions, and G are normalised with respect to E, S is also the channel efficiency achieved on the cable. T, shown in Figure 1, also normalised with respect to E, is the load applied by *BSPlib* on the transport layer. Our objective is to pace the injection of messages into the transport layer such that T, on average, is equal to a steady-state value of S without much variance. The value of T determines the position on the S–G curve of Figure 2 in a *steady state*; in particular, T can be chosen to maximise S. If the applied load is to the right of the maximum throughput in Figure 2, then small increases in the mean load lead to a decrease in channel efficiency which in turn increases the backlog in terms of retries and further increases the load. Working to the right of the maximum therefore exposes the system to these instabilities which manifest themselves in variances in the communication bandwidth—a metric we try to minimise [4, 1]. In contrast, when working to the left of the maximum, small increases in the applied load are accompanied by increases in the channel efficiency which helps cope with the increased load and therefore instabilities are unlikely. As an aside, the Ethernet exponential backoff handles the instabilities towards the right by rescheduling failed transmissions further and further into the future, which decreases the applied load.

In *BSPlib*, the mechanism of pacing the transport layer is achieved by using a form of statistical time-division multiplexing that works as follows. The frame size and the number of processors involved in the communication are known. As the processors' clocks are not necessarily synchronised, it is not possible to allow the processors access in accordance with some permutation, a technique applied successfully in more tightly-coupled architectures [6]. Thus the processors choose a *slot*, q, uniformly at random in the interval $[0 \ldots Q-1]$ (where Q is the number of processors communicating at the end of a particular superstep), and schedule their transmission for this slot. The choice of a random slot is important if the clocks are not synchronised as it ensures that the processors do not repeatedly choose a bad communication schedule. Each processor waits for time $q\varepsilon$ after the start of the *cycle*, where ε is a slot time, before passing another packet to the transport layer. The length of the slot, ε, is chosen based on the maximum time that the slot can occupy the physical medium, and takes into account collisions that might occur when good throughput is being achieved. The mechanism is designed to allow the medium to operate at the steady state that achieves a high throughput. Since the burstiness of communication has been smoothed by this slotting protocol, the erratic behaviour of the low-level protocol is avoided, and a high utilisation of the medium is ensured.

An alternative protocol, not considered here, would be to implement a deterministic token bus protocol in which each station can only send data whilst holding a "token". This scheme was not considered viable as it is inefficient for small amounts of traffic due to the need for the explicit token pass when the token holding processor has no message for the "next to-go" processor. In the worst case this would double the communication time. Also, token mechanisms protect shared resources such as a single Ethernet bus, however a network may be partioned into several independent segments, or processors may be connected via a switch. In this case, the token bus protocol would ensure only a single processor has access to the medium at any time, therefore wasting bandwidth. In contrast, the parameters used in the slotting mechanism can be trivially adjusted to take advantage of a switch based medium. For example, for a full-duplex cross-bar switch, Q can be assumed to be 1 ($Q = 2$ for half-duplex), and ε encapsulates the rate at which the switch and protocol stacks of sender and receiver can absorb messages in a steady state. If the back-plane capacity of the switch is less than the capacity of the sum of the links, then Q and ε can be adjusted accordingly. Therefore, the randomised slotting mechanism is superior to a deterministic token bus scheme, as Q and ε can be used to model a large variety of LAN interconnects.

3 Determining the Value of ε

In any steady state, $T = S$ because, if this were not the case, then either unbounded capacity for the protocol stacks would be required, or the stacks would dispense packets faster than they arrive, and hence contradict the steady-state assumption. Since ε is the slot time, packets are delivered with a mean rate of $1/\varepsilon$ packets per μs. Normalising this with respect to the frame size E gives a value for $T = E/\varepsilon$ packets per unit frame-time. We therefore choose an S value from the curve and infer a value of the slot size $\varepsilon = E/S$ as $S = T$ in a steady state. Choosing a value of $S = 80\%$ and $E = 1200\mu s$ gives a slot size of $1500\mu s$.

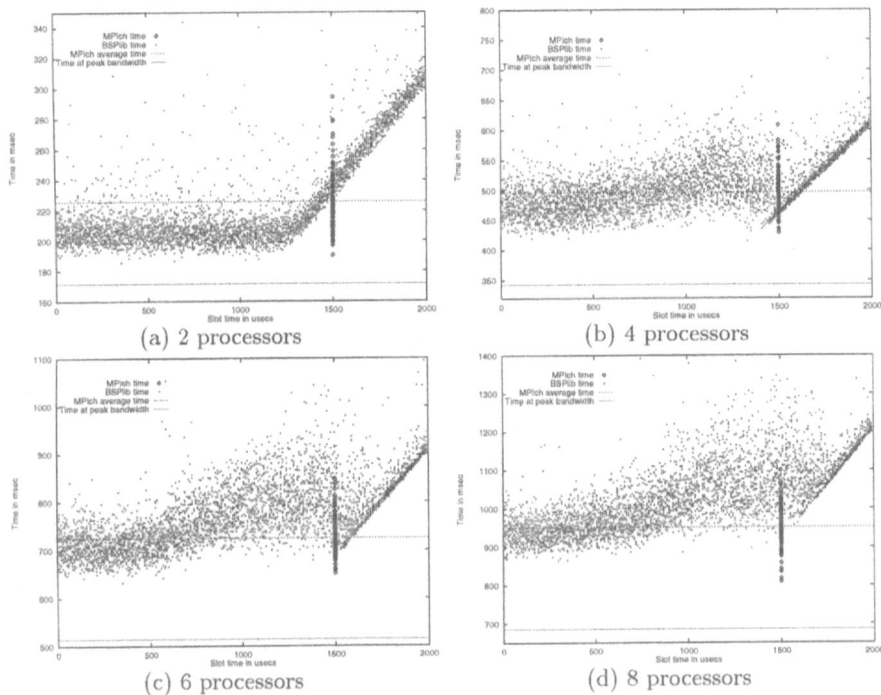

Fig. 4. Delivery time as a function of slot time for a cyclic shift of 25,000 words per processor, $p = 2, 4, 6, 8$, data for mpich shown at 1500 although slots are not used

In practice, while the maximum possible value for τ is known, the end-to-end propagation delay of the particular network segment is not, and this influences the slot size via the contention interval modelled in Figure 2. The analytic model assumes a Poisson arrival process, whereas for a finite number of stations, the arrival process is defined by independent Bernoulli trials (the limiting case of this process, as the number of processors increases, is Poisson, and approximates the finite case after the number of processors reaches about 20 [3]). More complicated topologies could also be considered where more than one segment is used.

The slot size ε can be determined empirically by running trials in which the slot size is varied and its effect on throughput measured. The experiments involved a 10Mbps Ethernet networked collection of workstations. Each workstation is a 266MHz Pentium Pro processor with 64MB of memory running Solaris 2. The experiments were carried out using the TCP/IP implementation of *BSPlib*. The machines and network were dedicated to the experiment, although the Ethernet segment was occasionally used for other traffic as it was a subset of a teaching facility. Figure 4(a) to Figure 4(d) plot the time it takes to realise a cyclic-shift communication pattern (each processor bsp_hputs a 25,000 word message into the memory of the processor to its right) for various slot sizes ($\varepsilon \in [0, 2000]$) and for 2, 4, 6 and 8 processors. The figures show the delivery time as a function

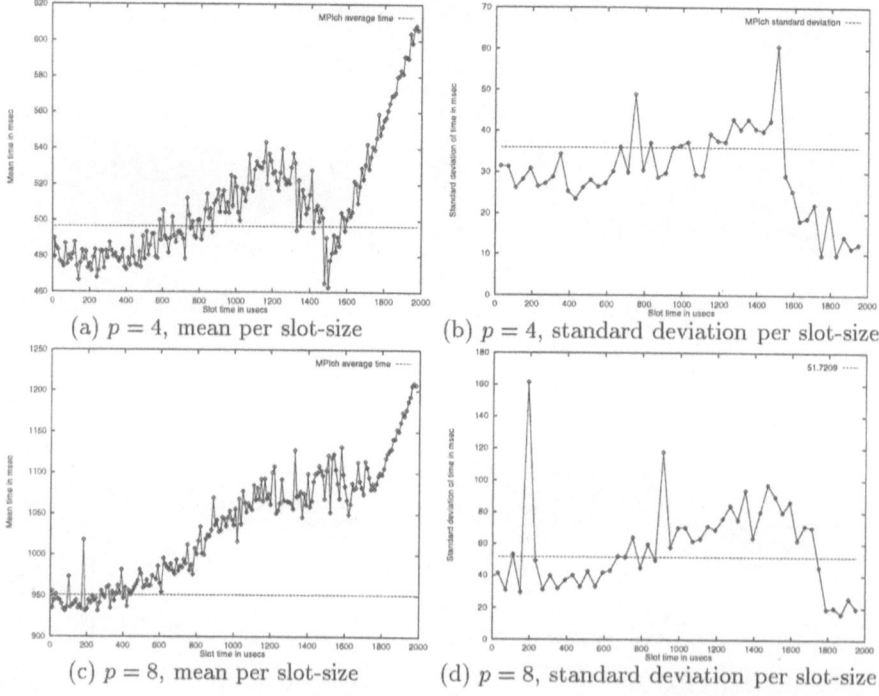

(a) $p = 4$, mean per slot-size

(b) $p = 4$, standard deviation per slot-size

(c) $p = 8$, mean per slot-size

(d) $p = 8$, standard deviation per slot-size

Fig. 5. Mean and standard deviation of delivery times of data from Figures 4(a,b)

of slot size, oversampled 10 times. The horizontal line towards the bottom of each graph gives the minimum possible delivery time based on bits transmitted divided by theoretical bandwidth.

Results for an MPI implementation of the same algorithm running on top of the Argonne implementation of MPI (mpich) [2] are also shown on these graphs. In this case, the data is presented at a slot size of 1500 μs (even though mpich does not slot), so only one (oversampled) measurement is shown. The dotted horizontal line in the centre of these figures is the mean delivery time of the MPI implementation.

The BSP slot time should be chosen to minimise the mean delivery time. Choosing a small slot time gives some good delivery times, but the scatter is large. In practice, a good choice is the smallest slot time for which the scatter is small. For $p = 2$ this is 1200 μs, for $p = 4$ it is 1450 μs, for $p = 6$ it is 1650 μs, and for $p = 8$ it is 1700 μs. Notice that these points do not necessarily provide the minimum delivery times, but they provide the best combination of small delivery times and small variance in these times.

Figure 4(b) shows a particularly interesting case at $\varepsilon = 1500$, as both the mean transfer rate and standard deviation of the *BSPlib* benchmark is much smaller than those of the corresponding mpich program. This slot-size can be clearly seen in Figure 5(c) and Figure 5(d) where the scatter caused by the

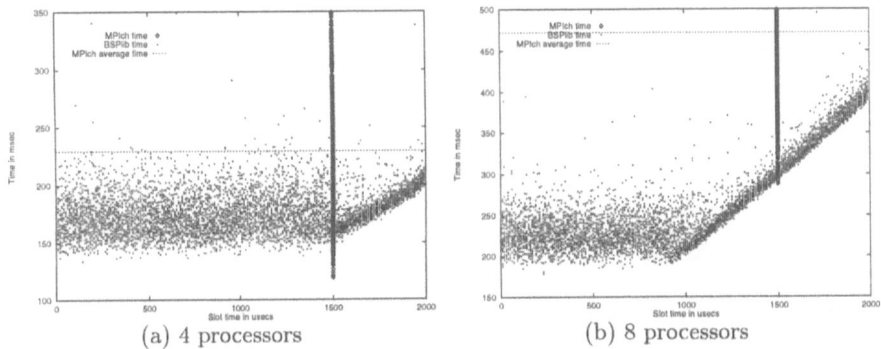

(a) 4 processors   (b) 8 processors

Fig. 6. Delivery time as a function of slot time for a cyclic shift of 8,300 words per processor, $p = 4, 8$, data for mpich shown at 1500 although slots are not used

oversampling at each slot size in Figure 4(b) has been removed by only displaying the mean and standard deviation of the oversampling. In contrast, the mean and largest outlier of the mpich program in Figure 4(d) is clearly lower than the corresponding *BSPlib* program when a slot size of 1500 is used. For larger configurations, the slot size that gives the best behaviour increases and the mean value of g for *BSPlib* quickly becomes worse than that for mpich.

An increase in the best choice of slot size from Figure 4(a) to Figure 4(d) should be expected as the probability $P(n)$ of n processors choosing a particular slot is binomially distributed. Thus as p increases, so does the expectation $E\{X \geq 2\}$ of the amount of contention for the slot, where

$$P(n) = \binom{p}{n} (1/p)^n (1 - 1/p)^{p-n} \text{ and } E\{X \geq 2\} = \sum_{i=2}^{p} iP(i)$$

Figure 3 shows that for $p \approx 20$ and greater, the dependence on p is minimal, and therefore the increase in slot size reaches a fixed point. Below twenty processors the dependence varies by at most 26%. The limit as $p \to \infty$ gives $E\{X \geq 2\} \to 1 - 1/e \approx 0.63$, as shown in the figure. The same is true of the probability of contention, but the range is very small, from 0.25 at $p = 2$, and as $p \to \infty$, $P\{x \geq 2\} \to 1 - 2/e \approx 0.26$.

In the mpich implementation [2] of MPI, large communications are presented to the socket layer in a single unit. However, in the *BSPlib* implementation all communications are split into packets containing at most 1418 bytes, so that we can pace the submission of packets using slotting. For this benchmark, each *BSPlib* process sends 71 small packets in contrast to mpich's single large message. Therefore, when p is small we would expect *BSPlib* to perform worse than mpich due to the extra passes through the protocol stack, and for larger values of p we would expect that the benefits of slotting out-weigh the extra passes through the protocol stack. Figures 4(a) to 4(d) show an opposite trend.

As can be seen from the Figures 4(b)–(d), as p increases, there is a noticeable "hump" in the data as the slot size increases. This phenomenon is not explained

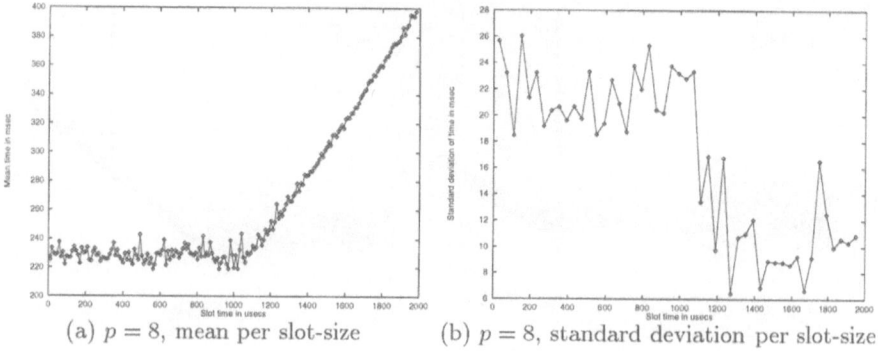

(a) $p = 8$, mean per slot-size   (b) $p = 8$, standard deviation per slot-size

Fig. 7. Mean and standard deviation of delivery times of data from Figure 6(b)

by the discussion above. The problem arises because we are modelling the communication as though it were directly accessing the Ethernet, without taking into account the TCP/IP stack. What we are observing is the TCP acknowledgement packets, which interfere with data traffic as they are not controlled by our slotting mechanism. The effect of this is to increase the optimum slot size to a value that ensures that there is enough extra bandwidth on the medium such that the extra acknowledgement packets do not negatively impact the transmission of data packets.

Implementations of TCP use a delayed acknowledgement scheme where multiple packets can be acknowledged by a single acknowledgement transmission. To minimise delays, a $200ms$ timeout timer is set when TCP receives data [11]. If during this $200ms$ period data is sent in the reverse direction then the pending acknowledgement is piggy-backed onto this data packet, acknowledging all data received since the timer was set. If the timer expires, the data received up to that point is acknowledged in a packet without a payload (a 512 bit packet).

In the benchmark program that determines the optimal slot size, a cyclic shift communication pattern is used. When $p > 2$ there is no reverse traffic during the data exchange upon which to piggy-back acknowledgements. If the entire communication takes less than $200ms$ then only p acknowledgement packets will be generated for each superstep; as the total time exceeds $200ms$, considerably more acknowledgement packets are generated. In Figure 4(a) the communication takes approximately $200ms$ and a minimal number of acknowledgements are generated as can be seen by the lack of a hump. In Figures 4(b)–(d), the size of the humps increases in line with the increased number of acknowledgements. The mpich program does not suffer as severely from this artifact as *BSPlib*. When slotting is not used (for example in mpich) there is potential for a rapid injection of packets onto the network by a single processor for a single destination, which means that it is likely that more packets arrive at their destination before the delayed acknowledgement timer expires. This reduces the number of acknowledgement packets. When slotting is used, packets are paced onto the network with a mean inter-packet time between the same source-destination pair of $p\varepsilon$. This drastically decreases the possibility of accumulated delayed acknowledgements. For example, in Figure 4(c), as the total time for communication is approximately 800ms, and as the slot size steadily increases, the number of acknowledgements increases. This in turn steadily increases the standard deviation and mean of the communication time. From the figure it can be seen that this suddenly drops off when the slot size becomes large as the probability of collision decreases due to the under-utilisation of the network.

The global nature of BSP communication means that data acknowledgement and error recovery can be provided at the superstep level as opposed to the packet by packet basis of TCP/IP. By moving to UDP/IP, we can implement acknowledgements and error recovery within the framework of slotting. This lower-level communication scheme is under development, although the hypothesis that it is the acknowledgements limiting the scalability of slotting can be tested by performing a benchmark on a dataset size that requires a total communication time that is less than $200ms$. Figures 6(a)–(d) shows the slotting benchmark for an 8333-relation where there are no obvious humps. In all configurations the mean and standard deviations of the *BSPlib* results are considerably smaller than `mpich`. Also, as can be seen from Figure 7 the optimal slot size at $p = 8$ is approximately $1200\mu s$.

4   Conclusions

We have addressed the ability of the BSP runtime system to improve the performance of shared-media systems using TCP/IP. Using BSP's global perspective on communication allows each processor to pace its transmission to maximise throughput of the system as a whole. We show a significant improvement over MPI on the same problem.

The approach provides high throughput, but also stable throughput because the standard deviation of delivery times is small. This maintains the accuracy of the cost model, and ensures the scalability of systems.

Acknowledgements

The work of Jonathan Hill was supported in part by the EPSRC Portable Software Tools for Parallel Architectures Initiative, as Research Grant GR/K40765 "A BSP Programming Environment", October 1995-September 1998. David Skillicorn is supported in part by the Natural Science and Engineering Research Council of Canada.

References

1. S. R. Donaldson, J. M. D. Hill, and D. B. Skillicorn. Communication performance optimisation requires minimising variance. In *High Perfomance Computing and Networking (HPCN'98)*, Amsterdam, April 1998.
2. W. Gropp and E. Lusk. A high-performance MPI implementation on a shared-memory vector supercomputer. *Parallel Computing*, 22(11):1513–1526, Jan. 1997.

3. J. L. Hammond and P. J. P. O'Reilly. *Performance Analysis of Local Computer Networks*. Addison Wesley, 1987.

4. J. M. D. Hill, S. Donaldson, and D. B. Skillicorn. Stability of communication performance in practice: from the Cray T3E to networks of workstations. Technical Report PRG-TR-33-97, Oxford University Computing Laboratory, October 1997.

5. J. M. D. Hill, B. McColl, D. C. Stefanescu, M. W. Goudreau, K. Lang, S. B. Rao, T. Suel, T. Tsantilas, and R. Bisseling. BSPlib: The BSP Programming Library. *Parallel Computing*, to appear 1998. see www.bsp-worldwide.org for more details.

6. J. M. D. Hill and D. B. Skillicorn. Lessons learned from implementing BSP. *Journal of Future Generation Computer Systems*, 13(4–5):327–335, April 1998.

7. P. J. B. King. *Computer and Communication Systems Performance Modelling*. International series in Computer Science. Prentice Hall, 1990.

8. D. B. Skillicorn, J. M. D. Hill, and W. F. McColl. Questions and answers about BSP. *Scientific Programming*, 6(3):249–274, Fall 1997.

9. S. Tasaka. *Performance Analysis of Multiple Access Protocols*. Computer Systems Series. MIT Press, 1986.

10. L. G. Valiant. A bridging model for parallel computation. *Communications of the ACM*, 33(8):103–111, August 1990.

11. G. R. Wright and W. R. Stephens. *TCP/IP Illustrated, Volume 2*. Addison-Wesley, 1995.

Adaptive Routing Based on Deadlock Recovery

Nidhi Agrawal[1] and C.P. Ravikumar[2]

[1] Hughes Software Systems, Sector18, Electronic City, Gurgaon, INDIA
[2] Deptt. of Elec. Engg., Indian Institute of Technology, New Delhi, INDIA

Abstract. This paper presents a deadlock recovery based fully adaptive routing for any interconnection network topology. The routing is simple, adaptive and is based on calculating the probabilities of routing at each node to neighbors, depending upon the static and dynamic conditions of the network. The probability of routing to the ith neighbor at any node is a function of the traffic and distance from the neighbor to the destination. Since with our routing algorithm deadlocks are rare, deadlock recovery is a better solution. We also propose here two deadlock recovery schemes. Since deadlocks occur due to cyclic dependencies, these cycles are broken by allowing one of the messages involved in deadlock to take an alternate path consisting of buffers reserved for such messages. These buffers can be centralized buffers accessible to all neighboring nodes or can be set of virtuals. The performance of our algorithm is compared with other recently proposed deadlock recovery schemes. The *2-Phase* routing is found to be superior compared to the other schemes in terms of network throughput and mean delay.

1 Introduction

Interprocessor communication is the bottleneck in achieving high performance in Message Passing Processors (MPPs). Various routing algorithms exist in the literature for wormhole routed[4] interconnection networks. Most routing algorithms achieve deadlock-freedom by restricting the routing, for example, Turn Model[5] router for hypercubes and meshes ensures deadlock-freedom by prohibiting certain turns in the routing algorithm. Many routing techniques achieve deadlock-freedom and adaptivity through the use of virtual channels [3,4]. Not only do virtual channels increase hardware complexity, they also reduce the speed of the router. Chien[2] has shown that virtual channels adversely affect the router performance, and multiplexing more than 2 to 3 virtual channels over one physical channel results in poor router performance. From the discussion above, it is clear that achieving deadlock-freedom and adaptivity at the cost of increased router complexity and increased delays is not a good solution.

In this paper we propose fully adaptive routing based on deadlock recovery. The algorithm breaks deadlocks without adding significant hardware. We introduce the notion of *Routing Probability* at each node. $P_i(j,k)$ denotes the probability of routing a message at node i to a neighboring node k, destined for j. Entirely chaotic routing can be achieved by selecting $P_i(j,k) = 1/d_i$. On the other hand, a deterministic routing algorithm can be modeled by selecting $P_i(j,l) = 1$ for some $1 \leq l \leq d_i$, and $P_i(j,k) = 0$, $\forall k \neq l$; here d_i is the degree of node i. In this paper we propose a simple heuristic to calculate the routing probabilities at any node depending on the distance from the destination and/or congestion at the neighboring node. Our algorithm dynamically computes the probabilities, thus providing adaptivity. A message is declared to be deadlocked at any node i, if the message header has to wait at node i for more than a pre-defined time called *TimeOut*. Since deadlocks are infrequent, it is more logical to recover from deadlocks rather than avoiding. Various schemes for deadlock recovery are available in the literature[1, 6]. Compressionless routing[6] is a deadlock recovery scheme based on killing the deadlocked packets and transmitting again. This scheme requires *padding flits* to be attached with the message when the message length is less than the network diameter. Moreover, messages that are killed and rerouted suffer large delays. A deadlock recovery scheme called *Disha*[1] uses a single flit buffer at each node. The disadvantage of this scheme is that at any time only one message can be routed onto deadlock-free buffers, thus only one of the deadlocks we can recover from at a time; furthermore the token based mutual exclusion on the central virtual network is implemented by adding additional hardware.

We propose here two schemes for deadlock recovery. Our first scheme makes use of multiple central buffers; on the other hand our second scheme, also called *2-Phase Routing* divides the network into two virtual networks. Both the schemes eliminate the disadvantages associated with the previously proposed recovery schemes. Rest of the paper is organized as follows : Next section describes the routing algorithm. Section 3 describes the proposed deadlock recovery schemes. The results of simulation are given in Section 4. Section 5 concludes the paper.

2 Adaptive Routing

2.1 Algorithm

Various heuristics may be used to calculate the routing probabilities $P_i(j,k)$ introduced in Section 1. In this section we propose a simple heuristic as shown below :

$$\forall k \in Nbr(i), \quad P_i(j,k) = \frac{\frac{1}{W_k}}{\sum_k \frac{1}{W_k}} \tag{1}$$

Here $Nbr(i)$ is the set of immediate neighbors of node i and W_k is a weight assigned to the neighbor k which is a linear function of the distance $\delta(j,k)$ and the congestion at the neighboring node k denoted as Q_k. The weight W_k is assigned as follows :

$$W_k = \alpha * \delta(j,k) + \beta * Q_k \tag{2}$$

Each node is assumed to have the knowledge of the fault-status of its neighbor and the congestion at each neighboring node. The congestion at any node k, also denoted by Q_k, is the average amount of time a header has to spend at k normalized against a timeout period $TimeOut$. To measure the value of Q_k, node k makes use of as many timers as there are neighbors. If a header waits for more than $TimeOut$ period, the message is declared to be deadlocked.

For most regular interconnection networks, the normalized distance function $\delta()$ is easy to evaluate. The scaling factors α and β are non-negative real constants, which dictate the nature of the routing algorithm; by selecting $\beta = 0$, we obtain distance-optimal routing. Similarly, by selecting $\alpha = 0$, we obtain a hot-potato routing algorithm. We found optimum values of the ratio $\frac{\alpha}{\beta}$, by simulation, for uniform and bit-reversal traffic (see Section 4).

2.2 Detection of deadlocks

The selection of $TimeOut$ interval greatly affects the router performance. If the $TimeOut$ period is very large, the deadlocked messages remain in the network and blocking many other messages; on the other hand if the $TimeOut$ period is small, false deadlocks are declared. Whenever node i receives a message header H, a timer $T(i, H)$ starts counting the number of clock cycles the header has to wait before being forwarded. If for any header H, $T(i, H)$ exceeds $TimeOut$ at node i, then header is declared to be *deadlocked*. The optimum $TimeOut$ interval can be found by simulations (see Section 4). We also measured the frequency of deadlocks due to the proposed routing algorithm. It is observed that under maximum load less than 5% messages are deadlocked for a 64 node hypercube and $TimeOut = 16$ clocks.

3 Deadlock-Recovery

3.1 Recovery Scheme 1

After detecting a deadlock, the deadlock cycle is broken by switching one of the messages involved in the deadlock cycle to the central buffers kept aside at each node for routing deadlocked messages. Each node say i has $K + 1$ central buffers numbered $0, 1, 2..., K$. These buffers are accessible by all neighboring nodes of i. The kth central buffers of all the nodes form a central virtual network VN_k, $0 \le k \le K$. On any central virtual network, only one message is allowed to travel at a time. Thus, at most K deadlocks can be recovered simultaneously.

The permission to break deadlocks is given in a "round-robin" fashion to all the nodes. This may be implemented using K *tokens* corresponding to each central virtual network. Token i corresponds to the central virtual network VN_i. These tokens are nothing but packets of one flit size, which rotate on VN_0 along a cycle including all healthy nodes of the network. Each token carries the address of the next node in the cycle. Further details on token management and implementation are given below:

- Token synchronization: The tokens are synchronized with the router clock i.e. each token remains at a node for one clock cycle while circulating. The worst case waiting time to capture a token is $(N * (Avg.Dist.)/K)$ clock cycles, where N is the number of nodes and $Avg.Dist.$ is the average distance of the network.

- Capturing a token: When a node detects a deadlock and concurrently receives a token j along the central buffer 0 carrying the address of the node, the token is captured by the node.
- Regenerating a token: In order to make sure that messages do not get deadlocked on central virtual networks, a captured token j at a node N_i is regenerated after $\delta(N_i, D)$ clock cycles, ensuring that there is at most one packet on the network, where D is the destination node.

3.2 Recovery Scheme-2

The recovery scheme-2 , also called *2-phase routing* discussed in this section, provides simultaneous recovery from all the deadlocks. Each physical channel is divided into two virtual channels. The network can be viewed as two virtual networks : *the Adaptive Virtual Network*(AVN) and *the Deadlock-free Virtual Network*(DVN). All the messages are initiated in AVN. Routing is carried out without any restrictions following the routing algorithm of Section 2. Whenever any message gets deadlocked, the routing enters phase 2 and the message is switched over to DVN. The routing procedure followed on DVN can be any deadlock-free routing for example, e-cube routing for hypercube, X-Y routing and negative-first routing for meshes and k-ary n-cubes.

We describe here briefly, a deadlock-free, Hamiltonian path based routing to route on DVN, which is partially adaptive and works for various topologies like hypercubes, k-ary n-cubes, Star graphs, meshes etc. The nodes of DVN are numbered according to Hamiltonian number denoted as $Hamilt(i)$. The DVN is further partitioned into two subnetworks *the Higher Virtual Subnetwork* (HSN) and *the Lower Virtual Subnetwork* (LSN). In HSN, there is an edge from any node i to any other node j, iff $Hamilt(i) < Hamilt(j)$. There is an edge from i to j in the lower network iff $Hamilt(i) > Hamilt(j)$. Each of these subnetworks is acyclic, thus routing in each of these networks is deadlock-free [7]. The routing function $\Re_1 (\Re_2)$ of Equation3(4) corresponds to routing in LSN(HSN) from i to j, where $E_L(E_H)$ is edge set of LSN(HSN).

$$\Re_1(i,j) = k : (i,k) \in E_L, Hamilt(i) > Hamilt(k) \geq Hamilt(j) \qquad (3)$$

$$\Re_2(i,j) = k : (i,k) \in E_H, Hamilt(i) < Hamilt(k) \leq Hamilt(j) \qquad (4)$$

The routing functions defined by Equations 3 and 4 are deterministic. In a modification of the routing function of Equation 3, if we allow a message to switch from LSN to HSN, when a faulty or congested node is encountered, without permitting the message to return to LSN, the routing function $\Re_1$ continues to remain deadlock-free (see Theorem 1). A similar argument holds for the function $\Re_2$. The resulting partially adaptive routing is called *Adapt-Hamilt*.

Theorem 1. *The adaptive routing* Adapt-Hamilt *is deadlock-free.*

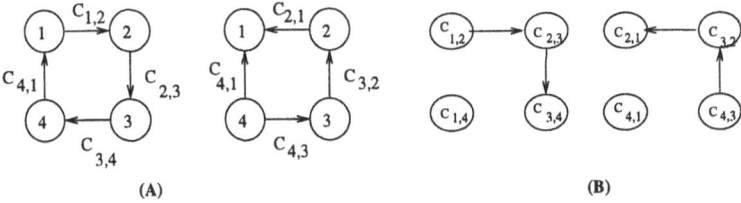

Fig. 1. (A) HSN and LSN for 2-dimensional Hypercube (B) CDG for $\Re_1$ and $\Re_2$

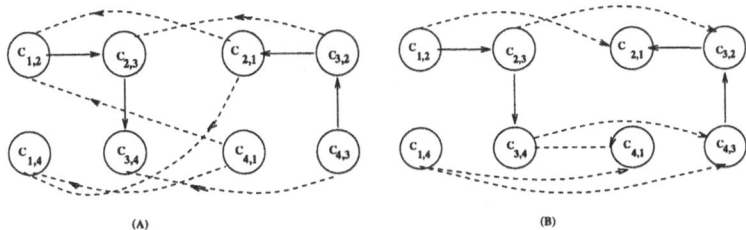

Fig. 2. CDG for adaptive routing in a 2-d Hypercube

Proof. We have seen that the routing function $\Re_1$ of Equation 3 is deadlock-free, implying that the channel dependency graph(CDG) induced by this function is acyclic [4](see Figure 1). The modified routing introduces additional edges in the CDG. These additional edges are of the form (x_h, y_l) as shown in Figure 2, where x_h is a node of the higher network CDG and y_l is a node of the lower network CDG. The adaptive routing function *Adapt-Hamilt*, if allows the additional edges of the form (x_h, x_l) then edges (x_l, y_h) are not permitted. Hence the theorem. □

4 Performance Study

We conducted experiments to measure the performance of our adaptive routing with various deadlock recovery schemes on a 256 node hypercube. The performance metric considered are the *throughput* and the *mean delay*. *Throughput* of the network is defined as the mean number of messages going out of the network per node per clock cycle. *Mean delay* is the mean number of clock cycles elapsed from the time the first flit of the message enters the network to the time the last flit leaves the network. It may be noted that throughput and latency are measured for various applied loads, thus throughput-latency graphs do not represent a function.

Each node generates messages with a poisson distribution. Two kinds of traffic patterns are generated: uniform traffic and bit-reversal traffic. In uniform traffic the destinations are generated uniformly with each node being equally probable; on the other hand in bit-reversal traffic pattern, the destination node address is obtained by reversing the bits of the source node address, for example source $s_1, s_2, ..., s_n$ sends a message to the destination node $D = s_n, s_{n-1}, ..., s_2, s_1$. Messages are 8 flits long. The results are taken for 4000 messages out of which the information for first 2000 messages is discarded (warm-up period). It takes one clock cycle time to transfer a flit across a physical channel. The header is assumed to be one flit long. The experiment is repeated several times by varying the seed and the average value is plotted.

4.1 Finding optimum *TimeOut* and $\frac{\alpha}{\beta}$

For fine tuning *TimeOut* period for each type of traffic, we conducted experiments. The network performance is measured for various *TimeOut* periods. Figure 3 shows the *throughput* versus *latency* curves for various *TimeOut* periods under bit-reversal traffic pattern. It may be observed that latency increases slowly initially upto throughput value of 0.02 messages per node per clock cycle and then there is a sharp increase. Since the highest throughput is achieved with *TimeOut* = 16, it is chosen as the optimum value. Similarly, for uniform traffic the optimum *TimeOut* value is also found to be 16. The plot is not shown due to the lack of space.

For both type of traffic model, we found optimum value of the ratio $\frac{\alpha}{\beta}$. Figure 4 shows the performance of the routing algorithm for various values of $\frac{\alpha}{\beta}$ for uniform traffic. The optimum value of $\frac{\alpha}{\beta}$ is the value which gives highest peak performance. The throughput reaches its maximum value upto a certain applied load then starts reducing; this is the unstable condition of the network. From the Figure 4 it can be noticed that the optimum value is $\frac{\alpha}{\beta} = 10$, because for this value the highest throughput is achieved before saturation. Similarly for bit-reversal traffic the optimum value of $\frac{\alpha}{\beta}$ is found to be 5, as shown in figure 5.

4.2 Comparison of various schemes

We compared the performance of the proposed routing algorithm for various deadlock recovery schemes. Both the proposed recovery schemes are compared with the existing deadlock recovery scheme *disha*. Figure 6 shows a comparison of the three recovery schemes under uniform traffic conditions. The throughput value saturates at 10% with both *disha* and scheme-1 and at 21% with *2-phase* routing. It is also observed that by adding more and more central virtual buffers (or tokens) the performance of scheme-1 improves. In figure 6, the performance of the scheme-1 is measured for 3 central virtual buffers at each node.

We analyze the performance of the proposed routing in the presence of node faults. It is observed that there is a negligible degradation in the performance routing for various fault conditions. The plots are not shown here due to the lack of space.

The performance is also compared for the three recovery schemes under bit-reversal traffic. Figure 7 shows the comparison. It may be observed that the recovery scheme-1 outperforms the recovery scheme *disha*, and the *2-phase* routing performs far better than both the scheme-1 and *disha*. Figure 7 shows that after saturation the mean delay increases sharply, however the throughput value does not increase. In this case saturation occurs only at 6%.

5 Conclusion

We have presented a fully adaptive, restriction-free routing algorithm. Since deadlocks are rare we considered deadlock recovery instead of traditional *deadlock avoidance*. We have presented two deadlock recovery schemes, the recovery scheme-1 and *2-phase* routing. The recovery scheme-1 provides simultaneous recovery from K deadlocks, where K is the number of central buffers at each node; on the other hand *2-phase* routing provides the simultaneous recovery from all the deadlocks. The performance of both the proposed recovery schemes is compared with the recovery scheme disha [1]. It is observed that the recovery scheme-1 outperforms the recovery procedure of [1], under both uniform and bit-reversal traffic. Also the *2-phase* routing performs far superior than both disha and the recovery scheme-1.

References

1. K.V.Anjan and T.M.Pinkston, "Disha: A deadlock Recovery Scheme for Fully Adaptive Routing", *Proc. of the 19th Int. Parallel Processing Symposium*, 1995.
2. A.A.Chien, "A Cost and Speed Model for k-ary n-cube Wormhole Routers", *IEEE Transactions on Parallel and Distributed Systems*, Vol.9, No.2, pages $150 - 162$, 1998.
3. W.Dally and H.Aoki, "Deadlock-free Adaptive Routing in Multicomputer Networks using Virtual Channels", *IEEE Transactions on Parallel and Distributed Systems*, Vol.4, No.4,1993, pages $466 - 475$.
4. W.Dally and C.Seitz, "Deadlock-free message routing in multiprocessor interconnection networks", *IEEE Transactions on Computers*, C-36(5), 1987, pages $547 - 553$.
5. C.J.Glass and L.M.Ni, "The turn model for Adaptive routing", *Proc. of the 19th Int. Symp. on Comp. Architecture*, 1992, pages $278 - 287$.
6. J.Kim, Z.Liu and A.Chien, "Compressionless Routing: A Framework for Adaptive and Fault-tolerant Routing", *IEEE Transactions on Parallel and Distributed Systems*, Vol.8, No.3, pages $229 - 244$, 1997.
7. X.Lin, P.K.McKinley and L.M.Ni, "Deadlock-free multicast wormhole routing in 2D mesh multicomputers", *IEEE Trans. on Parallel and Distributed Systems*, Vol.5, no.8, 1994, pages $793 - 804$.

988

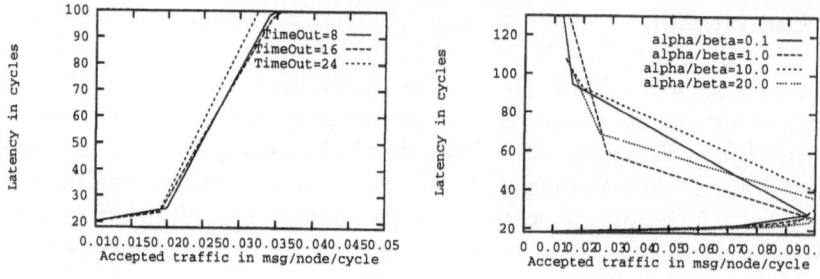

Fig. 3. Tuning *TimeOut* interval for bit-reversal traffic

Fig. 4. Tuning $\frac{\alpha}{\beta}$ for unifrom traffic

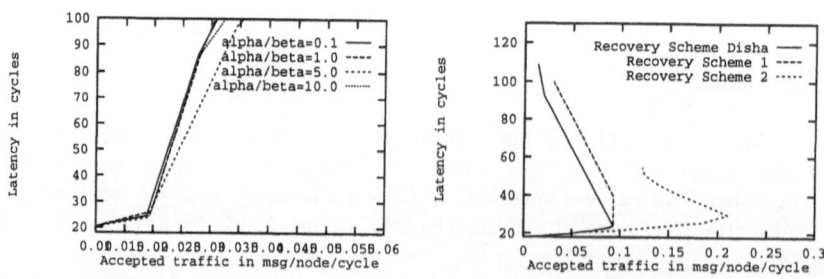

Fig. 5. Tuning $\frac{\alpha}{\beta}$ for bit-reversal traffic

Fig. 6. Throughput Vs. mean delay for the three recovery schemes under uniform traffic

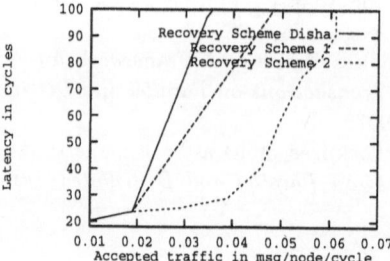

Fig. 7. Throughput Vs. mean delay for the three recovery schemes under bit-reversal traffic

On the Optimal Network for Multicomputers: Torus or Hypercube?

Mohamed Ould-Khaoua

Department of Computer Science, University of Strathclyde, Glasgow G1 1XH, UK.

Abstract. This paper examines the relative performance merits of the torus and hypercube when adaptive routing is used. The comparative analysis takes into account channel bandwidth constraints imposed by VLSI and multiple-chip technology. This study concludes that it is the hypercube which exhibits the superior performance, and therefore is a better candidate as a high-performance network for future multicomputers with adaptive routing.

1 Introduction

The hypercube and torus are the most common instances of k-ary n-cubes [5]. The former has been used in early multicomputers [3, 10] while the latter has become popular in recent systems [4, 8]. This move towards the torus has been mainly influenced by Dally's study [5]. When systems are laid out on a VLSI-chip, Dally has shown that under the constant wiring density constraint, the 2 and 3-D torus outperform the hypercube due to their higher bandwidth channels.

Abraham [1] and Agrawal [2] have argued that the wiring density argument is applicable where a network is implemented on a VLSI-chip, but not in situations where it is partitioned over many chips. In such circumstances, they have identified that the most critical bandwidth constraint is imposed by the chip pin-out. Both authors have concluded that it is the hypercube which exhibits better performance under this new constraint.

Wormhole routing [11] has also promoted the use of high-diameter networks, like the torus, as it makes latency independent of the message distance in the absence of blocking. In wormhole routing, a message is broken into flits for transmission and flow control. The header flit governs the route, and the remaining data flits follow in a pipeline. If the header is blocked, the data flits are blocked in situ.

Most previous comparative analyses of the torus and hypercube [1, 2, 5, 6] have used deterministic routing, where a message always uses the same network path between a given pair of nodes. Deterministic routing has been widely adopted in practice [3, 8, 10] because it is simple and deadlock-free. However, messages cannot use alternative paths to avoid congested channels. Fully-adaptive routing has often been suggested to overcome this limitation by enabling messages to explore all the available paths in the network. Duato [6] has recently proposed a fully-adaptive routing, which achieves deadlock-freedom with minimal hardware requirement. The Cray T3E [4] is an example of a recent machine that uses Duato's routing algorithm.

The torus continues to be a popular topology even in multicomputers which employ adaptive routing. However, before adaptive routing can be widely adopted in practical systems, it is necessary to determine which of the competing topologies are able to fully exploit its performance benefits. To this end, this paper re-assesses the relative performance merits of the torus and hypercube in the context of adaptive routing. The study compares the performance of the 2 and 3-D torus to that of the hypercube. The analysis uses Duato's fully-adaptive routing [6]. The present study uses queueing models developed in [9] to examine network performance under uniform traffic. Results presented in the next section reveal that it is the hypercube which provides the optimal performance under both the constant wiring density and pin-out constraints, and thus is the best candidate as a high-performance network for future multicomputers with fully-adaptive routing.

2 Performance Comparison

The torus has higher bandwidth channels than its hypercube counterpart under the constant wiring density and pin-out constraints. The detailed derivation of the exact relationship between the channel width of the torus in terms of that of the hypercube under both constraints can be found in [1, 2, 5].

The router's switch in the hypercube is larger than that in the torus due to its larger number of physical and virtual channels. As a consequence, the switching delay in the hypercube should be higher due to the additional complexity. Comparable switching delays in the two networks can be obtained if the routers have comparable switch sizes. This can be achieved by normalising the total number of channels in the routers of the two networks.

When mapped in the 2D plane, the hypercube ends up with longer wires, and thus higher wire delays than the torus. However, delays due to long wires can be reduced by using pipelined channels as suggested in [12]. The performance of the networks is examined below when both the wire delay is taken into account and when it is ignored. For illustration, network sizes of N=64 and 1024 nodes are examined. The channel width in the hypercube is one bit. The channel width in the torus is normalised to that of the hypercube. The message length (M) is 128 flits. A physical channel in the hypercube has V=2 virtual channels. The total number of virtual channels per router in the torus is normalised to that in the hypercube.

Figures 1-a and b depict latency results in the torus and hypercube under the constant wiring density constraint and when the effects of wire delays are taken into account in the 64 and 1024 node systems respectively. The figures reveal that the torus is able exploit its wider channels to provide a lower latency than the hypercube under light to moderate traffic. However, as traffic increases its performance degrades as message blocking rises, offsetting any advantage of

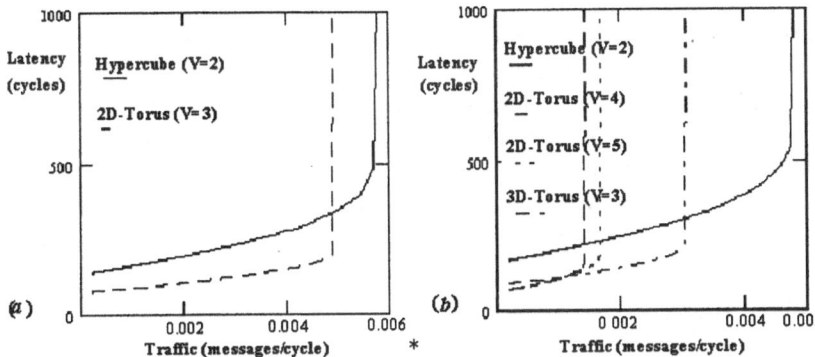

Fig. 1. The performance of the torus and hypercube including the effects of wiring delays. *(a) N=64, (b)N=1024.*

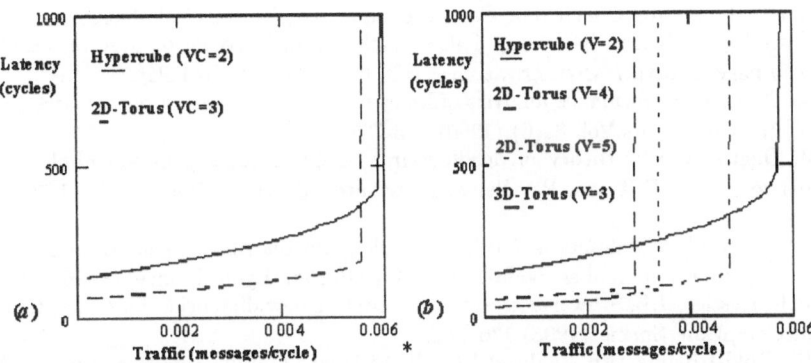

Fig. 2. The performance of the torus and hypercube ignoring the effects of wiring delays. *(a) N=64, (b)N=1024.*

having wider channels. Figures 2-a and b show latency when the effects of wire delays are ignored. The torus outperforms the hypercube under light to moderate traffic, but loses edge to the hypercube under heavy traffic. The difference in performance between the two networks increases in favour of the hypercube for larger network sizes. Figures 1 and 2 together reveal an important finding about the torus, and that is even though wires are longer in the hypercube, the torus is more sensitive to the effects of wire delays. This is because a message in the torus crosses, on average, a larger number of routers, and therefore require a longer service time to reach its destination. Since the ratio in channel width in the hypercube and torus decreases under the constant pin-out constraint, we can conclude that the hypercube is even more favourable when the networks are subjected to this condition.

3 Conclusion

This paper has compared the performance merits of the torus and hypercube in the context of adaptive routing. The results have revealed that the hypercube has superior performance characteristics to the torus, and therefore is a better candidate as a high-performance network for future multicomputers, that use adaptive routing.

References

1. S. Abraham, Issues in the architecture of direct interconnection networks schemes for multiprocessors, Ph.D. thesis, Univ. of Illinois at Urbana-Champaign(1991).
2. A. Agarwal., Limits on interconnection network performance, IEEE Trans. Parallel & Distributed Systems,Vol. 2(4), (1991) 398–412.
3. R. Arlanskas, iPSC/2 system: A second generation hypercube, Proc. 3rd ACM Conf. Hypercube Concurrent Computers and Applications,Vol. 1 (1988).
4. Cray Research Inc., The Cray T3E scalable parallel processing system, on Cray's web page at http://www.cray.com/PUBLIC/product-info/T3E/.
5. W.J. Dally, Performance analysis of k-ary n-cubes interconnection networks, IEEE Trans. Computers,Vol. 39(6) (1990) 775-785.
6. J. Duato, A New theory of deadlock-free adaptive routing in wormhole routing networks, IEEE Trans. Parallel & Distributed Systems, Vol. 4 (12) (1993) 320–331.
7. J. Duato, M.P. Malumbres, Optimal topology for distributed shared-memory multiprocessors: hypercubes again?, Proc. EuroPar'96, Lyon, France, (1996) 205–212.
8. R.E. Kessler, J.L. Schwarzmeier, CRAY T3D: A new dimension for Cray Research, in CompCon, Spring (1993) 176–182.
9. M. Ould-Khaoua, An Analytical Model of Duato's Fully-Adaptive Routing Algorithm in k-Ary n-Cubes, to appear in the Proc. 27th Int. Conf. Parallel Processing, 1998.
10. C.L. Seitz, The Cosmic Cube, CACM, Vol. 28 (1985) 22–33.
11. C.L. Seitz, The hypercube communication chip, Dep. Comp. Sci., CalTech, Display File 5182:DF:85 (1985).
12. S.L. Scott, J.R. Goodman, The impact of pipelined channel on k-ary n-cube net works, IEEE Trans. Parallel & Distributed Systems, Vol. 5(1) (1994) 2-16.

Constant Thinning Protocol for Routing h-Relations in Complete Networks

Anssi Kautonen[1], Ville Leppänen[2], and Martti Penttonen[1]

[1] University of Joensuu, Department of Computer Science, P.O.Box 111, 80101 Joensuu, Finland,
{Anssi.Kautonen,Martti.Penttonen}@cs.joensuu.fi
[2] University of Turku, Department of Computer Science and Turku Centre for Computer Science, Lemminkäisenkatu 14 A, 20520 Turku, Finland,
Ville.Leppanen@cs.utu.fi

Abstract. We propose a simple protocol, called *constant thinning protocol*, for routing in a complete network under OCPC assumption, analyze it, and compare it with some other routing protocols.

1 Introduction

Parallel programmers would welcome the "flat" shared memory, because it would make PRAM [15] style programming possible and thus make easier to utilize the rich culture of parallel algorithms written for the PRAM model. As large memories with a large number of simultaneous accesses do not seem feasible, the only possible way to build a PRAM type parallel computer appears to be to build it from processors with local memories. The fine granularity of parallelism and global memory access, what makes the PRAM model so desirable for algorithm designers, sets very high demands for data communication. Fortunately, memory accesses at the rate of the processor clock are not necessary, but due to the parallel *slackness* principle [16], latency in access does not imply inefficiency. It is enough to route an *h-relation* efficiently. By definition, in h-relation there is a complete graph, where each node (processor) has at most h packets to send, and it is the target of at most h packets. We assume the *OCPC* (Optical Communication Parallel Computer) or *1-collision* assumption [1]: If two or more packets arrive at a node simultaneously, all fail. An implementation of an h-relation is *work-optimal* at *cost c*, if all packets arrive at targets in time ch.

The first attempt to implement an h-relation is to use a *greedy* routing algorithm. By greedy principle, one tries to send packets as fast as one can. The fatal drawback of the greedy algorithm is the *livelock*: packets can cause mutual failure of sending until eternity. In a situation, when each of two processors has one packet targeted to the same processor, they always fail.

Another routing algorithm was proposed by Anderson and Miller [1], and it was improved by Valiant [15]. They realize work-optimally an h-relation for $h \in \Omega(\log p)$, where p is the number of processors. Other algorithms with even lower latency were provided by [6, 7, 2, 3, 11]. Contrary to these, the h-relation algorithm of Geréb-Graus and Tsantilas [5], GGT for short, has the advantage of being *direct*, i.e. the packets go to their targets directly, without intermediate nodes. For other results related with direct or indirect routing, see also [4, 14, 8, 9, 12, 13]. Actually our new algorithm was inspired by [13], although the latter deals with the continuous routing problem, not the h-relation.

proc GGT(h,ϵ,α)
for $i = 0$ **to** $\log_{1/(1-\epsilon)} h$ **do** Transmit($(1 - \epsilon)^i h,\epsilon,\alpha$)

proc Transmit(h,ϵ,α)
for all processors P **pardo**
   for $\frac{e}{1-\epsilon}(\epsilon h + max\{\sqrt{4\epsilon\alpha h \ln p}, 4\alpha \ln p\})$ times **do**
      choose an unsent packet x at random
      attempt to send x with probability $\frac{\#\ unsent\ packets}{h}$

2 Thinning Protocols

The throughput of the greedy routing of randomly addressed packets is characterized by

$$(1 - \frac{1}{h})^{h-1} \geq \frac{1}{e}$$

where $1/h$ is the probability that one of the $h-1$ competing processors is sending to the same processor at the same time. (For all $x > 0$, $(1-1/x)^{x-1} \geq e^{-1}$.) This would be the throughput if all processors would create and send a new randomly addressed packet, which is not the case in routing an h-relation. When failed packets are resent again and again, the risk of livelock grows. A solution is to decrease the sending probability. In GGT algorithm, the sending probability of packets varies between 1 and $1 - \epsilon$ ($0 < \epsilon < 1$). The transmission of packets is thus 'thinned' by factor 1 to $1/1 - \epsilon$, preventing the livelock. We now propose a very simple routing protocol, where thinning is more explicit.

proc CT(h,h_0,t,t_0)   %$1 \leq h_0 \leq h, 1 \leq t_0 \leq t$
for all processors **do**
   while packets remain **do**
      Transmit h randomly chosen packets (if so many remain)
         within time window $[1..[th]]$
      $h := max\{(1 - e^{-1/t_0})h, h_0\}$

A drawback of thinning is that a processor cannot successfully send a packet at those moments, when it does not even try to send. Thinning by factor t would thus imply inefficiency by factor t. But this is somewhat balanced by the success probability, which increases from $1/e$ to

$$(1 - \frac{1}{th})^{h-1} \geq \frac{1}{e^{1/t}}.$$

The expected throughput with thinning is thus characterized by the function $te^{1/t}$ whose growth is very modest: for $t = 1, 1.2, 1.5, 2.0$ and 3.0 the values of $te^{1/t}$ are $2.7, 2.8, 2.9, 3.3$, and 4.2. Still, even small thinning increases greatly the robustness against livelock. For very small numbers of packets constant thinning alone does not guarantee high success probability. For that reason, CT has a minimum size $h_0 \in \Omega(\log p)$ for thinning window, to prevent repeated collisions.

3 Analysis

The analyses for GGT and CT are rather similar. One can prove that

1. the number of unsent packets decreases geometrically from h to $\log p$
2. the rest of the packets can be routed in time $O(\log p \log \log p)$

Theorem 1. *For $h \in \Omega(\log p \log \log p)$, CT routes any h-relation in time $O(h)$ with high probability.*

Proof. Consider a round of the while loop, where $h' \geq k_0 \log p = h_0$ for a suitable constant k_0. We shall show that if in the beginning of a round the routing task is an h'-relation, it will be a ch'-relation after the round with high probability.

To ease the calculation of probabilities, we worsen the routing task a little by completing the initial h'-relation to a full h'-relation. We do this by adding 'dummy packets' with proper destination to those processors not having initially h' packets. We assume that the dummy packets participate in routing as the other packets. In the analysis below, the dummy packets can collide with normal packets as well as with other dummy packets, but in the actual algorithm attempting to route a dummy packet corresponds to an unallocated time slot (unable to cause any collisions). Thus the number of successful normal packets is always better in the actual situation.

We first show that, with high probability, the number of packets decreases to ch' or below. As a packet has the probability $1/th'$ of being sent at a given moment of time (from the interval $[1 \ldots \lceil th' \rceil]$), and by the h'-relation assumption, at most h' packets have the same target, the probability of success for a packet at a given moment of time is at least

$$\frac{1}{th'} \times (1 - \frac{1}{th'})^{h'-1} \geq \frac{1}{e^{1/t} \cdot th'},$$

since

$$(1 - \frac{1}{th'})^m \geq (1 - \frac{1}{th'})^{h'-1}$$

for any $m \leq h' - 1$ (consider m as the actual number of other packets with the same target) and

$$(1 - \frac{1}{th'})^{h'-1} = \left((1 - \frac{1}{th'})^{th'-t}\right)^{1/t} \geq \left((1 - \frac{1}{th'})^{th'-1}\right)^{1/t} \geq e^{-1/t}$$

for $h' > 1$ and $t \geq 1$. For each packet, the probability of success for a packet during the procedure call is at least

$$\frac{1}{e^{1/t} \cdot th'} \times th' = e^{-1/t}.$$

Hence, the expected number of successful packets of a processor within these th' units of time is $E_t = h'e^{-1/t}$. By applying Chernoff bound [10]

$$Pr(N < (1 - \epsilon)E_t) \leq e^{-\frac{1}{2}\epsilon^2 E_t}$$

and choosing $(1 - \epsilon)E_t = E_{t_0}$, we get $\epsilon = 1 - E_{t_0}/E_t \approx 1/t_0 - 1/t$, and

$$Pr(N < h'e^{-1/t_0}) \leq e^{-0.5 \times \epsilon^2 \times h'e^{-1/t}} \leq \frac{1}{p^{2.5}}$$

for $h' > h_0 = k_0 \log p$ for some constant k_0.

Hence, in a processor the number of outgoing packets decreases by compression factor $c = 1 - 1/e^{1/t_0}$ with probability $1 - 1/p^{2.5}$, and in all p processors and in all $\log_{1/c} p < \sqrt{p}$ rounds with probability $1 - 1/p$. Guaranteeing that the number of incoming packets is at most ch' for each processor is analyzed analogously. Clearly, the full h'-relation (completed with dummy packets) decreases to ch'-relation. Removing the dummy packets from the system can only decrease the degree of the relation. Within $\log_{1/c} h$ geometrically converging phases, or $O(ht)$ time, there remain no more than $O(\log p)$ packets in each processor.

Observe that by choosing a larger h_0 in the analysis above, we can easily show the same progression in the degree of the relation with probability $1 - p^{-\alpha}$ for any positive constant α.

For the rest of the algorithm, when $h' \leq h_0$ packets remain, the success probability of a packet is

$$\frac{h'}{th_0} \times (1 - \frac{1}{th_0})^{h_0-1} \geq \frac{h'}{h_0} \frac{1}{te^{1/t}}.$$

Thus the expectation of sending times for this packet is $te^{1/t}h_0/h'$. The sum of all these expectations, until all packets have been sent, is

$$te^{1/t}h_0\left(\frac{1}{h_0} + \frac{1}{h_0 - 1} + \ldots + \frac{1}{2} + 1\right) = E \in O(h_0 \log h_0).$$

By another form of Chernoff bound [10]

$$Pr(T > r) \leq \frac{1}{2^r} \quad \text{for } r > 6E$$

we see that

$$Pr(T > k_1 h_0 \log h_0) \leq \frac{1}{p^{2.5}}$$

for some k_0 and therefore all packets can be transmitted to their targets in time $O(\log p \log \log p)$ with high probability. By combining the two phases we see that all packets can be routed in time $O(h + \log p \log \log p)$ with high probability. By choosing $h \in \Omega(\log p \log \log p)$ we complete the proof. $\qquad\square$

4 Experiments

We ran some experiments to get practical experience of the new algorithm, see Table 1. The results are averages of 500 experiments.

Table 1. Routing cost of the CT algorithm with $h = \log_2^2 p$ and $h_0 = \log_2 p$. Thinning parameters $t = 1.2, 1.5, 2.0$ were tried. For comparison, results of GGT with $h = \log_2^2 p$, $\epsilon = 0.25$, $\alpha = 0.1$ are presented.

p	$t = 1.2$	$t = 1.5$	$t = 2.0$	GGT
128	3.74	3.82	4.20	4.86
256	3.73	3.77	4.16	4.92
512	3.76	3.77	4.13	4.95
1024	3.81	3.74	4.10	4.97
2048	3.78	3.72	4.07	4.99
4096	3.79	3.70	4.05	5.00

By the results in Table 1, the constant thinning algorithm CT achieves the cost 3.7 with slackness $h = \log^2 p$. With a suitable combination of ϵ and α GGT comes below 5 but is clearly slower than CT. Note that $e = 2.7$ is the lower bound of the cost.

In addition to mere numbers, our routing simulator shows the progress of routing graphically (see Figure 1). In CT the throughput of packets is constant for most of the time, while GGT follows a saw blade pattern. The "tail" of the graph is very critical. Too little thinning or too small h_0 grows the tail, as the algorithm approaches to the greedy algorithm.

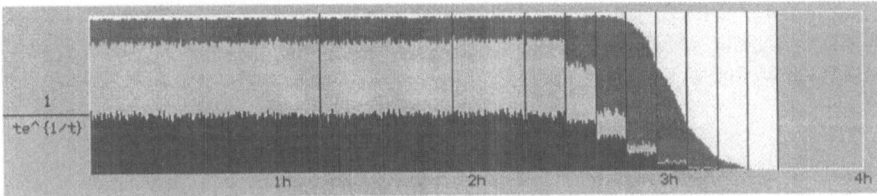

Fig. 1. Graphical output of a CT simulation in case $p = 1024$, $h = 128$, $h_0 = 16$, $t = t_0 = 1.2$. The picture is divided in three horizontal bands by ragged borderlines at $1/t = 83\%$ and at $1/te^{1/t} = 36\%$ of all processors. The bottom band represents successful processors, the middle band failed processors, and the top band passive processors. The vertical lines at intervals of cth, c^2th, ..., until $c^i th < h_0 = 16$ ($c = 1 - 1/e^{1/t_0}$) separate the phases of the algorithm. The total time in picture is $456 = 3.56h$.

References

1. R.J. Anderson and G.L. Miller. Optical communication for pointer based algorithms. Technical Report CRI-88-14, Computer Science Department, *University of Southern California*, LA, 1988.
2. A. Czumaj and F. Meyer auf der Heide, and V. Stemann. Shared memory simulations with triple-logarithmic delay. *Proc. ESA95*, 46–59, 1995.
3. M. Dietzfelbinger and and F. Meyer auf der Heide. Simple, efficient shared memory simulations. *Proc. SPAA93*, 110–119.
4. J. Håstad, T. Leighton, and B. Rogoff. Analysis of backoff protocols for multiple access channels. *SIAM J. Comput.* 25:740–774, 1996
5. M. Geréb-Graus and T. Tsantilas. Efficient optical communication in parallel computers. In *Proc. SPAA'92*, pp. 41 – 48, June 1992.
6. L.A. Goldberg, M. Jerrum, T. Leighton, and S. Rao. A doubly logarithmic communication algorithm for the completely connected optical communication parallel computer. In *Proc. SPAA'93*, pp. 300 – 309, June 1993.
7. L.A. Goldberg, Y. Matias, and S. Rao. An optical simulation of shared memory. In *Proc. SPAA'94*, pp. 257 – 267, June 1994.
8. L.A. Goldberg and P.D. MacKenzie, Analysis of Practical Backoff Protocols for Contention Resolution with Multiple Servers. *Proc. SODA'96*, pp. 554–563.
9. L.A. Goldberg and P.D. MacKenzie, Contention Resolution with Guaranteed Constant Expected Delay, *Proc. FOCS'97*, pp. 213–222.
10. T. Hagerup, C. Rüb. A guided tour of Chernoff bounds. *Information Processing Letters* 33:305–308, 1989.
11. R.M. Karp, M. Luby, and F. Meyer auf der Heide. Efficient PRAM simulation on a distributed memory machine. *Proc. STOC'92*, pp. 318–326.
12. A. Kautonen and V. Leppänen and M. Penttonen. Simulations of PRAM on Complete Optical Networks. In *Proc. EuroPar'96*, LNCS 1124:307–310.
13. M. Paterson and A. Srinivasan. Contention resolution with bounded delay. In *Proc. FOCS'95*, pp. 104–113.
14. P. Raghavan, and E. Upfal. Stochastic Contention Resolution With Short Delays. In *Proc. STOC'95*, pp. 229–237.
15. L.G. Valiant. General purpose parallel architectures. In *Handbook of Theoretical Computer Science, Vol. A*, 943–971, 1990.
16. L.G. Valiant. A bridging model for parallel computation. *Communications of the ACM*, 33:103-111, 1990.

NAS Integer Sort
on Multi-threaded Shared Memory Machines*

Thomas Grün[1] and Mark A. Hillebrand[1]

Computer Science Department, University of the Saarland
Postfach 151150, Geb. 45, 66041 Saarbrücken, Germany
(gruen@cs.uni-sb.de, mah@studcs.uni-sb.de)

Abstract. Multi-threaded shared memory machines, like the commercial Tera MTA or the experimental SB-PRAM, have an extremely good performance on the Integer Sort benchmark of the NAS Parallel Benchmark Suite and are expected to scale. The number of CPU cycles is an order of magnitude lower than the numbers reported of general purpose distributed memory or shared memory machines; even vector computers are slower. The reasons for this behavior are investigated. It turns out that both machines can take advantage of a fetch-and-add operation and that due to multi-threading no time is lost waiting for memory accesses to complete. Except for non-scalable vector computers, the Cray T3E, which supports fetch-and-add but not multi-threading, is the only parallel computer that could challenge these machines.

1   Introduction

One of the first programs that ran on a node processor of the Tera MTA (multi-threaded architecture) [2,3] was the Integer Sort (IS) from the Parallel Benchmark suite [8,14] of the Numerical Aerospace Simulation Facility (NAS) at NASA Ames Research Center. According to press releases of Tera Corporation [17], a single node processor of the Tera machine was able to beat a one-processor Cray T90, which hitherto held the record for one processor machines, by more than 30 percent. The SB-PRAM [1,9,12,15] has many features in common with the Tera MTA, although it is inspired by a totally different idea, viz realizing the PRAM model from theoretical computer science [18]. Both machines differ from most of todays shared memory parallel computers in two aspects: First, they implement the UMA (uniform memory access) model instead of the NUMA (non uniform memory access) scheme found in todays scalable shared memory machines like SGI Origin, Sun Starfire or Sequent Numa-Q. In other words, they do not employ data caches with a cache coherence protocol, but hide memory latency by means of multi-threading. Second, they provide special hardware for fetch-and-add instructions in the memory system. Commercial microprocessors, which are employed in most of todays parallel computers do not offer this kind

---

* This work was partly supported by the German Science Foundation (DFG) under contract SFB 124, TP D4.

of hardware support for running efficient parallel programs. An exception is the Cray T3E, which is a UMA machine with fetch–and–add support but without multi–threading.

The IS benchmark is part of the NAS (Numerical Aerospace Simulation Facility) Parallel Benchmark Suite (NPB) [8]. It models the ranking step of a counting sort (kind of bucket sort) application, which occurs for instance in particle simulations. The IS benchmark takes a list L of small integers as an input and computes for every element $x \in L$ the rank $r(x)$ as its position in the sorted list. On this benchmark, the best program on a 4–processor SB–PRAM spends 3.3 cycles per element (CPE) on the average, and the single–node Tera machine needs 2.4 cycles. Both machines are expected to scale without significant loss of efficiency. A Cray T90 node approaches 5 CPE and the numbers reported for the MPI–based IS sample code on various distributed and shared memory computers (including the Cray T3E) are greater than 25 CPE. Besides, scalability is a problem on these machines. In this article we explain, why this numbers differ in such a wide range and investigate whether there are better implementations than the sample code for CC–NUMA machines and the Cray T3E.

The paper is organized as follows. Section 2 discusses the IS benchmark specification. Section 3 addresses multi–threaded shared memory machines and Section 4 discusses a fast algorithm for vector computers. Section 5 presents results of the MPI–based message–passing sample code and Section 6 investigates CC–NUMA machines or the Cray T3E UMA machine could improve on these results. Section 7 concludes.

2 The NAS Integer Sort Benchmark

The Integer Sort (IS) program is part of the NAS Numerical Parallel Benchmark suite [8]. Despite its name it does not sort but rank an array of small integers. The first revision (NPB 1) was a "paper and pencil" benchmark specification, so as not to preclude any programming tricks for a particular machine. The benchmark specifies that an array keys [] of N keys is filled using a fully specified random number generator that produces Gaussian distributed random integers in the range $[0, B_{max} - 1]$. This key generation and a final verification step, which checks the results, are excluded from the timing. There are several "classes" of the benchmark, which define the parameter values N and B_{max}. We concentrate on class A ($N = 2^{23}, B_{max} = 2^{19}$); the parameters for class B and C are higher by a factor 4, and 16 respectively. The timed part of the benchmark consists of 10 iterations of (a) a modification of two keys, (b) the actual ranking step, and (c) a partial verification that tests if the well known rank of 5 keys has been computed correctly. A typical implementation of the inner loops that perform the actual ranking is listed in Table 1.

The NPB 2 revision of the benchmark intends to supplement the original benchmark description with sample implementations that can be used as a starting point for machine–specific optimizations. There is a serial implementation (NPB 2.3–serial) and an implementation based on the message passing standard

Table 1. Inner loops of the NAS IS benchmark

(1) for i = 1 to B: count[i] = 0	// clear count array
(2) for i = 1 to N: count[key[i]]++	// count keys
(3) for i = 1 to B: pos[i] = $\sum_{k=0}^{i-1}$ count[i]	// calc start position
(4) for i = 1 to N: rank[i] = pos[key[i]]++	// ONLY NPB 1

MPI (NPB 2.3b2). Unfortunately, the fourth loop of the benchmark, which was present in earlier sample implementations of NPB 1, has been omitted in NPB 2. However, the fourth loop is identical to the second loop except for storing the result, and does not require a new programming technique. Therefore, we use the NPB 2 variant of IS, class A throughout this paper.

Performance measures. The original performance measure of NAS IS is the elapsed time for 10 iterations of the inner loops. The main performance measure in our investigation is the number of CPU cycles per key element (CPE), because it emphasizes the architectural aspect rather than technology. We will point out situations in which CPE is an unfair measure.

3 Multi–threaded SMMs

First, we sketch the SB–PRAM design and highlight the Tera MTA differences. Then we describe the result of the best SB–PRAM implementation and indicate the algorithmic changes in the Tera MTA implementation.

3.1 Hardware Design

The SB–PRAM [1, 9, 12] is a shared memory computer with up to 128 processors and an equal number of memory modules, which are connected by a butterfly network. Processors send their requests to access a memory location via the butterfly network to the appropriate memory modules and receive an answer packet if the request was of type LOAD. There are no caches in the memory system. Three key concepts are employed to yield a uniform load distribution on the memory modules and to hide memory latency: (a) synchronous multi–threading of 32 virtual processors (VP) with a delayed load; (b) hashing of subsequent logical addresses to physical addresses that are spread over all memory modules; (c) butterfly network with input buffers and combining of packets. The combining facility is extended to do parallel prefix computations, which, due to synchronous execution, look to the programmer as if the VPs executed the operations one after another in a predefined order (sequential semantics).

The design ensures that the VPs are never stalled in practice, neither by network congestion on the way to the memory modules nor by answer packets that arrive too late. Although this issue has been intensively investigated in [7] and [19] using different and detailed simulators, the main criticism on the SB–PRAM is that these investigations were based on simulations only. Therefore, a 64 processor machine is being built as a proof of concept. At the time of this writing, the re–design of an earlier 4–processor prototype [4] has been completed. The 64 processor machine is expected to be running in summer 1998.

In [10] a variant of the SB–PRAM, called High Performance PRAM (HPP), has been sketched. Due to modest architectural changes and using top 1995 technology, the HPP is expected to achieve the tenfold performance of the SB–PRAM on average compiler generated programs.

Tera MTA. The design goal of the Tera MTA [2, 3, 17] was to build a powerful multi–purpose parallel computer with special emphasis on numerical programs. In order to meet this goal, it was determined early in the design phase to employ GaAs technology and liquid cooling. A processor chip runs at ≥ 294 MHz and has a power dissipation of 6 KW per processor.

Similar to the SB–PRAM, the Tera MTA hides memory latency by means of multi–threading. Unlike the SB–PRAM, it does not schedule a fixed number of VPs round robin, but it can support up to 128 threads, all of which can have up to 8 active memory requests. New threads can be created using a low overhead mechanism; the penalty for an instruction that attempts to use a register that is waiting for a memory request to arrive is only one cycle. The memory system of the Tera MTA is completely different from that of the SB–PRAM: the network topology is a kind of 3–dimensional torus. The messages are processed by a randomized routing scheme that may detour packets if the output link in the right direction is either overloaded or out of order. To our knowledge there is no detailed, technical description or a correctness proof of the network available to the public. The machine supports fetch–and–add instructions; the requests are not combined in the network, but serviced sequentially at the memory modules. Because the Tera threads may become asynchronous due to race conditions in the network, the machine does not offer the sequential semantics of the SB–PRAM.

3.2 Benchmark Implementation

Due to some limitations of our `gcc` compiler port for the SB–PRAM, we have hand–coded the loop bodies in assembly language and manually unrolled the inner loops of the benchmark. For details we refer to [11]. We now present how many instructions are necessary at least to implement the IS benchmark on the SB–PRAM. Hence, we hand–code the loop bodies in assembly language and neglect loop overhead.

As on every other machine, the count array can be cleared with a single "store with auto–increment" instruction, if every VP is assigned a different but contiguous portion of the count array. The arrays `count[]` and `pos[]` are used one at a time and can be coalesced into a single `cp[]` array, which saves a few address computations. Because there are no data dependencies between different iterations of loop 2, all `PROCS` virtual processors are mapped round robin onto the `keys` and `cp` arrays and synchronously execute loops 2 and 3. Two successive

(a) elem1 = M(k_ptr1+=2*PROCS);	elem2 = M(k_ptr2+=2*PROCS);
cp_ptr1 = elem1 + cp_base;	cp_ptr2 = elem2 + cp_base;
syncadd(cp_ptr1, 1);	syncadd(cp_ptr2, 1);
(b) elem1 = M(cp_ptr1+=2*PROCS);	elem2 = M(cp_ptr2+=2*PROCS);
elem1 = mpadd(ranksum, elem1);	elem2 = mpadd(ranksum, elem2);
M(cp_ptr1) = elem1;	M(cp_ptr2) = elem2;

Fig. 1. Interleaved loop bodies: (a) Loop 2, and (b) Loop 3.

loop bodies are interleaved in order to fill all load delay slots (see assembly code in Figure 1.a). The first instruction combines the incrementing of a properly initialized, private key pointer k_ptr by 2·PROCS with the loading of the next key. In the second instruction of the loop body, the pointer cp_ptr is computed as cp[] indexed by key, and the syncadd operation (mpadd without result) increments this entry. Figure 1.b lists the assembly code for the third loop, which also employs the interleaving technique. In the first instruction, the pointer cp_ptr is adjusted and an entry of cp[] is read. Then, the accumulated rank is computed in the multi-prefix add on ranksum, and the result is written back to the cp array. This loop relies on synchronous execution, because the order of mpadd instructions among VPs is crucial.

Loop 2 contributes most instructions to the timed portion of the code, because it is executed N times whereas loops 1 and 3 are executed only $B_{max} = \frac{N}{16}$ times. Thus, CPE(ranking) = $\frac{1}{16} + 3 + \frac{3}{16} = 3.25$ is optimal for the SB–PRAM. The benchmark implemented using assembly language macros and 64–fold unrolled loop achieves 3.30 CPE. A similar tuned C version is slower by roughly one cycle, because the compiler does not use "load with auto–add of 2*PROCS" but generates two separate instructions instead.

We have run the benchmark on our instruction–level simulator as well as on the real 4 processor machine. The run times differ by less than 0.1%, a fact that validates our network simulations which predicted that memory stalls are extremely unlikely. The run time obtained by a 128 processor simulation is only 1% slower than $\frac{1}{128}$th of the one processor run time, i.e. speedup is nearly linear. Because no caches are employed in the SB–PRAM, the run times of class B and C can be simply and accurately predicted; they are higher by a factor of 4, respectively 16.

The High–Performance–PRAM HPP would gain a factor of 5.6 in absolute speed, but the CPE value would be worse, because the machine can issue memory requests at only a third of the instruction rate. For a fairer comparison, the CPU clock should be replaced by the memory request rate.

Tera MTA. The Tera MTA system has a sophisticated parallelizing compiler, which processes the sequential source code and automatically generates threads when needed. The assembler output of loop 2, which is listed in [2], shows a 2 instruction loop body that is 5-fold unrolled. The memory operations are "load next key" and "fetch & increment count". An integer addition plus the loop

assume: V1 ← 0..63	
(a) 1 : V2 ← key[V1 + k]	(b) 2.1 : count[V2] ← V1
2 : V3 ← count[V2] + 1	2.2 : V4 ← count[V2] - V1
3 : count[V2] ← V3	2.3 : if(V4 ≠ 0) check

Fig. 2. Loop 2 vector code: (a) straight forward, (b) correction

overhead are hidden in the non–memory operations of the ten instructions of the loop body. Note, that the CPU clock and the memory request rate are equal in the super–scalar Tera MTA design.

Since the Tera MTA lacks synchronous execution, the third loop cannot be parallelized as on the SB–PRAM. Instead, every processor works on a contiguous block of the count array, computes the local prefix sums and outputs the global sum of its elements. In a second step, the prefix sum of the global sums is computed. Then, each processor adds the appropriate global prefix sum to its local elements. This procedure requires 3 instructions plus a small logarithmic term. For the class A benchmark it is less than 4 instructions. The Tera MTA requires less than $\frac{1}{16} + 2 + \frac{4}{16} = 2.3125$ CPE. The improvement compared to the SB–PRAM comes from the ability to execute one memory operation and two other (integer, jump, ...) operations per instruction.

We do not have access to a Tera MTA ourselves. The performance figures contained in this section are derived from [2,5] and personal communication with P. Briggs (Tera Computer Corporation) and L. Carter (SDSC). The numbers reported in [2] are 1.53 seconds and $5.25 < \text{CPE} < 5.3125$ for NPB 1. Assuming a CPE value of 5.25, the 294 MHz machine has an overhead of at least $\frac{1.53 \cdot 294 \cdot 10^6}{5.25 \cdot 10 \cdot 2^{23}} - 1 = 2.14$ %, i.e. hiding memory latency works well for the one processor prototype. The fourth inner loop of the benchmark, which is not present in NPB 2, accounts for 3 CPE. If we attribute all overhead to the three loops of IS(NPB 2) and conservatively assume CPE(NPB 2)=2.3125, we end up with a run time of < 0.68 seconds. Carter [5], who ran the NPB 2 benchmark on a 145 MHz machine, achieved a run time of 2.05 seconds, which corresponds to an overhead of 50 percent. He believes that on the early machine flipped bits in the network packets induced retransmissions, because the memory overhead figures for other benchmarks dropped as the hardware became more stable.

4 Vector Computers

In [5], a Tera MTA node is compared to a Cray T90 vector processor node on the NPB 2–serial sample code of IS. Vector processors have instructions for computing prefix sums on vector registers. Thus, the third benchmark loop can be parallelized in the same way as on the Tera MTA. The difficult part is loop 2, where even a single vector processor encounters problems.

Figure 2.a lists the straightforward, but wrong, implementation of loop 2. Successive, contiguous stripes of key[] are read into vector register V2. Then,

Table 2. IS (class A) benchmark results

Machine Name	CPU clock [MHz]	minimum			maximum		
		#procs	time	CPE	#procs	time	CPE
SB-PRAM	7	1	39.55	3.30	4	9.89	3.30
SB-PRAM simulation					128	0.31	3.33
Tera MTA	294	1	0.68	2.32			
Cray T90	440	1	1.11	5.82			
SB-PRAM (MPI)	7	2	160.8	26.85			
IBM SP2 WN	66	2	29.1	27.38	128	0.6	60.42
SGI Origin 2000	195	2	20.5	95.30	32	1.6	119.02
Cray T3E-900	450	2	12.7	136.26	256	0.4	258.15

V3 is filled by reading the corresponding count[] entries and incrementing them. Afterwards V3 is written back. If a specific key value v is contained twice in V2 at positions k_1 and k_2, count[v] is incremented only once, because V3[k_2], which finally ends up in count[v], contains count$_{old}$[v]+1 and not the incremented V3[k_1] value.

The Cray compiler can cure the situation using a patented algorithm [6] invented by Booth.[1] Before writing V3 back in line 3, the code listed in Figure 2.b tests whether duplicates occur, and, if necessary, processes them in the scalar unit. The run time of IS(NPB 2), as reported in [5], is 1.11 s, which corresponds to 5.82 CPE. The NPB 2.3–serial sample code contains an unnecessary store that possibly has not been detected by the compiler. Hence, the optimal CPE for a single vector computer could be lower by 1 CPE.

The above code for loop 2 can be easily parallelized by maintaining a private copy count$_i$[] of the count array on each node computer. When p denotes the number of participating processors, the third loop becomes a parallel prefix computation pos[i] $= \sum_{k=0}^{i-1} \sum_{l=0}^{p-1}$ count$_l$[k]. Memory consumption as well as the execution time of loop 3 scale with p. If p becomes very large, a two pass radix sort, like described by Zagha [20], can improve performance.

5 Comparison with MPI–based sample code

Table 2 lists the benchmark results discussed so far and compares them with run times of MPI–based sample code. We implemented the necessary MPI library functions on the SB–PRAM and optimized the sample code to the extent that an optimizing compiler could also achieve [11]. The times for the other machines have been taken from benchmark results published by NAS. The IBM SP2 is a distributed memory machine (DMM) with an extremely fast interconnection network, the SGI Origin 2000 a cache–coherent NUMA, and the Cray T3E a UMA shared memory machine (SMM) without cache coherence.

---

[1] Personal Communication with Larry Carter and Max Dechantsreiter.

Sample code implementation. The keys are distributed on all participating p computers before the timing starts. In a first local step, each computer sorts its keys using only the b most significant bits into 2^b buckets. Because the keys are put into a sorted order, this step alone requires more instructions than IS(NPB 1). Then, a mapping of buckets to processors, which balances the load, is computed. Afterwards, all local bucket parts are sent to their destinations in a single, global, all–to–all communication step. Now, every processor owns a distinct, contiguous key range and can count its keys locally. The pos[] entries can be computed from count[] like in the Tera MTA implementation.

The sample code algorithm is guided by the idea of doing one, central all–to–all communication. The price for this procedure is to rearrange the keys locally, such that keys that are sent to the same destination are stored contiguously in memory. A plus of the implementation is the good cache efficiency: the key arrays are accessed sequentially, so that the initial cache miss time is amortized over the complete cache line, and the count arrays fit into second level cache.

With less than $p = 16$ processors another strategy with fewer communication overhead would be possible: every processor could count its keys in a local count array and pos[] could be computed like on vector processors after transposing the $p \times B_{max}$ count matrix. If $p < 16$, then only $p \cdot B_{max} < 16 \cdot \frac{N}{16}$ data elements had to be sent over the network. However, the sample code is written for a large number of processors and slow communication networks.

The CPE of the SB–PRAM is 26.85, of which 9 cycles belong to memory instructions operating on key elements. The SB–PRAM issues one instruction per cycle and encounters no memory waits. The other machines have super-scalar processors, but suffer from memory waits. The impact of wait states on the CPE is the higher, the higher the clock rate of the specific machine is. Table 2 highlights this fact, which limits the benefit of cache–based architectures from future technological improvements compared to vector computers or multi–threaded machines.

6 Other Algorithms

The MPI–based sample code is tailored for distributed memory machines. We now investigate, if there are better algorithms for CC–NUMA machines or the Cray T3E.

CC–NUMA. Cache–coherent non–uniform–memory–access SMMs [13], like the SGI Origin, have overcome the argument "SMMs do not scale" by other means than the SB–PRAM or the Tera MTA. While multi–threading relies on having enough parallelism available to keep the processors busy, CC–NUMA machines try to minimize the *average* memory latency by caching, at the risk of running idle on cache misses.

We now investigate, if the direct approach of locking the count[] elements for updating could improve performance so much, that the results of multi–threaded SMMs could be reached. Therefore, we calculate the average memory latency introduced by accessing count[keys[i]] in loop 2. We assume that on a p-processor machine $\frac{1}{p}$ of the count[] elements are cached on every processor and the miss penalty is 20 cycles (i.e. 100 ns on a 200 MHz computer). Thus, the average memory latency is

$$t = \underbrace{\frac{1}{p} \cdot 1}_{\text{cache hit}} + \underbrace{\frac{p-1}{p} \cdot 20}_{\text{cache miss}}$$

On machines with more than $p = 16$ processors, t is greater than $18\frac{13}{16}$. In other words, the exchange of count[] entries between the coherent caches is rather expensive. If the remaining cycles for the program are taken into account, the CPE is at least an order of magnitude higher than on multi–threaded SMMs with fetch–and–add.

Cray T3E. A Cray T3E node computer consists of a DEC Alpha processor with up to 2 GByte local RAM. Additionally, the system logic provides a notion of a logically addressable shared memory through the E–registers [16]. Up to 512 E–registers can be accessed by the user to trigger shared memory accesses. In addition to load and store, E–registers also support fetch & add and an especially fast fetch & increment operation.

Although the machine does not support multi–threading in hardware, memory wait conditions can be avoided by using multiple E–registers to simulate multi–threading in software. With this technique, alternate key load and count increment operations can be performed at the maximal network clock of 75 MHz. The third loop of IS can also be parallelized in the same way as on the Tera MTA. Thus, less than 3 network cycles per key element are possible.

7 Conclusion

We have investigated the performance of the IS benchmark on the SB–PRAM and the Tera MTA, two scalable SMMs which employ multi–threading to hide memory latency instead of caching like most of todays computers. Besides multi–threading, both machines take advantage of a fetch–and–add instruction to update shared data structures conflict–free. These two properties lead to a performance that is an order of magnitude higher compared to published results of other scalable parallel computers, both DMMs and SMMs.

In particular, the experimental 7 MHz SB–PRAM prototype can compete with modern parallel computers on this benchmark. The Tera MTA makes use of expensive top technology, and achieves better performance figures than non-scalable top vector processors. Although CC–NUMA SMMs could possibly improve performance compared to DMMs, they are still an order of magnitude slower than multi–threaded machines. Alone the Cray T3E, which supports and atomic fetch & increment operation could challenge the performance of the Tera MTA by emulating multi–threading in software.

Acknowledgments. The authors would like to thank Larry Carter, Max Dechants-reiter, and Preston Briggs for helpful discussions and all people who contributed to the SB–PRAM project [15].

References

1. F. Abolhassan, R. Drefenstedt, J. Keller, W. J. Paul, and D. Scheerer. On the Physical Design of PRAMs. *Computer Journal*, 36(8):756–762, December 1993.
2. R. Alverson, P. Briggs, S. Coatney, S. Kahan, and R. Korry. Tera Hardware–Software Cooperation. In *Proc. of Supercomputing '97*. San Jose, CA, November 1997.
3. R. Alverson, D. Callahan, D. Cummings, B. Koblenz, A. Porterfield, and B. Smith. The TERA Computer System. In *Proc. of Intl. Conf. on Supercomputing*, June 1990.
4. P. Bach, M. Braun, A. Formella, J. Friedrich, T. Grün, and C. Lichtenau. Building the 4 Processor SB–PRAM Prototype. In *Proc. of the 30th Hawaii International Conference on System Sciences*, pages 14–23, January 1997.
5. J. Boisseau, L. Carter, K.S. Gatlin, A. Majumdar, and S. Snavely. NAS Benchmarks on the Tera MTA. In *Proc. of Workshop on Multi-Threaded Execution, Architecture and Compilation (M-TEAC 98)*, Las Vegas, February 1998.
6. M. Booth. US Patent 5247696. see http://www.patents.ibm.com, September 1993.
7. C. Engelmann and J. Keller. Simulation–based comparison of hash functions for emulated shared memory. In *Proc. PARLE (Parallel Architectures and Languages Europe)*, pages 1–11, 1993.
8. D. Bailey et al. The NAS Parallel Benchmarks. RNR Technical Report RNR-94-007, NASA Ames Research Center, March 1994. see also http://science.nas.nasa.gov/Software/NPB.
9. A. Formella, T. Grün, and C.W. Kessler. The SB–PRAM: Concept, Design and Construction. In *Draft Proceedings of 3rd International Working Conference on Massively Parallel Programming Models (MPPM-97)*, November 1997. see also http:// www-wjp.cs.uni-sb.de/~formella/ mppm.ps.gz.
10. A. Formella, J. Keller, and T. Walle. HPP: A High–Performance–PRAM. In *Proceedings of the 2nd Europar*, volume II of *LNCS 1124*, pages 425–434. Springer, August 1996.
11. T. Grün and M. A. Hillebrand. NAS Integersort on the SB–PRAM. Manuscript, available via http://www-wjp/~tgr/NASIS, May 1998.
12. T. Grün, T. Rauber, and J. Röhrig. Support for Efficient Programming on the SB–PRAM. *International Journal of Parallel Programming*, 26(3):209–240, June 1998.
13. D.E. Lenoski and W.-D. Weber. *Scalable Shared-Memory Multiprocessing.* Morgan Kaufmann Publishers, 1995.
14. NAS Parallel Benchmarks Home Page. http:// science.nas.nasa.gov/Software/NPB/.
15. SB–PRAM Home Page. http:// www-wjp.cs.uni-sb.de/sbpram.
16. S. L. Scott. Synchronization and Communication in the T3E Multiprocessor. In *Proc. of the VII ASPLOS (Architectural Support for Programming Languages and Operating Systems) Conference*, pages 26–36. ACM, October 1996.
17. Tera Computer Corporation Home Page. http://www.tera.com/.

18. J. van Leeuwen, editor. *Handbook of Theoretical Computer Science*, volume A, pages 869–941. Elsevier, 1990.

19. T. Walle. *Das Netzwerk der SB–PRAM*. PhD thesis, University of the Saarland, 1997. in German.

20. M. Zagha and G.E. Belloch. Radix Sort for Vector Multiprocessors. In *Proc. of Supercomputing '91*, pages 712–721, New York, NY, November 1991.

Analysing a Multistreamed Superscalar Speculative Instruction Fetch Mechanism

Rafael R. dos Santos* and Philippe O. A. Navaux**

Informatics Institute
CPGCC/Federal University of Rio Grande do Sul,
P.O. Box 15064 91501-970 Porto Alegre - RS, Brazil

Abstract. This work presents a new model for multistream speculative instruction fetch in superscalar architectures. The performance evaluation of a superscalar architecture with this feature is presented in order to validate the model and to compare its performance with a real superscalar architecture. This model intends to eliminate the instruction fetch latency introduced by branch instructions in superscalar pipelines. Finally, some considerations about the model are presented as well as suggestions and remarks to future works.

1 Introduction

Even using accurated branch prediction mechanisms, current superscalar architectures offer less performance than an ideal architecture.

When a branch is encountered, the branch predictor can predict it as to be taken or not taken. In the first case, contiguous instructions are already into the fetch stage when the prediction is made, and these instructions do not take part of the predicted path. For the second case, nothing occurs if the prediction is correct. But for both cases, we are assuming that the branch prediction mechanism is efficient.

The problem usually is that branches are predicted as to be taken and each time this occur a flow interruption also occurs. The time requiered to refill the fetch buffer and put the correct instructions into the instruction queue is greater than the necessary time to execute these instructions. If there are many functional units and flow interruptions occur frequently, it should be expected that the instruction queue will be empty for many cycles.

This means that the instruction fetch mechanism must be designed to keep the instruction queue with instructions that can be scheduled for execution on free functional units.

A superscalar architecture could not execute instructions faster than it can fetch instructions from the instruction cache. So, design an efficient fetch mechanism is more important than increase the number of resources, because it is not

---

* Ph.D. Student – E-mail: rrsantos@inf.ufrgs.br - UNISC - University of Santa Cruz do Sul - Santa Cruz - RS - Brazil

** Ph.D. – E-mail: navaux@inf.ufrgs.br - CPGCC/Federal University of Rio Grande do Sul

possible to increase the performance only by increasing the number of functional units.

The question is how many instructions could be fetched per cycle? Where are these instructions comming from? An alternative approach to branch prediction is instead of predict whether the branch is taken or not taken, is execute speculatively both the taken and not-taken paths and cancel the execution of the incorrect path as soon as the branch result is known [3].

2 Fetching instructions from multiple streams

Talcott [9] reffers a technique called *Fetch Taken and not-Taken Paths* to reduce the branch problem. In [1] was presented a new model based on this scheme. In this model, both paths of a conditional branch are fetched and put into the fetch buffer.

When the branch prediction stage transfers instructions, from the fetch buffer to the instruction queue, and reaches a branch into the fetch buffer, both possible paths of this branch and its instructions are already into the fetch buffer. In this case, no delay is introduced when the branch is predicted as taken. The transfer is redirected to the predicted path whithout delay.

To enable this operation, the fetch stage must detect the branch instruction just when it is fetched. In the next cycle, it must also start to fetch from both paths. When the branch is predicted, the instructions transfer to instruction queue is not interrupted.

In the multistreamed superscalar architecture proposed in [6], the fetch stage was modified to enable to fetch both paths of a branch instruction.

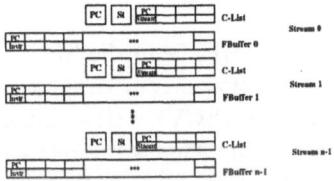

Fig. 1. Fetch Buffer Structure (e.g. fetch depth equal to 4 streams)

The figure 1 suggests a new buffer structure in the multistream pipeline. The number of stream buffers defines the fetch depth. Each stream has four independent elements: Program Counter, Status Bit, Children List and Fetch Buffer.

In the *Fetch Buffer* are stored the fetched instructions. The *PC* is used to point the instructions that are in the stream. The *status bit* indicates whether the stream structure is busy and the *Children-List* stores the identification of the children streams.

The *Children-List* also stores the branch address and the identifier of its children streams allowing the predict stage to start the transfer of instructions of the new stream when a branch is predicted, without any delay.

2.1 Multistream Architecture Operation

The *Fetch* stage fetches instructions to put them into the fetch buffer. When conditional branch instruction is detected, a new stream is generated and initialized, as if there were available resources.

The stream generation consist of *Children-List* updating operation with the branch address and the identification of a new stream structure that will store the instructions related to the new stream. For each branch detected there are two possible paths. The instructions that are in the not taken path are fetched and stored into the same stream structure where is stored the conditional branch. But, the taken path and their instructions will be stored in this new stream structure.

The stream structure initialization consists of the *PC* initialization with the target address and the setting of the *status bit*, to indicate that the structure was allocated.

The predict stage transfers instructions, from the fetch buffer to the instruction queue (like conventional superscalar architectures) looking for branch instructions. However, when it finds a branch instruction, it also makes a prediction. When the prediction is to be taken, this stage just concatenates the instructions which are in the children stream of this branch, discarding the closest instructions.

When the prediction is not taken, the children stream is discarded and the neightbouring instructions continue to be transfered to the instruction queue. When a stream is discarded, all children streams originated by this stream are also discarded, through a recursive operation.

3 Experimental Framework

For this work, we used 4 benchmarks from the SPECint95 suite (compress, go, ijpeg, li). Also, we used a execution-driven simulator to generate traces. This simulators acomplish with the SPARCV7 instruction set and simulates the execution in a scalar pipelined fashion. This is the reference machine.

For the superscalar simulations, we used 2 trace-driven simulators which executed the choosed benchmarks based on the traces generated by the first simulator. These 2 simulators are respectively called *Real* and *Mulflux* as we will describe below:

- Real Simulator: Simulates the execution of programs using two-level branch prediction [4, 5].
- Mulflux Simulator: Simulates the execution of programs using the proposed speculative instruction fetch mechanism [6].

4 Performance Analysis

In this section, we analize the performance of the multistreamed model based on the data extracted through several simulations. The experiments simulate different configurations of the real machine and the multistreamed machine. The tests started with a fetchwidth equal to 2 up to 8 instructions per cycle.

We can point out that in all situations the cycles with no dispatch are less than for the real machine. The no dispatch decreases with the increase of fetchwidth in the real machine. The percentage decreases from 40.50%, for fetchwidth equal to 2, to 39.30% in configurations with a fetchwidth equal to 8 instructions, as showed in the figure 2.

In the Mulflux machine, this percentage increases join to the increase of fetchwidth. This occur because the mispredictions and resources conflict increase too. Other studies have been developed to research more accurated branch predictors and the ideal balancing of the architectures resources to support a new and more powerfull parallelism through the pipeline. The sum of the three components, which causes no dispatch in both machines, correspond to in the percentage of cycles with no dispatch.

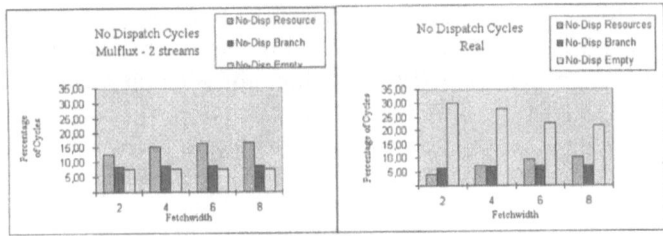

Fig. 2. No Dispatch in Mulflux with 2 Streams and Real Architecture

The divergency between the components is important in the Real machine. The occurency of an empty queue is the critical component that contributed to the existency of no dispatch cycles. This is not the case for the Mulflux machine. It is predictable that when there are more instructions ready to dispatch, the number of functional units becomes the main problem. This is true in the Mulflux architecture.

The occurency of empty queue in the Real machine decrease from 30.05%, with fetchwidth equal to 2 to 21.79%, with fetchwith equal to 8 instructions per cycle. In the Mulflux, the results obtained are 7.73% and 7.80% for the same configurations perhaps applying multiple streams.

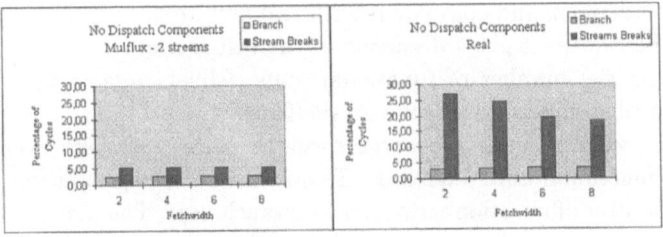

Fig. 3. Components that Causes Emptying Queue in Mulflux with 2 Streams and Real Architecture

The figure 3 presents the efficiency of the multistream model to reduce the empty queue occurrency. In the Mulflux, the stream interruptions are up to 5.24%, but in the Real architecure this percentage reach 27.07%.

The multistreamed model enables a reduction around 74.24% of empty queue occurrency. The effect of stream interruptions were reduced by 80.64% in the Mulflux machine. Another important aspect is that resource conflict and speculation depth must be considered with more attention when the multistream model is used.

4.1 Impact Analysis of Multistream Implementation

Before the implementation of the multistream model, we must consider the impact in the close blocks, and also the fetchwidth, resource conflicts and speculation depth.

In the previous results we did not consider the instruction cache misses. The use of the multistream model could generate more cache misses and the fetch latency can be important. So, the potential of the mechanism can be reduced. Then, we performed some experiments using a instruction cache with the same delay cycles as those of the Intel Pentium to observe the performance under real conditions. In this cache, the miss latency is equal to 3 cycles for the L1 cache, and 13 cycles when the miss causes an access to the L2 cache.

Also, for the last results we consider that the fetchwidth was multiplied by the number of valid stream structures. If the fetchwidth is equal to 8 instructions and the number of valid streams (initialized structures) is equal to 4 then the total and the real fetchwidth is equal to 32 instructions per cycle. This was made in the last experiments.

To avoid problems with the cache size and its configuration we have proposed a new strategy called *Dynamic Split Fetch - (DSF)* which consists in splitting the total fetchwidth between the valid stream structures [6]. In this case, if there are 4 valid streams and the fetchwidth is equal to 8 instructions per cycle, will be fetched 2 instructions for each valid stream and the fetchwidth is kept up to 8 instructions.

4.2 Overall Performance

In this section we discusse the overall performance delivered by the multistreamed mechanism and compare it with real architectures performance. Thus, we could observe the aspects discussed in the last section.

Increasing the number of functional units delivers more speed up for the multistreamed architecture as show in the figure 4.

However, we can observe that no dispatch cycles increases even when the number of functional units increases. This is due to the speculation depth that was kept for all configuration with a single branch unit. The machine considered has n generic functional units and only one branch unit that could stall the dispatch of instructions when saturated. This is showed in the graphic 5.

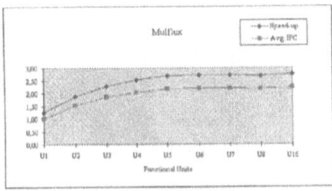

Fig. 4. Multistream Speed up when the Number of Functional Units Increase

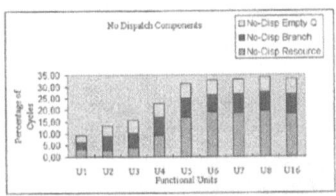

Fig. 5. Components of No Dispatch when Functional Units Increases

We made several simulations variating the number of functional units for the multistreamed architecture. We wanted to show that the resource conflict is the main factor in the limitation of the performance in this machine like in an ideal machine.

We performed experiments with 5 machines consisting of: a Multistream with perfect icache, a Multistream with normal icache, a Multistream with normal icache and DSF (*Dynamic Split Fecth*), a Real machine with perfect icache and a Real machine with normal icache.

The results that will be presented comes from the same configurations for each machine [6]. We used 2 streams structures to the multistream machines. All machines have a fetchwidth equal to 8 instructions per cycle, and a fetch buffer and instruction queue with 16 and 32 entries respectively; a dispatchwidth equal to 8 instructions; 8 functional units, each one with 8 reservation stations; 1 branch unit with 8 reservation stations; 8 bus results and reorder buffer with 64 entries.

The figure 6 shows the no dispatch cycles expended for each machine. The best case is the multistream machine with perfect icache and using a total fetchwidth that consists in multiplying the fetchwidth by the number of valid stream structures. We could observe that the Mulflux with normal icache has a percentage of no dispatch cycles similar to the Mulflux with normal icache using DSF. The division of the fetchwidth by each valid stream do not harm the performance of the considered cases. The Real machines expend more cycles with no dispatch.

The figure 7 shows the components that cause no dispatch cycles in each machine. In the mulstistream machines, the worst component is the resource conflict (around 45.00% of the no dispatch cycles). In the Real machines the emptying queue result around 60% of the occurency of no dispatch. In the second

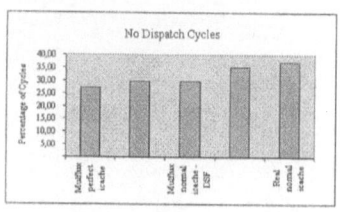

Fig. 6. Percentage of No Dispatch

figure 7, we could notice the reduction of the stream interruptions in multistream machines.

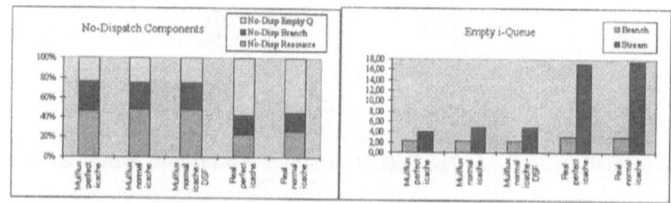

Fig. 7. No Dispatch Components and Emptying Queue Components

5   Conclusions and Future Works

The superscalar speed up is directly proportional to its IPC (*Instructions per Cycle*). It is desirable to obtain a speed up proportional to the number of functional units present in the architecture. However, the constant flow interruptions flush the instruction queue and decrease the number of ready instructions which could be dispatched. Thus, the IPC is reduced drastically because of the instruction queue flushing and a desirable speed up could not be achieved.

The multistreamed model allow a reduction around 74.24% of empty queue occurrency. The effect of stream breaks was reduced by 80.64% in the Mulflux machine. Another important aspect is that resource conflict (structural hazards) and speculation depth must be considered carefully when the multistream model is used.

In our experiments, the instruction cache performed similar performance in both cases: multistream and real architectures. The use of *Dynamic Split Fetch* brings a worthwhile strategy that allows to keep the number of icache buses.

Even if we reduce the occurency of empty instruction queue, we have verified that the decrease of no dispatch cycles do not decrease as we wanted. In the *Mulflux* machine the no dispatch cycles is between 28.99% and 33.11%, while in the *Real* machine it is between 39.30% and 40.50%. The increase of instructions flow in the instruction queue generate a major resource conflict like in the ideal architecture. Increasing the number of functional units but keeping the speculation depth did not allow good results in our experiments. Because of this, we are looking for resource balancing and ideal speculation depth in multistreamed architectures. The saturation of branch unit could come late the instructions execution.

Remarks to the next step could be pointed here. We are looking to introduce new instruction cache mechanisms in our simulations. Such mechanisms have been proposed and we believe that they will be used in next generations of superscalar microprocessors. Also, after to get a good configuration of our architecture we plan to compare it with other alternatives like trace processors [8], simultaneous multithreading [2], multiscalar [7] and other alternatives which certainly will be suggested.

Such comparison is very important to get an idea about the potential of such architecture and its complexity of implementation. The fourth generation of microprocessors can not be predicted at the moment but many new schemes have been proposed. The question is how to increase the performance of current microprocessors and how to obtain more machine parallelism using efficiently the increasingly chip density ?

References

1. Chaves Filho,Eliseu M. et al. A Superscalar Architecture with Multiple Instruction Streams. In: SBAC-PAD, VIII. Recife, Agosto 1996. **Proceedings**.... SBC/UFPE, 1996, pp 67-77 (in portuguese).
2. Eggers, Susan J. et al. Simultaneous Multithreading: A Plataform for Next-Generation Processors. *IEEE Micro*, V.17, n.5, Sep/Oct 1997.
3. Fromm; Richard. Branching on Superscalar Machines: Speculative Execution of Multiple Branch Paths Project Final Report. December 11, 1995. CS 252:Graduate Computer Architecture.
4. Yeh, Tse-Yu; Patt, Yale N. Two-Level Adaptive Training Branch Prediction. In: ANNUAL INTERNATIONAL SYMPOSIUM ON MICROARCHITECTURE, 24., 1991. **Proceedings...** New York: ACM, 1991. p. 51-61.
5. Yeh, Tse-Yu; Patt, Yale N. A Comparison of Dynamic Branch Predictors that use Two Levels of Branch History. In: ANNUAL INTERNATIONAL SYMPOSIUM ON COMPUTER ARCHITECTURE, 20., 1993. **Proceedings...** New York: ACM, 1993. p. 257-266.
6. Santos, Rafael R. dos. *A Mechanism for Multistreamed Speculative Instruction Fetch.* Porto Alegre: CPGCC/UFRGS, 1997 (M.Sc. Thesis - in portuguese).
7. Sohi, G. S.; Breach, S. E.; Vijaykumar, T. N. Multiscalar Processors. *Computer Architetcure News*, New York, v.23, n.2, p. 414-425, 1995.
8. Smith, James E,; Vajapeyam, Sriram. Trace Processors: Moving to Fourth-Generation. **Computer**, Los Alamitos, v.30, n.9, p.68-74, Sep. 1997.
9. Talcott, Adam R. *Reducing the Impact of the Branch Problem in Superpipelined and Superscalar Processors.* Santa Barbara: University of California, 1995 (Ph. D. Thesis).

Design of Processor Arrays
for Real-Time Applications

Dirk Fimmel   Renate Merker

Department of Electrical Engineering *
Dresden University of Technology
fimmel,merker@iee1.et.tu-dresden.de

Abstract. This paper covers the design of processor arrays for algorithms with uniform dependencies. The design constraint is a limited latency of the resulting processor array. As objective of the design the minimization of the costs for an implementation of the processor array in silicon is considered.

Our approach starts with the determination of a set of proper linear allocation functions with respect to the number of processors. It follows the computation of a uniform affine scheduling function. Thereby, a module selection and the size of partitions of a following partitioning is determined. A proposed linearization of the arising optimization problems permits the application of integer linear programming.

1 Introduction

Processor arrays represent an appropriate kind of co-processors for time consuming algorithms. Especially in heterogeneous systems a dedicated hardware for parts of algorithms, e.g. for the motion estimation of MPEG, is required. An admissible latency L_g for the selected part of the algorithm can be derived from the time constraint of the entire algorithm. The objective of the design is a processor array with minimal hardware costs that matches the admissible latency L_g.

In this paper, we consider algorithms with uniform dependencies. First, a set of linear allocation functions that lead to a small number of processors is computed. Then, for each allocation function a uniform affine scheduling function is determined. We assume that a LSGP (locally sequential and globally parallel) -partitioning can be applied to the processor array. The size of the partitions is derived with respect to the constraint that the latency of the resulting processor array is less than L_g. Furthermore, the processor functionality, i.e. kind and number of modules that have to be implemented in each processor, is determined. The objective of the design is a minimization of hardware costs. We measure hardware costs by the chip area needed to implement modules and registers in silicon.

---

* The research was supported by the "Deutsche Forschungsgemeinschaft", in the project A1/SFB358.

Related works handling the allocation function cover (1) a limited enumeration process to determine a linear allocation function leading to a minimal number of processors of the processor array [15], (2) the determination of a set of linear allocation functions that match a given interconnection network of the processor array [16], (3) the inclusion of a limited radius of the interconnections into the determination of the allocation function [14] and (4) the minimization of the chip area of the processor array by consideration of the processor functionality [3]. An approach to compute a variety of linear allocation and scheduling functions is proposed in [11]. Some notes to the determination of unconstrained minimal scheduling functions for algorithms with uniform dependencies can be found in [1]. Resource constraint scheduling for a given processor functionality is presented in [13]. An approach to minimize the throughput by consideration of the chip area is proposed in [9]. In [2] the approach [13] is extended to determine additionally the processor functionality in order to minimize a chip area - latency product.

The paper is organized as follows. Basics of the design of processor arrays are given in section 2. In section 3 hardware constraints considered in this paper are introduced. A linear program to determine a set of linear allocation functions is presented in section 4. Section 5 covers the determination of scheduling functions. A linear programming approach is presented in detail. Finally, a short conclusion is given in section 6.

2 Design of Processor Arrays

In this paper we restrict our attention to the class of algorithms that can be described as systems of uniform recurrence equations (SURE) [5].

Definition 1 (System of uniform recurrence equations). *A system of uniform recurrence equations is a set of equations S_i of the following form:*

$$S_i: \quad y_i[\mathbf{i}] = F_i(\cdots, y_j[f_{ij}^k(\mathbf{i})], \cdots), \quad \mathbf{i} \in \mathcal{I}, \quad 1 \le i, j \le m, 1 \le k \le m_{ij}, \quad (1)$$

where $\mathbf{i} \in \mathbb{Z}^n$ is an index vector, $f_{ij}^k(\mathbf{i}) = \mathbf{i} - \mathbf{d}_{ij}^k$ are index functions, the constant vectors $\mathbf{d}_{ij}^k \in \mathbb{Z}^n$ are called dependence vectors, y_i are indexed variables and F_i are arbitrary single valued operations. All equations are defined in the index space $\mathcal{I}$ being a polytope $\mathcal{I} = \{\mathbf{i} \mid \mathbf{H}_i \mathbf{i} \ge \mathbf{h}_{i0}\}$, $\mathbf{H}_i \in \mathbb{Q}^{m_i \times n}, \mathbf{h}_{i0} \in \mathbb{Q}^{m_i}$.

We suppose that the SURE has a single assignment form (every instance of a variable y_i is defined only once in the algorithm) and that there exists a partial order of the instances of the equations that satisfies the data dependencies.

Next, we introduce a graph representation of the data dependencies of the SURE.

Definition 2 (Reduced dependence graph (RDG)). *The equations of the SURE build the m nodes $v_i \in \mathcal{V}$ of the reduced dependence graph $\langle \mathcal{V}, \mathcal{E} \rangle$. The directed edges $e = (v_i, v_j) \in \mathcal{E}$ are the data dependencies weighted by the dependence vectors $\mathbf{d}(e) = \mathbf{d}_{ij}^k$. Source and sink of an edge $e \in \mathcal{E}$ are called $\sigma(e)$ and $\delta(e)$ respectively.*

The main task of the design of processor arrays is the determination of the time and the processor when and where each instance of the equations of the SURE has to be evaluated. In order to keep the regularity of the algorithm in the resulting processor array, we apply only uniform affine mappings [8] to the SURE.

Definition 3 (Uniform affine scheduling). *A uniform affine scheduling function* $\tau_i(\mathbf{i})$ *assigns an evaluation time to each instance of the equations:*

$$\tau_i : \mathbb{Z}^n \to \mathbb{Z} : \quad \tau_i(\mathbf{i}) = \boldsymbol{\tau}^T \mathbf{i} + t_i, \quad 1 \leq i \leq m, \tag{2}$$

where $\boldsymbol{\tau} \in \mathbb{Z}^n, t_i \in \mathbb{Z}$.

Definition 4 (Linear processor allocation). *A linear allocation function* $\pi(\mathbf{i})$ *assigns an evaluation processor to each instance of the equations:*

$$\pi : \mathbb{Z}^n \to \mathbb{Z}^{n-1} : \quad \pi(\mathbf{i}) = \mathbf{S}\mathbf{i}, \tag{3}$$

where $\mathbf{S} \in \mathbb{Z}^{n-1 \times n}$ *is of full row rank. Since* $\mathbf{S}$ *is of full row rank, the vector* $\mathbf{u} \in \mathbb{Z}^n$ *which is coprime and satisfies* $\mathbf{S}\mathbf{u} = \mathbf{0}$ *and* $\mathbf{u} \neq \mathbf{0}$ *is uniquely defined and called projection vector.*

Because of the lack of space we refer to [2, 3] for a treatment of uniform affine allocation functions $\pi_i : \mathbb{Z}^n \to \mathbb{Z}^{n-1} : \quad \pi_i(\mathbf{i}) = \mathbf{S}\mathbf{i} + \mathbf{p}_i, 1 \leq i \leq m$.

The importance of the projection vector is due to the fact that those and only those index points of an index space lying on a line spanned by the projection vector $\mathbf{u}$ are mapped onto the same processor. Due to the regularity of the index space and the uniform affine scheduling function, the processor executes the operations associated with that index points one after each other if $\boldsymbol{\tau}^T \mathbf{u} \neq 0$ with a constant time distance $\lambda = |\boldsymbol{\tau}^T \mathbf{u}|$ which is called iteration interval.

The application of a scheduling and an allocation function to a SURE results in a so called fullsize array.

3 Hardware Description

We consider a given set $\mathcal{M}$ of modules which are responsible to evaluate the operations of a processor. Instead of assuming given processors we want to determine modules which realize the operations of the processors. First, we introduce some measures needed to describe the modules. To each module $m_l \in \mathcal{M}$ we assign an evaluation time $d_l \in \mathbb{Z}$ in clock cycles needed to execute the operation of module m_l, a necessary chip area $c_l \in \mathbb{Z}$ needed to implement the module in silicon and the number $n_l \in \mathbb{Z}$ of instances of that module which are implemented in one processor. If a module $m_l \in \mathcal{M}$ has a pipeline architecture we assign a time offset o_l to that module which determines the time delay after that the next computation can be started on this module, otherwise $o_l = d_l$. Some modules are able to compute different operations, i.e. a multiplication unit is likewise able to compute an addition. To such modules different delays d_{li} and offsets o_{li} depending on the operations F_i are assigned.

The assignment of a module $m_l \in \mathcal{M}$ to an operation F_i is denoted as $m(i)$, and the set of modules which are able to perform the operation F_i is $\mathcal{M}_i$. The addressing of the instance of the module $m(i)$ which performs the operation F_i is given by $u_i \in \mathbb{Z}$.

4 Determination of Allocation Functions

The allocation function maps each index vector $\mathbf{i} \in \mathcal{I}$ to a processor $\mathbf{p}$ of the processor space $\mathcal{P} = \{\mathbf{p} \mid \mathbf{p} = \mathbf{Si} \wedge \mathbf{i} \in \mathcal{I}\}$. Our aim is the determination of a set of proper linear allocation functions that lead to processor spaces with a small number of processors.

In our approach we approximate the number of processors of the processor space $\mathcal{P}$ by the number of processors of the enclosing constant bounded polytop (cb-polytop) $\mathcal{Q} = \{\mathbf{p} \mid \mathbf{p}_{min} \leq \mathbf{p} \leq \mathbf{p}_{max}\}$ of $\mathcal{P}$, where $\mathcal{P} \subseteq \mathcal{Q}$, and each face of $\mathcal{Q}$ intersects $\mathcal{P}$ at least in one point. The consideration of the cb-polytop $\mathcal{Q}$ allows the formulation of the search for allocation functions as a linear optimization problem.

Program 1 (Determination of allocation functions)
for $j = 1$ to $n - 1$

$$
\begin{aligned}
& minimize\ v_j^1 - v_j^2 + 1, && v_j^1, v_j^2 \in \mathbb{Q}, \\
& subject\ to\ v_j^2 \leq \mathbf{s}_j^T \mathbf{w}_l \leq v_j^1, && 1 \leq l \leq |\mathcal{W}|, \mathbf{s}_j \in \mathbb{Z}^n, \\
& \qquad \mathbf{s}_j^T \mathbf{b}_k + (1 - r_k)R \geq 1, && 1 \leq k \leq n - j + 1, \quad (4.1) \qquad (4) \\
& \qquad \sum_{k=1}^{n-j+1} r_k = 1, && r_k \in \{0, 1\},
\end{aligned}
$$

end for
where $\mathcal{W}$ is the set of vertices $\mathbf{w}_l$ of $\mathcal{I}$, R is a sufficiently large constant and the vectors $\mathbf{b}_k$, $1 \leq k \leq n - j + 1$, are spanning the right null space of the matrix $(\mathbf{s}_1, \cdots, \mathbf{s}_{j-1})^T$. Constraint (4.1) is replaced by $\mathbf{s}_1 \neq \mathbf{0}$ for $j = 1$.

Constraint (4.1) ensures that the vectors $\mathbf{s}_j$, $1 \leq j \leq n - 1$, are linearly independent, i.e. that $rank(\mathbf{S}) = n - 1$. Constant R has to fulfill $R \geq \max_{1 \leq k \leq n-j+1} \{|\mathbf{s}_j^T \mathbf{b}_k|\}$.

The number of processors of the cb-polytop $\mathcal{Q}$ is $\prod_{j=1}^{n-1} (v_j^1 - v_j^2 + 1)$.

A motivation of our approach is given in the following theorem.

Theorem 1 (Number of processors of an enclosing cb-polytop). *If N (N') is the number of processors of the enclosing cb-polytop of the processor space resulting after application of the allocation function defined in program 1 (of another arbitrary linear allocation function), then $N \leq N'$.*

Since we are interested in several allocation functions we replace constraint (4.1) in program 1 for $j = 1$ by $\mathbf{s}_1^T \mathbf{u}_l \neq 0$, $1 \leq l < i$, in order to determine the i-th allocation function, where $\mathbf{u}_l$ are the projection vectors of the previous determined allocation functions.

Example 1.
We consider a part of the GSM speech codec system as example. The considered algorithm consists of two equations.

$$\begin{aligned}
&\text{I.} &&y_1[i,k] = y_1[i-1,k] + r[i]y_2[i-1,k-1], &&(i,k)^T \in \mathcal{I}, \\
&\text{II.} &&y_2[i,k] = y_2[i-1,k-1] + r[i]y_1[i-1,k], &&(i,k)^T \in \mathcal{I},
\end{aligned}$$

$$\mathcal{I} = \{(i,k)^T \mid 1 \le i \le 8, 1 \le k \le 120\}.$$

The index space with the data dependencies as well as the reduced dependence graph are depicted in Fig. 1.

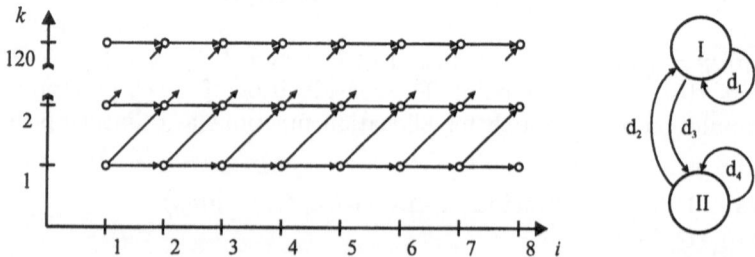

Fig. 1. *Index spaces with data dependencies and reduced dependence graph*

The dependence vectors are: $\text{I} \to \text{I} : \mathbf{d}_1 = (1,0)^T$, $\quad \text{II} \to \text{I} : \mathbf{d}_2 = (1,1)^T$, $\text{I} \to \text{II} : \mathbf{d}_3 = (1,0)^T$, $\quad \text{II} \to \text{II} : \mathbf{d}_4 = (1,1)^T$. The set of vertices $\mathcal{W}$ of the index space $\mathcal{I}$ is $\mathcal{W} = \{(1,1),(8,1),(1,120),(8,120)\}$.
Application of program 1 leads to the matrices $\mathbf{S}_1 = (1,0)$ and $\mathbf{S}_2 = (0,1)$, and hence to the projection vectors $\mathbf{u}_1 = (0,1)^T$ and $\mathbf{u}_2 = (1,0)^T$ respectively.

The next section covers the determination of a scheduling function and the processor functionality with respect to each projection vector $\mathbf{u}_i$ computed in program 1.

5 Determination of a Scheduling Function

5.1 General Optimization Problem

In this section we propose an approach to determine a scheduling function as well as a module selection and a conflict free assignment of the modules to the operations of the SURE. Our objective is the minimization of the costs for a hardware implementation of the processor array subject to the condition that a given latency L_g is satisfied.
A scheduling function is determined for each linear allocation function resulting after application of program 1. First, we present a general description of the optimization problem and go into more detail in the next paragraphs.

Program 2 (Determination of a scheduling function)

$$minimize \quad chip \; area \; C \; of \; the \; processor \; array$$
$$subject \; to \quad latency \; L \; of \; the \; processor \; array: \; L \leq L_g$$
$$causal \; scheduling \; function$$
$$selection \; of \; modules$$
$$conflict \; free \; access \; to \; the \; modules$$

5.2 Objective Function

The number of processors N_p is given exactly since the allocation function is fixed at this step. In each processor of the fullsize array we have to implement n_l instances of module m_l. The chip area needed to implement the modules of all processor of the fullsize array is therefore $C_f = N_p \sum_{l=1}^{|\mathcal{M}|} n_l c_l$.

Additional chip area is needed to implement (1) registers to store intermediate results, (2) combinatorial logic to control the behaviour of the processors and (3) interconnections between the processors. The costs for the control logic are assumed to be negligible. The number of registers depends on the number of processors as well as the number of data produced in each processor. We approximate the chip area C_r needed to implement the registers in silicon by $C_r = N_p \, m \, c_r$, where c_r is the chip area of one register and m is the number of equations of the SURE. Since the number of processors is independent of the scheduling function and the module selection, C_r can be treated as a constant. An assessment of the effort for the implementation of the interconnections is difficult. Instead of measuring the chip area of interconnections we propose a minimization of the length or a limitation of the radius of the interconnections respectively while determining the allocation function.

As a consequence of the above discussion we conclude that it is sufficient to consider the chip area C_f needed to implement the modules as measure for the hardware costs.

Now, we assume that the admissible latency L_g enables a slow down of the fullsize array. Hence, we can decrease the hardware costs by partitioning the fullsize array. We apply the LSGP-partitioning [4] (tiling) by a factor of K, i.e. each partition contains K processors of the fullsize array. Each partition represents a processor of the resulting processor array. We assume that the partitioning matches exactly the fullsize array and we neglect the additional hardware needed to control the partitions. Hence, we get the hardware costs of the resulting processor array $C_p = C_f/K$. Since the number of processors N_p is fixed it is sufficient to consider the chip area C_f' of only one processor of the fullsize array. Our objective is to minimize $C_p' = C_f'/K$, where $C_p = N_p C_p'$ and $C_f = N_p C_f'$.

5.3 Admissible Latency L_g

The latency L_f of the fullsize array is given by $L_f = t^{max} - t^{min}$, where

$$t^{min} \leq \tau^T \mathbf{w} + t_i \leq t^{max} - d_{m(i),i}, \quad \forall \mathbf{w} \in \mathcal{W}, \quad 1 \leq i \leq m,$$

where $t^{min}, t^{max} \in \mathbb{Z}$ and $\mathcal{W}$ is the set of vertices $\mathbf{w}$ of the index space $\mathcal{I}$. The consideration of only the vertices of $\mathcal{I}$ is justified by the fact that linear functions have their extreme values at extreme points of a convex space [10]. The relaxation to integer values has only small effort for sufficiently large index spaces $\mathcal{I}$.

The assumption that $L_g \gg L_f$ enables a partitioning of the fullsize array by a factor K, where $KL_f \le L_g$. This inequality is a sufficient condition since we can presume that a partitioning exists where the latency L of the resulting processor array satisfies $L \le KL_f$. An intuitive justification yields the opposite case where the distribution of a sequential program to K processors leads to a speed up less or equal to K.

Next, we linearize the constraints $C'_f = KC'_p$ and $KL_f \le L_g$. The minimal latency L_{min} of a fullsize array is easy to determine by assumption of unlimited resources and assignment of the fastest possible module to each operation of the SURE. Using L_{min} we can limit K by $1 \le K \le K_{max} = \lfloor L_g/L_{min} \rfloor$. The lower bound of K can be increased by consideration of the minimal hardware costs for one processor $C'_{min} = \min\{\sum_{l=1}^{|\mathcal{M}|} n_l c_l\}$ satisfying $\exists m_l \in \mathcal{M}_i.n_l \ge 1, 1 \le i \le m$.

The application of a resource constraint scheduling [13] with the modules leading to C'_{min} yields L_{max} and $K_{min} = \max\{1, \lfloor L_g/L_{max} \rfloor\}$.

The introduction of $K_{max} - K_{min} + 1$ binary variables $\gamma_j \in \{0,1\}$ enables an equivalent formulation of the constraints $C'_f = KC'_p$ and $KL_f \le L_g$ as follows:

$$
\begin{aligned}
C'_f &\le jC'_p + (1-\gamma_j)C'_{max}, & K_{min} \le j \le K_{max}, \\
jL_g &\ge jK_{min}L_f + \gamma_j(j - K_{min})L_g, & K_{min} \le j \le K_{max}, \\
\sum_{j=K_{min}}^{K_{max}} \gamma_j &= 1,
\end{aligned}
\tag{5}
$$

where $C'_{max} = \sum_{l=1}^{|\mathcal{M}|} n_l^{max} c_l$, and n_l^{max} is the maximal number of instances of module m_l which can be implemented in one processor of the fullsize array.

A reduction of the number of variables γ_i from $K_{max} - K_{min} + 1$ to $\lceil (K_{max} - K_{min} + 1)/z \rceil$, $z \in \mathbb{Z}$, is possible by solving the optimization problem iteratively. In the first iteration only such j, $K_{min} \le j \le K_{max}$, are considered that satisfy $j \bmod z = a$, where $a = K_{min} \bmod z$. The second iteration of the optimization problem is solved for $j_{max} - z \le j \le j_{max} + z$, where j_{max} is the solution of the first iteration.

5.4 Causality Constraint

In order to ensure a valid partial order or the equations preserving the data dependencies the scheduling function has to satisfy the causality constraint.

Definition 5 (Causality constraint). *A scheduling function $\tau_i(\mathbf{i}) = \tau^T \mathbf{i} + t_i$ has to satisfy the following constraint:*

$$
\tau^T \mathbf{d}(e) + t_{\delta(e)} - t_{\sigma(e)} \ge d_{m(\sigma(e)),\sigma(e)}, \quad \forall e \in \mathcal{E}.
\tag{6}
$$

5.5 Module Selection and Prevention of Access Conflicts

The assignment of a module to the operation F_j is described by $|\mathcal{M}_j|$ binary variables $r_j^l \in \{0, 1\}$, where $\sum_{m_l \in \mathcal{M}_j} r_j^l = 1$ and $r_j^k = 1 \leftrightarrow m(j) = m_k$.

The following resource constraint prevents access conflicts to the module.

Definition 6 (Resource constraint). *For a given projection vector* $\mathbf{u}$ *a uniform affine scheduling function* $\tau_i(\mathbf{i}) = \tau^T \mathbf{i} + t_i$ *has to satisfy the following constraint:*

$$
\left.
\begin{array}{r}
(t_j \bmod \lambda) - (t_k \bmod \lambda) \geq o_{m(j),k}, \\
\lambda - (t_j \bmod \lambda) + (t_k \bmod \lambda) \geq o_{m(j),j}, \\
(t_k \bmod \lambda) - (t_j \bmod \lambda) \geq o_{m(j),j}, \\
\lambda - (t_k \bmod \lambda) + (t_j \bmod \lambda) \geq o_{m(j),k},
\end{array}
\right\}
\begin{array}{l}
if\ (t_j \bmod \lambda) > (t_k \bmod \lambda), \\[2mm]
if\ (t_j \bmod \lambda) \leq (t_k \bmod \lambda),
\end{array}
\tag{7}
$$

for all $j \neq k$, $1 \leq j, k \leq m$, *with* $m(j) = m(k)$ *and* $u_j = u_k$, *where* $\lambda = |\tau^T \mathbf{u}|$.

We refer to [2] for an explanation and a linearization of (7).

5.6 Result of the Optimization Problem

The result of the optimization problem in program 2 is a set of modules have to be implemented in each processor and the number of processors of the fullsize array building one partition and hence one processor of the resulting processor array. Program 2 is solved for each projection vector computed in program 1. Then we select the projection vector $\mathbf{u}_i$ leading to minimal hardware costs

$$
C = C_p + C_r = N_p \left(\frac{1}{K} \sum_{i=l}^{|\mathcal{M}|} n_l c_l + m\, c_r \right).
$$

Example 2. (Continuation of example 1)
The admissible latency L_g is measured in clock cycles and supposed to be $L_g = 1200$. The considered set of modules is listed in table 1. We solve the

Table 1. Set of modules

	operation	evaluation time d_i in clock cycles	time offset o_i in clock cycles	chip area c_i normalized
m_1	add/mult	3	3	297
m_2	add/mult	4	4	169
m_3	add/mult	8	8	44
m_r	reg			1

optimization problem of program 2 separately for the projection vectors $\mathbf{u}_1$ and $\mathbf{u}_2$ determined in program 1. In order to justify our approach we present some interesting solutions in table 2 instead of only the best one given by the solver of the optimization problem.

Table 2. Results of the optimization problem

projection vector	N_p	$C_r = N_p \, m \, c_r$	module selection	K	$C_p = \frac{N_p}{K} \sum\limits_{i=1}^{|M|} n_i c_i$	$C = C_r + C_p$
$\mathbf{u}_1 = (0,1)^T$	8	16	$1 \times m_3$	1	704	720
"	8	16	$1 \times m_2$	1	1352	1368
"	8	16	$2 \times m_1$	3	1584	1600
$\mathbf{u}_2 = (1,0)^T$	120	240	$1 \times m_3$	9	586	826
"	120	240	$2 \times m_2$	37	1096	1336
"	120	240	$1 \times m_1$	25	1425	1665

Minimal hardware costs occur by using projection vector $\mathbf{u}_1$ and implementing one instance of module m_3 in each processor of the resulting processor array. Finally, we want to give some short notes to the computational effort of our example. The worst case is program 2 with respect to the projection vector $\mathbf{u}_1$. The program consists of 168 constraints with 51 binary, 14 integer and one rational variable. The solution takes 11.3 seconds of CPU time on a SUN SPARC Station 10.

5.7 Selection of Projection Vectors

An alternative approach to the separate computation of a scheduling function for each projection vector consists in consideration of the projection vectors $\mathbf{u}_i$ as parameters in program 2. We assume that the iteration interval $\lambda \leq \lambda_{max}$. A linearization of the constraint $\lambda = |\tau^T \mathbf{u}|$ is achievable using four inequalities and one binary variable v.

$$
\begin{aligned}
\lambda - 2v\lambda_{max} &\leq \quad \tau^T \mathbf{u} \leq \lambda, \quad \lambda \in \mathbb{Z}, \\
\lambda - 2(1-v)\lambda_{max} &\leq -\tau^T \mathbf{u} \leq \lambda, \quad v \in \{0,1\}.
\end{aligned}
\tag{8}
$$

Suppose, that we have to select one of P projection vectors $\mathbf{u}_i$, $1 \leq i \leq P$. Using P binary variables α_i, $1 \leq i \leq P$, we replace (8) by:

$$
\begin{aligned}
\lambda - 2v\lambda_{max} - 2(1-\alpha_i)\lambda_{max} &\leq \quad \tau^T \mathbf{u}_i \leq \lambda + (1-\alpha_i)\lambda_{max}, \quad 1 \leq i \leq P, \\
\lambda - 2(1-v)\lambda_{max} - 2(1-\alpha_i)\lambda_{max} &\leq -\tau^T \mathbf{u}_i \leq \lambda + (1-\alpha_i)\lambda_{max}, \quad v \in \{0,1\},
\end{aligned}
$$

$$
\sum_{i=1}^{P} \alpha_i = 1, \quad \alpha_i \in \{0,1\}, \quad 1 \leq i \leq P.
$$

The projection vector $\mathbf{u}_i$ is selected if $\alpha_i = 1$.

Furthermore, we have to include the hardware costs for the registers. The objective function is changed to minimize the hardware costs C of the resulting processor array. New constraints have to be introduced to take the different number of processors N_p^i of the fullsize arrays with respect to the projection vectors $\mathbf{u}_i$ into account:

$$
C \geq N_p^i(C_p' + m \, c_r) - (1-\alpha_i)N_p^i(C_{max}' + m \, c_r), \quad 1 \leq i \leq P.
$$

We do not recommend this approach since the condition of the integer linear program deteriorates strongly.

6 Conclusion

The presented approach is suitable to derive cost minimal processor arrays for algorithms with the requirement of an admissible latency. The arising optimization problems are given in a linearized form which permits the use of standard packages to solve the problems.

Possible extensions to our approach are the inclusion of the power consumption and the inclusion of a rough approximation of the effort needed to implement the interconnections.

References

1. A. Darte, Y. Robert: "Constructive Methods for Scheduling Uniform Loop Nests", *IEEE Trans. on Parallel and Distributed Systems*, Vol. 5, No. 8, pp. 814-822, 1994
2. D. Fimmel, R. Merker: "Determination of the Processor Functionality in the Design of Processor Arrays", *Proc. Int. Conf. on Application-Specific Systems, Architectures and Processors*, pp. 199-208, Zürich, 1997
3. D. Fimmel, R. Merker: "Determination of an Optimal Processor Allocation in the Design of Massively Parallel Processor Arrays", *Proc. Int. Conf. on Algorithms and Parallel Processing*, pp. 309-322, Melbourne, 1997
4. K. Jainandunsing: "Optimal Partitioning Scheme for Wavefront/Systolic Array Processors", *IEEE Proc. Symp. on Circuits and Systems*, 1986
5. R.M. Karp, R.E. Miller, S. Winograd: "The organization of computations for uniform recurrence equations", *J. of the ACM*, vol.14, pp. 563-590, 1967
6. D.I. Moldovan: "On the Design of Algorithms for VLSI Systolic Arrays", *Proceedings of the IEEE*, pp. 113-120, January 1983
7. P. Quinton: "Automatic Synthesis of Systolic Arrays from Uniform Recurrent Equations", *IEEE 11-th Int. Symp. on Computer Architecture*, Ann Arbor, pp. 208-214, 1984
8. S.K. Rao: "Regular Iterative Algorithms and their Implementations on Processor Arrays", *PhD thesis*, Stanford University, 1985
9. J. Rossel, F. Catthoor, H. De Man: "Extension to Linear Mapping for Regular Arrays with Complex Processing Elements", *Proc. Int. Conf. on Application-Specific Systems, Architectures and Processors*, pp. 156-167, Princeton, 1990
10. A. Schrijver: *Theory of Linear and Integer Programming*, John Wiley & Sons, New York, 1986
11. A. Schubert, R. Merker: "Systolization of Recursive Algorithms with DESA", in *Proc. 5th Int. Workshop Parcella '90*, Mathematical Research, G. Wolf, T. Legendi, U. Schendel (eds.), vol. 2, Akademie-Verlag Berlin, 1990, pp. 267-276, 1994
12. J. Teich: "A Compiler for Application-Specific Processor Arrays", *PhD thesis*, Univ. of Saarland, Verlag Shaker, Aachen, 1993
13. L. Thiele: "Resource Constraint Scheduling of Uniform Algorithms", *Int. Journal on VLSI and Signal Processing*, Vol. 10, pp. 295-310, 1995
14. Y. Wong, J.M. Delosme: "Optimal Systolic Implementation of n-dimensional Recurrences", *Proc. ICCD*, pp. 618-621, 1985

15. Y. Wong, J.M. Delosme: "Optimization of Processor Count for Systolic Arrays", *Research Report* YALEU/DCS/RR-697, Yale Univ., 1989
16. X. Zhong, S. Rajopadhye, I. Wong: "Systematic Generation of Linear Allocation Functions in Systolic Array Design", *J. of VLSI Signal Processing*, Vol. 4, pp. 279-293, 1992

Interval Routing & Layered Cross Product: Compact Routing Schemes for Butterflies, Mesh of Trees and Fat Trees

Tiziana Calamoneri[1] and Miriam Di Ianni[2]

[1] Dip. di Scienze dell'Informazione, Università di Roma "la Sapienza"
via Salaria 113, I-00198 Roma, Italy.
calamo@dsi.uniroma1.it
[2] Istituto di Elettronica, Università di Perugia
via G.Duranti 1/A, I-06123 Perugia, Italy.
diianni@istel.ing.unipg.it

Abstract. In this paper we propose compact routing schemes having space and time complexities comparable to a 2-Interval Routing Scheme for the class of networks decomposable as Layered Cross Product (LCP) of rooted trees. As a consequence, we are able to design a 2-Interval Routing Scheme for butterflies, meshes of trees and fat trees using a fast local routing algorithm. Finally, we show that a compact routing scheme for networks which are LCP of general graphs cannot be found by any only using shortest paths information on the factors.

1 Introduction

The information needed to route messages in parallel and distributed systems must be somehow stored in each node of the network. The simplest solution consists of a complete routing table – stored in each node u – that specifies for each destination v at least one link incident to u and lying on a path from u to v. The required space of such a solution is $\Theta(n \log \delta)$, where δ is the node degree and n the number of nodes in the network. Efficiency considerations lead to store *shortest paths* information. For better and fair use of network resources, storing, for each entry of the routing table, as many outgoing links as necessary to describe *all* shortest paths in the network should be aimed. Due to limited storage space at each processor, a linear increase of the routing table size in n is not acceptable. Research has then focused on identifying classes of network topologies whose shortest paths information can be succinctly stored, assuming that some "short" labels can be assigned to nodes and links at preprocessing time.

In the *Interval Routing Scheme* (in short, IRS) [7, 12], node-labels belong to the set $\{1, \ldots, n\}$, while link-labels are pairs of node-labels representing cyclic intervals of $[1, \ldots, n]$. A message with destination v arriving at a node u is sent by u onto an incident link whose label $[v_1, v_2]$ is such that $v \in [v_1, v_2]$. Such an approach allows one to achieve an efficient memory occupation. An IRS is said *optimum* if the route traversed by each message is a shortest path from its source to its destination. It is said *overall optimum* if a message can be routed along *any* shortest path. In [6, 12] optimum IRSs have been designed for particular network topologies. In [3, 11, 13] it has been proved the existence of networks that do not admit any optimum IRS. *Multi-label Interval Routing Schemes* were introduced [7] to extend the model in order to allow more than one interval to be associated to each link: a k-IRS is a scheme associating at most k intervals to each link. A message whose destination is node v is sent onto a link labeled $(I_1, \ldots, I_k)$ if $v \in I_i$ for some $1 \leq i \leq k$. In [2] a technique for proving lower bounds on the minimum k allowed was developed and in [4] it has been used to construct n-node networks for which any optimal k-IRS requires $k = \Theta(n)$. It was proved that for some well known interconnection networks, such as shuffle exchange, cube connected cycle, butterfly and star graph, each optimal k-IRS requires $k = \Omega(n^{1/2-\epsilon})$ – for proper values of ϵ – to store one shortest path for each pair [5]. Of course, this lower bound still holds to store *any* shortest path.

In this paper, after providing the necessary preliminary definitions (Section 2), we propose overall optimum compact routing schemes (Section 3) based on the same leading idea as the Multi-label Interval Routing for all networks which are Layered Cross Product (LCP) [1] of rooted trees (in short, T-networks). For many commonly used interconnection networks falling in this definition no overall optimum compact routing scheme was known. Among them we recall three widely studied topologies: butterflies, mesh of trees and fat trees. Our compact routing scheme requires as much space and time as those required by a 2-IRS. The achievement is particularly meaningful for butterflies because of the result in [5]. Finally, in Section 4, we give a negative result by proving that the knowledge of shortest paths on the factors of a network could be not enough to compute shortest paths on it.

2 Definitions and Preliminary Results

Point to point communication networks are usually represented by graphs, whose nodes stand for processors and edges for communication links. We always represent each edge $\{u, v\}$ by the pair of (oriented) arcs, (u, v) and (v, u).

An *l-layered graph*, $G = (V^1, V^2, \ldots, V^l, E)$ consists of l layers of nodes; V^i is the (non-empty) set of nodes in layer i, $1 \leq i \leq l$; every edge in E connects vertices of two adjacent layers. In particular a rooted tree T of height h is a h-layered graph, layer i defined either as the set of nodes having distance $i - 1$ from the root or as the set of nodes having distance $h - i$ from the root. From now on, we call T a *root-tree* or a *leaf-tree* according to whether the first or the second way of defining layers is chosen. In Fig. 1, T_1 is a root-tree, while T_2 is a leaf-tree.

Let $G_1 = (V_1^1, V_1^2, \ldots, V_1^l, E_1)$ and $G_2 = (V_2^1, V_2^2, \ldots, V_2^l, E_2)$ be two l-layered graphs. Their *Layered Cross Product* (LCP for short) [1] $G^1 \times G^2$ is an l-layered graph $G = (V^1, V^2, \ldots, V^l, E)$ where V^i is the cartesian product of V_1^i and V_2^i, $1 \leq i \leq l$, and a link $((a, \alpha), (b, \beta))$ belongs to E if and only if $(a, b) \in E_1$ and $(\alpha, \beta) \in E_2$.

Many common networks are LCP of trees [1]. Among theml: the *butterfly* with N inputs and N outputs is the LCP of two N-leaves complete binary trees (Fig. 1.a), the *mesh of trees* of size $2N$ is the LCP of two N-leaves complete binary trees with paths of length $\log N$ attached to their leaves (Fig. 1.b), the *fat tree* of height h [10] is the LCP of a complete binary tree and a complete quaternary tree, both of height h (Fig. 1.c).

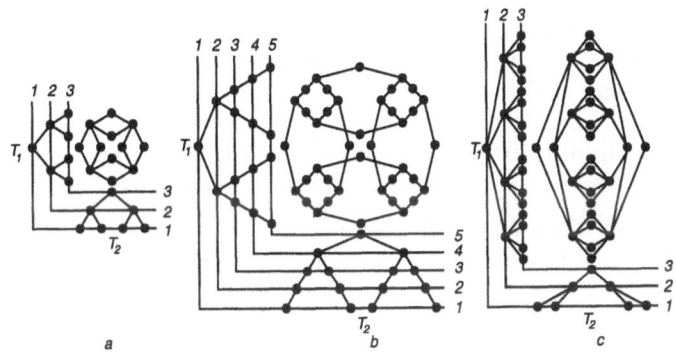

Fig. 1. Butterfly, Mesh of Trees and Fat-tree as LCP of rooted trees.

Fact 21 *Let $(a, \alpha), (b, \beta) \in V(G_1 \times G_2)$. Any shortest path from (a, α) to (b, β) is never shorter than a shortest path from a to b in G^1 and a shortest path from α to β in G^2.*

Fact 22 *If G is the LCP of either two root-trees or two leaf-trees then G is a tree.*

Observe that the LCP of two trees could be also not connected. This is not a restriction to our discussion, since we deal with connected networks that are the LCP of trees and not with *any* LCP of trees.

3   Designing Compact Routing Schemes for T-networks

Let $G = T_1 \times T_2$, where both T_1 and T_2 are trees. In [12] an IRS for trees has been shown. Thus, consider the two IRSs for T^1 and T^2 and let $\mathcal{L}_1$ and $\mathcal{I}_1$, $\mathcal{L}_2$ and $\mathcal{I}_2$ be the node- and link-labelings of an IRS for T^1 and T^2, respectively. A node $(u_1, u_2) \in V(G)$ is labeled with a triple (a, α, l) if $\mathcal{L}_1(u_1) = a$, $\mathcal{L}_2(u_2) = \alpha$ and l is the layer of both u_1 in T_1 and u_2 in T_2 (and of (u_1, u_2) in G). Similarly, a link $((u_1, u_2), (v_1, v_2)) \in E$ is labeled with a triple (I_1, I_2, l), where $I_1 = \mathcal{I}_1((u_1, v_1))$, $I_2 = \mathcal{I}_2((u_2, v_2))$ and l is the layer of (v_1, v_2). It is possible to rename all nodes according to their node-labeling $\mathcal{L}$; therefore, in the following we speak about *labeling* to mean link-labeling and we refer to nodes themselves to mean their node-labels.

To complete the definition of the compact routing scheme for G, we must describe algorithm $\mathcal{A}$ stored in each node (a, α, l_a) used to route a packet onto a shortest path connecting the current node (a, α, l_a) itself to the destination (t, τ, l_t). Informally, at each step $\mathcal{A}$ tries to take the greedy choice: if both shortest path factors $P_1(a, t)$ and $P_2(\alpha, \tau)$ move towards the same level then $\mathcal{A}$ moves in that direction, that is, it chooses link $((a, \alpha, l_a), (b, \beta, l_b))$. Otherwise, if $a = t$ (or $\alpha = \tau$), then P_1 (or P_2) is null, so the other path must be followed[1]. Finally, if $P_1(a, t)$ and $P_2(\alpha, \tau)$ go towards opposite levels, $\mathcal{A}$ follows the path going away from level l_t. We will prove that in this way a shortest path on G is always used. Now, we are ready to describe algorithm $\mathcal{A}$ formally.

Algorithm $\mathcal{A}$

(input: t, τ, l_t; output: one outgoing link labeled (I_1, I_2, l))

{ The labeling of the outgoing links, a, α and l_a are known constants in each node.}

if $a = t$ and $\alpha = \tau$ then extract the packet
else if there exists a link labeled (I_1, I_2, l) s.t. $t \in I_1$ and $\tau \in I_2$ then choose it
    else if there exists a link having label (I_1, I_2, l) s.t.
        $t \in I_1$ and $(\alpha = \tau$ or $|l_a - l_t| < |l - l_t|)$ then choose it
        else if there exists a link having label (I_1, I_2, l) s.t.
            $\tau \in I_2$ and $(a = t$ or $|l_\alpha - l_\tau| < |l - l_\tau|)$ then choose it;

Notice that algorithm $\mathcal{A}$, stored in each node, does not increase the asympthotic space complexity with respect to 2-IRS and runs in $\mathcal{O}(\delta)$ time, δ being the maximum node degree. As a consequence, if $\mathcal{A}$ is able to route packets to their destinations we have designed a compact routing scheme. In the following, we first show the optimality (Thm. 1) and then its overall optimality (Thm. 2).

Theorem 1. *If (a, α, l_a) transmits a packet, whose destination is (t, τ, l_t), to an adjacent node (b, β, l_b), then (b, β, l_b) belongs to a shortest path from (a, α, l_a) to (t, τ, l_t).*

Proof. If G is the LCP of either two root-trees or two leaf-trees the statement is trivially true because, from Fact 22, there exists a unique path between any pair of nodes. It must necessarily be the Layered Cross Product of the corresponding paths in the factors. Thus, links labeled (I_1, I_2, l) such that $t \in I_1$ and $\tau \in I_2$ can always be used. Therefore, from now on we shall always suppose that T_1 is a root-tree and T_2 is a leaf-tree.

The proof considers the truth of the **if**-conditions in $\mathcal{A}$.

1. There exists a link (I_1, I_2, l) such that $\tau \in I_2$ and $a = t$. Let $\langle \alpha, \beta, \beta_1, \ldots, \beta_k, \tau \rangle$ be the shortest path in T_2. Then, by definition of LCP, there exist $b, b_1, \ldots, b_k$ in T_1 such that $\langle a, b, b_1, \ldots, b_k, a \rangle$ is a path (crossing the same links more than

---

[1] When we say that $\mathcal{A}$ follows a path P_i (either $i = 1$ or $i = 2$), we mean that $\mathcal{A}$ follows an edge on G whose i-th factor belongs to P_i.

once) such that $\langle (a, \alpha, l_a), (b, \beta, l_b), (b_1, \beta_1, l_{b_1}), \ldots, (b_k, \beta_k, l_{b_k}), (a, \tau, l_a) \rangle$ is a path in G starting at (a, α, l_a) and ending at (a, τ, l_a). Fact 21 ensures that this is a shortest path in G. The case with $t \in I_1$ and $\alpha = \tau$ is symmetric.

2. There exists a link labeled (I_1, I_2, l_b) such that $t \in I_1$ and $\tau \in I_2$. W.l.o.g., suppose $l_a < l_t$: b is a child of a and β is the father of α (Fig. 2). Moreover, since b belongs to a shortest path from a to t, t is a descendant of b. Two cases are possible:

   – τ is an ancestor of α (Fig. 2.a). Then, the shortest paths from a to t in T_1 and from α to τ in T_2 have the same length $l_t - l_a$. It is easy to see that in G there exists a unique shortest path from (a, α, l_a) to (t, τ, l_t). Furthermore, all links in such path are labeled (I_1, I_2, l) such that $t \in I_1$ and $\tau \in I_2$. Thus, (b, β, l_b) belongs to a path of length $l_t - l_a$ and for Fact 21 it is the shortest path.

   – α and τ have a common ancestor (Fig. 2.b). Let γ be the nearest common ancestor and let l_c be its layer. By the definition of layers, and since T_1 is a root-tree and T_2 is a leaf-tree, it must be $l_t < l_c$.

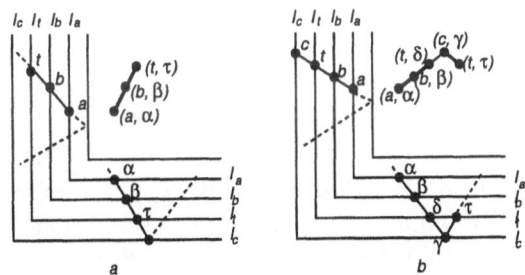

Fig. 2. $l_a < l_t$; a. τ is an acestor of α; b. α and τ have a common ancestor.

Let c be one of the descendants of t at layer l_c in T_1 and δ be the node at layer l_t belonging to the shortest path from α to γ in T_2. There exists a path from (a, α, l_a) to (t, δ, l_t) in G whose links are always labeled (I_1, I_2, l) such that $t \in I_1$ and $\tau \in I_2$ and having length equal to the length of the shortest path from α to δ in T_2 (this is easily proved by induction on the length). The path in G from (t, δ, l_t) to (t, τ, l_t) chosen by $\mathcal{A}$ is a shortest one because of case 1. Since the path from (a, α, l_a) to (t, τ, l_t) passing through (b, β, l_b) has the same length as the path from α to τ in T_2, it is a shortest one.

3. No outgoing link from (a, α, l_a) is labeled (I_1, I_2, l) such that $t \in I_1$ and $\tau \in I_2$; furthermore, neither $a = t$ nor $\alpha = \tau$. Then, one shortest path factor moves towards increasing layers while the other shortest path factor moves towards decreasing layers.

   Suppose first $l_t > l_a$, that is, t is closer than a to the leaves in T_1 while τ is closer than α to the root in T_2.

Notice that it is not possible to have both shortest paths factors moving towards the leaves. Thus, both shortest paths factors move towards the roots. Then, t and a have a nearest common ancestor c at layer l_c. On T_2, either τ is an ancestor of α (see Fig. 3) or τ and α have a nearest common ancestor δ at layer l_d (Fig. 4).

Consider the case of Fig. 3 first. Let b the father of a in T_1 and let I_1 be the label of (a, b) in T_1. Since $l_t - l_b > l_t - l_a$ and $t \in I_1$, $\mathcal{A}$ chooses one link labeled (I_1, I_2, l_b) and ending in (b, β, l_b). Let d be the node belonging to the shortest path from a to t at layer l_a in T_1. Node (b, β, l_b) lies on the path

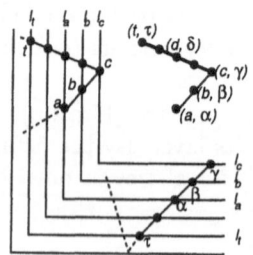

Fig. 3. The two shortest paths factors move towards the roots and τ is an ancestor of α while t and a have a common ancestor c.

from (a, α, l_a) to (d, α, l_a) – through a node (c, γ, l_c) – of the same length as the shortest path from a to d in T_1. From this last node, there is a single shortest path to (t, τ, l_t) constituted by links always labeled (I_1, I_2, l) such that $t \in I_1$ and $\tau \in I_2$. The length of such path from (a, α, l_a) to (t, τ, l_t) through (b, β, l_b) is equal to the length of the shortest path from a to t in T_1.

Consider now the situation in Fig. 4: both shortest paths factors move towards the roots, t and a have a nearest common ancestor c at layer l_c, and τ and α have a nearest common ancestor δ at layer l_d. To go from (a, α, l_a) to (t, τ, l_t) through a shortest path it is necessary to use either first a shortest path in T_1 till c or first a shortest path in T_2 till δ. Indeed, consider a path that alternates links whose first factor is from a shortest path in T_1 with links whose second factor is from a shortest path in T_2: let $(a, \alpha, l_a), (a_1, \alpha_1, l_1), \ldots, (a_k, \alpha_k, l_k)$ be the first fragment of such a path having the first factor as the shortest path in T_1, with $a_i \neq c$, $i = 1, \ldots, k$. If the next step is a link whose second factor belongs to a shortest path towards τ in T_2, such link ends necessarily in $(u, \alpha_{k-1}, l_{k-1})$, where u is a child of a_k in T_1. For the sake of symmetry, since u and a_{k-1} are at the same layer in the same subtree rooted at a_k (with $l_k > l_c$), the distance of $(u, \alpha_{k-1}, l_{k-1})$ from (t, τ, l_t) is equal to the distance of $(a_{k-1}, \alpha_{k-1}, l_{k-1})$ from (t, τ, l_t). Thus, in order to reach (t, τ, l_t) from (a, α, l_a) at least two steps have been wasted. Hence, necessarily one of the following is a shortest path in G:

- (Fig. 4.a) a path from (a, α, l_a) to (b, β, l_b) to (c, γ, l_c) to (h, α, l_a) to (t, η, l_t) to (f, δ, l_d) to (t, τ, l_t), where γ is a descendant of α at layer l_c in T_2, h and f are descendants of c at layer l_a and l_d, respectively, in T_1 and η is an ancestor of α at layer l_t in T_2. The length of such a path is $2(l_a - l_c) + (l_t - l_a) + 2(l_d - l_t) = 2(l_d - l_c) + l_a - l_t$

- (Fig. 4.b) a path from (a, α, l_a) to (d, δ, l_d) to (k, τ, l_t) to (a, θ, l_a) to (c, σ, l_c) to (h, θ, l_a) to (t, τ, l_t), where d and k are descendant of a at layers l_d and l_t, respectively, in T_1 and θ and σ are descendants of τ at layers l_a and l_c, respectively, in T_2. The length of such a path is $(l_d - l_a) + (l_d - l_t) + 2(l_t - l_c) = 2(l_d - l_c) + l_t - l_a$

Thus, the shortest path depends on the sign of $l_t - l_a$. Since in our hypothesis $l_t > l_a$, the shortest path is the first one, that is the path whose first step goes away from l_t. Since node (b, β, l_b) is such that $l_t - l_b > l_t - l_a$, then $\mathcal{A}$ chooses the first path.

When $l_t < l_a$, the same reasoning applies. Finally, the same discussion holds also if $l_t = l_a$. However, notice that in this case any of the two choices is possible.

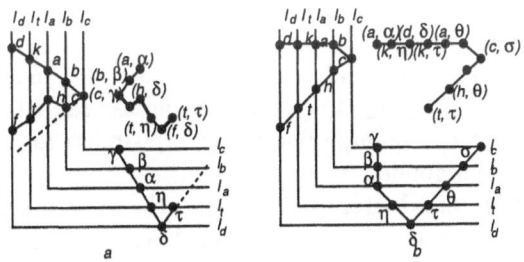

Fig. 4. The two shortest paths factors move towards the roots, t and a have a common ancestor c and τ and α have a common ancestor δ.

Theorem 2. *If a packet must be transmitted from node (a, α, l_a) to node (t, τ, l_t) and $\langle (a, \alpha, l_a), (a_1, \alpha_1, l_{a_1}), (a_2, \alpha_2, l_{a_2}), \ldots, (t, \tau, l_t) \rangle$ is any shortest path from (a, α, l_a) to (t, τ, l_t), then algorithm $\mathcal{A}$ can possibly use it.*

Proof. Again, thanks to Fact 22, we shall always suppose that T_1 is a root-tree and T_2 is a leaf-tree. The proof is divided according to the truth of the **if**-conditions in $\mathcal{A}$ and most of considerations done in the proof of Theorem 1 are used here.

1. There exists a link (I_1, I_2, l) such that $\tau \in I_2$ and $a = t$: then (cf. proof of Thm.1) G contains a path from (a, α, l_a) to (t, τ, l_t) of length equal to the path from α to τ in T_2. Let $k + 1$ be such length. Suppose that a path $\langle (a, \alpha, l_a), (a_1, \alpha_1, l_{a_1}), \ldots, (a_k, \alpha_k, l_{a_k}), (t, \tau, l_t) \rangle$ exists in G that is not found

by $\mathcal{A}$. Thus, α_1 does not belong to the shortest path from α to τ in T_2, that is, if link $((a, \alpha, l_a), (a_1, \alpha_1, l_{a_1}))$ is labeled (I_1, I_2, l_{a_1}), $\tau \notin I_2$. Hence, since by definition of LCP $p = \langle \alpha, \alpha_1, \ldots, \alpha_k, \tau \rangle$ must be a path in T_2, some α_i must be the same as some α_j, with $i \neq j$. But the distance between α and τ in T_2 is $k + 1$, then p must be longer than $k + 1$, an absurd. The case $t \in I_1$ and $\alpha = \tau$ is symmetric.

2. There exists a link labeled (I_1, I_2, l) such that $t \in I_1$ and $\tau \in I_2$. W.l.o.g., suppose $l_a < l_t$ and therefore b a child of a and β the father of α. Since b belongs to a shortest path from a to t, t is a descendant of b. Two cases are possible:

   $-$ τ is an ancestor of α. This case is trivially proved, since there is in G a unique shortest path from (a, α, l_a) to (t, τ, l_t) and $\mathcal{A}$ finds that path.
   $-$ α and τ have a nearest common ancestor γ at layer l_c. Recall that $l_t \leq l_c$ and a shortest path in G must be as long as the path from α to τ in T_2 (cf. proof of Thm.1). The proof of this case is similar to the one of case 1. of this theorem and thus omitted.

3. No outgoing link from (a, α, l_a) is labeled (I_1, I_2, l) such that $t \in I_1$ and $\tau \in I_2$. Suppose $l_t > l_a$. We have already proved that one of the following cases must occur:

   $-$ both shortest paths factors move towards the roots, τ is an ancestor of α while t and a have a nearest common ancestor c at layer l_c: algorithm $\mathcal{A}$ chooses a link labeled (I_1, I_2, l) such that $t \in I_1$ and such link belongs to a path from (a, α, l_a) to (t, τ, l_t) having the same length as the path from a to t in T_1. Again, a reasoning very similar to that one of case 1. of this theorem applies.
   $-$ both shortest paths factors move towards the roots, t and a have a nearest common ancestor c at layer l_c, τ and α have a nearest common ancestor δ at layer l_d. We have already proved that a shortest path from (a, α, l_a) to (t, τ, l_t) necessarily crosses either first a shortest path in T_1 to c or first a shortest path in T_2 to δ. This implies (cf. proof of Thm.1) that a shortest path in G must necessarily be a path from (a, α, l_a) to (c, γ, l_c) to (h, α, l_a) to (t, η, l_t) to (d, δ, l_d) to (t, τ, l_t), where γ is a descendant of α at layer l_c in T_2, h is a descendant of c at layer l_a in T_1 and η is an ancestor of α at layer l_t in T_2 (Fig. 4.a). Since $\mathcal{A}$ can choose any link $((a, \alpha, l_a), (b, \beta, l_b))$ labeled (I_1, I_2, l_b) such that $l_t - l_b > l_t - l_a$ and $t \in I_1$, then it is able to choose any path of this sort, and this proves the assertion.

   The same reasoning applies when $l_t < l_a$ and when $l_t = l_a$.

4   LCP of General Graphs: Driving some Conclusions

In the previous section we have proposed a method to compute a compact routing scheme for all T-networks. In the proofs of correctness, we have strongly used the properties of the factors and having a unique (shortest) path between any couple of nodes. It is immediate to wonder whether our technique can be extended to networks which are the LCP of more general graphs. In this section we show a negative result in this direction. Namely, we prove a property of the LCP allowing one to deduce that the knowledge of node- and link-labels on factors (and therefore the knowledge of their shortest paths) gives not enough information to find shortest paths in their LCP.

Theorem 3. *There exist a layered graph $G = (V, E)$ LCP of $G_1 = (V_1, E_1)$ and $G_2 = (V_2, E_2)$, a source node $(s, \sigma, l_s) \in V$, a destination node $(t, \tau, l_t) \in V$ and an edge $e \in E$ having the following properties: i. e is the layered cross product of two edges $e_1 \in E_1$ and $e_2 \in E_2$; ii. e belongs to a shortest path from (s, σ, l_s) to (t, τ, l_t); iii. neither e_1 belongs to a shortest path from s to t in G_1, nor e_2 belongs to a shortest path from σ to τ in G_2.*

Proof. The graph in the assertion is shown in Fig.5 in which the following convention is used: each edge in the drawing of G_1 and G_2 (and therefore of G) represents a simple chain whose length is determined by the difference of layers where its extremes lie.

Suppose the shortest path from s to t in G_1 passes through m and the shortest path from σ to τ passes through ν. Then, the following relations must hold:
$$|l_s - l_t| + 2|l_t - l_m| < 2|l_s - l_d| + |l_s - l_t| \text{ and } 2|l_s - l_n| + |l_s - l_t| < |l_s - l_t| + 2|l_t - l_r|$$
that is $|l_m - l_t| < |l_s - l_d|$ and $|l_s - l_n| < |l_t - l_r|$.

Whenever one of the following inequalities hold
$$|l_s - l_t| + 2|l_t - l_r| < 2|l_s - l_n| + |l_s - l_t| + 2|l_t - l_m|$$
$$2|l_s - l_d| + |l_s - l_t| < 2|l_s - l_n| + |l_s - l_t| + 2|l_t - l_m|$$
either the path through (m, μ_1, l_m) and (r, ρ, l_r) (first inequality) or the path through (n_2, ν, l_n) and (d, δ, l_d) (second inequality) are shorter than the path through (n_2, ν, l_n) and (m, μ_2, l_m). That is, the shortest path passes through an edge – either $((m, \mu_1, l_m), (r, \rho, l_r))$ or $((n_2, \nu, l_n), (d, \delta, l_d))$ – whose factors are not on a shortest path.

As a consequence of the previous theorem, we can state the following fact:

Fact 41 *Let G be a network that is the LCP of any two graphs G_1 and G_2. The only knowledge of the compact routing schemes (i.e. a node- and link-labeling scheme) on G_1 and G_2 may not be sufficient to deduce a compact routing scheme for G.*

Anyway, the previous claim does not forbid one to find some special cases in which the particular structure either of the network itself or of its factors helps in defining a compact routing scheme.

1038

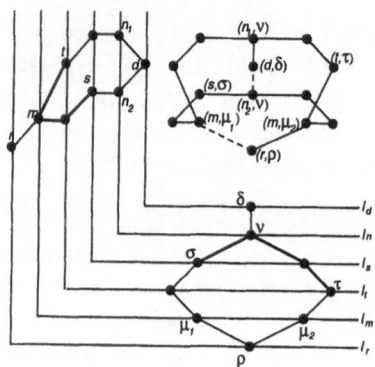

Fig. 5. Edges on a shortest path in G whose factors do not lie on any shortest path in G_1 and in G_2.

Acknowledgments: We thank Richard B. Tan for a careful reading of earlier drafts and for his helpful suggestions on improving the paper.

References

1. S. Even, A. Litman, "Layered Cross Product - A technique to construct interconnection networks", 4th *ACM SPAA*, 60-69, 1992.
2. M. Flammini, J. van Leeuwen, A. Marchetti-Spaccamela, "The complexity of interval routing in random graphs", *MFCS 1995*, LNCS 969, 37-49, 1995.
3. P. Fraigniaud, C. Gavoille, "Optimal interval routing", *CONPAR*, LNCS 854, 785-796, 1994.
4. C. Gavoille, S. Perennes, "Lower bounds for shortest path interval routing", *SIROCCO'96*, 1996.
5. R. Královič, P. Ružička, D. Štefankovič, "The complexity of shortest path and dilation bounded interval routing", *Euro-Par'97*, LNCS 1300, 258-265, 1997.
6. J. van Leeuwen, R.B. Tan, "Computer networks with compact routing tables", in: G. Rozemberg and A. Salomaa (Eds.) *The Book of L*, Springer-Verlag, Berlin, 298-307, 1986.
7. J. van Leeuwen, R.B. Tan, "Interval routing", *Computer Journal*, 30, 298-307, 1987.
8. J. van Leeuwen, R.B. Tan, "Compact routing methods: a survey", *Colloquium on Structural Information and Communication Complexity (SICC'94)*, 1994.
9. F.T. Leighton, *Introduction to Parallel Algorithms and Architectures: Arrays, Trees, Hypercubes.* Morgan Kaufmann Publishers, Inc.,1992.
10. C.E. Leiserson, "Fat-Trees: universal networks for hardware-efficient supercomputing", *IEEE Trans. on Comp.*, 34, 892-901, 1985.
11. P. Ružička, "On efficiency of Interval Routing algorithms", in: M.P. Chytil, L. Janiga, V. Koubeck (Eds.), *MFCS 1988*, LNCS 324, 492-500, 1988.
12. N. Santoro, R. Khatib, "Routing without routing tables", *Tech. Rep. SCS-TR-6*, School of Computer Science, Carleton University, 1982. Also as: "Labelling and implicit routing in networks", *Computer Journal*, 28, 5-8, 1985.

13. S.S.H. Tse, F.C.M. Lau, "A lower bound for interval routing in general networks", *Tech. Rep. 94-04*, Dept. of Computer Science, The University of Hong Kong, 1994 (to appear in *Networks*).

Gossiping Large Packets on Full-Port Tori

Ulrich Meyer and Jop F. Sibeyn

Max-Planck-Institut für Informatik
Im Stadtwald, 66123 Saarbrücken, Germany.
umeyer, jopsi@mpi-sb.mpg.de
http://www.mpi-sb.mpg.de/~umeyer, ~jopsi/

Abstract. Near-optimal gossiping algorithms are given for two- and higher dimensional tori. It is assumed that the amount of data each PU is contributing is so large, that start-up time may be neglected. For two-dimensional tori, an earlier algorithm achieved optimality in an intricate way, with a time-dependent routing pattern. In our algorithms, in all steps, the PUs forward the received packets in the same way.

1 Introduction

Meshes and Tori. One of the most thoroughly investigated interconnection schemes for parallel computation is the $n \times n$ *mesh*, in which n^2 processing units, *PUs*, are connected by a two-dimensional grid of communication links. Its immediate generalizations are d-dimensional $n \times \cdots \times n$ meshes. Despite their large diameter, meshes are of great importance due to their simple structure and efficient layout.

Tori are the variant of meshes in which the PUs on the outside are connected with "wrap-around links" to the corresponding PUs at the other end of the mesh, thus tori are node symmetric. Furthermore, for tori the bisection width is twice as large as for meshes, and the diameter is halved. Numerous parallel machines with two- and three-dimensional mesh and torus topologies have been built.

Gossiping. Gossiping is a fundamental communication problem used as a subroutine in parallel sorting with splitters or when solving ordinary differential equations: Initially each of the P PUs holds one packet, which must be routed such that finally all PUs have received all packets (this problem is also called *all-to-all broadcast*).

Earlier Work. Recently Šoch and Tvrdík [5] have analyzed the gossiping problem under the following conditions: Packets of unit size can be transferred in one *step* between adjacent PUs, *store-and-forward model*. In each step a PU can exchange packets with all its neighbors, *full-port model*.

The store-and-forward model gives a good approximation of the routing practice, if the packets are so large that the start-up times may be neglected. If an algorithm requires $k > 1$ packets per PU, then this effectively means nothing more than that the packets have to be divided in k chunks each.

In [5], it was shown that on a two-dimensional $n_1 \times n_2$ torus, the given problem can be solved in $\lceil (n_1 \cdot n_2 - 1)/4 \rceil$ steps if $n_1, n_2 \geq 3$. This is optimal. The algorithm is based on time-arc-disjoint broadcast trees: The action of each PU is time dependent, and thus, for every routing decision, a PU has to perform some non-trivial computation (alternatively these actions can be precomputed, but for a d-dimensional torus with P PUs this requires $\Theta(P)$ storage per processor).

New Results. In this paper, we analyze the same problem as Šoch and Tvrdík. Clearly, we cannot improve their optimality. Instead, we try to determine the minimal concessions towards simple time-independent algorithms: In our gossiping algorithms, after some $\mathcal{O}(d)$ precomputation on a d-dimensional torus, a PU knows once and for all, that packets coming from direction x_i have to be forwarded in direction x_j, $1 \leq i, j \leq d$. Time-independence ensures that the routing can be performed with minimal delay, for a fixed size network the pattern might even be built into the hardware. Also on a system on which the connections must be somehow switched, this is advantageous.

By routing the packets along d edge disjoint Hamiltonian paths, it is not hard to achieve the optimal number of steps if each PU initially stores $k = d$ packets. Here optimal means a routing time of $k \cdot P/(2 \cdot d)$ steps on a torus with P PUs. For $d = 2$, the schedule presented in Section 2 is particularly simple, and might be implemented immediately. Our scheme remains applicable for higher dimensions: the proof that edge-disjoint Hamiltonian cycles exist [3, 2, 1] is constructive, but these constructions are fairly complicted. Unfortunately, it is not possible to achieve the optimal number of steps using d fixed edge disjoint Hamiltonian paths and only one packet per PU which then circulates concurrently along all d paths. In [4] we show for the two-dimensional case that at least $P/40$ extra steps are needed due to multiple receives.

In Section 3, we investigate a second approach that allows an additional $o(P)$ routing steps. On a d-dimensional torus, it runs in $P/(2 \cdot d) + o(P)$ steps, with only one packet per PU. For $d = 2$, this is almost optimal and time independent. Therefore, this algorithm might be preferable over the one from [5]. More importantly, for higher dimensions, particularly for the practically relevant case $d = 3$, this gives the first simple and explicit construction that achieves close to optimally. In these algorithms, we construct *partial* Hamiltonian cycles: on a d-dimensional torus, we construct d cycles, each of which covers $P/d + o(P)$ PUs. These are such that for every cycle, every PU is adjacent to a PU through which this cycle runs.

2 Optimal-Time Algorithms

In this section we describe a simple optimal gossiping algorithm for two-dimensional $n_1 \times n_2$ tori assuming that initially each PU holds $k = 2$ packets.

Basic Case. The PU with index (i, j) lies in row i and column j, $0 \leq i < n_1$, $0 \leq j < n_2$. PU $(0, 0)$ is located in the upper-left corner. First PU (i, j) determines whether j is odd or even and sets its routing rules as follows:

$$j < n_2 - 1, j \text{ even} : T \leftrightarrow R; B \leftrightarrow L.$$
$$j < n_2 - 1, j \text{ odd} : T \leftrightarrow L; B \leftrightarrow R.$$

Here T, B, L, R designate the directions "top", "bottom", "left" and "right", respectively. By $T \leftrightarrow R$, we mean that the packets coming from above should be routed on to the right, and vice-versa. The other $\leftrightarrow$ symbols are to be interpreted analogously. Only in the special case $j = n_2 - 1$ we apply the rule $T \leftrightarrow R; B \leftrightarrow L$. The resulting routing scheme is illustrated in the left part of Figure 1.

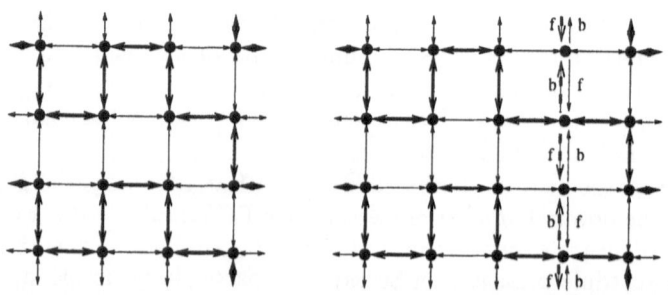

Fig. 1. Left: A Hamiltonian cycle on a 4×4 torus, whose complement (drawn with thin lines) also gives a Hamiltonian cycle. Right: Partial Hamiltonian cycles on a 4×5 torus. The PUs in column 3 lie on only one cycle. Such a PU passes the packets on this cycle that are running forwards to the PU below it, and those that are running backwards to the PU above it.

Odd n_i. Now we consider the case n_1 even and n_2 odd, the remaining cases can be treated similarly. Here we do not construct complete Hamiltonian cycles, but cycles that visit most PUs, and pass within distance one from the remaining PUs. Except for Column $n_2 - 2$, the rules how to pass on the packets are the same as in the basic case. In Column $n_2 - 2$ we perform $L \leftrightarrow R$.

PUs which do not lie on a given cycle, out-of-cycle-PUs, abbreviated OOC-PUs, are provided with the packets transferred along this cycle by their neighboring on-cycle-PUs, OC-PUs. With respect to different cycles, a PU can be both out-of-cycle and on-cycle. The packets received by an OOC-PU are not forwarded. This can be achieved in such a way, that a connection has to transfer only one packet in every step. The resulting routing scheme is illustrated in the right part of Figure 1.

Changing Direction. Each OOC-PU C receives packets from two different OC-PUs, A and B. Let A transfer the packets that are walking forwards, let B provide the packets in backward direction. If m denotes the cycle length and A and B are not adjacent on the cycle, then $m/2$ circulation steps are not enough: some packets are received from both directions, while others pass by without notice. We modify the algorithm as follows. Let l be the number of OC-PUs between A and B, then during the first l steps, A transfers to C the packets that are running forward, and during the last $m/2 - l$ steps those that run backward. B operates oppositely. In our case $l = 2 \cdot n_2 - 1$ for all PUs in Column $n_2 - 2$. Thus, each PU can still compute its routing decisions in constant time.

Theorem 1 *If every PU of an $n_1 \times n_2$ torus holds 2 packets, then gossiping can be performed in $\lceil n_1 \cdot n_2/2 \rceil$ steps.*

The considered schemes have the advantage that their precomputation can be performed in constant time. Alternatively, one might use the scheme in [3] for the construction of two edge-disjoint Hamiltonian cycles on any two-dimensional torus.

3   One-Packet Algorithms

In this section we give another practical alternative to the approach of [5]: we present algorithms which require only one packet per PU and are optimal to within $o(P)$ steps. The idea is simple: we construct d edge-disjoint cycles of length $P/d + o(P)$. Each of the cycles must have the special property, that if a PU does not lie on such a cycle, this PU must be adjacent to two other PUs belonging to this cycle. Each of these two PUs will transmit the packets from one direction of the cycle to the out-of-cycle-PU.

Two-Dimensional Tori. The construction can be viewed as an extension of the approach described in Section 2. There we interrupted the regular zigzag pattern for one column in order to cope with an odd value for n_2. Now we discard the zigzag in $n_2 - 2$ consecutive columns, the only two remaining zigzag columns connect the long stretched row parts to form two partial cycles:

$$j = 0 : T \leftrightarrow R; B \leftrightarrow L.$$
$$j = 1 : T \leftrightarrow L; B \leftrightarrow R.$$
$$2 \leq j < n_2 : L \leftrightarrow R.$$

Binding the out-of-cycle-PUs is done exactly as in the algorithm of Section 2; the case of odd n_1 and n_2 can be treated by inserting one special row.

Each of the partial cycles consists of $n_1 \cdot n_2/2 + n_1$ PUs, so using both directions in parallel one needs $n_1 \cdot n_2/4 + n_1/2$ steps to spread all the packets within the cycle. For every OOC-PU the two supplying OC-PUs are separated by $n_2 + 1$ other PUs on their cycle, thus we obtain a fully time-independent $n_1 \cdot n_2/4 + \mathcal{O}(n_1 + n_2)$ step algorithm. Applying the forward/backward switching as presented in Section 2 we get

Theorem 2 *If every PU of an $n_1 \times n_2$ torus holds 1 packet, then gossiping can be performed in $n_1 \cdot n_2/4 + n_1/2 + 1$ steps.*

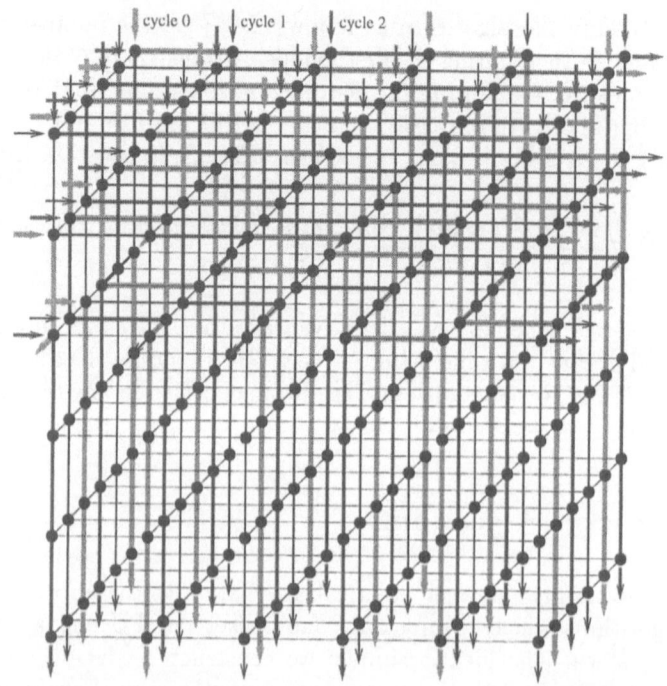

Fig. 2. Three edge-disjoint cycles used in the one-packet algorithm on a three-dimensional torus.

Three-Dimensional Tori.

For three-dimensional tori, we generalize the previous routing strategy, showing more abstractly the underlying approach. We assume that n_2 and n_3 are divisible by three. The constructed patterns are similar to the bundle of rods in a nuclear power-plant.

We construct three cycles. These are identical, except that they start in different positions: Cycle j, $0 \leq j \leq 2$, starts in PU $(0, j, 0)$. The cycles for two-dimensional tori were composed of $n_1/2$ *laps*: sections of a cycle starting with a zigzag and ending with the traversal of a wrap-around connection. The zigzags were needed to bring us two positions further in order to connect to the next lap. Here, we have three zigzags, bringing us three positions further. The sequence of directions in the zigzag pattern is given by $(2, 1, 2, 1, 2, 1)$. After these zigzags, moves in direction 1 are repeated until a wrap-around connection is traversed, the next lap can begin.

In this way we can fill up one plane (using $n_2/3$ laps), but in order to get to the next plane, there must be a second zigzag type consisting of the following direction pattern: $(2, 1, 2, 1, 3, 1)$. This type is applied every $n_2/3$ laps. The total schedule is illustrated in Figure 2.

If PU $P = (x, y, z)$, $x \geq 3$ belongs to cycle i, then PUs $P_r = (x, (y + 1) \bmod 3, z)$ and $P_d = (x, y, (z + 1) \bmod 3)$ belong to cycle $(i + 1) \bmod 3$, $P_l =$

$(x, (y-1) \bmod 3, z)$ and $P_u = (x, y, (z-1) \bmod 3)$ belong to cycle $(i-1) \bmod 3$. In this way, P on cycle i can be supplied with packets which are not lying on its own cycle: it receives packets running forward on cycle $(i+1) \bmod 3$ from P_r, backward-packets from P_d. P_l provides packets running backward on cycle $(i-1) \bmod 3$, forward-packets are sent by P_u. Each pair of supporting on-cycle-PUs is separated by $(n_1/3 + 1) \cdot n_2 - 1$ other PUs.

In the upper part of our reactor, exactly two cycles pass through each PU P, and the shifts are so, that the connections that are not used by cycle traffic lead to PUs that lie on the other cycle. At most $2 \cdot (n_1/3 + 1) \cdot n_2 - 1$ PUs lie between two supporting on-cycle-PUs. Multiple receives in the OOC-PUs can again be eliminated by the forward/backward switching of Section 2. Otherwise $\mathcal{O}(n_1 \cdot n_2)$ extra steps are necessary.

Theorem 3 *If every PU of an $n_1 \times n_2 \times n_3$ torus (n_2, n_3 integer multiples of three) holds 1 packet, then gossiping can be performed in $n_1 \cdot n_2 \cdot n_3/6 + n_1 \cdot n_2/2 + 1$ steps.*

Higher Dimensional Tori. The idea from the previous section can be generalized without problem for d-dimensional tori. Now, we construct d edge-disjoint cycles, each of them covering $P/d + o(P)$ PUs.

The cycles are numbered 0 through $d-1$. Cycle j starts in position $(0, j, 0, \ldots, 0)$. As before a cycle is composed of laps, starting with a zigzag and ending with moves in direction 1. Now there are $d-1$ types of zigzags, which are used in increasingly exceptional cases. The general pattern telling along which axis a (positive) move is made is as follows:

$$zigzag(2, 1) = (2, 1, 2, 1).$$

$$zigzag(3, 1) = (2, 1, 2, 1, 2, 1),$$
$$zigzag(3, 2) = (2, 1, 2, 1, 3, 1).$$

$$zigzag(i, 1) \doteq (zigzag(i-1, 1), 2, 1),$$
$$\vdots$$
$$zigzag(i, i-2) \doteq (zigzag(i-1, i-2), 2, 1),$$
$$zigzag(i, i-1) = (2, 1, 2, 1, 3, 1, 4, 1, 5, 1, \ldots, i-1, 1, i, 1).$$

Here $zigzag(d, j)$ indicates the j-th zigzag pattern for d-dimensional tori. During lap i, $1 \le i \le n_2/d \cdot n_3 \cdot \cdots \cdot n_d$, the highest numbered zigzag is applied for which the condition $i \bmod (n_2/d \cdot n_3 \cdot \cdots \cdot n_j) = 0$ is true. Namely, after precisely so many laps, the cycle has filled up the hyperspace spanned by the first j coordinate vectors.

Using induction over the number of dimension we can prove

Theorem 4 *If every PU of a d-dimensional $n_1 \times \cdots \times n_d$ torus ($n_2, \ldots, n_d$ integer multiples of d) with P PUs holds 1 packet, then gossiping can be performed in $P/(2 \cdot d) + P/(2 \cdot n_1) + 1$ steps.*

4 Conclusion

We have completed the analysis of the gossiping problem on full-port store-and-forward tori. In [5] only one interesting aspect of this problem was considered. We have shown that almost equally good performance can be achieved by simpler time-independent algorithms, and given explicit schemes for higher-dimensional tori as well.

References

1. Alspach, B., J-C. Bermond, D. Sotteau, 'Decomposition into Cycles I: Hamilton Decompositions,' *Proc. Workshop Cycles and Rays*, Montreal, 1990.
2. Aubert, J., B. Schneider, 'Decomposition de la Somme Cartesienne d'un Cycle et de l'Union de Deux Cycles Hamiltoniens en Cycles Hamiltonien,' *Discrete Mathematics*, 38, pp. 7–16, 1982.
3. Foregger, M.F., 'Hamiltonian Decomposition of Products of Cycles,' *Discrete Mathematics*, 24, pp. 251–260, 1978.
4. Meyer, U., J. F. Sibeyn, 'Time-Independent Gossiping on Full-Port Tori,' *Techn. Rep. MPI-98-1-014*, Max-Planck Inst. für Informatik, Saarbrücken, Germany, 1998.
5. Šoch, M., P. Tvrdík, 'Optimal Gossip in Store-and-Forward Noncombining 2-D Tori,' *Proc. 3rd International Euro-Par Conference*, LNCS 1300, pp. 234–241, Springer-Verlag, 1997.

Time-Optimal Gossip in Noncombining 2-D Tori with Constant Buffers[*]

Michal Šoch and Pavel Tvrdík

Department of Computer Science and Engineering
Czech Technical University, Karlovo nám. 13
121 35 Prague, Czech Republic
{soch,tvrdik}@sun.felk.cvut.cz
http://cs.felk.cvut.cz/pcg/

Abstract. This paper describes a time-, transmission-, and memory-optimal algorithm for gossiping in general 2-D tori in the noncombining full-duplex and all-port communication model.

1 Introduction

Gossiping, also called *all-to-all broadcasting*, is a fundamental collective communication problem: each node of a network holds a packet that must be delivered to every other node and all nodes start simultaneously. In this paper, we consider gossiping in the *noncombining store-and-forward all-port full-duplex* model. In such a model, a gossip protocol consists of a sequence of *rounds* and broadcast trees can be viewed as *levels* of arc sets, one level for one round. Given a broadcast tree $BT(u)$ rooted in vertex u of a graph G, the arcs crossed by packets in round t are denoted by $\mathcal{A}_t(BT(u))$. The *height* of a broadcast tree $BT(u)$, denoted by $h(BT(u))$, is the number of levels of BT, i.e., the number of rounds of a broadcast using this tree. Given $BT(u)$ and $i < h(BT(u))$, the *i-th level subtree* of $BT(u)$, denoted by $BT^{[i]}(u)$, is the subtree of $BT(u)$ consisting of the first i levels.

Given a graph G, every node of G has to receive $|\mathcal{V}(G)| - 1$ packets and in one round, it can receive $\delta(G)$ packets in the worst case, where $|\mathcal{V}(G)|$ is the size of G and $\delta(G)$ is the minimum degree of a node in G. The lower bound on the number of rounds of a gossip in G is therefore $\tau_g(G) = \left\lceil \frac{|\mathcal{V}(G)|-1}{\delta(G)} \right\rceil$.

If G is a vertex-transitive network, the gossip problem can be solved by constructing a *generic* broadcast tree, denoted by $BT(*)$, whose root can be placed into any vertex of G. All broadcast trees are therefore isomorphic.

2-D torus $T(m, n)$ is a cross product of 2 cycles of size m and n. Vertex set of $T(m, n)$ is the cartesian product $\{0, .., m - 1\} \times \{0, .., n - 1\}$.

Tori are vertex transitive and the isomorphic copies of the generic broadcast trees are made by *translation*.

---

[*] This research was supported by GAČR Agency under Grant 102/97/1055, FRVŠ Agency under Grant 1251/98, and CTU Grant 3098102336.

Definition 1. *Given nodes* $u = [u_x, u_y]$ *and* $v = [v_x, v_y]$ *of* $T(m, n)$, *a translation from* u *to* v, *denoted by* $\psi_{u \to v}$, *is induced by node mapping* $([w_x, w_y] \mapsto [w_x \oplus_m v_x \ominus_m u_x, w_y \oplus_n v_y \ominus_n u_y]$, *where the addition and subtraction is taken modulo* m *and* n, *respectively.*

A gossip in $T(m, n)$ is time- and transmission-optimal iff $h(BT(*)) = \tau_g(T(m, n))$ $= \lceil \frac{mn-1}{4} \rceil$ and all isomorphic copies of $BT(*)$ are pairwise *time-arc-disjoint*, i.e., two arcs at the same time-level of any two broadcast tree are never mapped on the same arc of $T(m, n)$ and the communication in every round is therefore contention-free. It is known that in tori, time-arc-disjointedness is equivalent to the distinctness of directions of arcs at every level of the generic tree. The set of directions of arcs $\mathcal{A}_t(BT(*))$ is denoted by $\text{dir}(\mathcal{A}_t(BT(*)))$. In a 2-D torus there exist four directions, usually denoted by N,E,W,S. N-S direction is vertical, W-E direction is horizontal. Hence, a sufficient condition for $BT(*)$ to guarantee a time-optimal gossip is that for any $t < \tau_g(T(m, n))$, every $\mathcal{A}_t(BT(*))$ is a set of 4 arcs of 4 distinct directions N, E, W, S.

In general, a gossip algorithm in $T(m, n)$ may require additional buffers in routers for packets which must wait at least one round before they can be sent out and the routers can get rid of them. Let $\beta(G)$ denote the maximum size of auxiliary buffers per router during gossiping in network G.

In [2], we have presented a time-optimal gossip protocol for general $T(m, n)$. However, this algorithm is not memory-optimal, it requires auxiliary buffers for $\Theta(\max(n, m - n))$ packets per router. The generic broadcast tree of $T(m, n)$ in [2] is built in two steps: filling up the maximal square submesh of odd side + informing the rest of nodes. Additional buffers for packets are required on the sides of the square submesh.

In this paper, we present a time- and memory-optimal gossip algorithm for $T(m, n)$ which requires $\beta(T(m, n)) = 3$. To keep the size of auxiliary buffers constant, the generic broadcast tree is built from vertical stripes of width 2. Only constant number of packets must be stored for dissemination in directions W and E, which allows to concatenate vertical stripes horizontally.

2 Generic Broadcast Trees for Optimal Gossip on $T(m, n)$.

Theorem 1. *For any* $m \geq n \geq 2$, *there exists a generic* $BT(*)$ *of* $T(m, n)$ *such that* $h(BT(*)) = \lceil \frac{mn-1}{4} \rceil$ *and* $|\mathcal{A}_t(BT(*))| = |\text{dir}(\mathcal{A}_t(BT(*)))|$ *for all* $1 \leq t \leq h(BT(*))$ *and* $\beta(T(m, n)) \leq 3$. *If* $n = 2$ *and* m *is even, one extra round is needed.*

Proof. For $m = n$, the algorithm is trivial and $\beta(T(m, m)) = 0$ if m is odd and $\beta(T(m, m)) = 1$ otherwise. Assume without losing generality $m > n$. The construction of time-arc-disjoint trees for time- and memory-optimal gossip depends slightly on the values of m and n, the number of different cases is 7, the basic idea is, however, the same in all of them, see [3] for details.

The memory requirements of our algorithm are stated in the following table.

	$n=2$	$n=3$, $m\leq5$	n odd	$n\geq4$ even, m odd	$n\geq4$ even, m even
$\beta(T(m,n))$	0	0	2	2	3

In this paper, we describe only one particular case when $m > n \geq 5$ are odd numbers. To make the construction of $BT(*)$ in $T(m,n)$ as simple as possible, the same patterns of arc sets $A_t(BT(*))$ are used repeatedly in various rounds t. The whole generic tree $BT(*)$ is then built using several arc patterns. Arc pattern i is depicted on Figures 1 and 2 as a quadruple of arcs labeled i. We associate with every pattern i a so called *expansion operator*, denoted by Γ_i. The broadcast tree is then specified by a regular expression over expansion operators. If $A_t(BT(*))$ has pattern i, an *attachment* of arcs at level t to $(t-1)$-level generic subtree is described as an *application* of the corresponding operator $BT^{[t]}(*) = BT^{[t-1]}(*)\Gamma_i$.

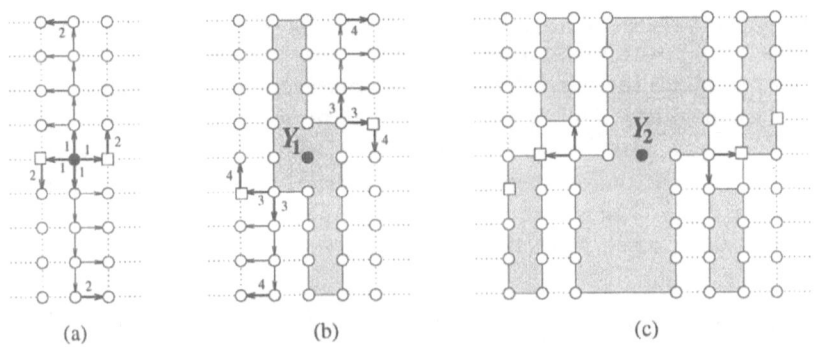

(a) (b) (c)

Fig. 1. The first phase of building $BT(*)$ in $T(m,n)$, if $m \geq 11$, $m > n \geq 5$ are odd. (a) $Y_1 = \Gamma_1^{\frac{n-1}{2}}\Gamma_2$. (b) $Y_2 = Y_1\Gamma_3^{\frac{n-3}{2}}\Gamma_4$. (c) Y_2^2.

The generic tree is built in two phases. Figure 1(a) depicts the first phase of constructing $BT(*)$ if $m \geq 11$. If $m \leq 9$, the first phase is void. The black circle is the root of $BT(*)$. The square symbols $\square$ denoted the nodes which have to store packets needed in later steps. Repeated application of Γ_1 in Figure 1(a) is followed by Γ_2 which produces $\frac{n+1}{2}$ level subtree $Y_1 = BT^{[\frac{n+1}{2}]}(*) = \Gamma_1^{\frac{n-1}{2}}\Gamma_2$. Note that the expansion by Γ_1 proceeds in N-S direction. Further $\frac{n-3}{2}$ levels are attached in direction N-S by repeated application of Γ_3 and after applying Γ_4, we get n-level subtree $Y_2 = BT^{[n]}(*) = Y_1\Gamma_3^{\frac{n-3}{2}}\Gamma_4$. If $m \geq 15$, the whole process can be repeated by expanding Y_2 in W-E direction, see Figure 1(c). The second phase depends on $m \bmod 4$. Let us describe the case of $m \equiv 1 \bmod 4$ (the other case is similar). Let $Y_3 = Y_2^{\frac{m-9}{4}}$. If $m = 9$, then Y_3 shrinks to the root.

Figure 2 depicts the solution. We interpret the use of expansion operators similarly as in the first phase. In $\frac{3n-1}{2}$ rounds, we expand Y_3 to $Y_4 = Y_3Y_1\Gamma_3^{\frac{n-3}{2}}\Gamma_5\Gamma_3^{\frac{n-3}{2}}\Gamma_4$ (see Figure 2(a)). Y_4 is then expanded in $3\lfloor\frac{n-1}{4}\rfloor$ rounds using Γ_6 and Γ_7 (see Figure 2(b)). Pattern $\Gamma_6^2\Gamma_7$ diffuses vertically until the

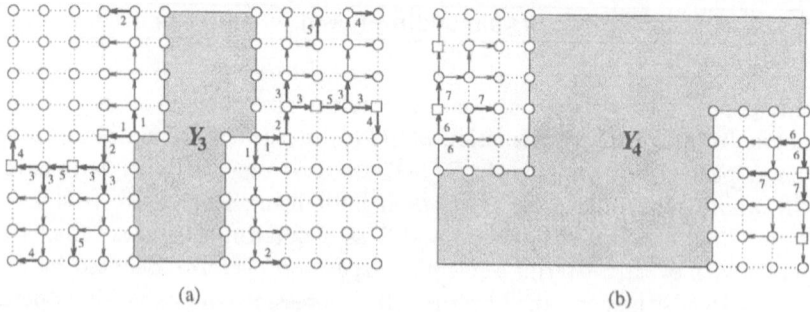

Fig. 2. The second phase of constructing $BT(*)$, if $m > n \geq 5$ are odd, $m \equiv 1 \bmod 4$. (a) Construction of Y_4. (b) Application of Γ_6 and Γ_7 patterns.

boundary is reached. Finally, if $n \equiv 1 \bmod 4$ (as shown on Figure 2(b)), one final round is needed, and if $n \equiv 3 \bmod 4$, after two more applications of Γ_6, only two uninformed nodes surrounded by informed ones remain.

The memory requirements for this case of m and n follow easily. In the first phase, 2 packets must be stored to apply Γ_2 (see the square symbols in Figure 1(a)). Then, these buffers can be reused to store packets for application of Γ_4 (see Figure 1(b)). The second phase is similar. In every round of the gossip, no more than two packets must be stored at a time. □

3 Conclusions

The minimal-height time-arc-disjoint trees in the proof of Theorem 1 provide time-, transmission-, and memory-optimal gossip algorithm in noncombining full-duplex all-port 2-D tori. The algorithm requires routers with buffers for at most 3 packets. An interesting problem is to find the exact lower bound on the size of additional buffers. Another open problem is to find a time- and memory-optimal gossip algorithm for 2-D meshes.

References

1. J.-C. Bermond, T. Kodate, and S. Perennes. Gossiping in Cayley graphs by packets. In M. Deza, et. al., editors, *Combinatorics and Computer Science*, LNCS 1120, pages 301–315. Springer, 1995.
2. M. Šoch and P. Tvrdík. Optimal gossip in store-and-forward noncombining 2-D tori. In C. Lengauer, et. al., editors, *Euro-Par'97 Parallel Processing*, LNCS 1300, pages 234–241. Springer, 1997. Research report at http://cs.felk.cvut.cz/pcg/.
3. M. Šoch and P. Tvrdík. Time-optimal gossip in noncombining 2-D tori with constant buffers. Manuscript at http://cs.felk.cvut.cz/pcg/.

Divide-and-Conquer Algorithms on Two-Dimensional Meshes*

Miguel Valero-García, Antonio González, Luis Díaz de Cerio and Dolors Royo

Dept. d'Arquitectura de Computadors - Universitat Politcnica de Catalunya
c/Jordi Girona 1-3, Campus Nord - D6, E-08034 Barcelona (Spain)
{miguel, antonio, ldiaz, dolors}@ac.upc.es

Abstract. The Reflecting and Growing mappings have been proposed to map parallel divide-and-conquer algorithms onto two - dimensional meshes. The performance of these mappings has been previously analyzed under the assumption that the parallel algorithm is initiated always at the same fixed node of the mesh. In such scenario, the Reflecting mapping is optimal for meshes with wormhole routing and the Growing mapping is very close to the optimal for meshes with store-and-forward routing. In this paper we consider a more general scenario in which the parallel divide-and-conquer algorithm can be started at an arbitrary node of the mesh. We propose and approach that is simpler than both the Reflecting and Growing mappings, is optimal for wormhole meshes and better than the Growing mapping for store-and-forward meshes.

1 Introduction

The problem of mapping divide-and-conquer algorithms onto two - dimensional meshes was addressed in [2]. First, a binomial tree was proposed to represent divide-and-conquer algorithms. Then, two different mappings (called the Reflecting and Growing mappings) were proposed to embed binomial trees onto two-dimensional meshes. It was shown that the Reflecting mapping is optimal for wormhole routing since the required communication can be carried out in the minimum number of steps (there are not conflicts in the use of links). On the other hand, the Growing mapping was shown to be very close to the optimal for the case of store-and-forward routing.

The communication performance of the Reflecting and the Growing mappings was analyzed in [2] under the assumption that the divide-and-conquer algorithm is always started at a fixed node of the mesh. In the following, we use the term fixed-root to refer to this particular scenario. In this paper, we consider a more general scenario in which a divide-and-conquer algorithm can be started at any arbitrary node of the mesh. The term arbitrary-root will be used to refer to this scenario. It will be shown that the Reflecting mapping is still optimal for wormhole routing in the arbitrary-root scenario but the performance of the Growing mapping for store-and-forward routing can be very poor in some common cases.

---

* This work was supported by the Ministry of Education and Science of Spain (CICYT TIC-429/95).

An alternative solution for the arbitrary-root scenario is proposed in this paper, which is inspired in a previous work on embedding hypercubes onto meshes and tori [1]. The proposed scheme, which will be called DC-cube embedding, has the following properties: a) it can be applied to meshes with either wormhole or store-and-forward routing, b) it is significantly simpler than the Reflecting and Growing mappings, c) it is optimal for wormhole routing, and d) it is significantly faster than the Growing mapping in some common cases, for store-and-forward routing. A more detailed explanation of the results presented in this paper can be found in [4].

The rest of this paper is organized as follows. Section 2 reviews the Reflecting and Growing mappings [2], which are the basis for our proposal. In section 3, these mappings are extended to the arbitrary-root scenario. Section 4 presents our proposal. Finally, section 5 presents a performance comparison of the different approaches and draws the main conclusions.

2 Background

Rajopadhye and Telle [2] propose the use of a binomial tree to represent a divide-and-conquer algorithm. Every node of the binomial tree represents a process that performs the following computations:

1. Receive a problem (of size x) from the parent (the host, if the node is the root of the tree).
2. Solve the problem locally (if 'small enough"), or divide the problem into two subproblems, each of size αx), and spawn a child process to solve one of these parts. In parallel, start solving the other part, by repeating step 2.
3. Get the results from the children and combine them. Repeat until all children's results have been combined.
4. Send the results to the parent (the host, if the node is the root).

Step 2 is referred to as the division stage. It has a number of phases corresponding to the different repetitions of step 2 (the number of levels of the binomial tree). The division stage is followed by the combining stage in which the results of the different subprobems are combined to produce the final result. For the sake of simplicity, as in [2], only the division stage will be considered from now on. Two different values for parameter α (see step 2) will be considered: $\alpha = 1$ (the problem is replicated) and $\alpha = 1/2$ (the problem is halved).

The execution of a binomial tree on a two-dimensional mesh can be specified in terms of the embedding of the tree onto the mesh. Two different embeddings were proposed in [2]: the Reflecting mapping and the Growing mapping. It was shown that the Reflecting mapping is optimal for wormhole routing since the communications are carried out without conflicts in the use of the mesh links. On the other hand, the Growing mapping is close to the optimal under store-and-forward routing. These performance properties were derived assuming a fixed-root scenario, that is, the root of the tree is always assigned to the same mesh node.

3 Extending the Reflecting and Growing Mappings to the Arbitrary-Root Scenario

In the following, we consider the case in which the divide-and-conquer algorithm can be started at an arbitrary node (a, b) of the mesh (this is called the arbitrary-root scenario).

We have first considered two straightforward approaches to extend the Reflecting and Growing mappings to the arbitrary-root scenario. In approach A we build a binomial tree which is an isomorphism of the binomial tree used in [2] for the fixed-root scenario (every label of the new tree is obtained by a fixed permutation of the bits in the old label). The isomorphism is defined so that the root of the binomial tree is mapped (by the corresponding embedding Reflecting or Growing) onto node (a, b). In approach B, the whole problem is first moved from node (a, b) to the starting node according to the original proposal (for the fixed-root scenario). These approaches will be compared with our proposal, that is described in the next section.

4 A New Approach for the Arbitrary-Root Scenario

Our approach to perform a divide-and-conquer computation on a mesh, under the arbitrary-root scenario is inspired in the technique proposed in [1] to execute a certain type of parallel algorithms (which will be referred to as CC-cube algorithms) onto multidimensional meshes and tori.

A d-dimensional CC-cube algorithm consists in 2^d processes which cooperate to perform a certain computation and communicate using a d-dimensional hypercube topology (the dimensions of the hypercube will be numbered from 0 to $d - 1$). The operations performed by every process in the CC-cube can be expressed as follows:

```
do i=0, d-1
    compute
    exchange information with neighbour in dimension i
enddo
```

The above case corresponds to a CC-cube which uses the dimensions of the hypercube in increasing order. However, any other ordering in the use of the dimensions is also allowed in CC-cubes.

A CC-cube can executed in a mesh multicomputer by using an appropriate embedding. In particular, the standard and xor embeddings were proposed to map the CC-cube onto a multidimensional meshes and tori respectively. The properties of both embeddings were extensively analyzed in [1, 3].

A divide-and-conquer algorithm can be regarded as a particular case of CC-cube algorithm. This particular case will be referred as DC-cube (from Divide-and-Conquer) and has the following peculiarities with regard to CC-cubes:

- a) In every iteration, only a subset of the processes are active (all the processes are active in every iteration of a CC-cube).

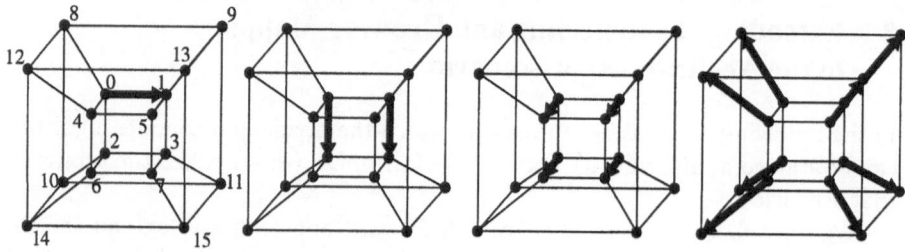

Fig. 1. The four iterations of a 4-dimensional DC-cube starting at process 0 and using the hypercube dimensions in ascending order.

– b) Communication between neighbour nodes is always unidirectional (instead of the bidirectional exchange used by CC-cubes).

A particular DC-cube is characterized by: (a) a process which is responsible for starting the divide-and-conquer algorithm, and (b) a certain ordering of the hypercube dimensions, which determine the order in which the processes of the DC-cube are activated. Figure 1 shows an example of a 4-dimensional DC-cube starting at process 0, and using the dimensions in ascending order. In this figure, the arrows represent the communications that are carried out in every iteration of the DC-cube. Finally, it can be shown that a binomial tree with the root labelled as l is equivalent to a d-dimensional DC-cube initiated at process l and using the hypercube dimensions in descending order.

The standard embedding has been proposed to map a $2k$-dimensional hypercube onto a $2^k \times 2^k$ mesh. The function S which maps a node i ($i \in [0, 2^{2k} - 1]$) of the hypercube onto a node (a, b) ($a, b \in [0, 2^k - 1]$) of the mesh is defined as[1]:

$$S(i) = \left(\left\lfloor \frac{i}{2^k} \right\rfloor, i \bmod 2^k \right) \ . \tag{1}$$

The properties of the standard embedding that are more relevant to this paper are: a) It is an embedding with constant distances (this property means that the neighbors in dimension i of the hypercube are found at a constant distance in the mesh, for any pair of neighbor nodes, and b) It has a minimal average distance. Property (a) is very attractive for the purpose of using the embedding on the arbitrary-root scenario. Property (b) is attractive from the performance point of view.

To start a divide-and-conquer algorithm from an arbitrary node (a, b) of the wormhole mesh, we use a DC-cube initiated at process $S^{-1}((a, b)) = a \cdot 2^k + b$, and using the dimensions of the hypercube in descending order. It is easy to see that the standard embedding of such a DC-cube is conflict free, and therefore the approach is optimal. The formal proof of this property is very similar to the

---

[1] The standard embedding can be easily extended to the general case of C-dimensional meshes. This extension is however out of the scope of this paper.

Table 1. Comparision of approaches

	t_s	t_e (general α)	t_e ($\alpha = 1$)	t_e ($\alpha = 0.5$)
G_A	$2^{k+1} - 2$	$(\alpha + \alpha^2)2^{k-1} + (\alpha^3 + \alpha^4)\frac{(2\alpha^2)^{k-1}-1}{2\alpha^2-1}$	$2^{k+1} - 2$	$\frac{3}{8}\left(2^k + 1 - \frac{1}{2^{k-1}}\right)$
G_B	$3 \cdot 2^{k-1} - 1$	$2^{k-1} - 1 + \alpha + \alpha^2 + (\alpha^3 + \alpha^4)\frac{(2\alpha^2)^{k-1}-1}{2\alpha^2-1}$	$3 \cdot 2^{k-1} - 1$	$2^{k-1} + \frac{1}{8} + \frac{3}{2^{k+2}}$
S	$2^{k+1} - 2$	$(\alpha + \alpha^2)\frac{2^k\alpha^{2k}-1}{2\alpha^2-1}$	$2^{k+1} - 2$	$\frac{3}{2}\left(1 - \frac{1}{2^k}\right)$

proof given in [2] for the case of the Reflecting mapping under the fixed-root scenario.

To start a divide-and-conquer algorithm from an arbitrary node (a, b) of the store-and-forward mesh, we use a DC-cube initiated at process $S^{-1}((a,b)) = a \cdot 2^k + b$, and using the dimensions of the hypercube in ascending order. In this way, the distances corresponding to the first iterations of the computation (involving larger messages) are smaller.

5 Comparison of Approaches and Conclusions

As a general consideration, it can be said that the standard embedding is significantly simpler than both Reflecting and Growing mapping.

When considering the wormhole arbitrary-root scenario, both the approach A for Reflecting mapping and the standard embedding of DC-cubes are optimal.

The comparison of approaches for the store-and-forward arbitrary-root scenario are summarized in table 1, in terms of the average communication cost (G_A for approach A, G_B for approach B and S for the standard embedding of DC-cube). The average cost G_A is defined as:

$$G_A = \frac{1}{2^{2k}} \sum_{a=0}^{2^k-1} \sum_{b=0}^{2^k-1} G^{a,b} \, , \tag{2}$$

where $G^{a,b}$ is the cost of the division stage when the computation is initiated at node (a, b). This cost is defined in terms of the startup cost incurred in every communication between neighbor nodes (denoted by t_s) and the transmission time per message size unit (denoted by t_e). The average costs G_B and S are defined in a similar way. Two particular cases of the term affecting t_e are distinguished, corresponding to $\alpha = 1$ and $\alpha = 0.5$.

The expressions in table 1 can be compared in two different cases. In the "small volume" case, the communication cost is assumed to be dominated by the term affecting t_s (this will happen when t_s/t_e is large and/or the problem is small). In the "large volume" case, the cost is assumed to be dominated by the term affecting t_e (this will happen when t_s/t_e is small and/or the problem is large).

The conclusions drawn from table 1 are:

- a) In the "small volume" case, approach B is the best, since: $G_A = S = (4/3)G_B$.

- b) In the "large volume" case and $\alpha = 1$, the conclusion is exactly the same as case (a).
- c) In the "large volume" case and $\alpha = 0.5$, approach S is significantly better than the rest, since:

$$G_A = \left(2^{k-2} + \frac{1}{2} \right) S \text{ and } G_B = \left(\frac{2^k}{3} + \frac{5}{12} \right) S . \tag{3}$$

Note that cases (a) and (c) are expected to be the most frequent since a parallel computer is targeted to solve large problems. Besides, they are also the most relevant since they may be very time consuming.

Note that in case (c) the improvement of the standard embedding of DC-cube over approaches A and B is proportional to the number of nodes and therefore it is very high for large systems.

References

1. González, A., Valero-García, M., Díaz de Cerio L.: Executing Algorithms with Hypercube Topology on Torus Multicomputers. IEEE Transactions on Parallel and Distributed Systems **8** (1995) 803–814
2. Lo, V., Rajopadhye, S., Telle, J.A.: Parallel Divide and Conquer on Meshes. IEEE Transactions on Parallel and Distributed Systems **10** (1996) 1049–1057
3. Matic, S.: Emulation of Hypercube Architecture on Nearest-Neighbor Mesh-Connected Processing Elements. IEEE Transactions on Computers **5** (1990) 698–700
4. Valero-García, M., González, A., Díaz de Cerio, L., Royo, D.: Divide-and-Conquer Algorithms on Two-Dimensional Meshes. Research Report UPC-DAC-1997-30, http://www.ac.upc.es/recerca/reports/INDEX1997DAC.html

All-to-all Scatter in Kautz Networks*

Petr Salinger and Pavel Tvrdík

Department of Computer Science and Engineering
Czech Technical University, Karlovo nám. 13
121 35 Prague, Czech Republic
{salinger,tvrdik}@sun.felk.cvut.cz

Abstract. We give lower and upper bounds on the average distance of
Kautz digraph. Using this result, we give a memory-optimal and asymptotically time-optimal algorithm for *All-to-All-Scatter* in noncombining
all-port store-&-forward Kautz networks.

1 Introduction

Shuffle-based interconnection networks, which include shuffle-exchange, de Bruijn,
and Kautz networks, are alternatives to orthogonal (meshes, tori, hypercubes)
and tree-based networks. Binary shuffle-based networks have been shown to have
similar properties as fixed-degree hypercubic networks (butterflies, CCCs), but
general de Bruijn and Kautz networks form a special family of interconnection
topologies whose properties are still poorly understood. Recently, the interest in
these networks has come from designers of optical interconnects [2, 3]. In this paper, we focus on d-ary Kautz digraphs. We give lower and upper bounds on the
average distance of Kautz digraph. Using this result, we give a memory-optimal
and asymptotically time-optimal algorithm for *All-to-All-Scatter* in noncombining all-port Kautz networks.

Vertex (arc) set of graph G is denoted by $V(G)$ ($A(G)$, respectively). The
alphabet of d letters, $\{0, 1, \ldots, d-1\}$, is denoted by Z_d. For $d \geq 2$ and $D \geq 2$,
Kautz digraph of *degree* d and *diameter* D, $K(d, D)$, has vertex set $V(K(d, D)) =$
$\{x_1 \ldots x_D; x_i \in Z_{d+1} \wedge x_i \neq x_{i+1} \ \forall i \in \{1, \ldots, D-1\}\}$ and there is an arc
from vertex $x = x_1 \ldots x_D$ to vertex $y = y_1 \ldots y_D$, $x, y \in V(K(d, D))$, iff
$x_{i+1} = y_i \ \forall i \in \{1, \ldots, D-1\}$. The arc from vertex $x_1 \ldots x_D$ to $x_2 \ldots x_D x_{D+1}$ is
denoted by $\langle x_1 \ldots x_D x_{D+1} \rangle$.

2 The Average Distance of Kautz Digraphs

Let $N = |V(G)|$. The *distance* between vertices $x, y \in V(G)$ is denoted by $d(x, y)$.
The *average distance* of $x \in V(G)$, $\overline{d}(x)$, is the average of distances between x
and all the vertices of G (including x): $\overline{d}(x) = \frac{1}{N} \sum_{y \in V(G)} d(x, y)$. The *average*

* This research has been supported by GAČR grant No. 102/97/1055 and FRVŠ grant
No. 1251/98.

distance of G is defined as the average of average distances of all vertices of G:
$\overline{d}(G) = \frac{1}{N} \sum_{x \in V(G)} \overline{d}(x)$.

In $K(d, D)$, given vertices $x = x_1 \ldots x_D$ and $y = y_1 \ldots y_D$, there is a unique shortest dipath from x to y. To calculate $d(x, y)$ and to construct the shortest path from x to y, we have to find the largest *overlap* of x and y, i.e., the smallest i such that $x_{i+1} \ldots x_D = y_1 \ldots y_{D-i}$. Then $d(x, y) = i$ and the shortest path is obtained by performing left shifts introducing successively $y_{D+1-i}, \ldots, y_D$. This fact allows to compute easily $\overline{d}(x)$ for any x and consequently $\overline{d}(K(d, D))$.

In this section, we give tight upper and lower bounds on $\overline{d}(K(d, D))$. Our analysis is based on shortest-path spanning trees, similarly to de Bruijn networks [1]. The upper (lower) bound on the average distance is reached when all vertices are as far (close) as possible from the root.

Lemma 1. *In $K(d, D)$, for any vertex x,*

$$D - \frac{1}{(d-1)(1 - \frac{1}{d^2})} + \frac{\frac{D}{d} + \frac{1}{d-1}}{d^{D-2}(d^2-1)} \leq \overline{d}(x) \leq D - \frac{1}{d-1} + \frac{2}{(d^2-1)d^{D-1}}.$$

There are $(d+1)d$ vertices matching the upper bound. These are all vertices formed by alternating a and b, $a \neq b \in Z_{d+1}$. There are $(d+1)d(d-1)^{D-2}$ vertices matching the lower bound. These are all vertices $x = x_1 \ldots x_D$ where $x_i \neq x_D$, for all i, $1 \leq i \leq D - 1$.

3 Memory-Optimal and Asymptotically Time-Optimal AAS in Kautz Digraphs

All-to-all scatter (AAS), also called *complete exchange*, is a fundamental collective communication pattern. Every vertex has a special packet for every other vertex and all vertices start scattering simultaneously. Most of the published AAS algorithms assume that in one communication step, vertices can combine individual packets arbitrarily into longer messages and can recombine incoming compound messages into new compounds. This assumption is not very realistic, since it implies substantial overhead and compound messages can become very large. Hence, we assume the *all-port noncombining lock-step* model. Every packet travels in the network as a separate entity. All vertices start the scattering simultaneously and transmit packets with the same speed in *steps*. A *router* receives packets from all input links in one step, those destined for the local processor stores into the local processor memory, and the remaining ones together with those ejected by the local processor sends out using all output links in the next step. We assume *store-and-forward* hardware: every input and output buffer of routers can store one packet. If a packet cannot be retransmitted in the next step, it must be stored in an auxiliary buffer of the router. An AAS algorithm is said to be *memory-optimal* if the routers require auxiliary buffers for $O(1)$ packets.

The AAS algorithm is based on lexicographic labeling of input and output arcs of vertices and on greedy routing. To describe these notions, we need three auxiliary functions.

Definition 1. *For any $a \in Z_{d+1}$, $b \in Z_d$:* $\sigma(a,b) = \begin{cases} b+1 & \text{if } a \leq b, \\ b & \text{otherwise.} \end{cases}$

For $a \neq b \in Z_{d+1}$: $\xi(b,a) = \mu(a,b) = \begin{cases} b & \text{if } a > b, \\ b-1 & \text{if } a < b. \end{cases}$

Kautz vertex $x_1 \ldots x_D$ is adjacent from d vertices $\sigma(x_1, i)x_1 \ldots x_{D-1}, i = 0, \ldots, d-1$. The arc $\langle \sigma(x_1, i)x_1 \ldots x_D \rangle$, denoted by $x{\downarrow}i$, has *input* label i. Similarly, vertex $x_1 \ldots x_D$ is adjacent to d vertices $x_2 \ldots x_D\sigma(x_D, j)$, $j = 0, \ldots, d-1$. The arc $\langle x_1 \ldots x_D\sigma(x_D, j) \rangle$, denoted by $x{\uparrow}j$, has *output* label j. Hence arc $\langle x_1 \ldots x_{D+1} \rangle$ of $K(d, D)$ has input label $\xi(x_1, x_2)$ and output label $\mu(x_D, x_{D+1})$ and therefore $\langle x_1 \ldots x_{D+1} \rangle = (x_2 \ldots x_{D+1}){\downarrow}\xi(x_1, x_2) = (x_1 \ldots x_D){\uparrow}\mu(x_D, x_{D+1})$.

Every Kautz vertex $x = x_1 \ldots x_D$ is the root of a "virtual" complete d-ary tree of height D, whose level l, $1 \leq l \leq D - 1$, consists of all Kautz vertices $x_{l+1} \ldots x_D y_1 \ldots y_l$, constructed from $x_1 \ldots x_D$ by l left shift operations, and whose level-D vertices are all Kautz vertices $y_1 \ldots y_D$, $y_1 \neq x_D$. This tree is called the *greedy routing virtual tree* rooted in x, denoted by $GRT(x)$. Obviously, some Kautz vertices appear more than once in $GRT(x)$, this depends on the periodicity of x. All $GRT(x)$ have the same output arc labeling.

The AAS algorithm consists of two similar phases. The first phase has d^{D-1} rounds. In each round r, $1 \leq r \leq d^{D-1}$, of the first phase, every root $x = x_1 \ldots x_D$ inserts a d-tuple of packets into $GRT(x)$, one packet per output arc, and the packets move down the tree using D-step greedy routing.

Let $y = y_1 \ldots y_D$ be a destination vertex, $y_1 \neq x_D$. The D-step greedy routing in $GRT(x)$ sends packets from x to y via vertices $x_2 \ldots x_D y_1$, $x_3 \ldots x_D y_1 y_2$, $\ldots$, $x_D y_1 \ldots y_{D-1}$. Transmission of packets from vertices $x_i \ldots x_D y_1 \ldots y_{i-1}$ to vertices $x_{i+1} \ldots x_D y_1 \ldots y_i$, $1 \leq i \leq D$, is said to be the i-th *step* of round r. (We put $x_{D+1} \equiv y_1$). The core of the AAS algorithm is functions Φ_1 and Φ_2.

Definition 2. *Let* $x = x_1 \ldots x_D \in V(K(d, D))$, $0 \leq r \leq d^{D-1}-1$, $0 \leq i \leq d-1$. *Then* $\Phi_1(x, r, i) = y_1^{r,i} \ldots y_D^{r,i}$, *where*

$\quad y_1^{r,i} = \sigma(x_D, i),$

$\quad y_{k+1}^{r,i} = \sigma(y_k^{r,i}, q_k^r), \qquad\qquad\qquad k = 1, \ldots, D-1,$

$\quad q_k^r = (\xi(x_k, x_{k+1}) + r \text{ div } d^{k-1}) \bmod d, \quad k = 1, \ldots, D-1.$

Let $0 \leq r \leq d^{D-2} - 1$. *Then* $\Phi_2(x, r, i) = z_1^{r,i} \ldots z_D^{r,i}$, *where*

$\quad z_1^{r,i} = x_D,$

$\quad z_{k+1}^{r,i} = y_k^{r,i}, \qquad k = 1, \ldots, D-1.$

Router in Kautz vertex $x = x_1 \ldots x_D$ executes in the first phase of the AAS protocol the following algorithm:

```
for r = 0 to d^{D-1} - 1 do_sequentially{
    for i = 0 to d - 1 do_in_parallel{
        take the packet destined for vertex Φ₁(x, r, i) from the local memory;
        apply the first step of the D-step greedy routing};
    for j = 2 to D do_sequentially
        for i = 0 to d - 1 do_in_parallel{
            receive a packet from arc x↓i;
```

      apply the j-th step of the D-step greedy routing};
for $i = 0$ to $d - 1$ **do_in_parallel**{
    receive a packet from arc $x{\downarrow}i$;
    store packet into the local processor memory}};

The second phase has d^{D-2} rounds and uses $(D-1)$-step greedy routing from $x = x_1 \ldots x_D$ to $x_D y_1 \ldots y_{D-1}$, which is defined similarly, using function Φ_2 instead of Φ_1.

Theorem 1. *The AAS algorithm is memory-optimal and contention-free. Every arc of $K(d, D)$ in every step of every round is used for transmitting one packet. The algorithm is asymptotically time- and transmission-optimal.*

Proof. It follows from Definition 2 that for all i and for any two rounds $r_1 \neq r_2$, $\Phi_1(x, r_1, i) \neq \Phi_1(x, r_2, i)$. Also $\Phi_1(x, r_1, i_1) \neq \Phi_1(x, r_1, i_2)$ for any $i_1 \neq i_2$. We need just to show that the communication is contention-free. Consider round r of the first phase and root x. The packet destined for $\Phi_1(x, r, i) = y_1^{r,i} \ldots y_D^{r,i}$ is issued via the i-th output arc of x. The first step is contention-free and every vertex receives one packet per input arc. Vertex $x_2 \ldots x_D y_1^{r,i}$, where $y_1^{r,i} = \sigma(x_D, i)$ gets the packet from its input arc with input label $\xi(x_1, x_2)$ and greedy routing forwards this packet to vertex $x_3 \ldots x_D y_1^{r,i} y_2^{r,i}$, where $y_2^{r,i} = \sigma(y_1^{r,i}, (\xi(x_1, x_2) + r \text{ div } d^0) \bmod d) = \sigma(y_1^{r,i}, (\xi(x_1, x_2) + r) \bmod d)$. Hence, vertex $x_2 \ldots x_D y_1^{r,i}$ will use output arc $\mu(y_1^{r,i}, \sigma(y_1^{r,i}, \xi(x_1, x_2) + r) \bmod d) = (\xi(x_1, x_2) + r) \bmod d$. This is clearly a permutation of input arcs to output arcs. The same argument applies to any step j, $j = 3, \ldots, D$, of round r. The same argument can be used to prove that the second-phase communication is also contention-free. The total number of steps of the AAS algorithm is clearly $\tau(d, D) = Dd^{D-1} + (D-1)d^{D-2}$. The lower bound on the number of steps of AAS in digraph G is

$$\tau_{\text{AAS}}(G) = \left\lceil \frac{\overline{d}(G)|V(G)|^2}{|A(G)|} \right\rceil.$$

Therefore

$$\frac{\tau(d, D)}{\tau_{\text{AAS}}(K(d, D))} = \frac{Dd^{D-1} + (D-1)d^{D-2}}{\overline{d}(K(d, D))(d+1)d^{D-2}} \leq 1 + \frac{2d-1}{D(d+1)(d-1)^2 - d^2}.$$

Let $D_{\min}(d)$ be the smallest diameter such that the slowdown of the AAS algorithm in $K(d, D)$ is at most 5 % for any $D \geq D_{\min}(d)$. Then $D_{\min}(2) = 22$, $D_{\min}(3) = 7$, $D_{\min}(4) = 4$, and for $d \geq 5$, $D_{\min}(d) = 2$.

4 Conclusions

We have given tight lower and upper bounds on the average distance in $K(d, D)$. They are very close to the diameter, they can be approximated with $D - \frac{1}{d-1}$. We have also designed a memory-optimal and asymptotically time- and transmission-optimal all-to-all scatter algorithm in noncombining all-port $K(d, D)$. It is based on greedy routing which treats the Kautz networks as if they were vertex-transitive. Every vertex becomes the root of a virtual complete d-ary tree. For $d \geq 5$, the slowdown of the algorithm is within 5% for any diameter D.

References

1. J.-C. Bermond, Z. Liu, and M. Syska. Mean eccentricities of de Bruijn networks. *Networks*, 30:187–203, 1997.
2. G. Liu, K. Y. Lee, and H. F. Jordan. TDM and TWDM de Bruijn networks and ShuffleNets for optical communications. *IEEE Transactions on Computers*, 46:695–701, 1997.
3. K. Sivarajan and R. Ramaswami. Lightwave networks based on de Bruijn graphs. *IEEE/ACM Trans. Networking*, 2(1):70–79, 1994.

Reactive Proxies: A Flexible Protocol Extension to Reduce ccNUMA Node Controller Contention

Sarah A. M. Talbot and Paul H. J. Kelly

Department of Computing
Imperial College of Science, Technology and Medicine
180 Queen's Gate, London SW7 2BZ, United Kingdom
{samt, phjk}@doc.ic.ac.uk

Abstract. Serialisation can occur when many simultaneous accesses are made to a single node in a distributed shared-memory multiprocessor. In this paper we investigate routing read requests via an intermediate proxy node (where combining is used to reduce contention) in the presence of finite message buffers. We present a *reactive* approach, which invokes proxying only when contention occurs, and does not require the programmer or compiler to mark widely-shared data. Simulation results show that the hot-spot contention which occurs in pathological examples can be dramatically reduced, while performance on well-behaved applications is unaffected.

1 Introduction

Unpredictable performance anomalies have hampered the acceptance of cache-coherent non-uniform memory access (ccNUMA) architectures. Our aim is to improve performance in certain pathological cases, without reducing performance on well-behaved applications, by reducing the bottlenecks associated with widely-shared data. This paper moves on from our initial work on proxy protocols [1], eliminating the need for application programmers to identify widely-shared data.

Each processor's memory and cache is managed by a node controller. In addition to local memory references, the controller must handle requests arriving via the network from other nodes. These requests concern cache lines currently owned by this node, cache line copies, and lines whose home is this node (*i.e.* the page holding the line was allocated to this node, by the operating system, when it was first accessed). In large configurations, unfortunate ownership migration or home allocations can lead to concentrations of requests at particular nodes. This leads to performance being limited by the service rate (occupancy) of an individual node controller, as demonstrated by Holt *et al.* [6].

Our proxy protocol, a technique for alleviating read contention, associates one or more proxies with each data block, *i.e.* nodes which act as intermediaries for reads [1]. In the basic scheme, when a processor suffers a read miss, instead

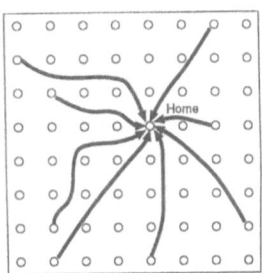 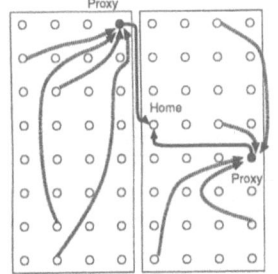 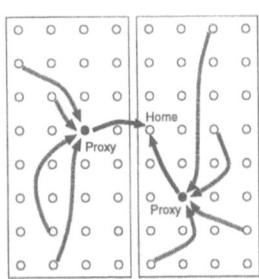

(a) Without proxies   (b) With two proxy clusters   (c) Read next data block
                              (read Line l)   (*i.e.* read Line $l + 1$)

Fig. 1. Contention is reduced by routing reads via a proxy

of directing its read request directly to the location's home node, it sends it
to one of the location's proxies. If the proxy has the value, it replies. If not, it
forwards the request to the home: when the reply arrives it can be forwarded to
all the pending proxy readers and can be retained in the proxy's cache. The main
contribution of this paper is to present a *reactive* version, which uses proxies only
when contention occurs, and does not require the application programmer (or
compiler) to identify widely-shared data.

The rest of the paper is structured as follows: reactive proxies are introduced
in Section 2. Our simulated architecture and experimental design are outlined in
Section 3. In Section 4, we present the results of simulations of a set of standard
benchmark programs. Related work is discussed in Section 5, and in Section 6
we summarise our conclusions and give pointers to further work.

2   Reactive Proxies

The severity of node controller contention is both application and architecture
dependent [6]. Controllers can be designed so that there is multi-threading of
requests (*e.g.* the Sun S3.mp is able to handle two simultaneous transactions [12])
which slightly alleviates the occupancy problem but does not eliminate it. Some
contention is inevitable, and will increase the latency of transactions. The key
problem is that queue lengths at controllers, and hence contention, are non-
uniformly distributed around the machine.

One way of reducing the queues is to distribute the workload to other node
controllers, using them as *proxies* for read requests, as illustrated in Fig. 1. When
a processor makes a read request, instead of going directly to the cache line's
home, it is routed first to another node. If the proxy node has the line, it replies
directly. If not, it requests the value from the home itself, allocates it in its own
cache, and replies. Any requests for a particular block which arrive at a proxy
before it has obtained a copy from the home node, are added to a distributed
chain of pending requests for that block, and the reply is forwarded down the
pending chain, as illustrated in Fig. 2. It should be noted that write requests are

(a) First request to proxy has to be forwarded to the home node:

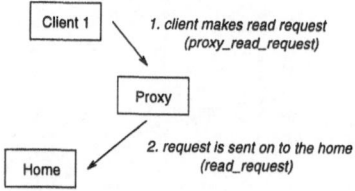

(b) Second client request, before data is returned, forms pending chain:

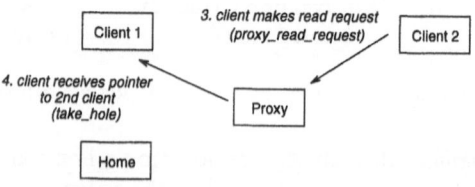

(c) Data is passed to each client on the pending chain:

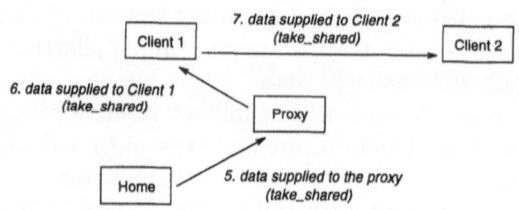

Fig. 2. Combining of proxy requests

not affected by the use of proxies, except for the additional invalidations that may be needed to remove proxy copies (which will be handled as a matter of course by the underlying protocol).

The choice of proxy node can be at random, or (as shown in Fig. 1) on the basis of locality. To describe how a client node decides which node to use as a proxy for a read request, we begin with some definitions:

- P: the number of processing nodes.
- $\mathcal{H}(l)$: the home node of location l. This is determined by the operating system's memory management policy.
- $\mathcal{NPC}$: the number of proxy clusters, *i.e.* the number of clusters into which the nodes are partitioned for proxying (*e.g.* in Fig. 1, $\mathcal{NPC}=2$). The choice of $\mathcal{NPC}$ depends on the balance between degree of combining and the length of the proxy pending chain. $\mathcal{NPC}=1$ will give the highest combining rate, because all proxy read requests for a particular data block will be directed to the same proxy node. As $\mathcal{NPC}$ increases, combining will reduce, but the number of clients for each proxy will also be reduced, which will lead to shorter proxy pending chains.

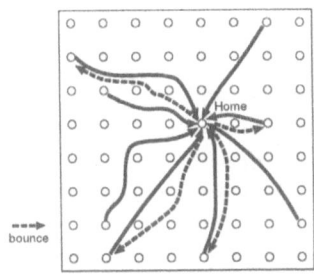 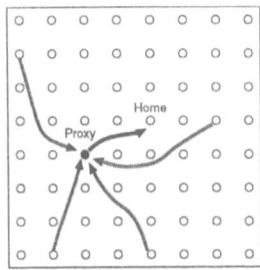

(a) Input buffer full, some read requests bounce (b) Reactive proxy reads

Fig. 3. Bounced read requests are retried via proxies

- $\mathcal{PCS}(C)$: the set of nodes which are in the cluster containing client node C. In this paper, $\mathcal{PCS}(C)$ is one of $\mathcal{NPC}$ disjoint clusters each containing $\mathcal{P}/\mathcal{NPC}$ nodes, with the grouping based on node number.
- $\mathcal{PN}(l, C)$ the proxy node chosen for a given client node (C) when reading location l. We use a simple hash function to choose the actual proxy from the proxy cluster $\mathcal{PCS}(C)$. If $\mathcal{PN}(l, C) = C$, or $\mathcal{PN}(l, C) = \mathcal{H}(l)$, then client C will send a read request directly to $\mathcal{H}(l)$

The choice of proxy node is, therefore, a two stage process. When the system is configured, the nodes are partitioned into $\mathcal{NPC}$ clusters. Then, whenever a client wants to issue a proxy read, it will use the hashing function $\mathcal{PN}(l, C)$ to select one proxy node from $\mathcal{PCS}(C)$. This mapping ensures that requests for a given location are routed via a proxy (so that combining occurs), and that reads for successive data blocks go to different proxies (as illustrated in Fig. 1(c)). This will reduce network contention [15] and balance the load more evenly across all the node controllers.

In the basic form of proxies, the application programmer uses program directives to mark data structures: all other shared data will be exempt from proxying [1]. If the application programmer makes a poor choice, then the overheads incurred by proxies may outweigh any benefits and degrade performance. These overheads include the extra work done by the proxy nodes handling the messages, proxy node cache pollution, and longer sharing lists. In addition, the programmer may fail to mark data structures that would benefit from proxying.

Reactive proxies overcome these problems by taking advantage of the finite buffering of real machines. When a remote read request reaches a full buffer, it will immediately be sent back across the network. With the reactive proxies protocol, the arrival of a buffer-bounced read request will trigger a proxy read (see Fig. 3). This is quite different to the basic proxies protocol, where the user has to decide whether all or selected parts of the shared data are proxied, and proxy reads are always used for data marked for proxying. Instead, proxies are only used when congestion occurs. As soon as the queue length at the destination node has reduced to below the limit, read requests will no longer be bounced and proxy reads will not be used.

The repeated bouncing of read requests which can occur with finite buffers leads to the possibility of deadlock: the underlying protocol has to detect the continuous re-sending of a remote read request, and eventually send a higher priority read request which is guaranteed service. Read requests from proxy nodes to home nodes will still be subject to buffer bouncing, but the combining and re-routing achieved by proxying reduce the chances of a full input buffer at the home node.

The reactive proxy scheme has the twin virtues of simplicity and low overheads. No information needs to be held about past events, and no decision is involved in using a proxy: the protocol state machine is just set up to trigger a proxy read request in response to the receipt of a buffer-bounced read request.

3 Simulated Architecture and Experimental Design

In our execution-driven simulations, each node contains a processor with an integral first-level cache (FLC), a large second-level cache (SLC), memory (DRAM), and a node controller (see Fig. 4). The node controller receives messages from, and sends messages to, both the network and the processor. The SLC, DRAM, and the node controller are connected using two decoupled buses. This decoupled bus arrangement allows the processor to access the SLC at the same time as the node controller accesses the DRAM. Table 1 summarises the architecture.

We simulate a simplified interconnection network, which follows the the $LogP$ model [3]. We have parameterised the network and node controller as follows:

- L: the latency experienced in each communication event, 10 cycles for long messages (which include 64 bytes of data, *i.e.* one cache line), and 5 cycles for all other messages. This represents a fast network, comparable to the point-to-point latency used in [11].
- o: the occupancy of the node controller. Like Holt *et al.* [6], we have adapted the LogP model to recognise the importance of the occupancy of a node controller, rather than just the overhead of sending and receiving messages. The processes which cause occupancy are simulated in more detail (see Table 2).
- g: the gap between successive sends or receives by a processor, 5 cycles.
- P: the number of processor nodes, 64 processing nodes.

We limit our message buffers to eight for read requests. There can be more messages in an input buffer, but once the queue length has risen above eight, all read requests will be bounced back to the sender until the queue length has fallen below the limit. This is done because we are interested in the effect of finite buffering on read requests rather than all messages, and we wished to be certain that all transactions would complete in our protocol. The queue length of $\sqrt{P}$ is an arbitrary but reasonable limit.

Each cache line has a home node (at page level) which: either holds a valid copy of the line (in SLC and/or DRAM), or knows the identity of a node which does have a valid copy (*i.e.* the owner); has guaranteed space in DRAM for the line; and holds directory information for the line (head and state of the sharing

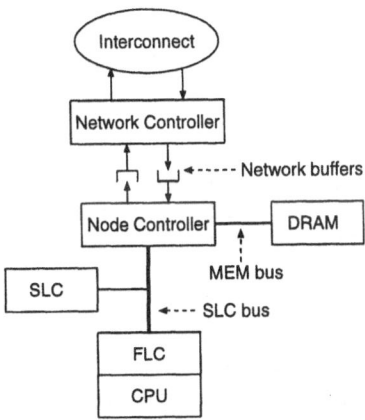

Fig. 4. The architecture of a node

Table 1. Details of the simulated architecture

CPU	CPI	1.0
	Instruction set	based on DEC Alpha
Instruction cache	All instruction accesses assumed primary cache hits	
First level data cache	Capacity	8 Kbytes
	Line size	64 bytes
	Direct mapped, write-through	
Second-level cache	Capacity	4 Mbytes
	Line size	64 bytes
	Direct mapped, write-back	
DRAM	Capacity	Infinite
	Page size	8 Kbytes
Node controller	Non-pipelined	
	Service time and occupancy	See Table 2
	Cycle time	10ns
Interconnection network	Topology	full crossbar
	Incoming message queues	8 read requests
Cache coherence protocol	Invalidation-based, sequentially-consistent ccNUMA, home nodes assigned to first node to reference each page (i.e. "first-touch-after-initialisation"). Distributed directory, using singly-linked sharing list Based on the Stanford Distributed-Directory Protocol, described by Thapar and Delagi [14]	

Table 2. Latencies of the most important node actions

operation	time (cycles)
Acquire SLC bus	2
Release SLC bus	1
SLC lookup	6
SLC line access	18
Acquire MEM bus	3
Release MEM bus	2
DRAM lookup	20
DRAM line access	24
Initiate message send	5

Table 3. Benchmark applications

application	problem size	shared data marked for basic proxying
Barnes	16K particles	all
CFD	64 x 64 grid	all
FFT	64K points	all
FMM	8K particles	f_array (part of G_Memory)
GE	512 x 512 matrix	entire matrix
Ocean-Contig	258 x 258 ocean	q_multi and rhs_multi
Ocean-Non-Contig	258 x 258 ocean	fields, fields2, wrk, and frcng
Water-Nsq	512 molecules	VAR and PFORCES

list). The distributed directory holds the identities of nodes which have cached a particular line in a sharing chain, currently implemented as a singly-linked list.

The directory entry for each data block provides the basis for maintaining the consistency of the shared data. Only one node at a time can remove entries from the sharing chain (achieved by locking the head of the sharing chain at the home node), and messages which prompt changes to the sharing chain are ordered by their arrival at the home node. This mechanism is not affected by the protocol additions needed to support proxies.

Proxy nodes require a small amount of extra store to be added to the node controller. Specifically we need to be able to identify which data lines have outstanding transactions (and the tags they refer to), and be able to record the identity of the head of the pending proxy chain. In addition, the node controller has to handle the new proxy messages and state changes. We envisage implementing these in software on a programmable node controller, e.g. the MAGIC node controller in Stanford's FLASH [9], or the SCLIC in the Sequent NUMA-Q [10].

The benchmarks and their parameters are summarised in Table 3. GE is a simple Gaussian elimination program, similar to that used by Bianchini and LeBlanc in their study of eager combining [2]. We chose this benchmark because it is an example of widely-shared data. CFD is a computational fluid dynamics application, modelling laminar flow in a square cavity with a lid causing friction [13]. We selected six applications from the SPLASH-2 suite, to give a cross-section of scientific shared memory applications [16]. We used both Ocean benchmark applications, in order to study the effect of proxies on the "tuned for data locality" and "easy to understand" variants. Other work which refers to Ocean can be assumed to be using Ocean-Contig.

4 Experimental Results

In this work, we concentrate on reactive proxies, but compare the results with basic proxies (which have already been examined in [1]). The performance results for each application are presented in Table 4 in terms of relative speedup with no proxying (i.e. the ratio of the execution time for 64 processing nodes to the execution time running on 1 processor), and percentage changes in execution time when proxies are used. The problem size is kept constant.

The relative changes results in Fig. 5 show three different metrics:

Table 4. Benchmark performance for 64 processing nodes

applications	relative speedup no proxies	proxy type	% change in execution time (+ is better, - is worse) for $\mathcal{NPC} = 1$ to 8							
			1	2	3	4	5	6	7	8
Barnes	43.2	basic	0.0	-0.1	0.0	0.0	-0.2	+0.3	-0.2	-0.1
		reactive	+0.3	+0.2	+0.3	+0.2	+0.1	-0.1	0.0	+0.3
CFD	30.6	basic	+6.6	+7.7	+6.4	+8.3	+6.7	+5.8	+6.0	+10.2
		reactive	+5.6	+5.4	+4.7	+4.5	+5.6	+4.1	+4.1	+4.8
FFT	47.4	basic	+9.3	+9.0	+9.8	+9.4	+9.2	+8.8	+8.9	+8.9
		reactive	+11.5	+11	+10.8	+10.8	+11.1	+11.6	+11.0	+10.6
FMM	36.1	basic	+0.2	+0.2	+0.2	+0.2	+0.2	+0.2	+0.2	+0.2
		reactive	+0.4	+0.3	+0.3	+0.4	+0.3	+0.4	+0.3	+0.3
GE	22.0	basic	+28.7	+28.7	+28.7	+28.7	+28.7	+28.8	+28.8	+28.7
		reactive	+23.3	+22.9	+22.3	+21.4	+21.5	+21.4	+21.7	+21.5
Ocean-Contig	48.9	basic	-3.2	+2.6	+0.1	-0.2	-0.8	-2.0	-2.1	-1.8
		reactive	-0.2	0.0	-0.1	-0.3	+0.2	+2	+1.3	+2.1
Ocean-Non-Contig	50.5	basic	-0.3	+1.6	-1.7	+4.1	+1.0	-0.4	+0.2	+4.2
		reactive	+3.1	+1.4	+1.4	+0.8	+5.1	+1.8	+1.2	+5
Water-Nsq	55.5	basic	-0.7	-0.6	-0.6	-0.5	-0.5	-0.5	-0.7	-0.5
		reactive	+0.2	+0.2	+0.2	+0.2	+0.2	+0.2	+0.1	+0.2

- *messages*: the ratio of the total number of messages to the total without proxies,
- *execution time*: the ratio of the execution time (excluding startup) to the execution time (also excluding startup) without proxies.
- *queueing delay*: the ratio of the total time that messages spend waiting for service to the total without proxies, and

The message ratios shown in Fig. 6 are:

- *proxy hit rate*: the ratio of the number of proxy read requests which are serviced directly by the proxy node, to the total number of proxy read requests (in contrast, a proxy miss would require the proxy to request the data from the home node),
- *remote read delay*: the ratio of the delay between issuing a read request and receiving the data, to the same delay when proxies are not used.
- *buffer bounce ratio*: the ratio of the total number of buffer bounce messages to read requests. This gives a measure of how much bouncing there is for an application. This ratio can go above one, since only the initial read request is counted in that total, *i.e.* the retries are excluded.
- *proxy read ratio*: the ratio of the proxy read messages to read requests - this gives a measure of how much proxying is used in an application.

The first point to note from Table 4 is that there is no overall "winner" between basic and reactive proxies, in that neither policy improves the performance of all the applications for all values of proxy clusters. Looking at the

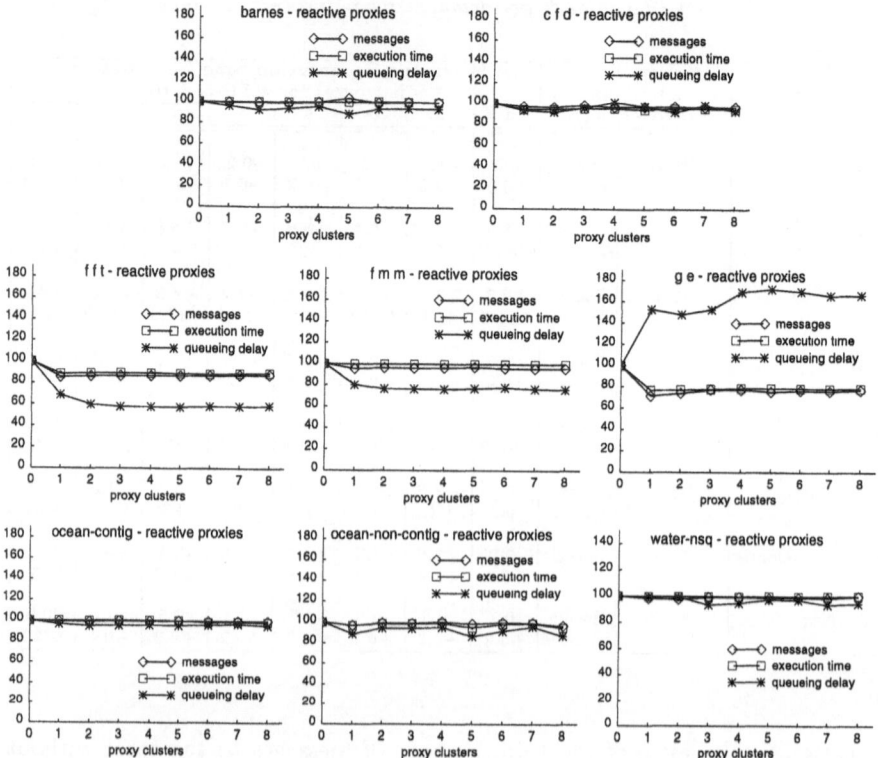

Fig. 5. Relative changes for 64 processing nodes with reactive proxies

results for different values of $\mathcal{NPC}$, for basic proxies there is no value which has a positive effect on the performance of all the benchmarks. However, for reactive proxies, there are two proxy cluster values that improve the performance of all the benchmarks, *i.e.* $\mathcal{NPC}=5$ and 8 achieve a balance between combining, queue distribution, and length of the proxy pending chains. Reactive proxies may not always deliver the best performance improvement, but by providing stable points for $\mathcal{NPC}$ they are of more use to system designers. It should also be noted that, in general, using reactive proxies reduces the number of messages, because they break the cycle of re-sending read messages in response to a finite buffer bounce (see Fig. 5).

Looking at the individual benchmarks:

Barnes. In general, this application benefits from the use of reactive proxies. However, changing the balance of processing by routing read requests via proxy nodes can have more impact than the direct effects of reducing home node congestion. Two examples illustrate this: when $\mathcal{NPC}=6$ for basic proxies, load miss delay is the same as with no proxies, store miss delay has increased slightly from the no proxy case, yet a reduction of 0.4% in lock and barrier delays results in an overall performance improvement of 0.3%. Conversely, for reactive proxies when $\mathcal{NPC}=6$, the load and store miss delays are the same as when proxies are

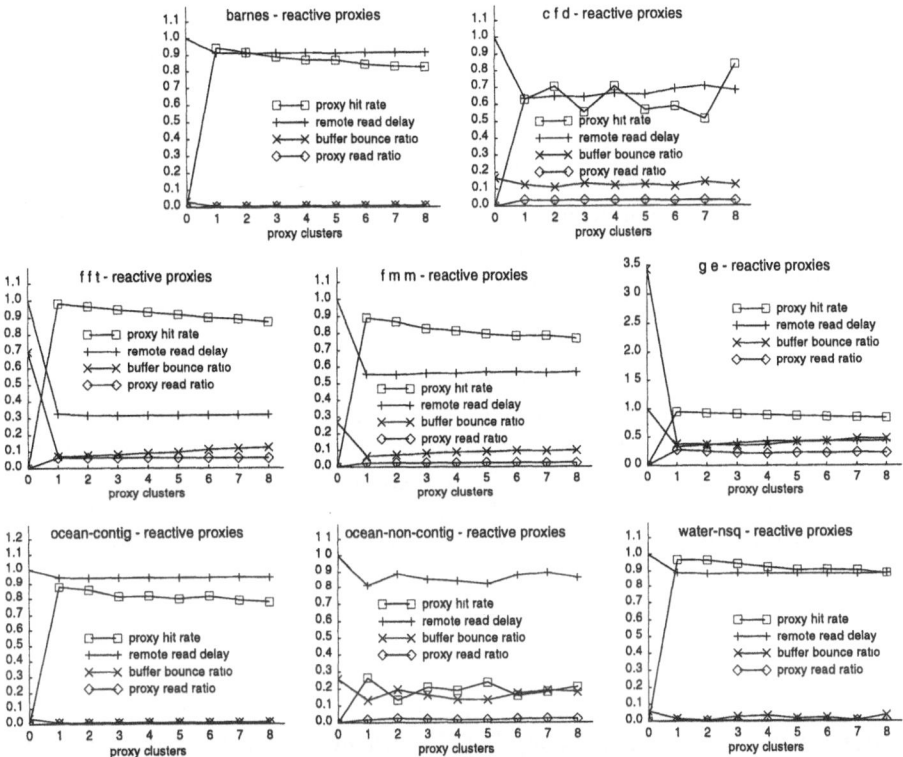

Fig. 6. Message ratios for 64 processing nodes with reactive proxies

not used, but a slight 0.1% increase in lock delay gives an overall performance degradation of -0.1%.

CFD. This application benefits from the use of reactive proxies, with performance improvements in the range 4.1% to 5.6%. However, the improvements are not as great as those obtained with basic proxies. The difference is attributable to the delay in triggering each reactive proxy read: for this application it is better to use proxy reads straight away, rather than waiting for read requests to be bounced. It should also be noted that the proxy hit rate oscillates, with peaks at $\mathcal{NPC}$=2,4,8 (see Fig. 6). This is due to a correspondence between the chosen proxy node and ownership of the cache line.

FFT. This shows a marked speedup when reactive proxies are used, of between 10.6% and 11.6%. The number of messages decreases with proxies because the buffer bounce ratio is cut from a severe 0.7 with no proxies. The mean queueing delay drops down as the number of proxy clusters increases, reflecting the benefit of spreading the proxy read requests. However, this is balanced by a slow increase in the buffer bounce ratio, because as more nodes act as proxy there will be more read requests to the home node, and these read requests will start to bounce as the number of messages in the home node's input queue rises to $\sqrt{\mathcal{P}}$ and above.

FMM. There is a marginal speedup compared with no proxies (between 0.3% and 0.4%). This is as expected given only the *f_array* (part of *G_Memory*) is known to be widely-shared, which was why it was marked for basic proxies. However, the performance improvement is slightly better than that achieved using basic proxies, so the reactive method dynamically detects opportunities for read combining which were not found by code inspection and profiling tools.

GE. This application, which is known to exhibit a high level of sharing of the current pivot row, shows a large speedup in the range 21.4% to 23.2%. However, the improvement is not as good as that obtained using basic proxies. This was to be expected, because proxying is no longer targeted by marking widely-shared data structures. Instead proxying is triggered when a read is rejected because a buffer is full, and so there will be two messages (the read and buffer bounce) before a proxy read request is sent by the client. It should also be noted that the execution time increases as the number of proxy clusters increases. As the number of nodes acting as proxy goes up, there will be more read requests (from proxies) being sent to the home node, and the read requests are more likely to be bounced, as shown by the buffer bounce ratio for GE in Fig. 6. Finally, the queueing delay is much higher when proxies are in use. This is because without proxies there is a very high level of read messages being bounced (and thus not making it into the input queues). With proxies, the proxy read requests are allowed into the input queues, which increases the mean queue length.

Ocean-Contig. Reactive proxies can degrade the performance of this application (by up to -0.3% at $\mathcal{NPC}=4$), but they achieve performance improvements for more values of $\mathcal{NPC}$ than the basic proxy scheme. Unlike basic proxies, reactive proxies reduce the remote read delay by targeting remote read requests that are bounced because of home node congestion. The performance degradation when $\mathcal{NPC}=1,3,4$ is attributable to increased barrier delays caused by the redistribution of messages.

Ocean-Non-Contig. This has a high level of remote read requests. These remote read requests result in a high level of buffer bounces, which in turn invoke the reactive proxies protocol. Unfortunately the data is seldom widely-shared, so there is little combining at the proxy nodes, as is illustrated by the low proxy hit rates. With $\mathcal{NPC}=4$, this results in a concentration of messages at a few nodes, overall latency increases, and the execution time suffers. For $\mathcal{NPC}=5$, the queueing delay is reduced in comparison to the no proxy case, and this has the best execution time. Given these results, we are carrying out further investigations into the hashing schemes suitable for the $\mathcal{PN}(l, C)$ function, and the partitioning strategy used to determine $\mathcal{PCS}(C)$, to obtain more reliable performance for applications such as Ocean-Non-Contig.

Water-Nsq. Using reactive proxies gives a small speedup compared to no proxies (around 0.2%). However, this is better than with basic proxies, where performance is always worse (in the range -0.5% to -0.7%, see Table 4). The extremely low proxy read ratios shows that there is very little proxying, but the high proxy hit rates indicate that when proxy reads are invoked there is a high level of combining. It is encouraging to see that the proxy read ratio is kept low:

this shows that the overheads of proxying (extra messages, cache pollution) are only incurred when they are needed by an application.

To summarise, the results show that for reactive proxies, when the number of proxy clusters ($\mathcal{NPC}$) is set to five or eight, the performance of all the benchmarks improves, *i.e.* they achieve the best balance between combining, queue length distribution, and the length of the proxy pending chains in our simulated system. This is a very encouraging result, because without marking widely-shared data we have obtained sizeable performance improvements for three benchmarks (GE, FFT, and CFD), and had no detrimental effect on the other well-behaved applications. By selecting a suitable $\mathcal{NPC}$ for an architecture, the system designers can provide a ccNUMA system with more stable performance. This is in contrast to basic proxies, where although better performance can be obtained for some benchmarks, the strategy relies on judicious marking of widely-shared data for each application.

5 Related Work

A number of measures are available to alleviate the effects of contention for a node, such as improving the node controller service rate [11], and combining in the interconnection network for fetch-and-update operations [4]. Architectures based on clusters of bus-based multiprocessor nodes provide an element of read combining since caches in the same cluster snoop their shared bus. Caching extra copies of data to speed-up retrieval time for remote reads has been explored for hierarchical architectures, including [5]. The proxies approach is different because it does not use a fixed hierarchy: instead it allows requests for copies of successive data lines to be serviced by different proxies.

Attempts have been made to identify widely-shared data for combining, including the GLOW extensions to the SCI protocol [8, 7]. GLOW intercepts requests for widely-shared data by providing agents at selected network switch nodes. In their dynamic detection schemes, which avoid the need for programmers to identify widely-shared data, agent detection achieves better results than the combining of [4] by using a sliding window history of recent read requests, but does not improve on the static marking of data. Their best results are with program-counter based prediction (which identifies load instructions that suffer very large miss latency) although this approach has the drawback of requiring customisation of the local node CPUs.

In Bianchini and LeBlanc's "eager combining", the programmer identifies specific memory regions for which a small set of server caches are pre-emptively updated [2]. Eager combining uses intermediate nodes which act like proxies for marked pages, *i.e.* their choice of server node is based on the page address rather than data block address, so their scheme does not spread the load of messages around the system in the fine-grained way of proxies. In addition, their scheme eagerly updates all proxies whenever a newly-updated value is read, unlike our protocol, where data is allocated in proxies on demand. Our less aggressive scheme reduces cache pollution at the proxies.

6 Conclusions

This paper has presented the reactive proxy technique, discussed the design and implementation of proxying cache coherence protocols, and examined the results of simulating eight benchmark applications. We have shown that proxies benefit some applications immensely, as expected, while other benchmarks with no obvious read contention still showed performance gains under the reactive proxies protocol. There is a tradeoff between the flexibility of reactive proxies and the precision (when used correctly) of basic proxies. However, reactive proxies have the further advantage that a stable value of $\mathcal{NPC}$ (number of proxy clusters) can be established for a given system configuration. This gives us the desired result of improving the performance of some applications, without affecting the performance of well-behaved applications. In addition, with reactive proxies, the application programmer does not have to worry about the architectural implementation of the shared-memory programming model. This is in the spirit of the shared-memory programming paradigm, as opposed to forcing the programmer to restructure algorithms to cater for performance bottlenecks, or marking data structures that are believed to be widely-shared.

We are currently doing work based on the Ocean-Non-Contig application to refine our proxy node selection function $(\mathcal{PN}(l, C))$. In addition, we are continuing our simulation work with different network latency (L) and finite buffer size values. We are also evaluating further variants of the proxy scheme: adaptive proxies, non-caching proxies, and using a separate proxy cache.

Acknowledgements

This work was funded by the U.K. Engineering and Physical Sciences Research Council through the CRAMP project GR/J 99117, and a Research Studentship. We would also like to thank Andrew Bennett and Ashley Saulsbury for their work on the ALITE simulator and for porting some of the benchmark programs.

References

1. Andrew J. Bennett, Paul H. J. Kelly, Jacob G. Refstrup, and Sarah A. M. Talbot. Using proxies to reduce cache controller contention in large shared-memory multiprocessors. In Luc Bougé et al, editor, *Euro-Par 96 European Conference on Parallel Architectures, Lyon*, volume 1124 of *Lecture Notes in Computer Science*, pages 445–452. Springer-Verlag, August 1996.
2. Ricardo Bianchini and Thomas J. LeBlanc. Eager combining: a coherency protocol for increasing effective network and memory bandwidth in shared-memory multiprocessors. In *6th IEEE Symposium on Parallel and Distributed Processing, Dallas*, pages 204–213, October 1994.
3. David E. Culler, Richard M. Karp, David Patterson, Abhijit Sahay, Eunice E. Santos, Klaus Erik Schauser, Ramesh Subramonian, and Thorsten von Eicken. LogP: a practical model of parallel computation. *Communications of the ACM*, 39(11):78–85, November 1996.

4. Allan Gottlieb, Ralph Grishman, Clyde P. Kruskal, Kevin P. McAuliffe, Larry Rudolph, and Marc Snir. The NYU Ultracomputer – designing a MIMD shared memory parallel computer. *IEEE Transactions on Computers*, C-32(2):175–189, February 1983.

5. Seif Haridi and Erik Hagersten. The cache coherence protocol of the Data Diffusion Machine. In E. Odijk, M. Rem, and J.-C Syre, editors, *PARLE 89 Parallel Architectures and Languages Europe, Eindhoven*, volume 365 of *Lecture Notes in Computer Science*, pages 1–18. Springer-Verlag, June 1989.

6. Chris Holt, Mark Heinrich, Jaswinder Pal Singh, Edward Rothberg, and John Hennessy. The effects of latency, occupancy and bandwidth in distributed shared memory multiprocessors. Technical Report CSL-TR-95-660, Computer Systems Laboratory, Stanford University, January 1995.

7. David V. James, Anthony T. Laundrie, Stein Gjessing, and Gurindar S. Sohi. Scalable Coherent Interface. *IEEE Computer*, 23(6):74–77, June 1990.

8. Stefanos Kaxiras, Stein Gjessing, and James R. Goodman. A study of three dynamic approaches to handle widely shared data in shared-memory multiprocessors. In *(to appear) 12th ACM International Conference on Supercomputing, Melbourne*, July 1998.

9. Jeffrey Kuskin. *The FLASH Multiprocessor: designing a flexible and scalable system*. PhD thesis, Computer Systems Laboratory, Stanford University, November 1997. Also available as a technical report, CSL-TR-97-744.

10. Tom Lovett and Russell Clapp. STiNG: a CC-NUMA computer system for the commercial marketplace. *23rd Annual International Symposium on Computer Architecture, Philadelphia, in Computer Architecture News*, 24(2):308–317, May 1996.

11. Maged M. Michael, Ashwini K. Nanda, Beng-Hong Lim, and Michael L. Scott. Coherence controller architectures for SMP-based CC-NUMA multiprocessors. *24th Annual International Symposium on Computer Architecture, Denver, in Computer Architecture News*, 25(2):219–228, June 1997.

12. Andreas Nowatzyk, Gunes Aybay, Michael Browne, Edmund Kelly, Michael Parkin, Bill Radke, and Sanjay Vishin. The S3.mp scalable shared memory multiprocessor. In *Proceedings of the International Conference on Parallel Processing Vol. 1*, pages 1–10, August 1995.

13. B. A. Tanyi. *Iterative Solution of the Incompressible Navier-Stokes Equations on a Distributed Memory Parallel Computer*. PhD thesis, University of Manchester Institute of Science and Technology, 1993.

14. Manu Thapar and Bruce Delagi. Stanford distributed-directory protocol. *IEEE Computer*, 23(6):78–80, June 1990.

15. Leslie G. Valiant. Optimality of a two-phase strategy for routing in interconnection networks. *IEEE Transactions on Computers*, C-32(8):861–863, August 1983.

16. Steven Cameron Woo, Moriyoshi Ohara, Evan Torrie, Jaswinder Pal Singh, and Anoop Gupta. The SPLASH-2 programs: characterization and methodological considerations. *Proceedings of the 22nd Annual International Symposium on Computer Architecture, in Computer Architecture News*, 23(2):24–36, June 1995.

Handling Multiple Faults in Wormhole Mesh Networks *

Tor Skeie

Department of Informatics, University of Oslo
Box 1080, Blindern, N-0316 OSLO, Norway.
mail: torsk@ifi.uio.no

Abstract. We present a fault tolerant method tailored for n-dimensional mesh networks that is able to handle multiple faults, even for two dimensional meshes. The method does not require existence of virtual channels. The traditional way of achieving fault tolerance based on adaptivity and adding virtual channels as the main mechanisms, has not shown the ability to handle multiple faults in wormhole mesh networks. In this paper we propose another strategy to provide high degree of fault-tolerance, we describe a technique which alters the routing function on the fly. The alteration action is always taken locally and distributed to a limited number of non-neighbor nodes.

1 Introduction

In recent years we have seen an increasing interest in multicomputers both in industry and academia. As the number of components in such computers grow the probability of failing components increase, and therefore the ability to adapt to errors becomes a significant issue in this field. In this paper we study wormhole routed interconnect networks with respect to fault tolerance.

The wormhole switching technique that was described by Dally and Seitz [5] is an improvement of the virtual cut through technique of Kermani and Kleinrock [15]. The concept of wormhole routing has been taken up in many interconnection networks, in particular in the domain of multiprocessor interconnection, see e.g. [18,20,21]. A survey of wormhole routing techniques can be found in [19].

Fault-tolerance in a communication network is defined as the ability of the network to effectively utilize its redundancy in the presence of faulty links. A failing link may destroy one path in use between two communicating nodes, but given that the failing link has not split the network into two unconnected parts, there will still be other paths that the two ends may communicate over. Much work on fault-tolerance in wormhole routed networks has been reported. Linder and Harden described an adaptive routing algorithm for k-ary n-cubes which requires up to 2^n virtual channels per physical channel [16]. Boppana and Chalasani present an algorithm to handle block faults in meshes where faulty links and nodes form rectangular regions [1]. The method requires 2 virtual

* This work is supported by Esprit under the OMI-ARCHES

channels per physical channel, and is able to handle any number of faults as long as the fault regions do not overlap. The technique requires that a fault free node knows the status of its own links and its neighbors' links. In [2] they generalize this method to also handle non-convex fault regions requiring 4 virtual channels per physical channel in the non-overlapping case. The method requires fault region shape-finding messages, and may in many cases mark more links as faulty than those that have actually failed. Several other methods rely on the virtual channel concept as well [3,4,6,8,9,11].

However, other approaches have been proposed. Glass and Ni have described a fault tolerant routing algorithm for meshes which does not require virtual channels [13]. Their method is based on the turn-model for generating adaptive routing algorithms [10,12] and needs extra control lines to agree upon allowed turns. Lysne, Skeie and Waadeland [17] also describe fault tolerant routing without using virtual channels. Basically, their methods rely on alteration of the routing function. Unfortunately, both these methods tolerate very few faults for low dimension meshes.

In this paper we present a method for fault tolerance which is different from the above mentioned work in terms of that it both *can handle a significant number of faults* and *does not need virtual channels*. The technique relies on a limited amount of non local status information, the action is taken locally and distributed to a limited number of nodes.

The remainder of the paper is organized as follows. Section 2 contains some basic notations. In section 3 we describe the fault tolerant routing method and finally in section 4 we conclude.

2 Preliminaries

The definitions used in this paper adhere to standard notation and definitions in wormhole routing [17]. The following theorem is due to Dally and Seitz [5] and Duato [7]. It will be used extensively within this paper.

Theorem 1. *A wormhole network is **free from deadlocks** if its channel dependency graph is acyclic.*

We shall in this paper consider bidirectional n-dimensional mesh networks, also called (k, n)-mesh networks, where k is the radix, n is the number of dimensions and $N = k^n$ is the number of nodes. A (k, n)-mesh is a (k, n)-torus without connecting wraparound links. To simplify presentation of the fault tolerant method, we will describe it in detail for the two-dimensional case and afterwards present generalization to three and higher dimensions. The four directions of a two-dimensional mesh are labeled *north, south, east* and *west*. Furthermore the channels of a horizontal (vertical) link l are denoted l_{east}/l_{west} (l_{south}/l_{north}), respectively.

2.1 The Fault Model

In this paper we will concentrate on link faults, and assume that if a channel fails then the entire link fails in both directions. The link failures are considered to be "hard" in the sense that faulty links will be down for quite a while (hours or even days). Furthermore we assume that each node in addition to knowing the status of its own links, also has status information regarding *east* and *north* links of some of the nodes in the row (column) that it is positioned in. For a more thorough definition we refer to section 3.2. To provide nodes with the link status information stated above we assume that the network has a control line structure where routers can send/receive specific control messages. The IEEE 1355 [14] STC104 [23] from SGS-Thomson is a router with such a control line structure.

3 A Distributed Fault Tolerant System for Positive-First Routing

In this section we present an algorithm on that handles multiple faults for *positive-first routing (PF)* without using virtual channels. *PF* is a variant of the *turn model* [12] which is a method for designing adaptive routing algorithms. The idea in the turn model is to start with a system where every packet can take all possible paths in every router. Then one prohibits just enough turns in order to break cycles in the channel dependency graph.

Positive first routing is defined by the turns showed in figure 1. We make the additional restriction that packets should be routed minimally in a fault free network, even if there are detours that do not introduce illegal turns. From the

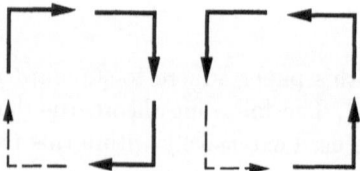

Fig. 1. The six allowed turns (solid arrows) for positive first routing in the fault free case.

above we observe that there is adaptivity in the south-west and north-east directions because in each of these directions one is allowed to make use of two turns. Routing in the south-east and north-west directions is, however, deterministic. The name *positive-first* comes from the hindrance of making turns from negative to positive directions.

The fault tolerant system is based on altering the routing function in some nodes, so that new paths are defined that avoid the faulty links. For most of the failure situations it will be nodes in two rows and/or columns that need a reprogrammed routing function, however, in some specific situations a cascading update might be necessary (see section 3.3). It will always be the node, where one or both of the *east* and *north* links recently failed, that initiates the necessary updating step to be taken. This means that the updating action to be provided is locally determined. Such a node will be called n_{master} throughout the remainder of the paper. Alteration messages are then distributed via the control line structure to nodes that need their routing function updated, henceforth the *distributed* view of the fault tolerant system. On the other hand the node where its *west-* and/or *south* links fail, stays passive concerning alteration of the routing function until reception of updating message. Such a node we denote for n_{slave}. We assume the nodes to have some amount of programmable logic so they can implement the alteration algorithm to respond to the failure situation, and furthermore that nodes on the fly can reprogram their routing function upon reception of updating messages.

Below we shall first present the necessary alteration of the routing function for different link failure situations in order to give all packets affected by a faulty link new paths. Next we show that the updated routing function will be deadlock free. Depending on the failure situation we divide the description in two parts:

1. *The situation when only one of the east and north links of a node are faulty.*
2. *The more complex situation where both these links are failing.*

In the first situation our methods do not use non-local link status information in order to perform the routing function alteration, while for the second case this becomes necessary. Recall that we denote the node with its east- and/or north links failing for n_{master}, and the nodes that are receiving a updating message from n_{master} are called n_{slave_i}. Furthermore the east and north links of n_{master} are designated l_e and l_n, respectively. For clarity, we shall in the methods proposed below assume that rerouting channels (links) are not faulty at the moment the updating action is initialized, in section 3.3 we also handle faulty rerouting channels.

3.1 The East or the North Link of a Node Fails

For this case we show that we can recover without introducing the two prohibited turns (figure 1).

The East Link of a Node Fails. The following method is used to alter the routing function:

Alteration 1: The packets that have paths through $l_{e_{east}}$ are rerouted northwards at n_{master} (*Alter 1* in figure 2(a)).

Alteration 2: The packets that have paths through $l_{e_{west}}$ that arrive (or originate) at the nodes in the row east of $l_{e_{west}}$ are rerouted northwards as well (*Alter 2* in figure 2(a)).

Alteration 3: The packets that have paths through $l_{e_{west}}$ that arrive at the nodes in the row above alteration 2 are rerouted westwards (*Alter 3* in figure 2(a)).

The *Alterations 2* and *3* are initialized by n_{master} distributing a update message to the two corresponding set of nodes. By altering the routing function to the nodes east of n_{slave} in the same row, we ensure that packets with original paths through $l_{e_{west}}$ are not sent towards n_{slave} (*Alter 2*). If they were sent towards n_{slave} they would be forced to take the prohibited west-north turn. Notice also that packets going in the adaptive south-west direction, that can possible use $l_{e_{west}}$ must for the same reason be stopped from going further south at the row right above the failing link (*Alter 3*).

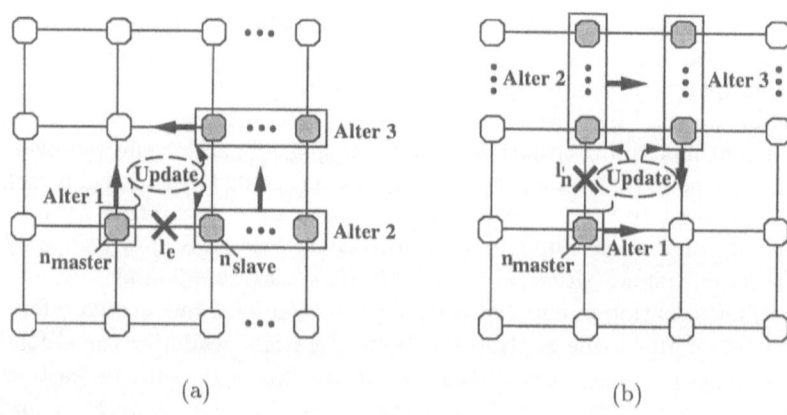

(a)   (b)

Fig. 2. Identifying the nodes (visualized as darkened boxes) that will get their routing function altered according to our method. The thick arrows indicate the redirections of the affected packets through the failing link. (a) The east link of n_{master} fails. (b) The north link of n_{master} fails.

The North Link of a Node Fails. Handling the north link failure situation of a node is symmetric to the east link one. Due to this symmetry we here only present the alteration of the routing function in a figurative manner (figure 2(b)).

Lemma 1. *If alteration of the routing function is done according to our methods when the east or the north link of a node fails, the new routing function is connected and deadlock free.*

Proof. From the described alteration methods it is obvious that the new routing function is connected, since nodes in the vicinity of the failing link are updated with respect to the affected packets in both channel directions. It also follows trivially from the *turn model* that the new routing function is deadlock free since none of the prohibited turns are used.

3.2 Both the East and North Links of a Node Fail

In this more complex failure situation the link status information has to be consulted in order for n_{master} to decide the proper alteration action. In addition it becomes necessary to introduce the two illegal turns, the south-east and west-north turn, respectively.

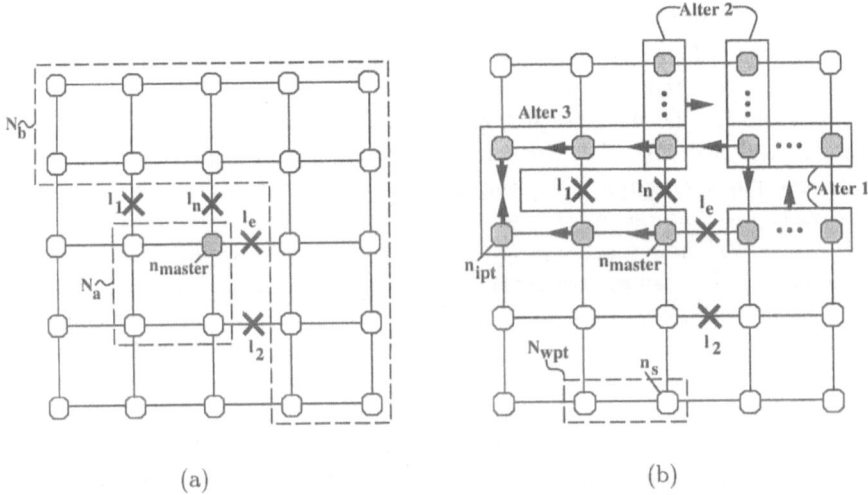

(a)   (b)

Fig. 3. (a) A failure situation where the two node sets N_a and N_b cannot communicate without introducing the prohibited *double turn*, the south-east and west-north dependencies, respectively. (b) Alteration of the routing function when both east and north links of n_{master} fail. The thick arrows indicate redirections of affected packets through the failing links (channels).

Let us first give some motivation for the algorithm proposed below; consider the failure situation showed in figure 3(a). The east and north connecting links (marked l_e and l_n, respectively) of node n_{master} have recently failed, in addition we assume that before this situation arises the links l_1 and l_2 have failed. Notice that since both l_e and l_n now are failing, packets originating in the node set N_a destined for the node set N_b and vice versa are not able to reach their destinations without introducing the prohibited *double turn*, south-east and west-north dependencies, respectively. Thus the key issue is to introduce the prohibited double turn in such a way that the two node sets can communicate with each other, and furthermore such that no cycles in the channel dependency graph are created as a consequence of the introduction of prohibited turns. We shall in the following introduce the prohibited double turn associated with operating east and north links of a node n_i, either positioned west of n_{master} in the same row or positioned south of n_{master} in the same column. We shall show that this reestablishes connectedness and preserves freedom from deadlocks. If there are several candidate nodes n_i in the same row (column) we pick the east (north) most of these, furthermore we denote this specific node for n_{ipt} (*Introduction of Prohibited Turns*). Henceforth, for n_{master} to be able to decide which node n_i is able to serve as n_{ipt}, it must have knowledge about the status of *east* and *north* links of all nodes in the same row (column) and that in addition are positioned west (south) of it. This defines the non local information required for our method.

Below we define the necessary alteration for this specific situation. We present it for the case when n_{ipt} is located west of n_{master}. The case when a node south of n_{master} is selected as n_{ipt} is symmetric.

Alteration 1: The packets that have paths through $l_{e_{west}}$ that arrive at the nodes in the row east of l_e are rerouted northwards. Furthermore the packets that have paths through $l_{e_{west}}$ that arrive at the nodes in the row above the previously specified one are rerouted eastwards (*Alter 1* in figure 3(b)).

Alteration 2: The rerouting proposed here consists of new paths for those packets via $l_{n_{south}}$ that can reach their destination without introducing the prohibited south-east turn. They could have been rerouted through n_{ipt} as well, however, we prefer eastwards rerouting in order to spread the traffic. If there are any nodes south of n_{master} in the same column that have their east link operating, let n_s be the up most node of these candidates. Furthermore let N_{wpt} (reachable *Without Prohibited Turn*) be the node set defined by the nodes located at the same vertical position as n_s or south of n_s, at the same horizontal position as n_s or west of n_s, and in addition positioned east of n_{ipt} (figure 3(b)).

The packets having paths through $l_{n_{south}}$ that arrive at the nodes in the column north of n_{master} and are destined for the node set N_{wpt}, are rerouted eastwards. Furthermore the packets having paths through $l_{n_{south}}$ that arrive at the nodes in the column east of the previously specified one and are destined for the node set N_{wpt}, are rerouted southwards (*Alter 2* in figure 3(b)).

Alteration 3: This alteration redirects those packets that have their detour through n_{ipt}. The packets having paths through $l_{e_{east}}$ and $l_{n_{north}}$ that arrive at nodes in the part of the row between n_{ipt} and n_{master} (inclusive), are rerouted westwards. In addition the packets having paths through $l_{e_{west}}$ and $l_{n_{south}}$ that arrive at nodes in the row above the previously defined one, are rerouted westwards as well. Furthermore the packets with paths through $l_{e_{east}}$ and $l_{n_{north}}$ are rerouted northwards at n_{ipt}, and the packets with paths through $l_{e_{west}}$ and $l_{n_{south}}$ are rerouted southwards in the node north of n_{ipt} (*Alter 3* in figure 3(b)).

Lemma 2. *If alteration of the routing function is done according to the method above for the situation when both east and north links of a node fail, the new routing function is connected and deadlock free.*

Proof. The proof is divided in two:

1. Connected: It follows trivially from the description of the methods that the new routing function is connected.

2. Deadlock free: Assume that a prohibited double-turn DT_i is introduced as a consequence of east and north link failures at n_{master_i}. We first prove that DT_i alone cannot form a cycle in the updated channel dependency graph. Secondly we prove that DT_i cannot be involved in a cycle together with other prohibited double-turns either. We assume that DT_i is introduced west of n_{master_i}, the case when DT_i is south of n_{master_i} is similar.

(i) If DT_i is involved in a cycle the cyclic dependencies must go through the channels connected to n_{master_i}. Since both channel directions of n_{master_i}'s east and north links are failing, a potential cycle must come via the south link of n_{master_i}. This means that another prohibited south-east or west-north turn must be involved to form a cycle in one of the directions. Therefore, DT_i cannot form a cycle if is the only prohibited turn introduced.

(ii) Now assume that there exists another prohibited double-turn DT_j south of n_{master_i}. We know that DT_j is introduced as a consequence of the failure of both the east and north links of n_{master_j} and consider DT_j to be located west of n_{master_j}. From point (i) above the cyclic dependencies must come through the south link of n_{master_j}, thus we need yet another prohibited double-turn to form a cycle, and so forth. The case when DT_j is introduced south of n_{master_j} is treated analogously. This means that no cycle can be formed even by two or more prohibited double-turns when they are introduced according to our method.

Our system is able to handle a significant number of link failure situations. The situations that the proposed method cannot handle are those where both east and north links of n_{master} fail simultaneously, and in addition one of the following two conditions apply:

1. There does not exist any node neither west of n_{master} in the same row, nor south of n_{master} in the same column that have both its east and north links operating.
2. Any candidate node n_i west (south) of node n_{master} in the same row (column) with both its east and north links operating, does not have a horizontal (vertical) path to n_{master}.

3.3 Rerouting Channel is Faulty

In the above presentation of alteration algorithms we assumed that the specified rerouting channels (links) were operating. In this section we identify what impact faulty rerouting channels will have on alteration of the routing function. When rerouting is requested along earlier failed channels we know there *already exist* rerouting paths for these faults, thus we only need to reupdate the routing function along these redirections to also include affected packets associated with the fault currently handled. Moreover, the additional nodes to be visited are *exactly defined* by the alteration methods specified in sections 3.1 and 3.2. To demonstrate this we give the three possible situations for requesting rerouting along a faulty channel, and refer them to the previous defined alterations:

(i) **Requesting rerouting along a faulty east or north channel when only one of these has failed:** According to *Alteration 1* specified in section

3.1, reupdate the routing function in this node (which must be a $n_{master_{old}}$) by letting affected packets from the original fault keep their rerouting northwards if it is the east channel that is failing. With respect to rerouting requested for the north channel direction, let affected packets keep their rerouting eastwards (refer to *Alter 1* in figure 2(b)).

(ii) **Requesting rerouting along a east or north channel when both of these have failed:** From *Alteration 3* in section 3.2, if $n_{ipt_{old}}$ is located west of $n_{master_{old}}$, reupdate the routing function in nodes between $n_{ipt_{old}}$ and $n_{master_{old}}$ (inclusive) to also include affected packets from the current fault (figure 3(b)). There will be a symmetric action for the case when $n_{ipt_{old}}$ is located south of $n_{master_{old}}$.

(iii) **Requesting rerouting along a faulty west or south channel:** Recall from *Alteration 2, 3* (in section 3.1) and figure 2(a), that if rerouting is requested for a failing *west channel direction* the nodes in the row west of this channel and nodes in the row above the previous one need to be updated with respect to the affected packets from current fault. Symmetric updating is necessary when rerouting is requested for a failing south channel (refer to figure 2(b)).

The three situations defined above require some additional updating to be initiated. Now, perhaps we in this reupdating see yet another faulty channel. Therefore, the natural question to ask is: will this repeated alteration process terminate successfully?

Lemma 3. *The alteration process initialized as a consequence of a broken rerouting channel will terminate successfully if it is done as specified above.*

Proof. Since there are a limited number of earlier faults to visit, it means that the only possible way to achieve a non terminating alteration process must be as a consequence of a "loop". Let us denote the failing link starting the updating for $l_{current}$. The rest of the proof is divided in three:

(i) First, when rerouting is requested along a faulty east or north channel, the additional updating is only done locally at the source node of this channel. It is therefore obvious that this situation cannot contribute to a "loop".

(ii) Second, the only possibility to construct a reupdating "loop" (revisiting $l_{current}$) is when the following apply: 1. The node $n_{master_{old}}$ and $l_{current}$ are located in the same row (column) and 2. $l_{current}$ is between $n_{master_{old}}$ and $n_{ipt_{old}}$. However, the last requirement could not entail correctness since it will mean that the path between $n_{master_{old}}$ and $n_{ipt_{old}}$ does not exist, which contradicts the assumption of the existence of such a path between these two nodes.

(iii) Third, if rerouting is requested for a broken west channel additional rerouting is only needed for channels located *east* and *north* of this one, which is even further east and north of $l_{current}$. Hence, $l_{current}$ cannot be revisited. The same reasoning can be used for a broken south channel. Therefore, we can conclude that a repeated alteration process will converge.

3.4 Three- and Higher-Dimensional Meshes

This section illustrates how the positive-first fault tolerant method can be extended to three- and higher-dimensional meshes. The modification follows naturally from the method presented above. In figure 4 we define the prohibited turns

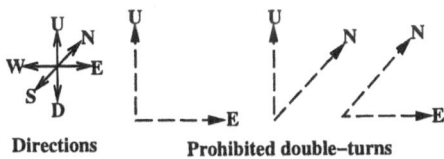

Directions   **Prohibited double-turns**

Fig. 4. The prohibited double-turns for fault free routing in three-dimensional meshes. The referred directions denote the following: U - Up, D - Down, N - North, S - South, E - East and W - West, respectively.

(shown as double-turns) for fault free routing in three-dimensional meshes. We have prohibited just enough turns to avoid cycles in the channel dependency graphs, thus it follows from [12] that the routing function is deadlock free.

In presence of faults we have two main routing function alteration actions depending of the failure situation in the three dimensional case as well:

1. *The situations where at least one of the east, north and up links of a node are operating.* Affected packets of a newly failing link are redirected through one that still is present. This means that no prohibited turns need to be introduced. This is analogous to the situations handled in section 3.1.
2. *The more complex situation when the east, north and up links are faulty simultaneously.* This means that affected packets have to be redirected via a node (n_{ipt}) either down, south or west of n_{master}. In this case it becomes necessary to introduce the prohibited double-turns, which corresponds to the situation defined in section 3.2.

Note that in the three-dimensional case each link participates in two planes, for example an east link is visible in both the *east-north* and *east-up* planes, respectively. Henceforth, nodes in the two planes that a failing link is located in, must have altered routing function. For each plane nodes in two rows/columns require to be updated, as in the two dimensional case. Regarding point 2 above, the additional nodes along the detour through n_{ipt} (the node where the prohibited turns are introduced) need to be updated, refer Alteration 3 in section 3.2.

Lemma 4. *If alteration of the routing function is performed as stated above in presence of faulty links in three-dimensional meshes, the new routing function is connected and deadlock free.*

Proof. The proof can be fully elaborated as the proofs in sections 3.1 and 3.2. Regarding the deadlock issue for the situation when all the *east, north and up* links of a node are faulty simultaneously, creation of any cycle requires several prohibited turns. Since we for each of these situations only associate the prohibited turns with one single node (n_{ipt}), this cannot form cyclic dependencies.

Generalization to n-dimensions follows the same pattern. For each new dimension one prohibits just enough turns to avoid cycles in the channel dependency graph in the fault free case. Then two main routing function alteration processes will be defined along the same lines as described above.

3.5 Simulation Experiments

In order to evaluate the performance of fault tolerant system we have done a series of experiments. We have simulated a 16×16 mesh in presence of 1%, 3% and 5% faulty links. The results show graceful degradation in performance, especially for non-uniform traffic and low number of faults where the degradation is hardly noticeable. The results are not presented here due to space constraints, but can be found in [22].

4 Conclusion

We have presented a method for achieving fault tolerance in wormhole routed (k, n)-mesh networks. Unlike other techniques we do not require existence of virtual channels and tolerate multiple faulty links, even for two-dimensional meshes.

The proposed concept relies on that the network is able to alter the routing function on the fly. The initiation for alteration is always taken locally and distributed to small number of non-neighbor nodes via control lines. The IEEE 1355 STC104 [23] router is an example of such a router with a control line structure. For most of the failure situations, the needed alteration of the routing function is fixed and simple. In order to handle more complex situations, such as when both the east and north links of a node (n) fail, we assume knowledge of link status information from some of the other nodes in the same row/column as node n. This non-local information is used by the local master node responding to a failure. The link status information is gathered via the control lines.

The implementation cost of our method is basically that it requires each node to have an amount of programmable logic, making it possible to implement the proposed alteration algorithms. Secondly we assume the network to have a control line structure where nodes can send/receive specific control messages. Extra logic in the nodes and some form of control lines are, however, usual mechanisms for achieving fault tolerance [2, 13].

Our investigations indicate that it is relatively easy to incorporate node faults into our method, where such faults will be modeled as a failure of all its links. An other interesting path of further work is to investigate if our technique also can be extended to handle (k, n)-tori networks.

Acknowledgments I would like to thank Associate Professor Olav Lysne and Associate Professor Øystein Gran Larsen for their valuable comments and assistance in preparing this paper.

References

1. R. V. Boppana and S. Chalasani. Fault-tolerant wormhole routing algorithms for mesh networks. *IEEE Transactions on Computers*, 44(7):848–864, 1995.

2. S. Chalasani and R. V. Boppana Communication in Multicomputers with Non-convex Faults. *IEEE Transactions on Computers*, 46(5):616–622, 1997.

3. A. A. Chien and J. H. Kim. Planar-adaptive routing: Low-cost adaptive networks for multiprocessors. *Journal of the Association for Computing Machinery*, 42(1):91–123, 1995.

4. W. J. Dally and H. Aoki. Deadlock-free adaptive routing in multicomputer networks using virtual channels. *IEEE Transactions on Parallel and Distributed Systems*, 4(4):466–475, 1993.

5. W. J. Dally and C. L. Seitz. Deadlock-free message routing in multiprocessor interconnection networks. *IEEE Transactions on Computers*, C-36(5):547–553, 1987.

6. B.V. Dao, J. Duato, and S. Yalamanchili. Configurable flow control mechanisms for fault-tolerant routing. In *Proceedings of the 22nd International Symposium on Computer Architecture*, pages 220–229. ACM Press, 1995.

7. J. Duato. A necessary and sufficient condition for deadlock-free adaptive routing in wormhole networks. *Int. Conf. on Parallel Processing*, I:142–149, Aug. 1994.

8. J. Duato. A theory to increase the effective redundancy in wormhole networks. *Parallel Processing Letters*, 4:125–138, 1994.

9. P. T. Gaughan and S. Yalamanchili. A family of fault-tolerant routing protocols for direct multiprocessor networks. *IEEE Transactions on Parallel and Distributed Systems*, 6(5):482–497, 1995.

10. C. J. Glass and L. M. Ni. The turn model for adaptive routing. In *Proceedings of the 19th International Simposium on Computer Architechture*, pages 278–287. IEEE CS Press, California, 1992.

11. C. J. Glass and L. M. Ni. Fault-tolerant wormhole routing in meshes. *In Twenty-Third Annual Int. Symp. on Fault-Tolerant Computing*, pages 240-249, 1993.

12. C. J. Glass and L. M. Ni. The turn model for adaptive routing. *Journal of the Association for Computing Machinery*, 41(5):874–902, 1994.

13. C. J. Glass and L. M. Ni. Fault-tolerant wormhole routing in meshes without virtual channels. *IEEE Transactions on Parallel and Distributed Systems*, 7(6):620–636, June 1996.

14. IEEE 1355-1995. IEEE standard for Heterogeneous InterConnect (HIC) (Low cost, low latency scalable serial interconnect for parallel system construction), 1995.

15. P. Kermani and L. Kleinrock. Virtual cut-through: A new computer communication switching technique. *Computer Networks*, 3:267–286, 1979.

16. D. H. Linder and J. C. Harden. An adaptive and fault tolerant wormhole routing strategy for k-ary n-cubes. *IEEE Transactions on Computers*, 40(1):2–12, 1991.

17. O. Lysne, T. Skeie and T. Waadeland. One-Fault Tolerance and Beyond in Wormhole Routed Meshes. *Microprocessors and Microsystems, Elsevier*, 21(7-8):471–481, 1998.

18. M. D. May, P. W. Thompson, and P. H. Welch, editors. *Networks, routers and transputers: function performance and application.* IOS Press, 1993.
19. L. M. Ni and P.K. McKinley. A survey of wormhole routing techniques in direct networks. *Computer,* 26:62–76, 1993.
20. Paragon XP/S product overview. *Intel Corp., Supercomputer Systems Div,* 1991.
21. C. L. Seitz, W. C. Athas, C. M. Flaig, A. J. Martin, J. Seizovic, C. S. Steele, and W.-K. Su. The architecture and programming of the Ametek series 2010 multicomputer. In *Proceedings of the Third Conference Hypercube Concurrent Computers and Applications, Pasadena (California),* volume I, pages 33–36, 1988.
22. T. Skeie. Topics in Interconnect Networking. *PhD Thesis, ISBN 82-7368-190-4, Dept. of Informatics, University of Oslo,* 1998.
23. P. W. Thompson and J. D. Lewis. The STC104 asynchronous packet switch. *VLSI Design,* 2(4):305–314, 1995.

Shared Control – Supporting Control Parallelism Using a SIMD-like Architecture

Nael B. Abu-Ghazaleh[1] and Philip A. Wilsey[2]

[1] Computer Science Dept.,
State University of New York
Binghamton, NY 13902-6000
[2] Department of ECECS, PO Box 210030
University of Cincinnati,
Cincinnati, Ohio 45221–0030

Abstract. SIMD machines are considered special purpose architectures chiefly because of their inability to support control parallelism. This restriction exists because there is a single control unit that is shared at the thread level; concurrent control threads must time-share the control unit. We present an alternative model for building centralized control architectures that better supports control parallelism. This model, called *shared control*, shares the control unit(s) at the instruction level — in each cycle the control signals for the supported instructions are broadcast to the PEs. In turn, a PE receive its control by synchronizing with the control unit responsible for its current instruction. There are a number of architectural issues that must be resolved. This paper identifies some of these issues and suggests solutions to them. An integrated shared-control/SIMD architecture design (**SharC**) is presented and used to demonstrate the performance relative to a SIMD architecture.

1 Introduction

Parallel architectures are classified according to their control organization as Multiple Instruction streams Multiple Data streams (MIMD), or Single Instruction stream Multiple Data streams (SIMD) machines. MIMD machines have a *distributed control organization*: each Processing Element (PE) has a control unit and is able to sequence a control thread (program segment) locally. Conversely, SIMD machines have a *centralized control organization*: the PEs share one control unit. A single thread executes on the control unit, broadcasting instructions to the PEs for execution. Because the control is shared and the operation is synchronous, SIMD PEs are small and inexpensive.

In a centralized control organization (*e.g.*, SIMD [11], [14] and MSIMD [5], [18] machines), an arbitrary number of PEs share a fixed number of control units. Traditionally, sharing of control has been implemented at the *thread level*; the PEs following the same thread concurrently share a control unit. The presence of application-level control-parallelism causes the performance of this model to drop (proportionately to the degree of control parallelism). This drop occurs because the control units are time-shared among the threads, with only the set of PEs requiring the currently executing control thread actively engaged in computation. General parallel applications contain control parallelism [3], [7] and, therefore, perform poorly on SIMD machines. Accordingly, SIMD machines have fallen out of favor as a platform for general-purpose parallel processing [10], [15].

This paper presents *Shared Control*: a model for constructing centralized control architectures that better supports control-parallelism. Under shared control, the control units are shared at the instruction (or atomic function) level. Each PE is assigned a local program and PEs executing the same instruction, but not necessarily the same thread, receive their control from the same control unit. A control unit is assigned to each instruction, or group of similar instructions, in the instruction set and, thus, broadcasts the microinstruction sequences to implement that instruction repeatedly to the PEs. Each PE receives its control by synchronizing with the control unit corresponding to its current instruction. Thus, all the PEs are able to advance their computation concurrently, regardless of the degree of control parallelism present in the application. The similarity of the hardware to the SIMD model allows the SIMD mode to be supported at little additional cost. With the ability to support control-parallelism efficiently, the major drawback of the SIMD model is overcome.

Shared control is a unique architectural paradigm; the classic association between control units and threads, present in all Von-Neumann based architectures, does not exist in this model. Therefore, it introduces several architectural issues that are unique to it. This identifies some of these issues and discusses solutions to them. The feasibility of the solutions is demonstrated using a SIMD/shared-control architecture design, **SharC**. Using a detailed simulator of **SharC**, the performance of the model is studied for some irregular problems. The remainder of this paper is organized as follows. Section 2 introduces the shared control model. Section 3 presents some architectural issues relating to a general shared control implementation. Section 4 presents a case study of a shared-control architecture. In Section 5, the performance of the architecture is studied. Finally, Section 6 presents some concluding remarks.

2   Shared Control

A shared control architecture is a centralized control architecture where the control units are shared at the operation level. PEs executing the same operation, but not necessarily the same thread, may share the use of the same control unit. An overview of a shared control machine is shown in Figure 1. The control program (microprogram) implementing the instruction set for the shared control mode is partitioned across a number of tightly coupled control units. This partitioning is static; it is carried out at architecture design time. Each control unit repeatedly broadcasts the microprogram sequence assigned to it to the PEs. A PE receives its control from the control unit associated with its current instruction. The PE synchronizes with the control unit by selecting the set

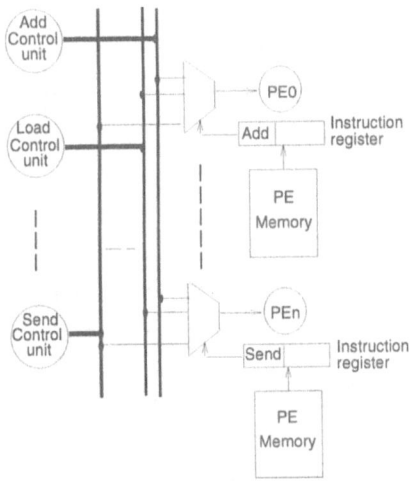

Fig. 1. Overview of a Shared Control Organization

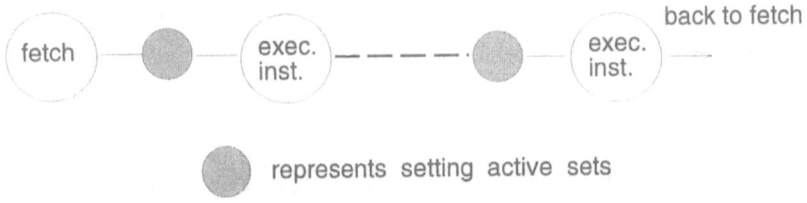

Fig. 2. Implementation of Several Instructions on a Single Control Unit

of control signals broadcast by that control unit. PEs are able to advance their computation concurrently; thus, MIMD execution is supported.

We first consider the problem of implementing and optimizing shared control using a single control unit; this is a special case that is needed for the general solution. Figure 2 shows the single control unit implementation. The control unit must supply control for all the instructions in every cycle. More precisely, the control unit sequentially issues *all* the paths through the microprogram, and the PEs conditionally participate in the path corresponding to their current instruction. In the remainder of this section, some of the architectural issues involved in constructing a single-control unit shared control multiprocessor are discussed.

Managing the Activity Status: Before every instruction execution stage, the activity bits for the PEs that are interested in this stage must be set (represented by the shaded circles in Figure 2). On SIMD machines setting the activity status before a conditional region requires the following operations on the PEs: (i) save current active set, (ii) evaluate the condition, and (iii) set active bit if condition is true. In addition, at the end of the conditional region the active set saved in

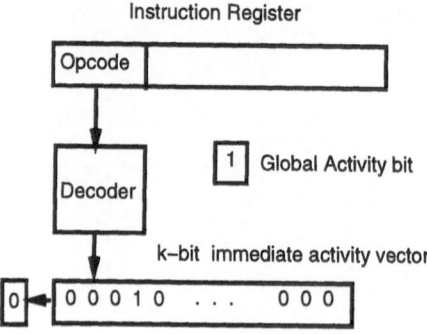

Fig. 3. Activity Set Management

step (i) is restored. The active bit can be tracked by saving the bit directly to an activity bit stack [13], or by using an activity counter that is incremented to indicate a deeper context [12]. Both schemes have a significant overhead (register space, as well as several execution steps). Implementing activity management using the SIMD model adds unacceptable overhead to the shared control cycle time since it is required several times per cycle in shared control.

The condition for the activity of a PE for instruction stage (called the *immediate* activity) is contained in the opcode field in the instruction register, allowing the following optimization to be made. The opcode field is decoded into a k-bit vector (as shown in Figure 3). At the beginning of every instruction segment, the bit vector is shifted into the immediate activity bit. Thus, instruction i is decoded into a bit vector consisting of 1 in the ith position and 0 elsewhere, mirroring the order of the execution regions. Only PEs with a high immediate activity bit participate in the current instruction segment. The register shift can be performed concurrently with the execution of each region at no additional cost; activity management cost is eliminated.

A compositional instruction set: Examining Figure 2), it can be observed that each PE receives the control for all k instructions, but uses only one. Compositional instruction sets relax this restriction by allowing the PEs to use any number of the k instruction segments in each cycle [4]. The output of each execution region is deposited in a temporary register that serves as an input to the next one. The output of the last stage is stored back to the final destination. Thus, an instruction is represented by the subset of the instruction segments that compose it. Composition can be easily incorporated into the activity management scheme in Figure 3.

3 General Shared Control

A general implementation of the shared control model uses multiple control units to implement the microprogram for the instruction set. There are a number of architecture issues that are introduced by this model (in addition to the ones present in the single control unit implementation). The discussion in this section will cover the range of possible solutions to each issue, rather than consider specific solutions in detail.

Control I/O pins: At first glance, this is the primary concern regarding the feasibility of shared control; because the model uses multiple control units, the width of the broadcast bus and the number of required pins at the PEs may become prohibitive. Fortunately, the increase in the number of pins is not linear with the number of control units because: (i) each control unit is responsible only for an instruction or a group of instructions; (ii) literal fields and register number fields are not broadcast; they are part of the instruction (they have to be broadcast in SIMD and MSIMD architectures); and (iii) pins that carry the same values on different control units throughout their lifetime are routed as a single physical pin.

The Control Broadcast Network: The control broadcast network is responsible for delivering the control signals from the control units to the PEs. Traditionally, control broadcast has been a bottleneck on centralized control machines; solutions to this problem include pipelining of the broadcast [3], and caching the control units closer to the PEs [16]. With advances in fabrication technology, there is a trend to move processing to memory [8]. For such systems, the control units may be replicated per chip, simplifying the control broadcast problem.

Control Unit Synchronization and Balance: The control stream required by each PE is not supplied *on demand* as per traditional computers. Rather, the control units have to be synchronized such that the control streams are available when a PE needs them (with minimal waiting time). A possible model to synchronize the control units is to force them to share a common cycle length, called *fundamental instruction cycle*, determined by the slowest control unit. However, the instruction cycle time may vary widely for the instructions in the instruction set, forcing long idle times for PEs with short instructions. Fortunately, there are a number of synchronization models that reduce PE idle time, including: (i) issuing the long instructions infrequently; (ii) allowing the short instructions to execute multiple times while the long instruction is executing; and (iii) breaking long instructions in a series of shorter instructions [2].

Support for Communication and I/O: Support of communication, I/O and other system functions poses the following problem: the system must provide support for both SIMD and MIMD operation at a cost that can be justified against the simple PEs. We focus this discussion on the communication subsystem. There are two options for the support communication. The first option restricts the support to the SIMD mode; the more difficult problem of supporting MIMD-mode communication is side stepped. Restricting communication to SIMD mode is inefficient because: (i) all the PEs have to synchronize and switch back to SIMD if any of them needs to communicate, (ii) because of SIMD semantics, all the PEs must wait for the PE with the worst path communication; a bottleneck of synchronous operation when irregular communication patterns are required [6]. Another alternative is to support MIMD operation using an inexpensive network [1].

4 A Case Study: SharC

In this section we present an overview of the architecture of *SharC*, a proof-of-concept shared control architecture [1]. We preface this discussion by noting that the *SharC* architecture was designed for possible fabrication within a very small budget (10,000 US dollars for a 64 PE prototype). The PE chip fits within the MOSIS small chip (a very small chip size with 100 I/O pads); much higher integration is possible with more aggressive technology. While *SharC* does not represent a realistic high-performance design using today's technology, it can still be used as an impartial model to investigate the shared-control performance relative to SIMD performance.

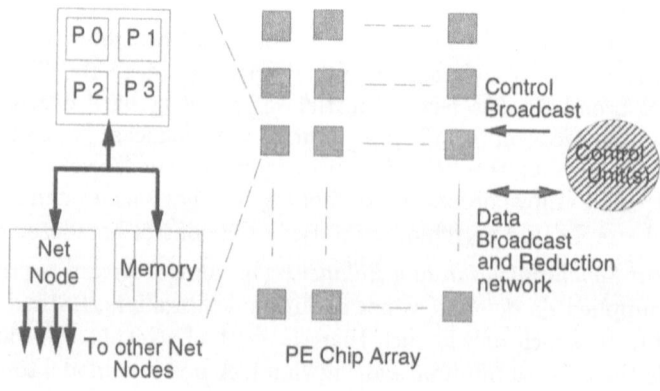

Fig. 4. System Overview

Figure 4 presents an overview of the system architecture. The shared control subsystem consists of 9 control units. The fundamental cycle for all the control units is 11 cycles, with the exception of the load/store control units which require 22 cycles. There are 4 PEs per chip sharing a single memory port. Communication occurs using the same memory port as well: 4 cycles of the 11-cycle fundamental cycle are reserved for the exchange of two message with the network processor (one each way). Most of the integer operations are mapped to the same control unit and implemented using composition. The fundamental cycle is 11 cycles long, constrained by the access time of the shared memory/communication port. If dedicated (non-shared) memory ports are supplied, this length can be reduced to 4 cycles for most instructions.

As was expected, the number of control pins for the shared control mode exceeded the number required by SIMD mode. However, the number of pins only increased from 112 pins for the SIMD mode to 118 pins for the shared control mode; the increase is small because of reasons discussed in Section 3. The average number of Clocks Per Instruction (CPI) for the shared control mode is higher than the CPI for SIMD mode; most SIMD instructions require less than

11 cycles to implement. The reason for the higher CPI is the time required for the additional fetch required for each instruction, as well as the idle cycles required to synchronize the control units. Thus, for a pure data-parallel applications, a SIMD implementation yields better performance than shared control. However, as the degree of control parallelism increases, the performance of shared control remains constant, while SIMD performance degrades.

The SHARC communication network is a packet-switched toroidal mesh network, with a chip (4 PEs) at each node sharing a network processor. The network processor is made simple by using an adaptive deflection routing strategy (no buffering is necessary) [9]. In many cases, deflection routing results in better performance than oblivious routing strategies (path taken by packet independent of other packets), because excess traffic is deflected away from congested areas; creating better network load balance.

5 Performance Analysis

A detailed simulator of a scalable configuration of the **SharC** design is used to study its performance. The model incorporates a structural description of the control subsystem at the microcode level; it also serves to verify the correctness of the microcode. The test programs were written using assembly language, and assembled using a macro assembler. The assembler supports SIMD/SPMD programming model; an algorithm is written as a SIMD/SPMD application, and any region can be executed in either mode (allowing mixed-mode programming). A transition from SIMD to MIMD operation is initiated explicitly by the SIMD program (using a special `c_mimd` instruction). A transition from MIMD back to SIMD occurs when all the PEs have halted (using a `halt` instruction); implementing a full barrier synchronization.

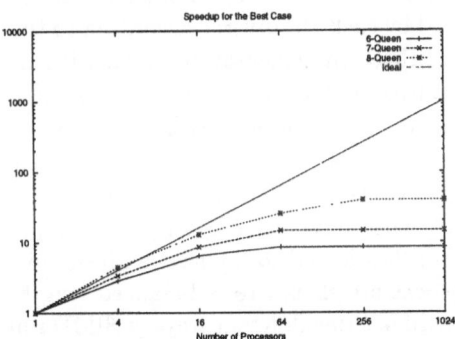

Fig. 5. Speedup for the N-Queen Problem

Our first example is the N-Queen problem: a classic combinatorial problem that is easy to express but difficult to solve. The N-Queen problem has recently

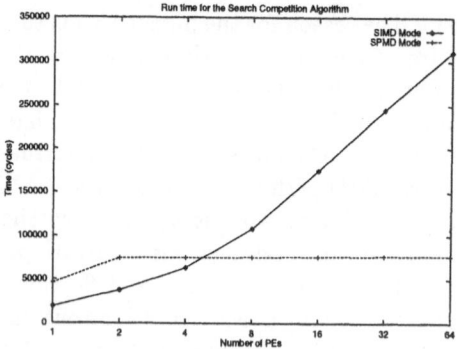

Fig. 6. Search Competition Algorithm Performance

found applications in optical computing and VLSI routing [17]. We considered the exhaustive case (find all solutions, rather than find a solution). The N-Queen problem is highly irregular and not suited to SIMD implementation. Figure 5 shows the performance of the shared control implementation normalized to that of the SIMD implementation. The SIMD solution was worse than a sequential solution because it failed to extract any parallelism, but incurred the activity management overhead. The speedup leveled because the of parallelism present at the small scales of the problem that were studied is limited.

The second application is a database search competition algorithm; an algorithm characteristic of massively parallel database operations. Each PE is initialized with a segment of a database (sorted by keys), and the PEs search for keys in their database segment. As the number of PEs is scaled, the size of the database is scaled (the size of the segment at each PE is kept the same). For a small number of PEs, there is little control parallelism, and SIMD performs better than shared control. As the number of PEs is increased, the paths through the database followed by each PE diverge according to their respective data. The performance of the SIMD implementation drops with the increased number of PEs while shared control performance remains constant.

Finally, we consider a parallel case statement with balanced cases (Figure 7). By varying the number of cases in the case statement, the degree of control parallelism is directly controlled. The instruction mix within the SIMD blocks generated randomly, with a typical RISC profile comprised of 30% load/store instructions with branches forced to the next address (allowing for the pipelined instruction). The blocks are chosen to of balanced lengths (10% variation). The program was simulated in three different ways: a SIMD implementation, a shared control (SPMD) implementation, and a mixed mode implementation. The mixed mode implementation executes the switch statement in the shared control mode, but executes block A and block B in SIMD.

Figure 8 shows execution times as a function of the number of cases. Not surprisingly, the performance of the SIMD mode degrades (linearly) with the increased control parallelism injected by increasing the number of cases. Both

```
plural p = p_random(1, n);

[SIMD BLOCK A]
switch(p) {
    case 1: [SIMD BLOCK 1];
            break;
        .   .
    case n: [SIMD BLOCK n];
    }
[SIMD BLOCK B]
```

Fig. 7. Parameterized Parallel Branching Region

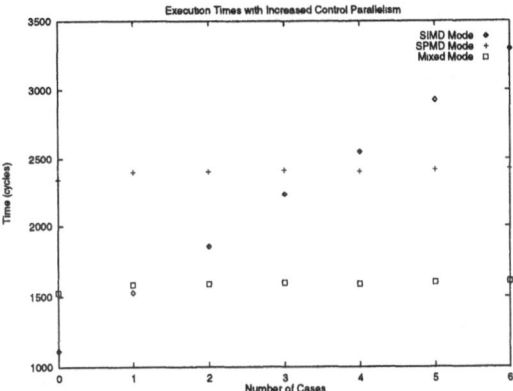

Fig. 8. Simulation results for the architecture on a parameterized branching region

the mixed mode and MIMD execution times did not change significantly with the added control parallelism. The mixed mode implementation is more efficient than the full shared control implementation because of its superior performance on the leading and trailing SIMD blocks.

6 Conclusions

SIMD machines offer an excellent price to performance alternative for applications that fit their restricted control organization. Commercial SIMD machines with superior price to performance ratio continue to be built for specific applications. Unfortunately, centralized control architectures (like the SIMD model) cannot support control parallelism in applications; the control unit has to sequentially broadcast the different control sequences required by each of the control threads. In this paper, we present a model for building centralized control architectures that is capable of supporting control parallelism efficiently. The shared control model is less efficient than the SIMD model on regular code sequences (it requires an additional instruction fetch, and memory space to hold the task instructions). When used in conjunction with the SIMD model, irregular regions of applications can be executed in shared control, extending the utility of the SIMD model to a wider range of applications.

The shared control model is fundamentally different from the SIMD model. Therefore, there are a set of architectural and implementation issues that must be addressed before an efficient shared control implementation can be realized. The performance of **SharC** was studied using a detailed RTL simulator. Applications were implemented in the SIMD mode, and in the shared control mode. The performance of the shared control implementation was compared to a pure SIMD implementation for several applications (the highly irregular N-queen, and massively parallel database search algorithms were used). Even at the small scales of the problems considered, shared control resulted in significant improvement in performance over the pure SIMD implementation. Unfortunately, larger applications (or problem sizes of the studied applications) could not be studied because: (i) simulating a massively parallel machine on a uniprocessor is inefficient: the simulation run time for the 1024 PE 8-queen problem was in excess of 20 hours on a SunSparc 100 workstation, and (ii) we do not have a compiler for the machine; all the examples were written and coordinated in assembly language. Finally, we used a parameterized conditional region benchmark to demonstrated that shared control is beneficial for applications where even a small degree of control parallelism is present.

References

1. Abu-Ghazaleh, N. B. *Shared Control: A Paradigm for Supporting Control Parallelism on SIMD-like Architectures.* PhD thesis, University of Cincinnati, July 1997. (in press).
2. Abu-Ghazaleh, N. B., and Wilsey, P. A. Models for the synchronization of control units on shared control architectures. *Journal of Parallel and Distributed Computing* (1998). (in press).
3. Allen, J. D., and Schimmel, D. E. The impact of pipelining on SIMD architectures. In *Proc. of the 9th International Parallel Processing Symp.* (April 1995), IEEE Computer Society Press, pp. 380–387.
4. Bagley, R. A., Wilsey, P. A., and Abu-Ghazaleh, N. B. Composing functional unit blocks for efficient interpretation of MIMD code sequences on SIMD processors. In *Parallel Processing: CONPAR 94 – VAPP VI* (September 1994), B. Buchberger and J. Volkert, Eds., vol. 854 of *Lecture Notes in Computer Science*, Springer-Verlag, pp. 616–627.
5. Bridges, T. The GPA machine: A generally partitionable MSIMD architecture. In *Proceedings of the 3rd Symposium on the Frontiers of Massively Parallel Architecutres* (1990), pp. 196–203.
6. Felderman, R., and Kleinrock, L. An upper bound on the improvement of asynchronous versus synchronous distributed processing. In *Proceedings of the SCS Multiconference on Distributed Simulation* (January 1990), vol. 22, pp. 131–136.
7. Fox, G. What have we learnt from using real parallel machines to solve real problems? Tech. Rep. C3P–522, Caltech Concurrent Computation Program, California Institute of Technology, Pasadena, CA 91125, March 1988.

8. Gokhale, M., Holmes, B., and Iobst, K. Processing in memory: The Terasys massively parallel PIM array. *IEEE Computer* (April 1995), 23–31.

9. Greenberg, A., and Goodman, J. Sharp approximate models of deflection routing in mesh networks. *IEEE Transactions on Communications 41* (Jan. 1993).

10. Hennesy, J. L., and Patterson, D. A. *Computer Architecture a Quantitave Approach, Second Edition.* Morgan Kaufman Publishers Inc., San Mateo, CA, 1995.

11. Hillis, W. D. *The Connection Machine.* The MIT Press, Cambridge, MA, 1985.

12. Keryell, R., and Paris, N. Activity counter: New optimization for the dynamic scheduling of SIMD control flow. In *1993 International Conference on Parallel Processing* (Aug. 1993), vol. 2, pp. 184–187.

13. MasPar Computer Corporation. *MasPar Assembly Language (MPAS) Reference Manual.* Sunnyvale CA, July 1991.

14. Nickolls, J. The design of the MasPar MP-1. In *Proceedings of the 35th IEEE Computer Society International Conference* (1990), pp. 25–28.

15. Parhami, B. SIMD machines: Do they have a significant future? *Computer Architecture News* (September 1995), 19–22.

16. Rockoff, T. SIMD instruction caches. In *Proceedings of the Symposium on Parallel Architectures and Algorithms '94* (May 1994), pp. 67–75.

17. Sosic, R., and Gu, J. Fast search algorithms for the N-queens problem. *IEEE Transactions on Systems, Man, and Cybernatics* (Nov. 1991), 1572–1576.

18. Weems, C., Riseman, E., and Hanson, A. Image understanding architecture: Exploiting potential parallelism in machine vision. *Computer* (Feb 1992), 65–68.

Workshop 23
ESPRIT Projects

Ron Perrott and Colin Upstill

Co-chairmen

The ESPRIT Programme has now been in existence for over a decade. One of the areas to benefit substantially from the injection of ESPRIT research and development funding has been the area of parallel computing or high performance computing. It is therefore appropriate that at this conference there should be two sessions dealing with HPC projects. A wide range of papers were submitted, and those selected represent state-of-the-art projects in various areas of interest. The projects break down into applications which can benefit from the use of HPC and those which deal with the software required for making high performance computers more efficient.

In the former category are papers like Parallel Crew Scheduling in Pharos. This is an application which considers the problem of scheduling airline crews and the use of a network of workstations. It is based on the widely known Carmen system which is used by most European airlines to schedule their crews among aeroplanes. The objective is to produce a parallel version of the Carmen system and execute it on a network of workstations and demonstrate the improvements and benefits which can be obtained. Essentially there are two critical parts in the application dealing with pairing generation and optimisation. The pairing generator distributes the numeration of pairings over the processors and the optimiser is based on an iterative Lagrangian heuristic. Initial results indicate the benefits the speed-up which can be obtained with this application on a network of workstations.

The second paper deals with CORBA (Common Object Request Broker Architecture) and develops a compliant programming environment for HPC. This is part of an ESPRIT Project known as PACHA. The main objective is to help the design of HPC applications using independent software components through the use of distributed objects. It proposes to harness the benefits of distributive and parallel programming using a combination of two standards, namely CORBA and MPI. CORBA provides transparent remote method invocations which are handled by an object request broker that provides a communication infrastructure independent of the underlying network. This project introduces a new kind of object which is referred to as a CORBA parallel object. It then allows the aggregation of computing resources to speed up the execution of a software component. The interface is described using an extended IDL to manage data distribution among the objects of the collection. CORBA is being implemented using Orbix from Iona Technologies and has already been tested using a signal processing application based on a client server approach.

The OCEANS project deals with optimising compilers for embedded applications. The objective of OCEANS is to design and implement an optimising compiler that utilises aggressive analysis techniques and integrates source level restructuring transformations with low level machine-dependent optimisations. This will then provide a prototype framework for iterative compilation where feedback from the low level is used to guide the selection of a suitable sequence of source level transformations and vice versa. The OCEANS compiler is centred around two major components, a high level restructuring system known as MT1 and a low level system for supporting language transformations and optimisations known as Salto. In order to validate the compiler, four public domain multimedia codes have been selected. Initial results indicate that the results using the initial prototype are satisfactory in comparison with the production compiler but their implementation still needs to be refined at both the high and low levels.

The fourth paper deals with industrial stochastic simulations on large-scale meta-computers, and reports the acheivements of the PROMENVIR (Probabilistic Mechanical Design Environment) project. This project culminated in a pan-European meta-computer demonstration of the use of PROMENVIR running a multibody simulation application. PROMENVIR is an advanced computing tool for performing stochastic analysis of generic physical systems. The tool provides the user with a framework for running a stochastic Monte Carlo simulation using a preferred deterministic solver and then provides sophisticated means to analyse the results. The European meta-computer consisted of a total of 102 processors geographically distributed among the members of the consortium. The paper details the problems that were encountered in establishing the meta-computer and carrying out the execution of the chosen application. It makes useful contributions on the feasibility and reliability of such an approach.

HIPEC stands for High Performance Computing Visualisation System Supporting Network Electronic Commerce applications. This is an ESPRIT Project whose main objective is to integrate advanced high performance technologies to form a generic electronic commerce application. The application is aimed at giving a large number of SMEs a configuration and visualisation tool to support the selling process within the show room and to enlarge their business using the same tool over the Internet. Computer graphics technology plays a primary role in the project as realistic images are required in such commerce. The particular experiment was based on the generic technology and applied to bathroom furniture application. The project was shown to be of benefit in this particular application area.

The final paper is the porting of the SEMC3D electromagnetics code to HPF. This was a project within the ESPRIT PHAROS Project, in which four industrial simulation codes are ported to HPF. The electromagnetic simulation code was ported onto two machines, namely the Meiko CS2 and the IBM SP2. The paper describes the application and the numerical computational features which are important and then analyses the porting of the code to a parallel machine. A series of phases such as code cleaning, translation to Fortran 90, and inserting HPF directives are described. This is followed by a series of measurements. The first stage in the port was the conversion of the Fortran 77 code to Fortran

90, for which converted code single processor times on the IBM SP2 where obtained. This was then followed by conversion to HPF code and performance times over a range of processors documented. Finally, a comparison with message passing code was carried out. The speed-ups obtained in the benchmarking of this particular code are described as encouraging by the users, and give evidence of the benefits of using HPF.

Parallel Crew Scheduling in PAROS*

Panayiotis Alefragis[1], Christos Goumopoulos[1], Efthymios Housos[1], Peter Sanders[2], Tuomo Takkula[3], Dag Wedelin[3]

[1] University of Patras, Patras, Greece
[2] Max-Planck-Institut für Informatik, Saarbrücken, Germany
[3] Chalmers University of Technology, Göteborg, Sweden

Abstract. We give an overview of the parallelization work done in PAROS. The specific parallelization objective has been to improve the speed of airline crew scheduling, on a network of workstations. The work is based on the Carmen System, which is used by most European airlines for this task. We give a brief background to the problem. The two most time critical parts of this system are the pairing generator and the optimizer. We present a pairing generator which distributes the enumeration of pairings over the processors. This works efficiently on a large number of loosely coupled workstations. The optimizer can be described as an iterative Lagrangian heuristic, and allows only for rather fine-grained parallelization. On low-latency machines, parallelizing the two innermost loops at once works well. A new "active-set" strategy makes more coarse-grained communication possible and even improves the sequential algorithm.

1 The PAROS Project

The PAROS (Parallel large scale automatic scheduling) project is a joint effort between Carmen Systems AB of Göteborg, Lufthansa, the University of Patras and Chalmers University of Technology, and is supported under the European Union ESPRIT HPCN (High Performance Computing and Networking) research program. The aim of the project is to generally improve and extend the use and performance of automatic scheduling methods, stretching the limits of present technology.

The starting point of PAROS is the already existing Carmen (Computer Aided Resource Management) System, used for crew scheduling by Lufthansa as well as by most other major European airlines. The crew costs are one of the main operating costs for any large airline, and the Carmen System has already significantly improved the economic efficiency of the schedules for all its users. For a detailed description of the Carmen system and airline crew scheduling in general, see [1]. See also [3, 6, 7] for other approaches to this problem.

At present, a typical problem involving the scheduling of a medium size fleet at Lufthansa requires 10-15 hours of computing, for large problems as much as 150 hours. The closer to the actual day of operation the scheduling can take

---

* This work has been supported by the ESPRIT HPCN program.

place, the more efficient it will be with respect to market needs and late changes. Increased speed can also be used to solve larger problems and to increase the solution quality. The speed and quality of the crew scheduling algorithms are therefore important for the overall efficiency of the airline.

An important objective of PAROS is to improve the performance of the scheduling process through parallel processing. There exist attempts to parallelize similar systems on high end parallel hardware (see [7]), but our focus is primarily to better utilize a network of workstations and/or multiprocessor workstations, allowing airlines such as Lufthansa to use their already existing hardware more efficiently.

In section 2 we give a general description of the crew scheduling process in the Carmen system. The following sections report the progress in parallelizing the main components of the scheduling process: the pairing generator in section 3, and the optimizer in section 4. Section 5 gives some conclusions and pointers for future work.

2 Solution Methodology and the Carmen System

For a given timetable of flight *legs* (non stop flights), the task of crew scheduling is to determine the sequence of legs that every crew should follow during one or several workdays, beginning and ending at a home base. Such routes are known as *pairings*. The complexity of the problem is due to the fact the crews cannot simply follow the individual aircraft, since the crews on duty have to rest in accordance with very complicated regulations, and there must always be new crews at appropriate locations which are prepared to take over. At Lufthansa, problems are usually solved as weekly problems with up to the order of 10^4 legs.

For problems of this kind there is usually some top level heuristic which breaks down the problem into smaller subproblems, often to different kinds of daily problems. Due to the complicated rules, the subproblems are still difficult to handle directly, and are solved using some strategy involving the components of pairing generation and optimization.

The main loop of the Carmen solution process is shown in Figure 1. Based on an existing initial or current best solution a subproblem is created by opening up a part of the problem. The subproblem is defined by a *connection matrix* containing a number of legal leg connections, to be used for finding a better solution. The connection matrix is input to the *pairing generator* which enumeratively generates a huge number of pairings using these connections. Each pairing must be legal according to the contractual rules and airline regulations, and for each pairing a cost is calculated. The generated pairings are sent to the optimizer, which selects a small subset of the pairings in order to minimize the total crew costs, under the constraints that every flight leg must receive a crew. In both steps the main algorithmic difficulty is the combinatorial explosion of possibilities, typical for problems of this kind. Rules and costs are handled by a special rule language, and is translated into runnable code by a rule compiler, which is then called by the pairing generator.

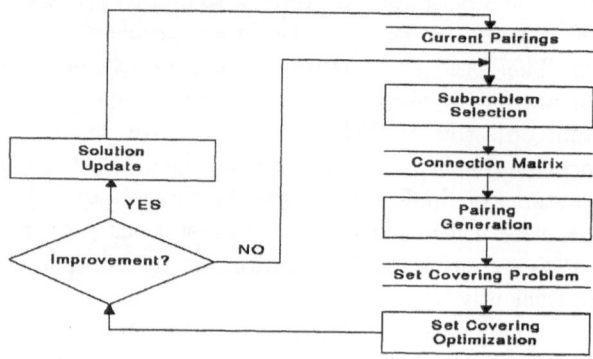

Fig. 1. Main loop in Carmen solution process.

A typical run of the Carmen system consists of 50-100 daily iterations in which 10^4 to 10^6 pairings are generated in each iteration. Most of the time is spent in the pairing generator and the optimizer. For some typical Lufthansa problems, profiling reveals that 5-10% of the time is consumed in the overall analysis and the subproblem selection, about 70-85% is spent in the pairing generator, and 10-20% in the optimizer. These figures can however vary considerably for different kinds of problems. The optimizer can sometimes take a much larger proportion of the time, primarily depending on the size and other characteristics of the problem, the complexity of the rules that are called in the inner loop of the pairing generator, and various parameter settings. As a general conclusion however, it is clear that the pairing generator and the optimizer are the main bottlenecks of the sequential system, and they have therefore been selected as the primary targets for parallelization.

3 Generator Parallelization

The pairing generation algorithm is a quite straightforward depth first enumeration that starts from a crew base, and builds a chain of legs by following possible connections as defined by the matrix of possible leg connections. The search is heuristically pruned by limiting the number of branches in each node, typically between 5-8. The search is also interrupted if the chain created so far is not legal.

The parallelization of the pairing generator is based on a manager/worker scheme, where each worker is given a start leg from which it enumerates a part of the tree. The manager dynamically distributes the start legs and the necessary additional problem information to the workers in a demand driven manner. Each worker generates all legal pairings of the subtree and returns them to the manager.

The implementation has been made using PVM, and several efforts have been made to minimize the idle and communication times. Large messages are sent whenever possible. Load balancing is achieved by implementing a dynamic workload distribution scheme that implicitly takes into account the speed and the current load of each machine. The number of start legs that are sent to each worker is also changing dynamically with a fading algorithm. In the beginning, a large number of start legs is given and in the end only one start leg per worker is assigned. This scheme reconciles the two goals of minimizing the number of messages and of good load balancing. Efficiency is also improved by pre-fetching. A worker requests the next set of start legs before they are needed. It can then perform computation while its request is being serviced by the manager.

The parallel machine can also be extended dynamically. It is possible to add a new host at any time to the virtual parallel machine and this will cause a new worker to be started automatically.

Another feature of the implementation is the suspend/resume feature which makes sure that only idle workstations are used for generation. Periodically the manager requests the CPU load of the worker, suspends its operation if the CPU load for other tasks is over a specified limit, and resumes its operation if the CPU load is below a specified limit. The performance overhead depends on the rate of the load checking operation which in any case is small enough ($\leq 1\%$).

3.1 Scalability

For the problems tested, the overhead for worker initialization depends on the size of the problem and consists mainly of the initialization of the legality system and the distribution of the connection matrix. This overhead is very small compared to the total running time.

The data rate for sending generated pairings from a worker to the manager is typically about 6 KB/s, assuming a typical workstation and Lufthansa rules. This is to be compared with the standard Ethernet speed of 1 MB/s, so after its start-up phase the generator is CPU bound for as many as 200 workstations. Given the high granularity and the asynchronous execution pattern we therefore expect the generator to scale well up to the order of 100 workstations connected with standard Ethernet, of course provided that the network is not too loaded with other communication.

3.2 Fault Tolerance

Several fault tolerance features have been implemented. To be reliable when many workstations are involved, the parallel generator can recover from task and host failures. The notification mechanism of PVM [4] is used to provide application level fault tolerance to the generator.

A worker failure is detected by the manager which also keeps the current computing state of the worker. In case of a failure the state is used for reassigning the unfinished part of the work to another worker. The failure of the manager can be detected by the coordinating process, and a new manager will be started. The recovery is achieved since the manager periodically saves its state to the filesystem (usually NFS) and thus can restart from the last checkpoint. The responsibility of the new manager is to reset the workers and to request only the generation of the pairings that have not been generated, or have been generated partially. This behavior can save a lot of computing time, especially if the fault appears near the end of the generation work.

3.3 Experimental Results

We have measured the performance of the parallel generator experimentally. The experiments have been performed on a network of HP 715/100 workstations of roughly equivalent performance, connected by Ethernet at the computing center of the University of Patras, during times when almost exclusive usage of the network and the workstations was possible. The results are shown in Table 3.3. What we see is the elapsed time for a single iteration of the generator, and the running time in seconds as a function of the number of CPUs. The speedup in all cases is almost linear to the number of CPUs. The generation time decreases in all cases almost linearly to the number of the CPUs used.

Table 1. Parallel generator times for typical pairing problems.

problem name	legs	pairings	1 CPU	2 CPUs	4 CPUs	6 CPUs	8 CPUs	10 CPUs
lh_dl_gg	946	396908	26460	13771	7061	4536	3466	2818
lh_dl_splimp	946	318938	20760	10797	5448	3686	2797	2181
lh_wk_gg	6196	594560	31380	16834	8436	5338	4312	3288
sj_dl_kopt	1087	159073	10860	5563	2804	1892	1385	1112

4 Optimizer Parallelization

The problem solved by the optimizer is known as the set covering problem with additional base capacity constraints, and can be expressed as

$$\min\{cx : Ax \geq 1, Cx \leq d, x \text{ binary}\}$$

Here A is a 0-1 matrix and C is arbitrary non-negative. In this formulation every variable corresponds to a generated pairing and every constraint corresponds to a leg. The size of these problems usually range from 50×10^4 to $10^4 \times 10^6$ (constraints × variables), usually with 5-10 non-zero A-elements per variable. In most cases only a few capacity ($\leq$) constraints are present.

This problem in its general form is NP-hard, and in the Carmen System it is solved by an algorithm [9], that can be interpreted as a Lagrangian based heuristic. Very briefly, the algorithm attempts to perturbate the cost function as little as possible to give an integer solution to the LP-relaxation. From the point of view of parallel processing it is worth pointing out that the character of the algorithm is very different, and for the problems we consider also much faster, compared to the common branch and bound approaches to integer programming, and is rather more similar to an iterative equation solver. This also means that the approach to parallelization is completely different, and it is a challenge to achieve this on a network of workstations.

The sequential algorithm can be described in the following simplified way that highlights some overall aspects which are relevant for parallelization. On the top level, the sequential algorithm iteratively modifies a Lagrangian cost vector $\bar{c}$ which is first initialized to c. The goal of the iteration is to modify the costs so that the sign pattern of the elements of $\bar{c}$ corresponds to a feasible solution, where $\bar{c}_j < 0$ means that $x_j = 1$, and where $\bar{c}_j > 0$ means that $x_j = 0$. In the sequential algorihm the iteration proceeds iteratively by considering *one constraint at a time*, where the computation for this constraint modifies the $\bar{c}$ values for the variables of that constraint. Note that this implies that values corresponding to variables not in the constraint are not changed by this update. The overall structure is summarized as

$\bar{c} = c$
reset all s^i to 0
$\kappa = 0$
repeat
      for every constraint i
         $r^i = \bar{c}^i - s^i$
         $s^i = $ function of r^i, b_i and some parameters
         $\bar{c}^i = r^i + s^i$
      increase κ according to some rule
until no sign changes in $\bar{c}$

In this pseudocode, $\bar{c}^i$ is the sub-vector of $\bar{c}$ corresponding to non-zero elements of constraint i. The vector s^i represents the contribution of constraint i to $\bar{c}$. This contribution is cancelled out before every iteration of constraint i where a new contribution is computed. The vectors r^i are temporaries.

A property relevant to parallelization is that when an iteration of a constraint has updated the reduced costs of its variables, then the following constraints have to use these new updated values. Another property is that the vector s^i is usually a function of the current $\bar{c}$-values for at most two variables in the constraint, and these variables are referred to as the *critical variables* of the constraint.

In the following subsections we summarize the different approaches to parallelization that have been investigated. In addition to this work, an aggressive optimization of the inner loops was performed, leading to an additional speedup of roughly a factor of 3.

4.1 Constraint Parallel Approach

A first observation is that if constraints have no common variables they may be computed independently of each other, and a graph coloring of the constraints would reveal the constraint groups that could be independent. Unfortunately, the structure of typical set covering problems (with many variables and few long constraints) is such that this approach is not possible to apply in a straightforward way. The approach can however be modifed to let every processor maintain a local copy of $\bar{c}$, and a subset of the constraints. Each processor then iterates its own constraints and updates its local copy of $\bar{c}$. The different copies of $\bar{c}$ must then be kept consistent through communication. Although the nature of the algorithm is such that the $\bar{c}$ communication can be considerably relaxed, the result for our type of problems did not allow a significant speedup on networks of workstations.

4.2 Variable Parallel Approach

Another way of parallelization is to parallelize over the variables and the inner loops of the constraint calculation. The main parts of this calculation that concern the parallelization are

1) collect the $\bar{c}$ for the variables in the constraint.
2) find the two least values of r^i.
3) copy the result back to $\bar{c}$.

When the variables are distributed over the processors, each processor is then responsible for only one piece of $\bar{c}$ and the corresponding part of the A-matrix. Some of the operations needed for the constraint update can be conveniently done locally, but the double minimum computation requires communication. This operation is associative, and it is therefore possible to get the global minimal elements through a reduction operation on the local critical minimum elements. To minimize communication it is also possible to group the constraints and perform the reduction operation for several independent constraints at a time.

This strategy has been successfully implemented on an SGI Origin2000, with a speedup of 7 on 8 processors. However, it cannot be used directly on a network of workstations due to the high latency of the network. We have therefore investigated a relaxation to this algorithm, that does a lazy update of $\bar{c}$. The approach is based on updating $\bar{c}$ only for the variables required by the constraint group that will be iterated next. The reduced cost vector is then fully updated based on the contributions of the previous constraint group, during the reduction operation of the current constraint group, and so overlaps computation with both communication and idle time. The computational part of the algorithm is slightly enlarged and the relaxation requires that constraint groups with a reasonable number of common variables can be created. The graph colouring heuristics that have been used are based on [8], which gives 10-20% fewer groups than a greedy first fit algorithm, although the computation times are slightly higher.

For numerical convergence reasons random noise is added in the solution process and even with minor changes in the parameters the running times can vary considerably, and also the solution quality can be slightly different. Tests are shown in Table 4.2 for pure set covering problems from the Swedish Railways, and the settings have been adjusted to ensure that a similar solution quality is preserved on the average (usually well within 1% of the optimum), and the problems shown here are considered as reasonably representative for the average case. Running times are given in seconds for the lazy variable parallel implementation for 1 to 4 processors. The hardware in this test has been 4 × HP 715/100 connected with 10 Mbps switched Ethernet. It can be seen that reasonably good speedup results are obtained even for medium sized problem instances. For networks of workstations a possibility could here also be to use faster networks such as SCI [5] or Myrinet [2] and streamlined software interfaces.

Table 2. Results for the lazy variable parallel code.

problem name	rows	columns	1 CPU	2 CPUs	3 CPUs	4 CPUs
sj_daily_17sc	58	1915	0.60	2.95	3.53	4.31
sj_daily_14sc	94	7388	17.30	9.45	11.78	11.82
sj_daily_04sc	429	38148	288.73	124.76	82.94	72.00
sj_daily_34sc	419	156197	951.53	634.54	365.90	259.32

4.3 Parallel Active Set Approach

As previously mentioned, the constraint update is usually a function of at most two variables in the constraint. If the set of these critical variables was known in advance, it would in principle be possible to ignore all other variables and receive the same result much faster. This is not possible, but it gives intuitive support for a variation of the sequential algorithm, where a smaller number of variables that are likely to be critical are selected in an *active set*. The idea is to make the original algorithm work only on the active set, and then sometimes do a scan over all variables to see if more variables should become active. For our problems this is especially appealing considering the close relation to the so called column generation approach to the crew scheduling problem, see [3]. The approach was first implemented sequentially and turns out to work well on real problems. An additional and important performance benefit is that the scan can be implemented by a columnwise traversal of the constraint matrix which is much more cache friendly and gives an additional sequential speedup of about 3 for large instances.

This modified algorithm also gives a new possiblity for parallelization. If the active set is small enough, then the scan will dominate the total execution time, even if the active set is iterated many times between every scan. In constrast to the original algorithm, all communication necessary for a global scan of all variables can be concentrated into a single vector valued broadcast and reduction operation. Therefore, the parallel active set approach is successful even on a network of workstations connected over Ethernet. Some results for pure set covering problems from Lufthansa are shown in Table 4.3. The settings have been adjusted to ensure that a similar solution quality is preserved on the average.

Running times are given in seconds for the original production code Probl, the new code with the active set strategy disabled, and with active set strategy for 1, 2 and 4 processors. The hardware in this test has been 4 × Sun Ultra 1/140 connected with a shared 100 Mbps Ethernet. From these results, and other more

Table 3. Results of parallel active set code.

problem name	rows	columns	Probl	no active set	1 CPU	2 CPUs	4 CPUs
lh_dl26_02	682	642613	9962	4071	843	412	294
lh_dl26_04	154	121714	1256	373	216	105	60
lh_dt1_11	5287	266966	1560	765	298	78	66
lh_dt58_02	5339	409350	2655	2305	924	406	207

extensive tests, it can be concluded that in addition to the sequential speedup, we can obtain a parallel speedup of about a factor of 3 using a network of four workstations. It does not make sense however to increase the number of workstations significantly, since it is then no longer true that the scan dominates the total running time.

5 Conclusions and Future Work

We have demonstrated a fully successful parallelization of the two most time critical components in the Carmen crew scheduling process, on a network of workstations.

For the generator, where a coarse grained parallelism is possible, a good speedup can be obtained also for a large number of workstations, when however issues of fault tolerance become very important. Further work includes the management of multiple APC jobs, a refined task scheduling scheme, preprocessing inside the workers (e.g. elimination of identical columns) and enhanced management of the virtual machine (active time windows of machines).

For the optimizer, which requires a much more fine grained parallelism, Figure 2 shows a theoretical model of the expected parallel speedup of the different parallelizations, for a typical large problem. Although the variable parallel approach scales better, the total speedup is higher with the active set since the sequential algorithm is then faster. On the other hand, the active set can be responsible for a lower solution quality for some instances. Further tuning of the parallel active set strategy and other parameters is therefore required to improve the stability of the code both with respect to quality and speedup. It may be possible to combine the two approaches but it is not clear if this would give a significant benefit. Also, the general capacity constraints are not yet handled in any of the parallel implementations.

The parallel components have been integrated in a first PAROS prototype system that runs together with existing Carmen components. Given that the

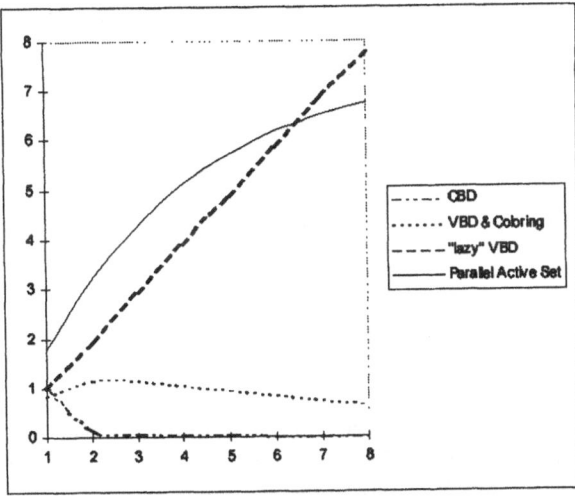

Fig. 2. Expected parallel speedup.

generator and the optimizer are now much faster, we consider parallelization also of some other components such as the generation of the connection matrix, which should be comparatively simple. Other integration issues are the communicaton between the generator and the optimizer, which is now done through files, and related issues such as optimizer preprocessing.

References

1. E. Andersson, E. Housos, N. Kohl, and D. Wedelin. *OR in the Airline Industry*, chapter Crew Pairing Optimization. Kluwer Academic Publishers, Boston, London, Dordrecht, 1997.
2. N. J. Boden, D. Cohen, R. E. Felderman, A. E. Kulawik, C. L. Seitz, J. N. Seizovic, and W.-K. Su. Myrinet: A gigabit-per-second Local Area Network. *IEEE Micro*, 15(1):29–36, Feb. 1995.
3. J. Desrosiers, Y. Dumas, M. Solomon, and F. Soumis. *Handbooks in Operations Research and Management Science*, chapter Time Constrained Routing and Scheduling. North-Holland, 1995.
4. A. Geist. Advanced programming in PVM. *Proceedings of EuroPVM 96*, pages 1–6, 1996.
5. IEEE. Standard for the scalable coherent interface (sci), 1993. IEEE Std 1596-1992.
6. S. Lavoie, M. Minoux, and E. Odier. A new approach for crew pairing problems by column generation with an application to air transportation. *European Journal of Operational Research*, 35:45–58, 1988.
7. R. Marsten. RALPH: Crew Planning at Delta Air Lines. *Technical Report. Cutting Edge Optimization*, 1997.
8. A. Mehrota and M. Trick. A clique generation approach to graph coloring. *INFORMS Journal of Computing*, 8, 1996.
9. D. Wedelin. An algorithm for large scale 0-1 integer programming with application to airline crew scheduling. *Annals of Operations Research*, 57, 1995.

Cobra: A CORBA-compliant Programming Environment for High-Performance Computing

Thierry Priol and Christophe René

IRISA -Campus de Beaulieu - 35042 Rennes, France

Abstract. In this paper, we introduce a new concept that we call a parallel CORBA object. It is the basis of the *Cobra* runtime system that is compliant to the CORBA specification. *Cobra* is being developed within the PACHA Esprit project. It aims at helping the design of high-performance applications using independent software components through the use of distributed objects. It provides the benefits of distributed and parallel programming using a combination of two standards: CORBA and MPI. To support CORBA parallel objects, we propose to extend the IDL language to support object and data distribution. In this paper, we discuss the concept of CORBA parallel object.

1 Introduction

Thanks to the rapid increase of performance of nowadays computers, it can be now envisaged to couple several high-intensive numerical codes to simulate more accurately complex physical phenomena. Due to both the increased complexity of these numerical codes and their future developments, a tight coupling of these codes cannot be envisaged. A loosely coupling approach based on the use of several components offers a much more attractive solution. With such approach, each of these components implements a particular processing (pre-processing of data, mathematical solver, post-processing of data). Moreover, several solvers are required to increase the accuracy of simulation. For example, fluid-structure or thermal-structure interactions occur in many field of engineering. Other components can be devoted to pre-processing (data format conversion) or post-processing of data (visualisation). Each of these components requires specific resources (computing power, graphic interface, specific I/O devices). A component, which requires a huge amount of computing power, can be parallelised so that it will be seen as a collection of processes to be ran on a set of network nodes. Processes within a component have to exchange data and have to synchronise. Therefore, communication has to be performed at different levels: between components and within a component. However, requirements for communication between components or within a component are not the same. Within a component, since performance is critical, low level message-passing is required whereas between components, although performance is still required, modularity/interoperability and reusability are necessary to develop cost effective applications using generic components.

However, till now, low level message-passing libraries, such as MPI or PVM, are used to couple codes. It is obvious to say that this approach does not contribute to the design of applications using independent software components. Such communication libraries were developed for parallel programming so that they do not offer the necessary support for designing components which can be reused by other applications. Solutions already exist to decrease the design complexity of applications. Distributed object-oriented technology is one of these solutions. A complex application can be seen as a collection of objects, which represent the components, running on different machines and interacting together using remote object invocations. Existing standard such as CORBA (Common Object Request Broker Architecture) aims at helping the design of applications using independent software components through the use of CORBA objects [1]. CORBA is a distributed software platform which supports distributed object computing. However, exploitation of parallelism within such object is restricted in a sense that it is limited to a single node within a network. Therefore, both parallel and distributed programming environments have their own limitations which do not allow, alone, the design of high performance applications using a set of reusable software components.

This paper aims at introducing a new approach that takes advantage of both parallel and distributed programming systems. It aims at helping programmers to design high performance applications based on the assembling of generic software components. This environment relies on CORBA with extensions to support parallelism across several network nodes within a distributed system. Our contribution concerns extensions to support a new kind of object we called a parallel CORBA object (or parallel object) as well as the integration of message-passing paradigms, mainly MPI, within a parallel object. These extensions exploit as much as possible the functionality offered by CORBA and requires few modifications to existing CORBA implementations. The paper is organised as follows. Section 2 gives a short introduction to CORBA. Section 3 describes our extensions to the CORBA specification to support parallelism within an object. Section 4 introduces briefly the *Cobra* runtime system for the execution of parallel objects. Section 5 describes some related works that share some similarities with our own work. Finally, section 6 draws some conclusions and perspectives.

2 An overview of CORBA

CORBA is a specification from the OMG (Object Management Group) [5] to support distributed object oriented applications. Such applications can be seen as a collection of independent software components or CORBA objects. Objects have an interface that is used to describe operations that can be remotely invoked. Object interface is specified using the Interface Definition Language (IDL). The following example shows a simple IDL interface:

---

[1] For the remaining of the paper, we will use simply object to name a CORBA object

```
interface myservice {
  void put(in double a);
  double myop(inout long i, out long j);
};
```

An interface contains a list of operations. Operations may have parameters whose types are similar to C++ ones. A keyword added just before the type specifies whether the parameter is an input or an output parameter or both. IDL provides an interface inheritance mechanism so that services can be extended easily. Figure 1 provides a simplified view of the CORBA architecture.

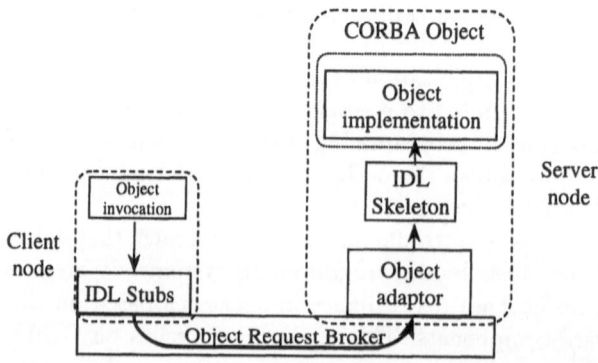

Fig. 1. CORBA system architecture

In this figure, an object located at the client side is bound to an implementation of an object located at the server side. When a client invokes an operation, communication between the client and the server is performed through the Object Request Broker (ORB) thanks to the IDL stub (client side) and the IDL skeleton (server side). Stub and skeleton are generated by an IDL compiler taking as input the IDL specification of the object. A CORBA compliant system offers several services for the execution of distributed object-oriented applications. For instance, it provides object registration and activation.

3 Parallel CORBA object

CORBA was not originally intended to support parallelism within an object. However, some CORBA implementations provide a multi-threading support for the implementation of objects. Such support is able to exploit simultaneously several processors sharing a physical memory within a single computer. Such level of parallelism does not require modification of the CORBA specification since it concerns only the object implementation at the server side. Instead of having one thread assigned to an operation, it can be implemented using several threads. However, the sharing of a single physical memory does not allow a large number of processors since it could create memory contention. One objective of our work was to exploit several dozen of nodes available on a network to carry out a parallel execution of an object. To reach this objective, we introduce the concept of parallel CORBA object.

3.1 Execution model

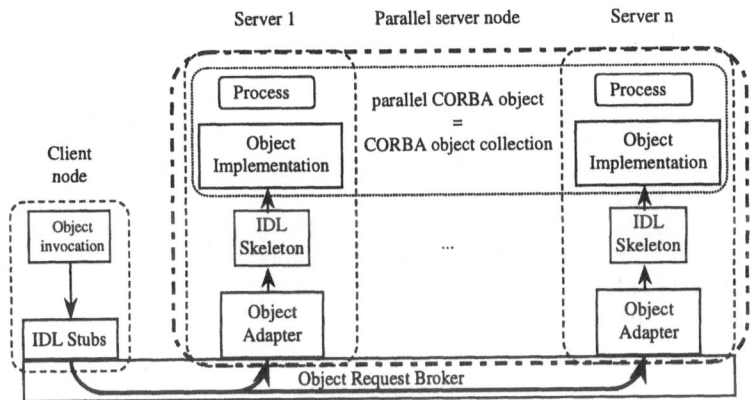

Fig. 2. Parallel CORBA object service execution model.

The concept of parallel object relies on a SPMD (Single Program Multiple Data) execution model which is now widely used for programming distributed memory parallel computers. A parallel object is a collection of identical objects having their own data so that it complies with the SPMD execution model. Figure 2 illustrates the concept of parallel object. From the client side, there is no difference when calling a parallel object comparing to a standard object. Parallelism is thus hidden to the user. When a call to an operation is performed by a client, such operation is executed by all CORBA objects belonging to the collection. Such parallel execution is handled by the stub that is generated by an Extended-IDL compiler, which is a modified version of the standard IDL compiler.

3.2 Extended-IDL

As for a standard object, a parallel object is associated with an interface that specifies which operations are available. However, this interface is described using an IDL we extended to support parallelism. Extensions to the standard IDL aim at both specifying that an interface corresponds to a parallel object and at distributing parameter values among the collection of objects. Extended-IDL is the name of these extensions.

Specifying the degree of parallelism The first IDL extension corresponds to the specification of the number of objects of the collection that will implement the parallel object. Modifications to the IDL language consist in adding two brackets to the IDL *interface* keyword. A parameter can be added within the two brackets to specify the number of objects belonging to the collection. Such parameter can be a "*", that means that the number of objects belonging to the collection is not specified in the interface. The following example illustrates the proposed extension.

```
interface[*] ComputeFEM {
   typedef double dmat[100][100];
   void initFEM(in dmat mat, in double p);
   void doFEM(in long niter, out double err);
};
```

In this example, the number of objects will be fixed at runtime depending on the available resources (i.e. the number of network nodes if we assume that each object of the collection is assigned to only one node). The implementation of a parallel object may require a given number of objects in the collection to be able to run correctly. Such number may be inserted within the two brackets of the proposed extension. The following example gives an example of a parallel object service which is made of 4 objects.

```
interface[4] ComputeFEM {
   ...
};
```

Instead of giving a fixed number of objects in the collection, a function may be added to specify a valid number of objects in the collection. The following example illustrates such possibility. In that case, the number of objects in the collection may be only a power of 2.

```
interface[n^2] ComputeFEM {
   ...
};
```

It is the responsibility of the runtime system, in our case *Cobra*, to check whether the number of network nodes has been allocated according to the specification of the parallel object. IDL allows a new interface to inherit from an existing one. Parallel interface can do the same but with some restrictions. A parallel interface can inherit only from an existing parallel interface. Inheritance from standard interface is forbidden. Moreover, inheritance is allowed only for parallel interfaces that could be implemented by a collection of objects for which the number of objects coincides. The following example illustrates this restriction.

```
interface[*] MatrixComponent {
   ...
};
interface[n^2] ComputeFEM : MatrixComponent {
   ...
};
```

In this example, interface *ComputeFEM* derives from interface *MatrixComponent*. The new interface has to be implemented using a collection having a power of 2 objects. In the following example, the Extended-IDL compiler will generate an error when compiling because inheritance is not valid :

```
interface[3] MatrixComponent {
  ...
};
interface[n^2] ComputeFEM : MatrixComponent {
  ...
};
```

Specifying data distribution Our second extension to the IDL language concerns data distribution. The execution of a method on a client side will provoke the execution of the method on every objects of the collection. Since, each object of the collection has it own separate address space, we must envisage how to distribute parameter values for each operation. Attributes and types of operation parameters act on the data distribution. Proposed extension of IDL for data distribution is allowed only for parameters of operations defined in a parallel interface. When a standard IDL type is associated with a parameter with an **in** mode, each object of the collection will receive the same value. When a parameter of an operation has either an **out** or a **inout** mode, as a result of the execution of the operation, stub generated by the Extended-IDL compiler will get a value from one of the objects of the collection.

The IDL language provides multidimensional fixed-size arrays which contains elements of the same type. The size along each dimension has to be specified in the definition. We provide some extensions to allow the distribution of arrays among the objects of a collection. Data distribution specifications apply for both **in**, **out** and **inout** mode. They are similar to the ones already defined by HPF (High Performance Fortran). The following example gives a brief overview of the proposed extension.

```
interface[*] MatrixComponent {
  typedef double dmat[100][100];
  typedef double dvec[100];
  void matrix_vector_mult(in dist[BLOCK][*] dmat, in dvec v,
                          out dist[CYCLIC] dvec u);
};
```

This extension consists in the adding of a new keyword (**dist**) which specifies how an array is distributed among the objects of the collection. For example, the 2D array *mat* is distributed by block of rows. Stubs generated by the Extended IDL compiler do not perform the same work when the parameter is an input or an output parameter. With an input parameter, the stub must scatter the distributed array so that each object of the collection received a subset of the whole array. With an output parameter, the stub must do the reverse operation. Indeed, each object of a collection contains a subset of the array. Therefore, the stub is in charge of gathering data from objects belonging to the collection.

Such gathering may include a redistribution of data if the client is itself a parallel object. In the previous example, the number of objects in the collection is not specified in the interface. Therefore, the number of elements assigned to a particular object can be known only at runtime. It is why a distributed array of a given IDL type is mapped to an unbounded sequence of this IDL type. Unbounded sequence offers the advantage that its length is set up at runtime. We propose to extend the sequence structure to store information related to the distribution.

4 A runtime system to support parallel CORBA objects

The *Cobra* runtime system [2] aims at providing resource allocation for the execution of parallel objects. It is targeted to a network of PCs connected together using SCI [6]. Resource allocation consists in providing network nodes and shared virtual memory regions for the execution of parallel objects. Resource allocation services are performed by the resource management service (RmProcess) of *Cobra*. It is used when a parallel service must be bind to a client. We propose to extend the _bind method provided by most of the CORBA implementations. Binding to a parallel object differs from the standard binding method. Indeed, a reference to a virtual parallel machine (**vpm**) is given as an argument of the *bind* method instead of a single machine. The **vpm** reference is obtained through the *Cobra* resource allocator. The following example illustrates how to use a parallel object service within the *Cobra* runtime:

```
...
// Obtain a reference from the RmProcess service
cobra = RmProcess::_bind("cobra.irisa.fr");
// Create a VPM
cobra->mkvpm(vpmname, NumberOfNodes, NORES);
// Get a reference to the allocated vpm
pap_get_info_vpm(&vpm, vpmname);
// Obtain a reference from the parallel object service: MatrixComponent
cs = MatrixComponent::_bind ( &vpm );
...
// Invoke an operation provided by MatrixComponent service
cs->matrix_vector_mult( a, b, &c);
...
```

The *bind* method may be called either by a single object, or by all objects belonging to a collection if the client is itself a parallel object.

5 Related works

Several projects deal with environments for high-performance computing combining the benefits of distributed and parallel programming. The RCS [1], NetSolve [3] and Ninf [8] projects provide an easy way to access linear algebra method libraries which run on remote supercomputer. Each method is described by a specific interface description language. Clients invoke methods thanks to specific functions. Arguments of these functions specify method name and method arguments. These projects propose some mechanisms to manage load balancing on different supercomputers. One drawback of these environments is the difficulty for the user to add new functions in the libraries. Moreover, they are not compliant to relevant standard such as CORBA. The Legion [4] project aims at creating a world wide environment for high-performance computing. A lot of principles of CORBA (such as heterogeneity management and object location) are provided by the Legion run-time, although Legion is not CORBA-compliant. It manipulates parallel objects to obtain high-performance. All these features are in common with our *Cobra* run-time. However, Legion provides others services such as load balancing on different hosts, fault-tolerance and security which are not present in *Cobra*. The PARDIS [7] project proposes a solution very close to our approach because it extends the CORBA object model to a parallel object model. A new IDL type is added: *dsequence* (for distributed sequence). It is a generalisation of the CORBA sequence. This new sequence describes data type, data size, and how data must be distributed among objects. In PARDIS, distribution of objects is let to the programmers. It is the main difference with *Cobra* for which a resource allocator is provided. Moreover in *Cobra*, extended-IDL allows to describe object parallel services in more details.

6 Conclusion and perspectives

This paper introduced the parallel CORBA object concept. It is a collection of standard CORBA objects. Its interface is described using an Extended-IDL to manage data distribution among the objects of the collection. *Cobra* is being implemented using Orbix from Iona Tech. It has already been tested for building a signal processing application using a client/server approach [2]. For such application, the most computing part of the application is encapsulated within a parallel CORBA object while the graphical interface is a Java applet. This applet, acting as a client, is connected to the server through the CORBA ORB. Current works are now focusing on the experiment of the coupling of numerical codes. Particular attention will be paid on the performance of the ORB which seems to be the most critical part of the software environment to get the requested performance. It is planned, within the PACHA project, to implement an ORB that fully exploits the performance of the SCI clustering technology while ensuring compatibility with existing ORB through standard protocols such as TCP/IP.

References

1. P. Arbenz, W. Gander, and M. Oettli. The Remote Computation System. In *HPCN Europe '96*, volume 1067 of *LNCS*, pages 662–667, 1996.
2. P. Beaugendre, T. Priol, G. Alleon, and D. Delavaux. A client/server approach for hpc applications within a networking environment. In *HPCN'98*, pages 518–525, April 1998.

3. H. Casanova and J. Dongara. NetSolve: A Network Server for Solving Computational Science Problems. *The International Journal of Supercomputer Applications and High Performance Computing*, 11(3):212–223, 1997.

4. A. S. Grimshaw, W. A. Wulf, and the Legion team. The Legion Vision of a Worldwide Virtual Computer. *Communications of the ACM*, 1(40):39–45, January 1997.

5. Object Management Group. The common object request broker: Architecture and specification 2.1, August 1997.

6. Dolphin Interconnect. Clustar interconnect technology. White Paper, 1998.

7. K. Keahey and D. Gannon. PARDIS: CORBA-based Architecture for Application-level Parallel Distributed Computation. In *Proceedings of Supercomputing '97*, November 1997.

8. M. Sato, H. Nakada, S. Sekiguchi, S. Matsuoka, U. Nagashima, and H. Takagi. Ninf: A Network Based Information Library for Global World-Wide Computing Infrastructure. In *HPCN Europe '97*, volume 1225 of *LNCS*, pages 491–502, 1997.

OCEANS: Optimising Compilers for Embedded ApplicatioNS*

Michel Barreteau[1], François Bodin[2], Peter Brinkhaus[3], Zbigniew Chamski[4],
Henri-Pierre Charles[1], Christine Eisenbeis[5], John Gurd[6], Jan Hoogerbrugge[4],
Ping Hu[5], William Jalby[1], Peter M. W. Knijnenburg[3], Michael O'Boyle[7],
Erven Rohou[2], Rizos Sakellariou[6], André Seznec[2], Elena A. Stöhr[6],
Menno Treffers[4], and Harry A. G. Wijshoff[3]

[1] Laboratoire PRiSM, Université de Versailles, 78035 Versailles, France.
[2] IRISA, Campus Universitaire de Beaulieu, 35042 Rennes, France.
[3] Department of Computer Science, Leiden University, P.O. Box 9512,
2300 RA Leiden, The Netherlands.
[4] Philips Research, Information and Software Technology, Prof. Holstlaan 4,
5656 AA Eindhoven, The Netherlands.
[5] INRIA, Rocquencourt, BP 105, 78153 Le Chesnay Cedex, France.
[6] Department of Computer Science, The University, Manchester M13 9PL, U.K.
[7] Department of Computer Science, The University, Edinburgh EH9 3JZ, U.K.

Abstract. This paper presents an overview of the activities carried out
within the ESPRIT project OCEANS whose objective is to investigate
and develop advanced compiler infrastructure for embedded VLIW processors. This combines high and low-level optimisation approaches within
an iterative framework for compilation.

1 Introduction

Embedded applications have become increasingly complex during the last few
years. Although sophisticated hardware solutions, such as those exploiting instruction level parallelism, aim to provide improved performance, they also create a burden for application developers. The traditional task of optimising assembly code by hand becomes unrealistic due to the high complexity of hard-
ware/software. Thus the need for sophisticated compiler technology is evident.

Within the OCEANS project, the consortium intends to design and implement an optimising compiler that utilises aggressive analysis techniques and
integrates source-level restructuring transformations with low-level, machine dependent, optimisations [1, 14, 16]. A major objective of the project is to provide a
prototype framework for iterative compilation where feedback from the low-level
is used to guide the selection of a suitable sequence of source-level transformations and *vice versa*. Currently, the Philips TriMedia (TM1000) VLIW processor
[8] is used for the validation of the system.

In this paper, we present the work that has been carried out during the
first 15 months since the project started (September 1996). This has largely

---

* This research is supported by the ESPRIT IV reactive LTR project OCEANS, under
contract No. 22729.

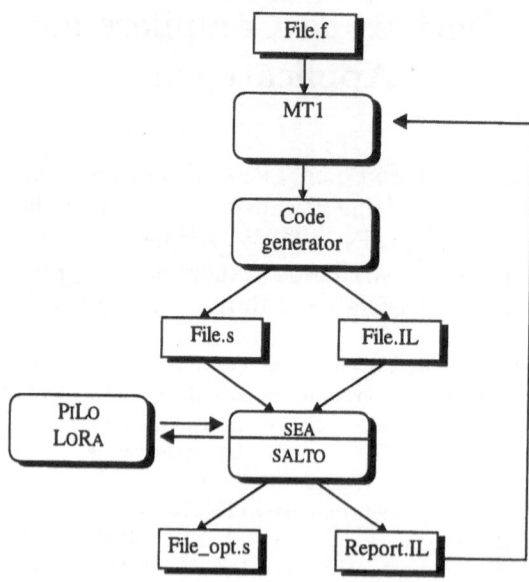

Fig. 1. The Compilation Process.

concentrated on the development of the necessary compiler infrastructure. An overall description of the system is given in Section 2. Sections 3 and 4 present the high-level and the low-level subsystems respectively, while the steps that have been taken towards their integration are highlighted in Section 5. Finally, some results from the initial validation of the system are shown in Section 6, and the paper is concluded with Section 7.

2   An Overview of the OCEANS Compiler System

The OCEANS compiler is centred around two major components: a high-level restructuring system, MT1, and a low-level system for supporting assembly language transformations and optimisations, SALTO, which is coupled with SEA, a set of classes that provides an abstract view of the assembly code, and tools for software pipelining (PILO) and register allocation (LORA). Their interaction is illustrated in Figure 1 which shows the overall organisation of the OCEANS compilation process. In particular, a program is compiled in three main steps:

- First, MT1 performs lexical, syntactical and semantic analysis of a source FORTRAN program (**File.f**). Also, a sequence of source program transformations can be applied.
- The restructured source program is then fed into the code generator which generates sequential assembly code that is annotated with instruction identifiers used to identify common objects in MT1 and SALTO, and a file written in an *Interface Language* (**File.IL**) that provides information on data dependences and control flow graphs.

- Finally, SALTO (coupled with SEA) performs code scheduling and register allocation. At this step guarded instructions are created and resource constraints are taken into account.

The above process is repeated iteratively until a certain level of performance is reached. Thus, different optimisations, both at the source-level and the low-level, are checked and evaluated. An important feature of the system is the existence of a client-server protocol that has been implemented in order to provide easy access to the compiler over the Internet, for all members of the consortium. MT1 and the code generator are located at Leiden, and SALTO, SEA, PILO and LORA are located at Rennes.

3 High-Level Transformations

Optimizing and restructuring compilers incorporate a number of program transformations that replace program fragments by semantically equivalent fragments to obtain more efficient code for a given target architecture. The problem of finding an optimum order for applying them is commonly known as the *phase ordering problem*. Within the MT1 compilation system [5] this problem is solved by providing a *Transformation Definition Language (TDL)* [3] and a *Strategy Specification Language (SSL)* [2]. Transformations and strategies specified in these languages can be loaded dynamically into the compiler.

3.1 Transformation Definition Language

The TDL is based on pattern matching. The user can specify an *input pattern*, a transformed *output pattern* and a *condition* when the transformation can be legally and/or beneficially applied. Patterns may contain *expression* and *statement variables*. When a pattern is matched against the code these variables are bound to actual expressions and code fragments, respectively. The expression and statement variables can be used in turn in the specification of the output pattern and the condition. This mechanism allows one to specify a large number of transformations, such as loop interchange, loop distribution or loop fusion. However, it is not powerful enough to express other important transformations, such as loop unrolling. Therefore, the TDL also allows for *user-defined functions* in the output pattern. User-defined functions are the interface to the internal data structures of the compiler. In this way, any algorithm for transforming and testing code can be implemented and made accessible to the TDL.

3.2 Strategy Specification Language

The order in which transformations have to be applied is specified using a Strategy Specification Language (SSL). It allows the specification of an optimising strategy at a more abstract level than the source code level. This language contains sequential composition of transformations, a choice construct and two repetitive constructs.

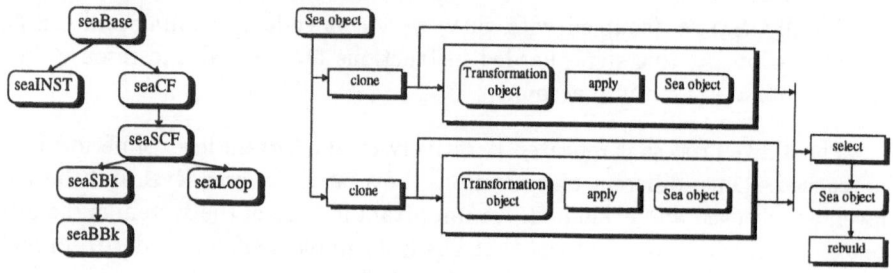

Fig. 2. SEA class hierarchy.   **Fig. 3.** Typical usage of SEA classes.

An IF statement consists of a transformation that acts as a condition, a THEN part and an optional ELSE part. The transformation in the condition can be applied successfully or not. If it is successful, the transformations in the THEN part are to be executed. Optionally, in the ELSE part a list of transformations can be given which should be executed in case the transformation matched but was not applied successfully due to failing conditions.

The two repetitive constructs consist of a transformation to be checked and a statement list to be executed if the condition is true or false, respectively. They consist of a WHILE-ENDWHILE and a REPEAT-UNTIL construct.

Examples of how to specify strategies in SSL can be found in [2].

4 Low-Level Optimisations

Low-level optimisations are built on the top of SALTO, a retargetable system for assembly language transformation and optimisation [15]. To facilitate the implementation of optimisations, a set of classes has been designed, SEA (SALTO Enhanced Abstraction), that provides an abstract view of the assembly code which is more pertinent to the code scheduling and register allocation problems. The most important features of SEA are that it allows the evaluation of various code transformations before producing the final code, and that it separates the implementation of the global low-level optimisation strategy from the implementation of individual optimisation sequences.

The SEA model contains two kinds of objects:

code fragments The following objects can be used. **seaINST**: an instruction object; **seaCF**, an unstructured set of code fragments; **seaSCF**, a structured subset of control flow graph with a unique entry point; **seaSBk**, a superblock; **seaBBk**, a basic block; and finally, **seaLoop**, a structured piece of code that has loop properties. Figure 2 illustrates the corresponding class hierarchy.

transformations to be applied to subgraphs. All transformations are characterized by the following main methods: **preCond()** returns the set of control flow subgraphs that qualifies for the transformation; **apply()** applies the transformation to a given subgraph, and finally, **getStatus()** that returns the status of the transformation after application (*success* or *failure*) and the reason for the failure.

The usage of the SEA objects is shown in Figure 3. A transformation is tried on a cloned piece of code, then according to performance or size criteria one of the solutions found is chosen and propagated to the low-level program representation using the `rebuild()` method.

The optimisations currently available within SEA are: *register renaming, superblock construction* [12], *guard insertion* [11], *loop unrolling* (also available at the high-level), *local/superblock scheduling* [12], *software pipeline*, and *register allocation*. The implementation of software pipeline is based on the tools PiLo [17] and LoRa [9] which generate a modulo scheduling of the loop body.

5 Integration

5.1 The Interface Language

In order to transmit information between the various components of the compiler, an *Interface Language* (IL) was designed. This allows the propagation of information, such as data dependences and loop control data, from MT1 to SALTO, as well as feedback information from the scheduled code back to MT1.

An IL description consists of three sections: *a list of keywords* that specifies the list of attributes that can apply to an object; *a default level setting* that indicates the type of code the objects belong to; and *a list of object references* which specify the nature, contents and attributes of an object. More details on the IL can be found in [7].

5.2 Information Forwarded and Feedback

Data dependence information is propagated from MT1 to SALTO and is used for memory disambiguation. The feedback from SEA to MT1 (file `Report.IL` in Figure 1) contains information on the code structure, the basic blocks, as well as a record of the transformations that were applied. Data related to each basic block include the total number of assembly instructions, the critical path for scheduling the code, the number of cycles of the scheduled code, and a grouping depending on the nature of the instructions. Examples can be found in [7].

MT1 uses the feedback from SALTO in order to build an internal data structure that can be accessed by the TDL and the SSL by means of user-defined functions in the condition that can check for the identity of a code fragment and suggestions made by SALTO. When such a transformation is used as the condition for an IF construct in the SSL, we are able to select the transformations we want to apply to this fragment.

Initially, MT1 compiles the program without performing any restructuring and the compiled program is scheduled by SALTO. SALTO identifies the code fragments that can be improved. It reports its diagnostics to a cost model that makes a decision on what kind of restructuring could be performed next. Then, MT1 reads the suggestions for restructuring and performs these. It is intended that a transformation sequence for a given program fragment is selected by following a systematic approach for searching through a domain of possible transformations. First, each different transformation is applied once and then the same follows for each branch of the tree. The search space is minimised by using a threshold condition for terminating branches that are not likely to yield an optimum result in their descendants. Some preliminary experiments using this strategy can be found in [10]; further work is in progress.

6   Validation of the Initial System

In order to validate the compiler, four public domain multimedia codes have been selected [4]. These are a low bit-rate encoder/decoder for H.263 bitstreams, an MPEG2 encoder/decoder, an implementation of the CCITT G.711, G.721 and G.723 voice compression standards, and the Persistence of Vision Ray-Tracer for creating 3D graphics.

At the high-level, initial experimentation aimed at identifying those transformations that appear to be the most crucial in optimising code scheduling. Inspection of the benchmarks revealed that they contain many imperfectly nested double or triple loops with much overhead due to branch delays. In order to deal with such loops, a transformation that converts the imperfectly nested loop into single loop has been suggested [13].

At the low-level, the initial validation of the system has been carried out by applying four different optimisation sequences:

- S_0 is the simplest sequence. First, the code is scheduled locally and then register allocation is performed.
- $S_1(u)$ is based on unrolling the loop body u times. The unrolled body is transformed into a superblock by guarding instructions. As in S_0, register allocation is performed after local scheduling.
- $S_2(u)$ is similar to $S_1(u)$ except that register allocation is performed before scheduling. This usually requires less registers, allowing this sequence to succeed when $S_1(u)$ fails due to a lack of registers.
- S_3 consists in applying a software pipelining algorithm.

The above optimisation sequences were validated and indicative results, using the most-time consuming loops of H263, are illustrated in Figure 4. Every optimisation sequence has been applied to each of the six selected loops and the size of the resulting VLIW code and the speed of the loop, i.e. the number of cycles per iteration, were computed. From the table, a well-known result is observed: the more we unroll a loop, the faster it runs — cf. columns $S_1(2)$, $S_1(3)$, $S_1(4)$ — but at the expense of a larger code size. As expected $S_2(2)$ yields too poor performance and large code because of the presence of false dependences. Finally, software pipelining (S_3) gives the best performance but at the expense of a very large increase in code size. Note that this transformation failed with the last loop, due to a lack of registers.

	Optimisation sequences						C code
	S_0	$S_1(2)$	$S_1(3)$	$S_1(4)$	$S_2(2)$	S_3	
speed	8	6	5	5	7	3	for (i=xa;i<xb; i++)
size	8	12	16	20	13	75	{ d[i]=s[i]*om[i]; }
speed	9	7	6	6	10	5	for (i=xa; i<xb; i++)
size	9	13	18	22	19	55	{ d[i]+=s[i]*om[i]; }
speed	12	8	8	7	12	6	for (i=xa; i<xb; i++)
size	12	16	24	28	24	121	{ dp[i]+=(((unsigned int)(sp[i] +sp2[i]+1))>>1)*om[i]; }
speed	15	10	9	9	16	6	for (i=xa; i<xb; i++)
size	15	20	28	34	31	172	{ dp[i]+=(((unsigned int)(sp[i]+ sp[i+1]+1))>>1)*OM[c][j][i]; }
speed	15	10	10	8	17	7	for (i=xa; i<xb; i++)
size	15	19	29	33	33	179	{ dp[i]+=(((uint)(sp[i]+sp2[i]+ sp[i+1]+sp2[i+1]+2))>>2)*om[i]; }
speed	19	13	12	11	30	–	for (k=0; k<5; k++)

size	19	25	36	44	59	–	{ xint[k]=nx[k]>>1; xh[k]=nx[k] & 1;
							yint[k]=ny[k]>>1; yh[k]=ny[k] & 1;
							s[k]=src+1x2*(y+yint[k])+x+xint[k];
							}

Fig. 4. Time consuming loops extracted from H263.

In most embedded applications, it is necessary to answer globally questions such as: "Given a maximum code size, what is the highest performance that can be achieved?", or "Given a performance goal, what is the smallest code size that can be achieved?". Within the OCEANS compiler this trade-off is evaluated quantitatively by applying a novel compiler strategy. Thus, the choice of the most suitable optimisation is made *a posteriori*, when the impact of each possible transformation is known. More details can be found in [6].

7 Conclusion and Future Work

The previous sections outlined the current status of the OCEANS compiler. Although the results obtained so far, using the initial prototype, are satisfactory (comparing with a production compiler), the implementation work still continues on both the high and low levels. A major part of the work during the next months and until the end of the project is devoted to the integration of the two levels, the development of a prototype framework for iterative compilation, and its experimental evaluation and tuning. Finally, it is intended that the system be made publically available in due time (at the moment SALTO is available on request).

References

1. B. Aarts, *et al.* OCEANS: Optimizing Compilers for Embedded Applications. In C. Lengauer, M. Griebl, S. Gorlatch (Eds.), *Proceedings of Euro-Par'97*, Lecture Notes in Computer Science 1300, Springer-Verlag, 1997, pp. 1351–1356.
2. R. A. M. Bakker, F. Bregt, P. M. W. Knijnenburg, P. Touber, and H. A. G. Wijshoff. Strategy Specification Language. OCEANS Deliverable D1.2a, 1997.
3. A. J. C. Bik, P. J. Brinkhaus, P. M. W. Knijnenburg, P. Touber, and H. A. G. Wijshoff. Transformation Definition Language. OCEANS Deliverable D1.1, 1997.
4. A. J. C. Bik, P. J. Brinkhaus, P. M. W. Knijnenburg, P. Touber, H. A. G. Wijshoff, W. Jalby, H.-P. Charles, M. Barreteau. Identification of Code Kernels and Validation of Initial System. OCEANS Deliverable D3.1c, 1997.
5. A. J. C. Bik and H. A. G. Wijshoff. MT1: A Prototype Restructuring Compiler. Technical Report 93-32, Department of Computer Science, Leiden University, 1993.
6. F. Bodin, Z. Chamski, C. Eisenbeis, E. Rohou, and A. Seznec. GCDS: A Compiler Strategy for Trading Code Size Against Performance in Embedded Applications. Research Report 1153, IRISA, 1997.
7. F. Bodin and E. Rohou. High-level Low-level Interface Language. OCEANS Deliverable D2.3a, 1997.
8. B. Case. Philips Hope to Displace DSPs with VLIW. *Microprocessor Report*, 8(16), 5 Dec. 1994, pp. 12–15. See also http://www.trimedia-philips.com/
9. C. Eisenbeis, S. Lelait, and B. Marmol. The meeting graph: a new model for loop cyclic register allocation. *Proceedings of PACT'95* (Cyprus, June 1995).
10. J. Gurd, A. Laffitte, R. Sakellariou, E. A. Stöhr, Y. T. Chu, and M. F. P. O'Boyle. On Compile-Time Cost Models. OCEANS Deliverable 1.2b, 1997.
11. W. Hwu, R. E. Hank, D. M. Gallagher, S. A. Mahlke, D. M. Lavery G. E. Haab, J. C. Gyllenhaal, and D. I. August. Compiler Technology for Future Microprocessors. *Proceedings of the IEEE*, 83(12), Dec. 1995, pp. 1625–1639.
12. W. Hwu, *et al.* The Superblock: An Effective Technique for VLIW and Superscalar Compilation. *The Journal of Supercomputing*, 7(1), May 1993, pp. 229–248.
13. P.M.W. Knijnenburg. Flattening: VLIW Code Generation for Imperfectly Nested Loops. *Proceedings CPC'98*, 1998. To appear.
14. OCEANS Web Site at http://www.wi.leidenuniv.nl/Oceans/
15. E. Rohou, F. Bodin, A. Seznec, G. Le Fol, F. Charot, F. Raimbault. SALTO: System for Assembly-Language Transformation and Optimization. Technical Report 1032, IRISA, June 1996. See also http://www.irisa.fr/caps/Salto/
16. R. Sakellariou, E. A. Stöhr, and M. F. P. O'Boyle. Compiling Multimedia Applications on a VLIW Architecture. *Proceedings of the 13th International Conference on Digital Signal Processing (DSP97)* (Santorini, July 1997), vol. 2, IEEE Press, 1997, pp. 1007–1010.
17. J. Wang, C. Eisenbeis, M. Jourdan, and B. Su. Decomposed Software Pipelining: a New Perspective and a New Approach. *International Journal on Parallel Processing*, 22(3), 1994, pp. 357–379.

Industrial Stochastic Simulations on a European Meta-Computer

Ken Meacham, Nick Floros, and Mike Surridge

Parallel Applications Centre, 2 Venture Road,
Chilworth, Southampton, SO17 7NP, UK.
{kem,nf,ms}@pac.soton.ac.uk
http://www.pac.soton.ac.uk/

Abstract. This paper outlines the experiences of running a large stochastic multi-body simulation across a pan-European meta-computer, to demonstrate the use of the PROMENVIR tool within such a large-scale WAN environment. We describe the meta-application management approach developed by PAC and discuss the technical issues raised by this experiment.

1 Introduction

PROMENVIR (PRObabilistic MEchanical desigN enVIRonment) is an advanced meta-computing tool for performing stochastic analysis of generic physical systems, which has been developed within the PROMENVIR ESPRIT project (No. 20189) [1, 2]. The tool provides a user with the framework for running a stochastic Monte Carlo (MC) simulation, using any preferred deterministic solver (e.g. NASTRAN), then provides sophisticated statistical analysis tools to analyse the results.

PROMENVIR facilitates investigations into the reliability of components, by analysing how uncertainties in their manufacture, deployment or environment affect key mechanical properties such as stresses. PROMENVIR does this by generating many hundreds of analysis "shots", in which the uncertain parameters are generated from statistical distributions selected by the user. Key output values are then extracted from each analysis shot, and can be analysed to determine the sensitivity of the component with respect to uncertain parameters, correlations between physical properties and behaviour, and clustering characteristics (e.g. failure set distributions). PROMENVIR is an open environment, and can generate analyses for essentially any solver.

Since solver runs are independent, the overall simulation is intrinsically parallel and is therefore an ideal candidate for exploiting parallel HPC resources, which may include heterogeneous clusters of workstations, MPP or SMP platforms. These resources may reside within a company Local Area Network (LAN), or may be accessible over a geographically-distributed Wide Area Network (WAN). The only restriction on the use of resources within a PROMENVIR simulation is the availability of hosts and solver licenses. The implication of this is that, by co-operating with partner sites across a WAN, a PROMENVIR stochastic simulation of many hundreds of solver runs may be executed within a few hours, rather than many days, hence providing a rapid turnaround time for analysis of new designs.

One of the main challenges within the PROMENVIR project was therefore to demonstrate the effectiveness of the PROMENVIR package within a pan-European meta-computing (WAN) environment, to solve a large stochastic problem of industrial significance. The Parallel Applications Centre (PAC) has successfully set up and run a series of "WAN experiments", involving partner sites within a European consortium. This paper outlines the experiments which took place, the technical issues which arose and summarizes the results.

2   Testcase for WAN Experiment

The most widely available solver within the PROMENVIR consortium was the Multi-Body Simulation code SIMAID, developed by CEIT in San Sebastian [3], so it was decided that a suitable demonstration would be set up, using this code as the basis for a large stochastic simulation.

The testcase chosen was a simulation of a satellite antenna deployment, during which the antenna unfolds to give a planar rim (see Figs 1 - 3). The antenna was modelled using a set of beams, with springs and dampers at the joints. A single deterministic simulation had already been carried out, which predicted that the outer rim of the antenna would be essentially flat. However, the antenna simulation model was constructed using components with identical properties, and did not take into account the manufacturing tolerances inherent within the components and their assembly.

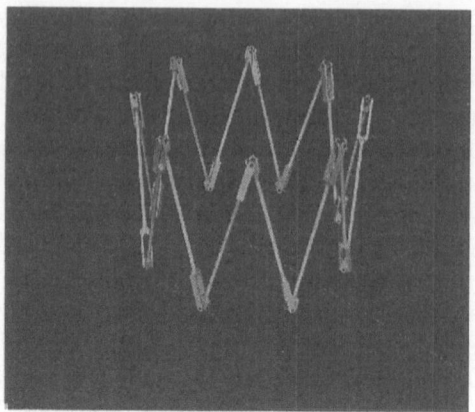

Fig. 1. SIMAID antenna model (folded)

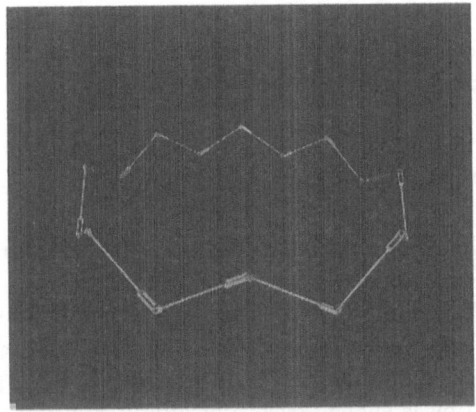

Fig. 2. SIMAID antenna model (partially deployed)

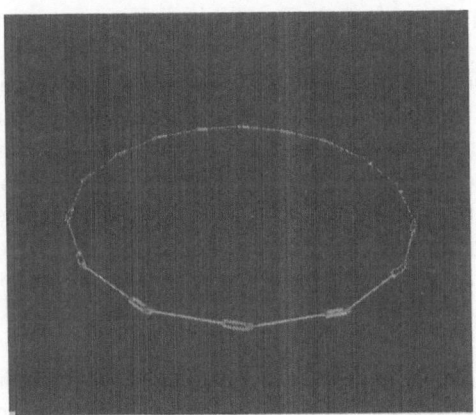

Fig. 3. SIMAID antenna model (fully deployed)

The PROMENVIR Monte Carlo simulation consisted of 500-1000 shots of the SIMAID solver, for the antenna testcase (at least 500 shots were required for convergence of results). For each shot, values for the critical component parameters (e.g. spring stiffness) were chosen from a normal distribution, which modelled the known tolerances. The results of the stochastic simulation could then be analysed to see the effect of manufacturing tolerances on the planarity of the deployed antenna, and hence the effectiveness of the antenna.

3   The European Meta-Computer

A total of 102 CPUs were provided for the experiment, from within the consortium (see Fig. 4). These were restricted to SGI architectures, since SIMAID was only available for IRIX or IRIX64 operating systems. However, many different types of SGI machines were made available, ranging from older Indigo R3000 workstations up to a 64-cpu Origin (at UPC, Barcelona). The topology of this hardware configuration is shown geographically in Fig 4. The master host was running at PAC, while slave hosts were used in UK, Spain, Germany and Italy, at various partner sites. We refer to such a meta-computer as a Parallel Virtual Computer (PVC).

SIMAID was installed and licensed on each machine, and access permissions set up to allow remote access from the master SGI Indy at PAC, on which the PROMENVIR front end and resource manager were running.

4   Technical Issues

4.1   Meta-Application Manager

A fundamental component of the package is the Advanced Parallel Scheduler (APS), which has been developed at the Parallel Applications Centre (PAC). This has been designed as a meta-application manager, which orchestrates the use of the PVC by an application such as PROMENVIR. The APS is capable of making intelligent resourcing decisions, based on performance models (developed at PAC) of the application and resources. These models are used to predict resource usage for a solver in terms of CPU, memory, disk space and I/O traffic, and hence allows capacity planning of the available resources, before submission of any jobs.

A hosts (PVC) database is set up, using the System Definition tool (SDM), which automatically obtains characteristics for each required host, within the LAN or WAN. Performance characteristics for hosts may also be entered here, to be used by the APS. The APS creates daemons on remote hosts, which are responsible for submitting jobs, copying files and communicating load information back to the master workstation. It is capable of initiating UNIX processes directly, but can also be used to control and submit meta-applications via conventional load-sharing software, such as LSF from Platform Computing [4].

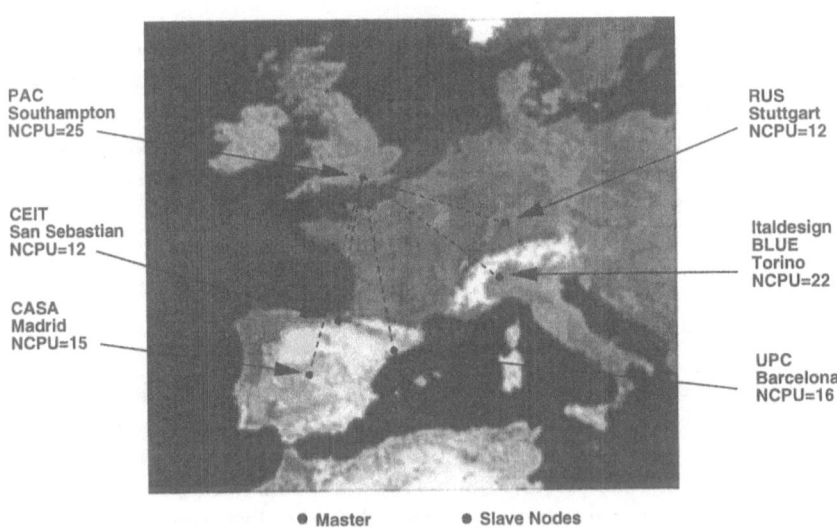

PROMENVIR Multi–National Meta–Computer (UK, D, E, I)
NCPU = 102

PAC
Southampton
NCPU=25

CEIT
San Sebastian
NCPU=12

CASA
Madrid
NCPU=15

RUS
Stuttgart
NCPU=12

Italdesign
BLUE
Torino
NCPU=22

UPC
Barcelona
NCPU=16

● Master   ● Slave Nodes

Fig. 4. Meta-computer (PVC) topology for WAN experiment

4.2   Remote Host Access and Security

A major part of the work involved in setting up the WAN experiment was ensuring remote host accessibility. For example, some partners were able to set
up, quite quickly, remote access to their hosts via rsh (remote shell); these sites
generally consisted of academic institutions, where access is not normally too
restricted. However, several partners (including PAC) had security features including firewall machines. This was a major obstacle, since rsh access is normally
restricted and, even where available, a firewall machine generally allows access
in one direction only. This makes it impossible to set up two-way rsh access
between two secure sites, through two firewalls (i.e. one at each end).

A further requirement on the meta-computer was to make any file transfers
between remote sites both robust and secure. For these reasons, we decided to
explore the use of SSH/SCP protocols, which turned out to solve most of our
problems and requirements.

SSH (Secure Shell) and SCP (Secure Remote Copy), developed by Data Fellows [5,6], provide secure TCP/IP connections between trusted sites. Features
include:

- strong authentication (via RSA keys)
- automatic data encryption and compression
- access though firewalls (via SSH officially registered port 22)

We were also able to exploit an advanced SSH facility called port forwarding, which enables TCP traffic to be forwarded to a remote host via the secure connection.

Various partner sites already had an SSH facility, while others needed to install it. By using SSH and SCP protocols for all TCP connections, and by configuring partner firewalls to allow SSH access, we were then able to submit tasks transparently to the PVC, via the APS, and receive back results from solver runs, for collection by the PROMENVIR application. This demonstrated sharing of resources between sites for a single meta-application, even between sites which both have firewall security.

5 Running the Experiment

Several smaller-scale tests of PROMENVIR were carried out, firstly using machines that were available and easily accessible (i.e. without firewalls). It was soon found that the unreliability of the network was the major factor in any lost or failed solver runs (shots). Due to the statistical nature of the PROMENVIR environment, it is not critical if some shots are lost, since results will still converge, as long as enough shots are completed in total.

However, we were finding that up to 10-20% of the remote tasks could fail, due to slow and unreliable connections to certain sites (particularly to Italy). This figure was unacceptable, particularly if the WAN was intended for use with other (non-stochastic) types of application.

We therefore made several improvements to the APS, to enhance robustness and optimise data transfer. For example, rather than attempting single file transfers, we set up the APS to perform multiple retries. If a host could not be contacted after this, it was assumed to be down, and no further tasks were sent to it.

As more and more machines were made available, we were able to increase the number of CPUs used for the experiments. Our target was to reach 100 CPUs across the meta-computer.

6 WAN Experiment Findings

6.1 Results

A summary of the most successful experiment is shown in Table 1. This shows a list of the partner sites, and their contributions to the meta-computer in terms of CPUs provided (as SGI workstations or SMP platforms).

Of the 8 sites represented in this table, 5 had firewall security.

6.2 Availability and Reliability

Of course, the fact that remote hosts are set up and installed does not necessarily imply that they will be available at the time of running the experiment. Of the

Table 1. PVC host usage statistics during WAN experiment

Partner	CPUs Nproc	Availability			Shot Statistics		
		In PVC	Access	Used	Failed	Successful	Total
PAC	15	15	15	14	1	150	151
So'ton University	10	10	9	6	0	40	40
UPC	16	16	16	16	1	275	276
RUS	12	12	11	9	0	104	104
CASA	15	15	15	14	15	184	199
CEIT	12	12	12	8	0	98	98
ItalDesign	11	11	6	5	2	63	65
Blue	11	11	11	7	6	61	67
Grand Totals	102	102	95	79	25	975	1000

Total cpus installed	102	Elapsed Execution Time:		4:39:16
Total cpus defined in PVC	102			
Total cpus available	95	Approx Single CPU Time:		250 hrs
Total cpus used in WAN	79			

102 CPUs defined in the PVC, only 95 could be contacted at the time of running the experiment. We found that this number fluctuated during the day; this was mainly due to the network load, but was sometimes due to machines being down, or turned off.

Of the 95 CPUs available to PROMENVIR via the APS (i.e. with daemons running), 79 were actually used during the simulation to run solver tasks. The reason for this lies in how the APS decides to allocate tasks to hosts. The CPU load is measured on each remote host (via the UNIX uptime command), and sent back to the APS. If this is below a certain threshold and the performance models indicate that it is advantageous to do so, a task will be submitted to that host; otherwise the task will be scheduled elsewhere.

We found that jobs were already running on certain hosts (generally those which had not been provided exclusively for the WAN experiment), so these were not used by the APS to run PROMENVIR tasks. However, we also found that some machines, which were apparently not running any jobs, were still not used by the APS; these hosts still appeared overloaded. We eventually traced this problem to certain SGI workstations which were running screen lock programs, which fully loaded the machines CPU resources, when no other (higher priority) jobs were running.

A total of 1000 SIMAID solver shots were submitted to the meta-computer, of which 975 completed successfully, and the total elapsed time of the PROMEN-VIR simulation was only 4h:39m:16s. This compares with an approximate time of 250 hours on a typical SGI workstation (a single SIMAID run taking around 15 mins). The 25 shot failures were still mainly due to network problems, which will always be present to some extent, within a large meta-computer, unless very fast and reliable links are used (e.g. ATM). A handful of shots were lost, due to solver problems. In certain extreme cases, combinations of random input parameters can cause the solver to crash, though this is rare.

6.3 Analysis of Results

The detailed analysis of the results is beyond the scope of this paper, and will be published elsewhere, in conjunction with results from further simulations which are being carried out using differing ranges of manufacturing tolerances.

However, preliminary results show that, when a stochastic approach is used, there is a broader variation in the deployed geometry, and hence a poorer planarity in the dish than predicted with the deterministic simulation.

By using the stochastic approach, further insights may be obtained than were previously available. For example, by examining the planarity variation according to the standard deviation of manufacturing tolerances, it is possible to work back and ask questions such as "what tolerances must be ensured in the manufacturing process, to produce an antenna of certain planarity". These types of questions are highly relevant to industry.

7 Conclusions

PROMENVIR has been demonstrated as a highly useful tool for running industrial stochastic simulations across a European meta-computer, showing that corporate-scale meta-computing resources really can be used to solve large problems for industry. The APS, developed at PAC, has facilitated management of meta-applications and meta-computing resources (including firewall-protected systems), and is also being used with other meta-computing applications [7,8].

Results of the WAN experiment have shown that, once a meta-computer has been set up, further experiments can be carried out routinely. However, the overheads of setting up remote sites can be quite large. Furthermore, it has been found to be difficult to obtain full (exclusive) access to machines, during a large-scale experiment, without the full cooperation from contributing partners.

Our experience suggests that, due to a combination of background loads, network performance and human factors (e.g. failure to log out of a machine), around 80% resource availability is the maximum which can realistically be expected for large-scale meta-applications.

Finally, there will always be some inherent unreliability within a large-scale meta-computing resource, mainly due to network problems. These must be eliminated as much as possible and, where uncontrollable problems arise, the system and application must be made as robust as possible, to ensure that any simulation does not suffer fatally from individual task failures.

Acknowledgements

The work reported here was part of the ESPRIT PROMENVIR project. We are greatful to our partners (CASA, UPC, CEIT, RUS, Atos, ItalDesign and Blue Engineering) for a most fruitful collaboration.

References

1. Marczyk, J.: Meta-Computing and Computational Stochastic Mechanics. Computational Stochastic Mechanics in a Meta-Computing Perspective (1997) 1–18 Ed. J. Marczyk
2. PROMENVIR product web page:
   http://www.atos-group.de/cae/index-promenvir.htm
3. CEIT home page: http://www.ceit.es/
4. Platform Computing (LSF) home page: http://www.platform.com
5. Data Fellows home page: http://www.datafellows.com/
6. Further information on SSH may be found at: http://www.cs.hut.fi/ssh/
7. Upstill C.: Multinational Corporate Metacomputing. Proceedings of the ACM/IEEE SC97 Conference, San Jose, CA, November 15-21, 1997
8. Hey, A.J.G., Scott, C.J., Surridge, M. and Upstill, C.: Integrating Computation and Information Resources - An MPP Perspective. 3rd Working Conference on Massively Parallel Programming Models (MPPM-97), London, November 12-14, 1997

Porting the SEMC3D Electromagnetics Code to HPF

Henri Luzet[1] and L.M. Delves[2]

[1] SEMCAP, 20 rue Sarinen,
Silic 270 RUNGIS 94578, France
[2] N.A Software Ltd, 62 Roscoe St,
Liverpool L1 9DW, UK
delves@nasoftware.co.uk

Abstract. Within the Esprit Pharos project, four industrial simulation codes were ported to HPF1.1 to provide a test of the suitability of this language for handling realistic Fortran77 applications, and to estimate the effort involved. We describe here the port of one of these: the SEMC3D electromagnetic simulation code. An outline of the porting process, and results obtained on the Meiko CS2 are given.

1 The Pharos Project

Pharos was an Esprit–funded project aimed at:

- Bringing together an HPF-oriented toolset from three vendors;
- Using this toolset to develop HPF versions of four industrial–strength codes owned by project partners to estimate the effectiveness of HPF1.1 and current compilers for handling such codes, and the effort involved in producing effective HPF ports.

The project ran from January 1996 to December 1997; partners included:

Tools Vendors :
 N.A. Software : HPF Compiler and source level debugger (*HPFPlus*);
 Simulog : Fortran77 to Fortran90 conversion tool (*FORESYS*);
 Pallas : Performance and communications monitor (*VAMPIR*).
Code Owners :
 SEMCAP : Electromagnetic Simulation code (*SEMC3D*);
 CISE : Finite Element Structural code (*CANT-SD*);
 Matra BAe Dynamics : Compressible Flow code (*Aerolog*);
 debis : Linear Elastostatics code (*DBETSY3D*).
HPF Experts :
 GMD : Developers of ADAPTOR, an early and highly efficient HPF research compiler
 University of Vienna : Developers of Vienna Fortran, a precursor to HPF; and of VFCS, an interactive research compilation system implementing Vienna Fortran and later modified to accept standard HPF.

We describe here the HPF port of *SEMC3D*, and give benchmark results obtained with this code.

2 The *SEMC3D* Code

2.1 The Application

The SEMC3D package is an Electromagnetic code developed to simulate EMC (ElectroMagnetic Compatibility) phenomena. It has a wide range of applications. For example, in the automotive industry electric and electronic components and systems must comply with increasingly tougher regulations. They must resist hostile electromagnetic environments and show reduced emissions.

The *SEMC3D* package computes the electromagnetic environment of complex structures made up of metal, dielectric, ferrite and multilayers or human tissues. These structures could be surrounded with ionised gas, laid on a realistic or perfect ground, or with antennas or cables. The electromagnetic environment should be taken as the knowledge of electric and magnetic fields inside and outside the analysed structures. Many source models can be used: NEMP (Nuclear ElectroMagnetic Pulse), plane wave, current injection, lightning and all analytical and digital signals.

The connection of the SEMC3D software with specialised Cable Networks codes (as CRIPTE) allows many applications in the field of EMC problems (emission and /or susceptibility) as well as test cases where EMI (ElectroMagnetic Interferences) phenomena occur.

2.2 Numerical and Computational Features

The original Fortran77 source code comprises approximately 3500 lines. The kernel of SEMC3D code is built around the Leap-Frog scheme in time and space. The Leap-Frog scheme uses only the components located within semi space-time away with regard to the updated component. This characteristic gives a strong locality in the solution of Maxwell's equations, that is to say the updating of variables is independent of the order of their appearance and each variable depends only on its closest neighbour through another variable.

For each cell 6 variables (Ex, Ey, Ez, Hx, Hy, Hz) are necessary to solve Maxwell's equations.

Of course, the high locality of the finite-difference method leads to a domain decomposition of the work space.

2.3 Message Passing Version

The usage of electronics in vehicles has increased at a phenomenal rate in the last decade, and continues to do so, for example to enhance automatization, driver comfort and vehicle safety. As a result, the computer power needs (CPU time and memory size) steadily increase as more complex and more accurate simulations are required to meet customer needs. The use of MPP systems is one way to satisfy these needs, and a message passing version of the code was developed in the early 90s, and tested on many platforms including Paragon, CS2, CM5 and Workstation and with NX, MPI, PVM and PARMACS libraries.

The design of parallelisation is founded on a standard domain decomposition approach based on nodal variables, namely there is a correspondence between one calculation node and one 3D element array (IJK representation). This correspondence is used both in the HPF version and in the message passing version, but the implementations differ.

The message passing version used a decomposition with overlap. At each step of the computation, the calculations are made in each domain and then the quantities on the boundaries of domains are updated. This is a fine grain parallelisation which requires very substantial effort in a message passing implementation. A major advantage of HPF is its intrinsic ability to provide a fine grained parallelisation with modest development effort; and this is what we sought to produce.

3 Outline Porting Procedure

The porting procedure had the same steps for each code; these steps are those foreseen as appropriate at project start:

Code Cleaning : The original Fortran77 code was tidied up to remove language features which were either:
- non-standard Fortran77 (some of these remain in many codes either as legacies from earlier Fortran versions or because they have been widely supported by compilers);
- or can be replaced even within Fortran77 by better constructs.

This stage proved to be well-supported by the *FORESYS* tool. However, some work was necessary by hand.

Translation to Fortran90 : This provides a better base for HPF. The *FOREST90* tool (part of the *FORESYS* suite) proved to give good support for this, providing an automatic translation which was then hand polished.

Improving the Fortran90 version : to make the source better suited to HPF by utilising array syntax where possible, and by removing further features (storage association, sequence association, assumed size and explicit shape array arguments) which cannot directly be parallelised in HPF. Static storage allocation was replaced by dynamic storage allocation, as part of this process. This stage made use of the FORESYS vectorising tool; but much of the work was done manually.

Inserting HPF Directives : was the final stage. It involved also a small amount of code restructuring:
- loops were restructured to be **INDEPENDENT** where possible;
- Procedures were marked as **HPF_SERIAL** where appropriate;
- Some I/O statements were moved to improve parallelisation.

4 Porting the *SEMC3D* Code

4.1 Translation to Fortran 90.

As with all of the codes, *FORESYS* was used to generate a Fortran90 version semi-automatically. The main features of the Fortran 90 version are :

- New Fortran 90 syntax used.
- All variables are declared.
- Interface modules are provided for every procedure.
- Sequence association is not used. COMMON blocks have been replaced by modules.
- Introduction of INTENT attribute for dummy arguments.
- Removal of obsolete features.

In the Fortran77 version, all arrays are defined in the main program with their exact dimensions specified via an include file. Fortran 90 is able to introduce dynamic arrays, so we have rewritten the main program using this feature.

The next stage was the introduction of array syntax or array operations when possible. This work was very easy because the initial version of SEMC3D was written in Fortran 80 for the Alliant platform.

In Fortran 77 the actual and the dummy arguments of subroutines can differ in rank. In addition, many "Fortran77" codes use vendor extensions which allow dummy and actual to differ also in type or kind. The original SEMC3D code uses all these facilities.

We used the capacity of Fortran 90 to create user–defined generic procedures to keep the same calls.

After the initial translation was completed we carried out portability tests, and final tuning on several platforms: IBM, Cray, HP with vendor–supplied compilers and on the Sun and Meiko CS2 platforms at Vienna with the NASL FortranPlus compiler. Some minor parts were rewritten to bypass various compiler bugs.

4.2   HPF Port

As noted above, the HPF port was performed using fine grain parallelisation. No major difficulties were encountered; the code was originally designed to vectorise well, and the majority of loops have no data dependency.

Choosing the partition *SEMC3D* works within a three dimensional rectangular envelope, which maps naturally onto a 3D processor grid:

```
!HPF$ PROCESSORS Grid_3d(number_of_processors(dim=1), &
!HPF$                    number_of_processors(dim=2), &
!HPF$                    number_of_processors(dim=3))
```

The user can specify the actual size and shape of the grid in the run command.

Using TEMPLATE The simulation space is a box which can be defined by the following TEMPLATE.

```
!HPF$ TEMPLATE,  DIMENSION(1:nx,1:ny,1:nz) :: BOX
```

where nx, ny and nz are the number of discretization points on each axis.

Following our message passing experience, we chose to distribute this TEMPLATE with the (BLOCK,BLOCK,BLOCK) attribute; this was extended to BLOCK(mx), BLOCK(my), BLOCK(mz) during tuning with mx, my, mz chosen to optimise load balancing. However, since all directives are set via a single INCLUDE file, the actual mapping used can be trivially changed at the cost of a recompilation.

Arrays: Alignment and Distribution We distinguish two cases: the main program and the subroutines. In the main program, all the arrays are aligned with the BOX template, and in the subroutines they are distributed according to the inherited mapping.

For example:

```
In main :
          !HPF$ ALIGN ex(I,J,K) WITH BOX(I,J,K)
In the subroutines :
          !HPF$ DISTRIBUTE ex  *(BLOCK,BLOCK,BLOCK)
```

Major Parallel Constructs Used The two major sources of parallelism in the HPF code are:

Use of Array Expressions : As noted, *SEMC3D* is highly vectorisable. Many of the Fortran77 DO loops translate naturally to Fortran90 array assignments (and *FORESYS* carries out many of these translations automatically). These are very efficiently handled by the NASL *HPFPlus* compiler.

Use of HPF_SERIAL Procedures : Computationally expensive code sections which could not readily be expressed using array syntax, were wrapped in HPF_SERIAL procedures. These are typically called within a loop; the NASL *HPFPlus* compiler places the calls intelligently by analysing the processor residency of the actual arguments, and the resulting loops run in parallel with very low overheads.

Note that this approach can be used for both fine grain and coarse grain parallelism; in this application the granularity is low.

4.3 Indirection

Some boundary conditions used in SEMC3D can only be handled using indirection. In the Fortran77 version, these conditions are located with an integer pointer array. In the first translation step we replicated these arrays on all processors, but at a later stage we recoded the one dimensional pointer array to reflect the 3-D nature of the problem, and used a masked WHERE to retain parallelism:

```
integer  iptstr(1:nstr)
do i=1,nstr
   e(ipstr(i)) = 0
enddo
```

became

```
      real,    dimension(1:,1:,1:), intent(inout) :: e
      logical, dimension(1:,1:,1:), intent(in)    :: iptstr
!HPF$ DISTRIBUTE e*(BLOCK,BLOCK,BLOCK)
!HPF$ ALIGN      iptstr(I,J,K) WITH *e(I,J,K)
      WHERE(iptstr(:,:,:))
         e(:,:,:) = 0
      ENDWHERE
```

5 Benchmarks

We have benchmarked the code using several datasets:

cube_07 : a small (44*44*44) test cube used primarily for single processor testing; primary data requirements 3MBytes

cube_08 : a modest (54*54*54) test cube; primary data requirements around 6MB.

cube_09 : An 80*80*80 test cube, with primary data requirements around 20MB. This provides a test problem small enough to be run on a single processor of the Meiko CS2, which has only limited storage (48MB user–accessible per processor), but large enough to provide a test of the parallelism.

CAR_1 : An industrial model of a complete car, with primary data requirements around 100MB. This represents a typical production problem for the code.

5.1 Single Processor Benchmarks

The first stage in the port was the conversion of the Fortran77 code to Fortran90. This conversion itself has major advantages: the resulting code is cleaner, and written at a much higher level; so that it is easier to maintain and develop.

Utilising these advantages in practice means relying on the Fortran90 compiler available on the chosen target system. The NASL HPFPlus compiler also has this reliance, since it targets Fortran90.

Fortran90 compilers however still vary in their optimisation level for specific targets. Table 1 shows the single processor performance of the original Fortran77 code; of the converted Fortran90 code; and of the final HPF code running on one processor of the Vienna Meiko CS2. The compilers used were:

Fortran77: F77 SC3.0.1
Fortran90: NASL FortranPlus Release 2.0beta
HPF: NASL HPFPlus Release 2.1

We see that the CS2 FortranPlus compiler has not been comparably optimised to the Sun F77 compiler. Using HPF on one processor introduces a factor of between 2.8 and 3.4 in the single processor times. Since (surprisingly perhaps) the HPF times are *better* than the F90 times, all of this is due to the performance of the backend compiler.

Table 1. Single Processor Times (seconds) on the CS2

Compiler	Cube_07	Cube_08	Cube_09
F77 -O2	26.0	46.2	147
F77 -O5	16.6	31	101
F90 -fast	79.4	106	348
HPF -O	56.5	88	281

5.2 Parallel Performance of the HPF Code

We have run the converted HPF code on the Meiko CS2 at the University of Vienna; this is one of the two project platforms for *Pharos*. Table 2 shows the performance achieved. The times recorded are for four simulation time periods for cube_08, cube_09; one time period for CAR_1.

Table 2. Performance of the HPF version of *SEMC3D* on the Meiko CS2. Times are in seconds.

Processors	Cube_08	Cube_09	CAR_1
1	89	281	
2	52	154	
4	35	76	268
8	13	39.7	152
16	8.6	22.7	86
32	6.2	10.7	45

The speedups achieved are very good; the code clearly scales well. The mildly anomalous time measured for cube_08 on eight processors was repeatable but not really understood. CAR_1 could not be run on less than four processors.

5.3 Comparison with a Message-Passing MPI Port

As noted above, an MPI/Fortran77 version of the code had already been developed prior to the project start. Interest in HPF stemmed from the cost of producing and maintaining this port; and especially the difficulty of keeping it in step with ongoing development of the serial code.

We ran the cube_09 benchmark with both MPI and HPF versions, on the Meiko CS2, with results shown in Table 3.

The single processor performance advantage of the Fortran77 compiler is maintained roughly constant over the range of processors used; that is, the speedups achieved by HPF and MPI codes are essentially the same. The MPI figures are roughly a factor 3 better than those achieved with HPF; comparison with Table 1 shows that this is attributable wholly to the lack of optimisation of the backend Fortran90 compiler.

Table 3. MPI (F77 -O5) vs HPF (HPFPlus -O) for cube_09. Times are in seconds.

Procs	MPI	HPF	Ratio
1	101	281	1:2.8
2	57.5	154	1:2.7
4	28.3	76	1:2.7
8	13.5	39.7	1:2.9
16	6.9	22.7	1:3.3
32	5.7	10.7	1:1.9

6 Discussion

The speedups obtained in the benchmarking are extremely encouraging; what do the results tell us about the use of HPF at this stage in compiler development?

6.1 Friendliness of the HPF Language

This has certainly been confirmed. The HPF version of the code is as easy to read as the Fortran90 version; and certainly much easier to maintain than either the Fortran77 or the message passing version. This last is particularly hard to maintain: the implementation of a fine grain message passing parallelisation obscures the structure of the original serial code, while this structure is cleanly maintained in the HPF version.

6.2 Requirements on the HPF Compiler

One important aspect of the initial porting process was to identify those requirements on the HPF compiler needed to produce a final scalable working version of the code. Here is a summary of the findings:

Fortran90 Compiler Optimisation Improved runtime efficiency in the backend Fortran90 compilers, would help the benchmarks. The NASL HPFPlus compiler targets Fortran90 directly, and hence can (and necessarily does) rely on the many optimisations available to such a compiler. In the medium term, Fortran90 compilers will generate better code for well written programs than the equivalent code passed through a Fortran77 compiler. This is already true for some Fortran90 compilers; but not yet on the CS2.

HPFPlus Efficiency Release 2.0 of *HPFPlus* has substantial optimisations which have reduced the HPF overheads in many codes to between zero and 20%; indeed, here they prove to be *negative overheads*. HPF overheads are certainly not a significant issue with *SEMC3D*.

Language Support The major language features used in this port were:

- Multidimensional `BLOCK(m)` distributions;
- `ALIGN`;
- `PROCESSOR` arrays with runtime determined size and shape;
- Efficient handling of distributed array arguments;
- Array expressions using distributed components;
- Autoparallelised `DO` loops with calls to `HPF_SERIAL` procedures.

These features were sufficient to produce an extremely effective parallel code with parallel efficiency as good as the pre-existing MPI version; the HPF source is however very much easier to maintain, and further develop, than the message-passing MPI code.

7 Acknowledgements

This work was supported in part via Esprit Project No P20162, Pharos. We are indebted to our colleagues in this project for many helpful discussions, and aid in running on a variety of systems. Especial thanks are due to Thomas Brandes, Junxian Liu, Ruth Lovely, John Merlin and Dave Watson.

HiPEC: High Performance Computing Visualization System Supporting Networked Electronic Commerce Applications*

Reinhard Lüling and Olaf Schmidt

University of Paderborn (Department of Computer Science)
Fürstenalle 11,D-33102 Paderborn, Germany
{rl,merlin}@uni-paderborn.de

Abstract. This paper presents the basic ideas and the technical background of the Esprit project HiPEC. The HiPEC Project aims at integrating HPCN technologies into an electronic commerce scenario to ease the interactive configuration of products and their photorealistic visualization by a high performance parallel rendering system.

1 Introduction

It is well accepted that electronic commerce systems will be of large benefit for the European industry if these technologies are actively used by companies and integrated into their business as soon as possible. Especially SMEs can benefit largely as market access and communication to suppliers and larger industries can be eased considerably using Internet technologies.

The main objective of the HiPEC project is to integrate advanced HPCN technologies to form a generic electronic commerce application. This application is aimed at giving a large number of SMEs a configuration and visualization tool to support the selling process within the showroom and to enlarge their business using the same tool also on the Internet. The enabling technology behind this tool is a high performance computing system as well as advanced networking technology. The HPC system is used to run the configuration and visualization system and networks give the end-users access to a HPC system to run the service from their showroom. It is the special aim of this project to make the service as general usable as possible. This means that open interfaces have to be used that allow a future integration of other components to customize the system to any special needs. The HPC equipment which is used for the electronic commerce application is provided as a service to the end-users in a way that allows the end-users to use these technologies via suitable communication networks. Thus, the end-users can have cost-efficient access to HPC technologies. The potential customers of these technologies are retailers of industrial products. The application range selected here is reselling of bathroom equipment. In

---

* This work is supported by the EU, ESPRIT project 27232 (Domain 6, High Performance Computing and Networking).

a typical scenario, the customer can configure the bathroom equipment in an interactive configuration session. To ease the decision process of the customer, the configuration of the final product is supported by presenting the customer a photorealistic animation of the composed bathroom. As such a presentation has to be computed in short time with highest quality it is necessary to use a high performance computing system for the realistic image synthesis.

In section 2 the basic ideas of HiPEC will be discussed and a more detailed view on the technical background of the project is given.

2   Approach of the HiPEC Project

The HiPEC project aims at integrating HPCN technologies into an electronic commerce scenario to ease the interactive configuration of products. In the project scenario a retailer is equipped with a PC that is connected to the Internet. This PC holds a local database of product information from different manufacturers of bathroom equipment. Product information that was previously only available in books and brochures is digitized and loaded into the local database that can be frequently updated by the manufacturer of the bathroom equipment via the Internet.

A customer that wants to configure his bathroom can do this using a CAD program on the PC. During this selling process an employee of the reseller guides the selection of products and shows the different scenarios using the local CAD program on the PC. Using the local database selections can easily be made and different scenarios can be presented to the customer. Configuration of products is only one aspect. The other and even more important aspect is the visualization of a product. Customers want to have a visual impression of their product. This is of special importance in areas where furniture and other consumer goods are configured and offered to customers. As a high quality animation of products can lead to an increased attractivity the HiPEC project integrates a rendering service that generates photorealistic pictures of the selected product. As these animations can only be generated using HPC computing systems, a parallel computing system is used to generate these pictures. To generate this in highest quality, the architecture of the bathroom is loaded to a remote parallel computer system where a database is installed that contains detailed graphical models of all products. The model of the complete bathroom is build here and forwarded to the parallel system that generates a picture of highest quality. The encoded picture is send back to the PC to be printed and handed over to the customer. Thus, the HiPEC electronic commerce system contains the following components (see Figure 1).

User Interface and Configuration System
The User Interface and configuration system is installed on a PC in the showroom of the seller of the product. Together with the seller the customer configures the final product and has the possibility of using the remote high-end rendering service installed on the parallel computer system for a high-quality visualization.

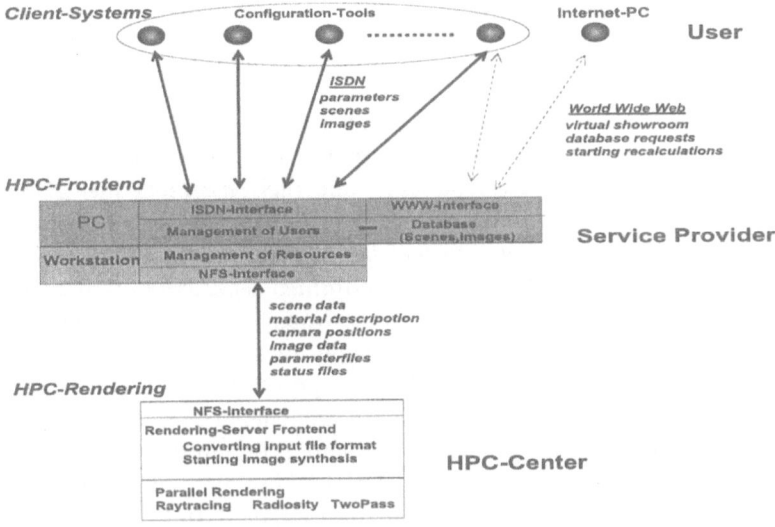

Fig. 1. Architecture of the HiPEC-System

This visualization might be only initiated at the end of the selling process or also in between to guide the user. The user is enabled to store generated product visualizations and configured products in a database. This database is used to set up a virtual showroom on the World Wide Web, which can be visited by potential customers from their PCs at home.

Communication network
The configuration and visualization service will be offered in the showroom of the retailers. Thus, communication networks will be used for different purposes. We will use ISDN dial-up lines to connect the PCs installed in the end-users showrooms to the parallel server. Using the database installed at the site of the parallel server, the bandwidth of these lines is large enough to support the application.

Service Broker
The electronic commerce application is achieved by a service broker which is installed on frontend computers to the HPC system. The service broker is responsible for managing the user accounts and for performing efficient job-scheduling of rendering jobs on the available hardware resources.

Database of product components
To generate a high quality animation, the complete product that is configured on the local PC has to be delivered to the parallel rendering server. To save communication bandwidth, the most important information about structure of the sub-components, structure of the surface of the object and other information are stored in a database that is located on a frontend PC connected to the parallel system. Using this database, the client system only has to communicate some

small information about the configuration. The complete scene that has to be animated is generated on the parallel computing system.

Parallel rendering system

Ray tracing and radiosity algorithms currently implemented in image synthesis systems provide the necessary rendering quality, but these methods are suffering from their extensive computational costs and their enormous memory requirements. Parallel computing and graphics are two fields of computer science that have much to offer each other. Efficient use of parallel computers requires algorithms where the same computation is performed repeatedly or where separate tasks are performed with little or no coordination, and most computer graphics tasks fit these needs well. On the other hand, most graphics applications have an insatiable appetite for raw computational power, and parallel computers are a good way to provide this power. It can be concluded that there is an obvious synergy between computer graphics and parallel computation. Several sophisticated methods for efficient parallel simulation of illumination effects in complex environments have been developed in the past [1][2] [4]. In the HiPEC approach the parallel rendering system is installed on a parallel computing platform that is located at the site of a service provider that provides the parallel computing service.

Over all, this application shows the benefits of networked HPC technologies in an electronic commerce application that goes beyond presenting products in the WWW. This technology is general, i.e. it can be applied to a large number of other applications and is presented here for a complete industrial area allowing to demonstrate the benefits of these technologies.

3 Conclusion

The HiPEC project integrates advanced HPC technology into a special electronic commerce application supporting the selling process within showrooms of retailers of bathroom furniture. So even SMEs can profit by the advantages of powerful HPC systems which are to expensive for a local installation. The design of the HiPEC system is kept as general as possible in order to allow the integration of other components to customize the system to the demands of additional electronic commerce applications.

References

1. Baum, D.R.; Winget, J.M.: Real time radiosity through parallel processing and hardware acceleration, Proc. of SIGGRAPH 90, March 1990, pp. 67-75
2. Chalmers, A.G.; Paddon, D.J.: Parallel processing of progressive refinement radiosity methods, Proc. 2nd Eurographics Workshod on Rendering, Mai 1991
3. Schmidt, O.; Lange, B.: Interaktive photorealistische Bildgenerierung durch effiziente parallele Simulation der Lichtenergieverteilung, Proc. of Informatik97,1997, pp. 476-485, Springer- Verlag
4. Schmidt, O.; Rathert, J.; Reeker, L.: Parallel Simulation of the Global Illumination, *accepted for PDPTA '98*, Las Vegas, July 1998

Index of Authors

Abdallah, Ali E.   165
Abu-Ghazaleh, Nael B.   1089
Agrawal, Nidhi   981
Akarsu, E.   55
Alefragis, Panayiotis   1104
Almeida, F.   234
Alur, Rajeev   191
Andonov, Rumen   480
Antonis, Konstantinos   352
Antonoiu, Gheorghe   545
Arbenz, Peter   771

Banerjee, P.   422
Barbosa da Silva, Fabricio Alves   367
Barreteau, Michel   1123
Bataller, Jordi   887
Beckmann, Olav   413
Ben-Asher, Yosi   933
Bernabéu-Aubán, José M.   887
Berrendorf, Rudolf   299
Bianchini, Ricardo   831
Bischof, Stefan   383
Biswas, Rupak   307
Bode, Arndt   193
Bodin, François   1123
Boeres, Cristina   337
Boku, Taisuke   244
Bono, Ida de   812
Borges, Leonardo   763
Boukerche, Azzedine   318,534
Boulet, Pierre   263
Bourgeois, J.   113
Brandes, Thomas   629,639
Brégier, Frédéric   639
Brent, Richard P.   1
Brinkhaus, Peter   1123
Broughton, P.   126
Brown, T.J.   102
Brunie, Lionel   503
Bull, J. Mark   377
Burger, A.   126

Burkhart, Helmar   75

Cahoon, Brendon   521
Calamoneri, Tiziana   1029
Campos, Luis Miguel   367
Cappello, Franck   216
Carpenter, Bryan   659
Catthoor, F.   923
Cavalheiro, Gerson G. H.   373
Cesati, Marco   892
Chakravarty, Manuel M.T.   709
Chamski, Zbigniew   1123
Chapman, Barbara   650
Charles, Henri-Pierre   1123
Chiola, G.   620
Choudhary, A.   422
Ciaccio, G.   620
Clark, A.F.   92
Clint, Maurice   747
Cole, Murray   625
Collard, Jean-François   411
Coloma, I.   539
Corradi, Antonio   625
Counilh, Christine Marie   639
Crookes, D.   102

Danelutto, M.   698
Dantas, M.A.R.   397
Das, Dibyendu   401
Das, Sajal K.   318,534
Datta, Ajoy   534
Davenport, Glorianna   47
Davy, John   136
De Man, H.   923
De Vito, Dominique   742
Delaitre, T.   113
Delgado, A.   539
Delves, L.M.   1140
Dempster, E.W.   126
Denneulin, Yves   373
Díaz de Cerio, Luis   1051

Dickens, Phillip M.   959
Diekmann, Ralf   347
Di Ianni, Miriam   892,1029
di Serafino, Daniela   812
Doallo, Ramón   224
Donaldson, Stephen R.   80,970
dos Santos, Rafael R.   1010
Downton, A.C.   92
Ducloux, Eric   812

Ebner, Ralf   383
Eisenbeis, Christine   1123
Eisenbiegler, Jörn   456,865
Erlebach, Thomas   383
Essah, Wissal   136
Evripidou, Paraskevas   181,463

Feautrier, Paul   470
Ferreira, A.   875
Fimmel, Dirk   1018
Fisher, M.D.   852
Fissgus, U.   273
Fleury, M.   92
Flocchini, P.   554
Floros, Nick   1131
Flynn Hummel, Susan   297
Fox, Geoffrey C.   55,659
Fraguela, Basilio B.   224
Frayssé, Valérie   751
Fuerle, Thomas   953
Furmanski, W.   55
Futatsugi, Kokichi   846

Gaber, J.   405
Gabow, Harold N.   307
Garofalakis, John   352
Gehrke, Thomas   733
Germain, Cécile   629
Geus, Roman   771
Giavitto, Jean-Louis   742
Giraud, Luc   751
González, Antonio   1051
Goumopoulos, Christos   1104
Grün, Thomas   999
Guérin Lassous, I.   875

Guerraoui, Rachid   513
Gurd, John   1123

Haber, Gady   933
Haupt, T.   55
Hayashi, Tatsuya   390
Heinrich-Litan, L.   273
Herley, Kieran   967
Hernández, Emilio   220
Hey, Tony   220
Hill, Jonathan M.D.   80,157,970
Hillebrand, Mark A.   999
Hirata, Hiromichi   846
Hockauf, Robert   206
Hofstedt, Petra   676
Hoogerbrugge, Jan   1123
Housos, Efthymios   1104
Hu, Ping   1123

Ioroi, Shigenori   846
Ishii, Naohiro   390
Itakura, Ken'ichi   244
Iwahori, Yuji   390

Jalby, William   1123
Jarvis, Stephen A.   157
Jensch, Jörg   944
Justo, G.R.   113

Kacsuk, Péter   842
Kandemir, M.   422
Kaniewski, Juri   798
Karl, Wolfgang   206
Kautonen, Anssi   993
Keane, J.A.   852
Keller, Gabriele   709
Kelly, Paul H.J.   413,1062
Kharraz-Aroussi, Hatim   751
Kiper, Ayşe   793
Knijnenburg, Peter M.W.   1123
Knoop, Jens   445
Koch, Povl T.   601
Koppler, Rainer   435
Kort, Iskander   255,940
Kreuchlin, Wolfgang   747

Krommer, Arnold R.   804
Kshemkalyani, Ajay D.   578
Kubota, Kazuto   244
Kulkarni, C.   923

Lachanas, Adrianos   463
Lanfear, Tim   80
Launay, Pascale   729
León, C.   539
Leberecht, Markus   206
LeMaster, Timothy E.   534
Leppänen, Ville   993
Li, Dingchao   390
Li, Xiaoming   659
Li, Xinying   659
Lodi, E.   554
Lorcy, Stéphane   738
Löwe, Welf   328,865
Lü, J.   126
Lu, Zhihong   521
Luccio, F.   554
Ludwig, Thomas   173,193
Lüling, Reinhard   944,1149
Lundberg, Lars   288
Luzet, Henri   1140

Mallet, Julien   688
Malony, Allen D.   191
Marcus, K.   875
Marín, Mauricio   897
Maslennikow, Oleg   798
May, David   591
Mayer, Ernst   503
McAleese, G.   102
McColl, Bill   863
McKinley, Kathryn S.   521
Meacham, Ken   1131
Mehrotra, Piyush   650
Melas, P.   570
Merker, Renate   1018
Meyer, Ulrich   1040
Middendorf, Martin   328
Miller, Barton P.   146
Molitor, P.   273
Morales, D.G.   234

Morrow, P.J.   102
Muller, Henk L.   591

Najjar, Faza   528
Nash, Jonathan   906
Navaux, Philippe O.A.   1010
Németh, Zsolt   842
Neophytou, Neophytos   181
Newhall, Tia   146
Nikolopoulos, Dimitrios S.   491
Nodera, Takashi   788

O'Boyle, Michael   1123
Oberhuber, Michael   206
Ogata, Kazuhiro   846
Oliker, Leonid   307
Oliveira, Suely   763
Orlando, Salvatore   356
Osoba, Femi O.   704
Ould-Khaoua, Mohamed   989
Ozawa, Kazufumi   780
Ozdemir, H.   55

Pagli, L.   554
Papatheodorou, Theodore S.   491
Pazat, Jean-Louis   729
Pedone, Fernando   513
Pelagatti, S.   698
Penttonen, Martti   993
Perego, Raffaele   356
Perrott, Ron   1101
Piquer, José M.   610
Plouzeau, Noël   738
Polychronopoulos, Eleftherios D.   491
Preis, Robert   347
Priol, Thierry   1114
Pua, C.S.   126
Puntigam, Franz   720

Rabenseifner, Rolf   563
Rabhi, Fethi A.   704
Rahayu, J. Wenny   505
Rajopadhye, Sanjay   480
Ramanujam, J.   422
Ramirez, Rafael   279

1156

Randoux, S.   113
Rapine, Christophe   322
Rau-Chaplin, A.   875
Rauber, Th.   273
Ravikumar, C.P.   981
Rebello, Vinod E.F.   337
Redon, Xavier   263
René, Christophe   1114
Richard, Olivier   216
Riley, Graham   297
Roantree, D.   102
Roch, Jean-Louis   373
Roda, J.L.   234
Röder, Christian   193
Rodríguez, C.   234,539
Rohou, Erven   1123
Roman, Jean   639
Rousset de Pina, Xavier   601
Royo, Dolors   1051

Sakellariou, Rizos   297,1123
Salinger, Petr   1057
Sande, F.   539
Sanders, Peter   1104
Sansonnet, Jean-Paul   742
Santoro, N.   554
Santos Costa, Vítor   831
Sarvan, N.   92
Sato, Mitsuhisa   244
Scherson, Isaac D.   322,367
Schikuta, Erich   953
Schimmler, Manfred   916
Schinkmann, F.   113
Schiper, André   513
Schlimbach, Frank   347
Schmidt, Bertil   916
Schmidt, Olaf   1149
Schröder, Heiko   916
Scott, R.I.   852
Sensen, Norbert   944
Seznec, André   1123
Shenoy, N.   422
Sibeyn, Jop F.   1040
Siniolakis, Constantinos   157
Sips, Henk   625

Skeie, Tor   1076
Skillicorn, D.B.   337,698,970
Slimani, Yahya   528
Snelling, David   967
Šoch, Michal   1047
Spence, I.T.A.   102
Spezzano, Giandomenico   669
Spies, F.   113
Spirakis, Paul   352
Srimani, Pradip K.   545
Stöhr, Elena A.   1123
Surridge, Mike   1131
Sutter, St.   273

Takkula, Tuomo   1104
Talbot, Sarah A.M.   1062
Talia, Domenico   669
Taniar, David   505
Taylor, H.   126
Thakur, Rajeev   959
Thorelli, Lars-Erik   682
Tomov, N.T.   126
Toursel, B.   405
Townsend, Paul D.   35
Treffers, Menno   1123
Trinitis, Jörg   173
Trystram, Denis   255,322,940
Tsuno, Naoto   788
Tvrdík, Pavel   1047,1057

Upstill, Colin   1101

Valero-García, Miguel   1051
Vanneschi, Marco   21
Vasilev, Vasil P.   157
Vekariya, P.   113
Visconti, Ivana   610
Vlassov, Vladimir   682

Wadsworth, Chris   75
Wagner, Michael   206
Wagner, Clemens   821
Walker, David   863
Walshaw, Chris   347
Wanek, Helmut   953

Wedelin, Dag   1104
Wen, Yuhong   659
Wijshoff, Harry A.G.   1123
Williams, M.H.   126
Wilsey, Philip A.   1089
Winter, S.C.   113
Wismüller, Roland   173
Wyrzykowski, Roman   798

Yamada, Susumu   780
Yanev, Nicola   480

Zaluska, E. J.   570
Zapata, Emilio L.   224
Zavanella, A.   698
Zemerly, M.J.   113
Zhang, Guansong   659
Zimmermann, Wolf   328,865

Springer
and the
environment

At Springer we firmly believe that an international science publisher has a special obligation to the environment, and our corporate policies consistently reflect this conviction.

We also expect our business partners – paper mills, printers, packaging manufacturers, etc. – to commit themselves to using materials and production processes that do not harm the environment. The paper in this book is made from low- or no-chlorine pulp and is acid free, in conformance with international standards for paper permanency.

 Springer

Lecture Notes in Computer Science

For information about Vols. 1–1397

please contact your bookseller or Springer-Verlag

Vol. 1398: C. Nédellec, C. Rouveirol (Eds.), Machine Learning: ECML-98. Proceedings, 1998. XII, 420 pages. 1998. (Subseries LNAI).

Vol. 1399: O. Etzion, S. Jajodia, S. Sripada (Eds.), Temporal Databases: Research and Practice. X, 429 pages. 1998.

Vol. 1400: M. Lenz, B. Bartsch-Spörl, H.-D. Burkhard, S. Wess (Eds.), Case-Based Reasoning Technology. XVIII, 405 pages. 1998. (Subseries LNAI).

Vol. 1401: P. Sloot, M. Bubak, B. Hertzberger (Eds.), High-Performance Computing and Networking. Proceedings, 1998. XX, 1309 pages. 1998.

Vol. 1402: W. Lamersdorf, M. Merz (Eds.), Trends in Distributed Systems for Electronic Commerce. Proceedings, 1998. XII, 255 pages. 1998.

Vol. 1403: K. Nyberg (Ed.), Advances in Cryptology – EUROCRYPT '98. Proceedings, 1998. X, 607 pages. 1998.

Vol. 1404: C. Freksa, C. Habel. K.F. Wender (Eds.), Spatial Cognition. VIII, 491 pages. 1998. (Subseries LNAI).

Vol. 1405: S.M. Embury, N.J. Fiddian, W.A. Gray, A.C. Jones (Eds.), Advances in Databases. Proceedings, 1998. XII, 183 pages. 1998.

Vol. 1406: H. Burkhardt, B. Neumann (Eds.), Computer Vision – ECCV'98. Vol. I. Proceedings, 1998. XVI, 927 pages. 1998.

Vol. 1408: E. Burke, M. Carter (Eds.), Practice and Theory of Automated Timetabling II. Proceedings, 1997. XII, 273 pages. 1998.

Vol. 1407: H. Burkhardt, B. Neumann (Eds.), Computer Vision – ECCV'98. Vol. II. Proceedings, 1998. XVI, 881 pages. 1998.

Vol. 1409: T. Schaub, The Automation of Reasoning with Incomplete Information. XI, 159 pages. 1998. (Subseries LNAI).

Vol. 1411: L. Asplund (Ed.), Reliable Software Technologies – Ada-Europe. Proceedings, 1998. XI, 297 pages. 1998.

Vol. 1412: R.E. Bixby, E.A. Boyd, R.Z. Ríos-Mercado (Eds.), Integer Programming and Combinatorial Optimization. Proceedings, 1998. IX, 437 pages. 1998.

Vol. 1413: B. Pernici, C. Thanos (Eds.), Advanced Information Systems Engineering. Proceedings, 1998. X, 423 pages. 1998.

Vol. 1414: M. Nielsen, W. Thomas (Eds.), Computer Science Logic. Selected Papers, 1997. VIII, 511 pages. 1998.

Vol. 1415: J. Mira, A.P. del Pobil, M.Ali (Eds.), Methodology and Tools in Knowledge-Based Systems. Vol. I. Proceedings, 1998. XXIV, 887 pages. 1998. (Subseries LNAI).

Vol. 1416: A.P. del Pobil, J. Mira, M.Ali (Eds.), Tasks and Methods in Applied Artificial Intelligence. Vol.II. Proceedings, 1998. XXIII, 943 pages. 1998. (Subseries LNAI).

Vol. 1417: S. Yalamanchili, J. Duato (Eds.), Parallel Computer Routing and Communication. Proceedings, 1997. XII, 309 pages. 1998.

Vol. 1418: R. Mercer, E. Neufeld (Eds.), Advances in Artificial Intelligence. Proceedings, 1998. XII, 467 pages. 1998. (Subseries LNAI).

Vol. 1419: G. Vigna (Ed.), Mobile Agents and Security. XII, 257 pages. 1998.

Vol. 1420: J. Desel, M. Silva (Eds.), Application and Theory of Petri Nets 1998. Proceedings, 1998. VIII, 385 pages. 1998.

Vol. 1421: C. Kirchner, H. Kirchner (Eds.), Automated Deduction – CADE-15. Proceedings, 1998. XIV, 443 pages. 1998. (Subseries LNAI).

Vol. 1422: J. Jeuring (Ed.), Mathematics of Program Construction. Proceedings, 1998. X, 383 pages. 1998.

Vol. 1423: J.P. Buhler (Ed.), Algorithmic Number Theory. Proceedings, 1998. X, 640 pages. 1998.

Vol. 1424: L. Polkowski, A. Skowron (Eds.), Rough Sets and Current Trends in Computing. Proceedings, 1998. XIII, 626 pages. 1998. (Subseries LNAI).

Vol. 1425: D. Hutchison, R. Schäfer (Eds.), Multimedia Applications, Services and Techniques – ECMAST'98. Proceedings, 1998. XVI, 532 pages. 1998.

Vol. 1427: A.J. Hu, M.Y. Vardi (Eds.), Computer Aided Verification. Proceedings, 1998. IX, 552 pages. 1998.

Vol. 1429: F. van der Linden (Ed.), Development and Evolution of Software Architectures for Product Families. Proceedings, 1998. IX, 258 pages. 1998.

Vol. 1430: S. Trigila, A. Mullery, M. Campolargo, H. Vanderstraeten, M. Mampaey (Eds.), Intelligence in Services and Networks: Technology for Ubiquitous Telecom Services. Proceedings, 1998. XII, 550 pages. 1998.

Vol. 1431: H. Imai, Y. Zheng (Eds.), Public Key Cryptography. Proceedings, 1998. XI, 263 pages. 1998.

Vol. 1432: S. Arnborg, L. Ivansson (Eds.), Algorithm Theory – SWAT '98. Proceedings, 1998. IX, 347 pages. 1998.

Vol. 1433: V. Honavar, G. Slutzki (Eds.), Grammatical Inference. Proceedings, 1998. X, 271 pages. 1998. (Subseries LNAI).

Vol. 1434: J.-C. Heudin (Ed.), Virtual Worlds. Proceedings, 1998. XII, 412 pages. 1998. (Subseries LNAI).

Vol. 1435: M. Klusch, G. Weiß (Eds.), Cooperative Information Agents II. Proceedings, 1998. IX, 307 pages. 1998. (Subseries LNAI).

Vol. 1436: D. Wood, S. Yu (Eds.), Automata Implementation. Proceedings, 1997. VIII, 253 pages. 1998.

Vol. 1437: S. Albayrak, F.J. Garijo (Eds.), Intelligent Agents for Telecommunication Applications. Proceedings, 1998. XII, 251 pages. 1998. (Subseries LNAI).

Vol. 1438: C. Boyd, E. Dawson (Eds.), Information Security and Privacy. Proceedings, 1998. XI, 423 pages. 1998.

Vol. 1439: B. Magnusson (Ed.), System Configuration Management. Proceedings, 1998. X, 207 pages. 1998.

Vol. 1441: W. Wobcke, M. Pagnucco, C. Zhang (Eds.), Agents and Multi-Agent Systems. Proceedings, 1997. XII, 241 pages. 1998. (Subseries LNAI).

Vol. 1442: A. Fiat. G.J. Woeginger (Eds.), Online Algorithms. XVIII, 436 pages. 1998.

Vol. 1443: K.G. Larsen, S. Skyum, G. Winskel (Eds.), Automata, Languages and Programming. Proceedings, 1998. XVI, 932 pages. 1998.

Vol. 1444: K. Jansen, J. Rolim (Eds.), Approximation Algorithms for Combinatorial Optimization. Proceedings, 1998. VIII, 201 pages. 1998.

Vol. 1445: E. Jul (Ed.), ECOOP'98 – Object-Oriented Programming. Proceedings, 1998. XII, 635 pages. 1998.

Vol. 1446: D. Page (Ed.), Inductive Logic Programming. Proceedings, 1998. VIII, 301 pages. 1998. (Subseries LNAI).

Vol. 1447: V.W. Porto, N. Saravanan, D. Waagen, A.E. Eiben (Eds.), Evolutionary Programming VII. Proceedings, 1998. XVI, 840 pages. 1998.

Vol. 1448: M. Farach-Colton (Ed.), Combinatorial Pattern Matching. Proceedings, 1998. VIII, 251 pages. 1998.

Vol. 1449: W.-L. Hsu, M.-Y. Kao (Eds.), Computing and Combinatorics. Proceedings, 1998. XII, 372 pages. 1998.

Vol. 1450: L. Brim, F. Gruska, J. Zlatuška (Eds.), Mathematical Foundations of Computer Science 1998. Proceedings, 1998. XVII, 846 pages. 1998.

Vol. 1451: A. Amin, D. Dori, P. Pudil, H. Freeman (Eds.), Advances in Pattern Recognition. Proceedings, 1998. XXI, 1048 pages. 1998.

Vol. 1452: B.P. Goettl, H.M. Halff, C.L. Redfield, V.J. Shute (Eds.), Intelligent Tutoring Systems. Proceedings, 1998. XIX, 629 pages. 1998.

Vol. 1453: M.-L. Mugnier, M. Chein (Eds.), Conceptual Structures: Theory, Tools and Applications. Proceedings, 1998. XIII, 439 pages. (Subseries LNAI).

Vol. 1454: I. Smith (Ed.), Artificial Intelligence in Structural Engineering. XI, 497 pages. 1998. (Subseries LNAI).

Vol. 1456: A. Drogoul, M. Tambe, T. Fukuda (Eds.), Collective Robotics. Proceedings, 1998. VII, 161 pages. 1998. (Subseries LNAI).

Vol. 1457: A. Ferreira, J. Rolim, H. Simon, S.-H. Teng (Eds.), Solving Irregularly Structured Problems in Prallel. Proceedings, 1998. X, 408 pages. 1998.

Vol. 1458: V.O. Mittal, H.A. Yanco, J. Aronis, R-. Simpson (Eds.), Assistive Technology in Artificial Intelligence. X, 273 pages. 1998. (Subseries LNAI).

Vol. 1459: D.G. Feitelson, L. Rudolph (Eds.), Job Scheduling Strategies for Parallel Processing. Proceedings, 1998. VII, 257 pages. 1998.

Vol. 1460: G. Quirchmayr, E. Schweighofer, T.J.M. Bench-Capon (Eds.), Database and Expert Systems Applications. Proceedings, 1998. XVI, 905 pages. 1998.

Vol. 1461: G. Bilardi, G.F. Italiano, A. Pietracaprina, G. Pucci (Eds.), Algorithms – ESA'98. Proceedings, 1998. XII, 516 pages. 1998.

Vol. 1462: H. Krawczyk (Ed.), Advances in Cryptology - CRYPTO '98. Proceedings, 1998. XII, 519 pages. 1998.

Vol. 1464: H.H.S. Ip, A.W.M. Smeulders (Eds.), Multimedia Information Analysis and Retrieval. Proceedings, 1998. VIII, 264 pages. 1998.

Vol. 1465: R. Hirschfeld (Ed.), Financial Cryptography. Proceedings, 1998. VIII, 311 pages. 1998.

Vol. 1466: D. Sangiorgi, R. de Simone (Eds.), CONCUR'98: Concurrency Theory. Proceedings, 1998. XI, 657 pages. 1998.

Vol. 1467: C. Clack, K. Hammond, T. Davie (Eds.), Implementation of Functional Languages. Proceedings, 1997. X, 375 pages. 1998.

Vol. 1468: P. Husbands, J.-A. Meyer (Eds.), Evolutionary Robotics. Proceedings, 1998. VIII, 247 pages. 1998.

Vol. 1469: R. Puigjaner, N.N. Savino, B. Serra (Eds.), Computer Performance Evaluation. Proceedings, 1998. XIII, 376 pages. 1998.

Vol. 1470: D. Pritchard, J. Reeve (Eds.), Euro-Par'98: Parallel Processing. Proceedings, 1998. XXII, 1157 pages. 1998.

Vol. 1471: J. Dix, L. Moniz Pereira, T.C. Przymusinski (Eds.), Logic Programming and Knowledge Representation. Proceedings, 1997. IX, 246 pages. 1998. (Subseries LNAI).

Vol. 1473: X. Leroy, A. Ohori (Eds.), Types in Compilation. Proceedings, 1998. VIII, 299 pages. 1998.

Vol. 1475: W. Litwin, T. Morzy, G. Vossen (Eds.), Advances in Databases and Information Systems. Proceedings, 1998. XIV, 369 pages. 1998.

Vol. 1477: K. Rothermel, F. Hohl (Eds.), Mobile Agents. Proceedings, 1998. VIII, 285 pages. 1998.

Vol. 1478: M. Sipper, D. Mange, A. Pérez-Uribe (Eds.), Evolvable Systems: From Biology to Hardware. Proceedings, 1998. IX, 382 pages. 1998.

Vol. 1479: J. Grundy, M. Newey (Eds.), Theorem Proving in Higher Order Logics. Proceedings, 1998. VIII, 497 pages. 1998.

Vol. 1480: F. Giunchiglia (Ed.), Artificial Intelligence: Methodology, Systems, and Applications. Proceedings, 1998. IX, 502 pages. 1998. (Subseries LNAI).

Vol. 1482: R.W. Hartenstein, A. Keevallik (Eds.), Field-Programmable Logic and Applications. Proceedings, 1998. XI, 533 pages. 1998.

Vol. 1483: T. Plagemann, V. Goebel (Eds.), Interactive Distributed Multimedia Systems and Telecommunication Services. Proceedings, 1998. XV, 326 pages. 1998.

Vol. 1487: V. Gruhn (Ed.), Software Process Technology. Proceedings, 1998. IX, 157 pages. 1998.

Vol. 1488: B. Smyth, P. Cunningham (Eds.), Advances in Case-Based Reasoning. Proceedings, 1998. XI, 482 pages. 1998. (Subseries LNAI).

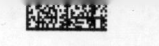